Isometric View
of Foot of Column.

Plan.

Beam Box

Plan.

The webs of stiffeners are sometimes taken from the corners if there is a large area of metal.

Distance pieces.

Section.

F

Plan of beams through Web.

Firring or Filling in piece.

Joist

Joist.

Iron Beam.

Isometric View (looking up)

showing rail Flooring here are Iron Beams. Floor Boards. Firring

Joist.

12"x6" Steel.

1" Metal.

12"x6" Steel.

Section.

12"x6" Steel.

feather.

1¾" Metal Bottom plate.

3'-0"x3'-0"x

Elevation at J.

Thimble joint.

Web ¾

1¾" Bed

Detail of Beams showing the way they are bolted to column to keep the work rigid.

Flange 11½" diam.

7" diam.

¾" Bolts.

12"x6" Steel

11½x4½

Early Structural Steel in London Buildings

Early Structural Steel in London Buildings

A discreet revolution

Jonathan Clarke

ENGLISH HERITAGE

Published by English Heritage, The Engine House, Fire Fly Avenue, Swindon SN2 2EH
www.english-heritage.org.uk
English Heritage is the Government's lead body for the historic environment.

First published 2014

ISBN 978–1-84802–103–7

Product code 51664

British Library Cataloguing in Publication data
A CIP catalogue record for this book is available from the British Library.

For more information about images from the English Heritage Archive, contact
Archives Services Team, The Engine House, Fire Fly Avenue, Swindon SN2 2EH;
telephone (01793) 414600.

Brought to publication by Jeremy Toynbee, Toynbee Editorial Services, and
Robin Taylor and Sarah Enticknap, Publishing, English Heritage.
Typeset in Charter Regular ITC 9.5pt
Edited by Jeremy Toynbee
Indexed by Caroline Jones, Osprey Indexing
Page layout by Pauline Hull
Printed in the UK by Butler Tanner & Dennis, Ltd.

Part 1 opening image
Detail from 'Eisenwalzwerk' The Iron Rolling Mill (Modern Cyclopes) by Adolph Menzel.
Oil on canvas, 1875.
[© bpk/National Gallery, Berlin]

Frontispiece
Interior of Birkbeck Bank; dome.
[BB65/00466]

CONTENTS

ACKNOWLEDGEMENTS

This book has had a long and discontinuous gestation, weathering institutional redesignations, reprioritisations and relocations, as well as the sometimes faltering zeal of its author. Its origins lie with a thematic report produced in 2000, which, like this, benefited from the advice, support and enthusiasm of many individuals, not least former colleagues at English Heritage. To those persons who I thanked then, and whose patience has been sorely tried in awaiting the manuscript's published appearance, I wish to express profound gratitude and obligation.

Foremost among these is Peter Guillery, a linchpin in the book's conception and development. Besides closely reading the manuscript, and doing his best to keep it on course, his unwavering, sagacious and multifarious support has been simply tremendous. Colin Thom, also of the Survey of London, has my heartfelt thanks for reading the text, offering invaluable comments and suggestions. Those who sat on the early steering group, or who otherwise had managerial responsibility or involvement with the project – Susie Barson, Sara Brown, Paul Barnwell, John Cattell, Martin Cherry, Keith Falconer, Pete Herring, Colum Giles, Chris Scull and Humphrey Welfare – have all been supportive in differing ways, in Colum and Keith's cases also reading and commenting on early chapters. Similarly, I am grateful to my former immediate colleagues for their assistance, particularly Jo Smith for support and advice and June Warrington for help with some picture acquisition. Andy Donald produced the great majority of new drawings, and I am grateful to him, and Helen Jones, for also making available copies of their drawings published elsewhere. Many of the recent photographs are by Derek Kendall and Sid Barker, with others by Mike Seaforth, Alun Bull, Steve Cole, Nigel Corrie, Damian Grady and other former and current staff within the photographic section of English Heritage. Charles Walker provided great support and advice on image acquisition and improvement. Colleagues in the English Heritage Archive helped to identify and produce illustrations and articles, with Nicky Cryer and Nigel Wilkins particularly and unwearyingly helpful, but also Clare Broomfield, Felicity Gilmour, Cynthia Howell, Helen Jurga, Alyson Rogers and Lucinda Walker. Thanks are also due to Jane Trodd and Karen Horn, former librarians at the Savile Row offices. Colleagues in the Survey of London, and from other offices, generously made available information or images from their own current or past work, including Colin Thom, Philip Temple, Andrew Saint, Steven Brindle, Aileen Reed, Kathryn Morrison, Alan Johnson, John Minnis, Ian Leith, Simon Taylor and the late Ian Goodall.

The contribution of people outside English Heritage has been equally important. Without their enthusiastic support, and specialist input, this book would not have been possible. Lawrance Hurst shared liberally his knowledge, sources and archive, chaired valuable discussions at the IStructE History Study Group, and offered valuable comment to the manuscript, which he kindly read. Robert Thorne also read the manuscript and made helpful suggestions. Malcolm Tucker read closely and commented on the chapter on industrial buildings, also generously sharing information, insights and images. Mike Chrimes has been consistently supportive. I am also grateful to the many people I have met or corresponded with who freely shared their knowledge or advice, including Michael Bussell, (the late) Colin Cunningham, Malcolm Dunkeld, John Earl, Clare Hartwell, David Lawrence, Anne Mayoh, Tom Ridge, Andrew Smith, Tim Smith, James Sutherland, Thom Swailes, Stuart Tappin, Wilma Tulloch, Isobel Watson, Sara Wermiel and David Yeomans. Other individuals, who helped with specific questions and references, are thanked in the endnotes. All the above help notwithstanding, any errors of fact and interpretation that may have crept in are entirely my own.

The staff of libraries, institutions, companies and record offices in London and beyond were invariably helpful and patient, with many imparting their specialist knowledge. Particular thanks are due to Rob Thomas (Institution of Structural Engineers), Carol Morgan and Annette Ruehlman (Institution of Civil Engineers), Peter O'Reilly (The British Library), Justine Sambrook (RIBA), Hugh Alexander (The National Archives), Doug Bartram-Weight (London Metropolitan Archives) and all their colleagues. Thanks also to the staff of other archive-holding institutions, and individuals, mentioned in the illustrations credits, Malcolm Dickson, Christine Frick, Clare Hackett, Dawn Humm, Nicole P K Lau, Rob Lloyd, Sheila O'hare, John Porter, Susan Scott, Jessica Talmage and Nicholas Webb. The book has also greatly benefited from the cooperation of owners and occupiers of interesting buildings, to whom I am most grateful. Lastly, I should like to thank, Sarah Enticknap, Pauline Hull, Robin Taylor and Jeremy Toynbee for steering the book through to publication.

Note on locations
All buildings referred to in the book are in (modern) Greater London, unless otherwise stated.

Notes on measurements
Imperial measurements are used in the text and in the great majority of original drawings.
Conversion table:

1in. = 25.4mm	1 acre = 0.4ha
1ft (12in.) = 304.8mm	1 sq ft = 0.09m^2
1 mile = 1.6km	1 cu ft = 0.0283m^3

There are a few measurements relating to Continental products that were originally expressed in metric units and for these an imperial conversion has been provided to allow for ease of comparison.

Note on titles
(Sir) in parentheses indicates a person later knighted, but not so at the time under discussion.

INTRODUCTION

The advent of the employment of rolled-iron joists into building art, with its later development of complete steel-framed construction, marks an era upon the threshold only of which we now stand ... Nothing in architectural history can compare in importance with this development since the rediscovery and application of the dome to buildings of magnitude by Brunelleschi and Michael Angelo at the Renaissance.

Arthur Beresford Pite, 1902[1]

A practising London architect and Professor of Architecture at the Royal College of Art, Arthur Beresford Pite had the ability 'to look with equal passion into the past and the future'.[2] His observation was not an overstatement. It was written at a pivotal point in British architectural history, when building technologies based on structural steel components were ascendant and exerting profound influence on building design. Mainstream architecture had finally caught up with the 'Steel Age',[3] and in London and other cities a revolution in construction technique was underway. Its most dramatic expression was the all-steel, skeleton-framed building, in which the rigid, orthogonal, skeletal framework carried the walls, floors, roofs, equipment and inhabitants. Conceptually analogous to timber-framed construction,[4] albeit far stronger, less combustible and on a much grander scale, skeleton construction was an efficient, adaptable building system ideally suited to an advanced capitalist society (Figs 0.1 and 0.2). For a growing number of building types it offered particular advantages, including speed and soundness of construction, fire-resistance, spatial openness, flexibility of planning and, in a commercially driven era, a sturdy mount for huge display windows. Buildings turned heads even as they arose: one Edwardian journalist exclaimed how 'Instead of the enormous confused mass of scaffolding formerly employed, one now sees a towering crane-like erection which hoists steel and iron, stone and brick, into its allotted place, and the great building

seems to be put together upon a plan something like that adopted by the child who from his "box of bricks" completes the toy edifice according to the coloured picture.'[5] Across the Atlantic, in Chicago, New York and other American cities, this system had already elevated offices to unprecedented heights of 20 storeys and more, provided rigid enclosures for electrical power-houses and even framed a soaring church. In Paris and Brussels it had fashioned wondrous, glass-filled shopping emporia, and even in Cape Town and Johannesburg steel-framed technologies were giving rise to the first generation of tall commercial buildings. Such was the reach of this change that in London, where the architectural press and national newspapers periodically vented concerns of an 'American Invasion', it seemed likely that the 'American System' of

Fig 0.1
Gloucester House Flats in course of erection at the corner of Old Park Lane and Piccadilly (1906; Collcutt and Hamp, architects). [RIBA46480, RIBA Library Photographs Collection]

Fig 0.2
Gloucester House Flats as it survives today, with its street-level motor-car showroom reused as a fast-food restaurant.
[Jonathan Clarke]

train, and, equally importantly, British designers were by then independently constructing buildings with steel framing. They failed to capture the level of publicity accorded this long-forgotten proposal, or indeed of better-known subsequent projects such as The Ritz Hotel, but undeniably they all signified the dawn of a new era.

Pite's pronouncement was not only prescient, it was also extremely perceptive. Most revolutions are preceded, precipitated or even accompanied by changes of a more gradual, evolutionary nature and such was the case with the appearance of steel-frame construction. By the close of the first decade of the 20th century, steel frameworks of varying comprehensiveness had become a standard means of structuring large or important buildings in London. On a purely technical level, this state of affairs represented the culmination and coalescence of building technologies that had been discreetly evolving since the Industrial Revolution. Those based on or around the use of structural metal components formed important technological lineages that stretched back as far as the late 17th century, when Christopher Wren used slender iron columns designed by Jean Tijou to help support the galleries enclosing three sides of the House of Commons.[7] By the mid-19th century, with the arrival of structural wrought iron, iron construction had spread far and wide within the metropolis. Spectacular railway sheds at Euston, Paddington and St Pancras, lofty market halls at Smithfield, Covent Garden and Islington, extravagant shopping arcades in the West End, and the Crystal Palace, a vast exhibition hall in Hyde Park that awed the world, all announced an 'Iron Revolution'.[8] The Crystal Palace and a handful of other trabeated mid-19th-century iron structures were particularly important milestones in the history of iron construction, anticipating fully load-bearing techniques of multi-storey framing in steel. Thereafter, in Britain at least, the line of development to steel-frame construction was anything but seamless and sequential. In fact, an apparent hiatus until the construction in 1903–5 of the Anglo-American designed Ritz Hotel, long regarded by many as Britain's first steel-framed building, has helped foster the impression that the technology was imported from America. This most iconic of all early British steel-framed buildings has great significance, but in many ways it has stolen the

construction might take root in its most towering 'skyscraper' guise. A few months before Pite's remark, readers of *The Times* and other broadsheets learned that an Anglo-American syndicate planned to build an enormous office block on the site of the proposed Aldwych 'crescent'. Costing an asserted £2,000,000, with a 750ft frontage to the Strand, parts of this seven-storey glass- and sandstone-clad steel skeleton were to rise to 10 storeys, the whole edifice served night and day by 30 lifts.[6] Nothing came of this, but other projects involving direct American expertise and labour were already in limelight, distorting our understanding both in terms of the assumptions we make about and the questions we ask of this fundamental epoch of architectural and constructional history. American influences and the uptake of steel-frame construction form part of a far broader, longer and largely untold story, one in which Pite's observation about rolled iron joists is key.

The focus of this book is evolution and change in London's buildings and architecture from about 1880 to about 1910. Its emphasis is unashamedly constructional. A great deal has been written about the shape, style and ornament of metropolitan buildings of the period, but comparatively little on their structural anatomy and physiology. Scholarship that has addressed the use of structural iron in British architecture has largely centred on the early or mid-19th century, when the most conspicuous technological advances were made, or on particular types of buildings, such as textile mills, railway sheds and market halls, where iron witnessed some of its most innovative, heroic and celebrated applications. The intention here is to pick up the story where the 'Iron Revolution' ended in *c* 1880 and, mindful of the key, often building-type-specific legacies in structural iron, pursue its various strands into the early 'Steel Age'. At its heart, this book is an examination of how a new structural material – mass-produced steel – came to be produced and first applied to the buildings of one of the world's great cities. Even as the 'Iron Revolution' reached its peak, the ground was being prepared for another. Steel, a superior form of iron, had hitherto only been produced in high quality but small quantity and, from the 1840s, in larger quantity but inferior quality.[9] The invention of the Bessemer and Siemens-Martin processes revolutionised steelmaking, enabling, in time, the mass production of a metal whose homogeneity, quality and physical properties outmatched those of both cast and wrought iron. Steel became the pillar of a new phase of industrialisation and urbanisation throughout the world, and London, where Henry Bessemer had conducted his initial steelmaking experiments, was one of the first cities to make use of it. The first structural steel beams manufactured for use in a building were used here, in a headquarters bank erected in 1860–1 in the city's emergent financial district. They were fabricated from puddled steel, the product of a curious application of wrought-iron-making technology, and showed well the advantages that could accrue from the use of steel. Compared to iron equivalents, they were half as deep yet equally strong, thus helping the architect extract every last inch of interior space from a tight plot. But this demonstration, and more importantly, the gradual uptake of the Bessemer process in these years, was premature, even inopportune. Bessemer's famous invention of 1856 did not, as is commonly supposed, significantly impact on architecture or structural engineering in Britain or America. Bessemer steel was a brittle, moderate-quality material not truly suited to structural applications. Experience began to show that open-hearth steel produced by the rival Siemens-Martin process was of a structural grade, although it was not until the 1880s that steel producers, seeking new markets, began rolling structural shapes for use in construction. London, with a long pedigree of iron construction and a voracious appetite for wrought-iron joists and girders, would be a key consumer. Change begat change. A discreet revolution had begun.

The next quarter century, between 1885 and 1910, saw significant changes in the design and construction of many building types in London. It was a period of considerable complexity, transition and overlap. Older structural techniques and forms developed in iron and masonry persisted, and in many buildings steel was used alongside both cast and wrought iron or load-bearing brick or stone. Constructional advances provoked changes in some building types more than others, but reciprocally, the demands of some types were more stimulating to innovation than others. It was also a period of increased territorial ambiguity within the design professions. As steel construction developed, and buildings became larger and more complex, specialised engineering contractors became crucial, and consulting engineers indispensable. Structure was forced back onto the architectural agenda, much to the dismay of many architects. But this transition was also marked by cooperation among the design professions, often resulting in buildings of great distinction that incorporated improved constructional components or techniques, and even new structural forms. Techniques of framing evolved to make buildings more open, better lit, more stable, or to give them stronger floors or wider roofs. Initially incorporating steel as a lesser component, but increasingly as the major constituent, framing technologies of varying sophistication, comprehensiveness and self-sufficiency were early on exploited in a great many buildings. This evolutionary thrust gained in momentum, and is perhaps the most significant characteristic of these developments. By the early 1900s, with the all-steel, multi-storey skeleton frame finding growing application for some of London's most important buildings, it was clear that building technology had outpaced building legislation. London's 1894 Building Act had not foreseen full steel-frame construction. It soon became clear that a new Act was needed. In 1909, after considerable debate that threw into focus the changing nature of architecture, and its relationship with engineering, such an Act was finally passed. London's opening chapter of building with steel had ended, and a new one had begun.

This then is a story of change, not purely in the buildings themselves, but also in the legislative and professional environments, including the attitudes and understandings of the designers concerned. Architects responded varyingly to steel, as they had done to iron; some used it modestly, secretly and guiltily, while others exploited it fully, openly and audaciously. These responses were themselves conditioned by specific factors and wider milieux; matters of cost, fire-resistance and corrosion, of building legislation and of architectural philosophy and aesthetics, all exerted varying pushes and pulls. Constructional firms and engineers, many little known, are as much a part of the story as the architects and their clients; without their specialised contributions, most of the buildings covered in this book would not have been possible in the forms they took. Yet the central role of these artisans and professionals was often overlooked in the architectural and building periodicals of the day, at least in the early years before increasing steel usage forced recognition. Indeed the steel (and before that iron) component of outwardly brick buildings was often as not ignored, or referred to only in passing, a partiality often reflected in English architectural histories of the period. A typical instance of this was the write-up of a large textile warehouse completed in 1884 in Wood Street in the City of London. *The Building News* paid tribute to architects Ford and Hesketh for 'the great skill' they 'displayed in obtaining the greatest amount of light without depreciating from the architectural effect' (Fig 0.3). The building's extensively glazed granite and Portland stone façade signalled a new look for such commercial warehouses, in which stone or brick pilasters were routinely diminished to sizes unexampled by those of an earlier generation. Constructionally, unassisted brickwork of such attenuated proportions would have been unsound, and this 170ft-long elevation must have been carried on structural iron columns and beams.[10] At this date it might just possibly have incorporated steel joists. Yet the write-up omitted all mention of the building's structural underpinnings that directly informed its appearance, and probably its interior arrangement. The contractor, William Brass, who during the previous year was busy supplying, fabricating and erecting the iron and steelwork for the Alhambra Theatre in Leicester Square, was, however, disclosed. In most instances, it was left to the engineering contractors or consulting

Fig 0.3
Seldom disclosed in building accounts, the visual effect of constructional iron and steel was nonetheless unmistakable in many of London's commercial buildings of the later 19th century.
[The Building News, 7 November 1884]

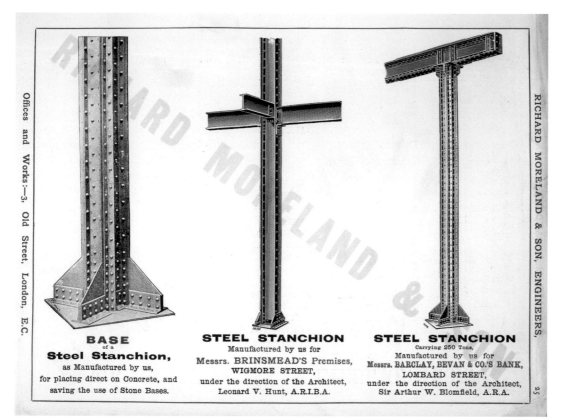

Fig 0.4
Trade catalogues such as this from Richard Moreland & Son suggest that by the early 1890s rolled-steel I-sections had come to be used for interior support in a range of London buildings.
[©The British Library Board (F60077-54)]

engineers to proclaim their involvement through specialist periodicals and trade catalogues (Figs 0.4 and 0.5). Such examples indicate the difficulties of actually recognising and researching construction that is both poorly documented and, where extant, often hidden from view for aesthetic or technical reasons. However, synthesised understandings derived from a number of sources, documentary and physical, do permit informed speculation, and hence, it is hoped, a more holistic account that reinstates some of the era's 'Cinderellas', namely the fabricators, builders and designers, and their works in iron and steel.

The book is structured into two parts. The first examines the technological developments and economic forces that brought structural steel into being. Chapter 1 looks at how production of structural-grade steel began, what its special qualities were, and why it took so long to achieve official and institutional credibility. Some engineers were prepared to use it early on, yet it was not until it had gained acceptance in other areas, notably shipbuilding, that restrictions placed on its use were lifted, and its full strength and utility could be exploited in engineering and architecture. Chapter 2 looks at the emergence of steel constructional components, which had a long pedigree in structural iron equivalents. Beginning with cast iron, which had many inherent limitations, it moves on to wrought iron, a metal similar in many respects to steel. It suggests that the introduction of the I-beam, which was not possible in the older, brittle metal, marked a fundamental advance in industrial production and structural efficiency, ultimately revolutionising constructional technique. It fostered the development of fire-resisting flooring technologies, as well as other new innovations, notably Julius Homan's system of compounding beams which helped overcome the problems associated with limited sectional sizes. Finally, it considers the great array of structural steel components that British firms, seeking new markets, began to produce in the 1880s. One company in particular, Dorman Long & Co., directed itself wholly to making a product that would win over the lucrative girder trade, dominated mostly by continental companies. The London market was a key objective in this, and setting up constructional works there in the 1890s ensured its leading role. With increasing numbers of structural sections produced by greater numbers of steel rollers, standardisation, following American and German precedent, became desirable.

Standardisation promoted greater rationalisation, and yet further consumption. Chapter 3 looks at a largely unsung stimulus for British structural steel production, the importation of continental rolled iron, especially from Belgium and Germany. Continental imports of iron and, to a lesser extent, steel, really helped kick-start British production by offering a competitive spur, and not until British rollers became competitive did the use of Belgian, French and German iron slacken.

Chapters 4 to 8 are a series of contextual essays that examine some of the most important factors, consequences and milieux pervading and surrounding early structural steel. The acceptance and uptake of steel construction, and of its gainful conclusion, the fully framed, multi-storey building, were affected and mediated by factors both specific and non-specific to London: its building regulations, the philosophical attitudes of leading theorists and practitioners within the architectural profession, and the transatlantic importation of ideas, expertise and men. The evolution of the fully framed building was a hugely significant advance, and given the historiographic interest and debate surrounding it in this country, as in America, it receives detailed consideration.

Fig 0.5
Framed steelwork like that illustrated in Fig 0.4 was employed in John Brinsmead & Sons' piano showroom and warehouse at 18–22 Wigmore Street, erected in 1892–3 to designs by Leonard V Hunt, architect. [DP134017]

Its fully fledged appearance on the streets of Edwardian London was emblematic of the productive cooperation and interdependency of the architect and the engineer but it was also symptomatic of a disjuncture between the professions, a polarisation of respective roles long underway, but one compounded by the advent of steel-frame architecture. The changed and changing dynamics of architect–engineer roles formed a subtext to the London County Council General Powers Act 1909 (the 'Steel Frame' Act or 'Engineers Act' as it was also dubbed) and helped set the tone for professional relations for the rest of the century and beyond.

The second part of the book is its heart, an examination of how structural steel was exploited in different types of London building before 1910. Theatres and music halls, clubs and hotels, banks and offices, industrial buildings, stores, houses, churches, pools, fire stations and tube stations are all considered. For the sake of comparison, and to gain clearer insight into the changes engendered by steel, forerunning developments in iron are reviewed first. In most buildings, early techniques of steel construction closely followed those of iron. However, in others, such as power stations, which were unknown until the late 19th century, there were few iron-using precedents for designers to turn to. Various factors have influenced the selection of the buildings discussed in this part of the book. Some, such as hotels, for instance, are central to the story, having played an important role in this country in the evolution of the full steel frame. The scale of steel usage, and the sophisticated nature of the construction influenced choice in these and other instances, such as theatres, where steel was put to highly specific, and demanding uses. Other types, such as power stations and tube stations, are included because they are among the more distinctive groups of buildings that London produced, with shared structural and architectural design features. One drawback of this typological approach is some repetition whereby proprietary fire-resisting floors, particular forms of columns and even generic forms of construction reappear in different contexts. This makes the first part of the book all the more important, in which the details of construction techniques are introduced in context. An approach of this nature has, however, the advantage that constructional techniques and structural forms endemic to particular types of buildings are better revealed. In the 19th century, as today and always, structure, form and function were closely interrelated. Constructional history should not be separated from functional understanding, let alone architectural or social history.

The final part of the book both sums up, and offers conclusions on the impact of structural steel in its opening era to London's buildings. The substitution of iron for masonry in certain load-bearing situations had initiated decisive changes in the formal and spatial composition of many building types through the 19th century, but the advent of structural steel from the 1880s had a lasting, more profound impact. The interior of a typical late-Victorian theatre, where steel-cantilevered balconies enabled, for the first time, sightlines fully unobstructed by supporting columns, is but one demonstration of this. Reaching maturity in the early 1900s, steelwork theory and practice laid the foundations for the rest of the century, with ultimately transformational effects on architecture. Yet in the period considered by this book, London was not obviously transformed in the way that other cities were. Truly the greatest 'world city', at the height of its economic, social and political might, it was also innately traditional, a condition that helps explain the discreet nature of its own steel revolution.

Charting the development of some of London's key building types of the late 19th and early 20th centuries from this perspective presents a rich and complex, yet largely untold, story. It is a story that it is hoped will lead to more rounded assessments and greater appreciation of the constructional aspect of London's – and England's – built heritage of this era. This book arose from publicly funded and institutionally supported architectural investigation of buildings threatened with alteration or demolition, and is therefore a step towards an appraisal of buildings the technological importance of which has been undervalued, complementing previous or current research into the subject. If this book enhances appreciation for an often hidden and comparatively neglected side to Victorian and Edwardian architecture, then an important purpose will have been achieved. This is not simply a preservationist argument, but a matter of bearing witness for posterity. Above all, it is written for all those who have ever paused to wonder what lies beneath the surface, and what happened on the British side of the pond during an exciting period in architectural and constructional history.

Part I
Technological preconditions and other contexts

Towards a structural steel: technological and official hurdles and professional diffusion

Recent improvements in the manufacture of steel have been of such a nature as to bring in a prominent manner before the attention of the [architectural] profession the question of the comparative qualities of steel and welded iron.

(1882)[1]

Steel is not only from 30 to 40 per cent stronger than iron, and four times as ductile, but it is a material which is capable of being heated with so exact precision in the open-hearth furnace that any given quality can be made to predominate, whether hardness or ductility, and to almost any extent.

(1897)[2]

Looking back, it is astonishing with what difficulties the use of steel for structural purposes was beset.

(1899)[3]

It was not until the early 1880s, after nearly three decades of technical snags, official and institutional uncertainty, and the obstructive torpor or hostility of ironmasters, that mass production of structural-grade steel really began. Virtually the same product that is used by engineers and architects today, mild steel, as it had come to be known, combined a degree of strength, toughness and ductility that set it apart from all other materials, including cast and wrought iron. With a compressive strength almost as great as cast iron, and a tensile strength appreciably in excess of wrought iron, it lent itself to use for all the cardinal structural members in buildings and bridges – beams, columns, struts, ties and connectors. Like wrought iron, its ductility was such that it could be riveted without shattering, and, subject to excessive stress, would fail gradually rather than abruptly without warning. Critically, mild steel was able to endure far more stress than wrought iron when approaching its elastic limit or yield point, and beyond its limit it had a greater propensity to deform plastically by buckling or elongating rather than by tearing or rupturing. Its elasticity, or ability to return to its original shape and dimensions once a load had been removed, gave it great resistance to fatigue, making it well suited to the most intensive of applications. In essence it was superior in a whole range of criteria, allowing, in theory, a new scale of building with a higher margin of safety. These properties were directly related to the precise quantity of constituent chemical compounds and elements, namely carbon, and the purity and homogeneity of the molten metal, neither of which had ever been satisfactorily resolved by (Sir) Henry Bessemer's method of producing bulk steel.

Unleashed on the world in 1856, Bessemer's revolutionary process may have held the promise of cheap, high-quality steel for a multitude of uses, but constructional application was never destined to be one of them (Fig 1.1). Beset by technical difficulties from the start,

Fig 1.1
Britain's only surviving Bessemer converter, built in 1934 for a steelworks in Workington, Cumbria, but now standing silent at Kelham Island Industrial Museum, Sheffield. Henry Bessemer's process for mass-producing steel was used until the 1970s, yet the quality of steel it produced in the 19th century was unsatisfactory for most structural purposes.
[AA022440]

including reliance on low-phosphoric ores, and viewed with suspicion and scepticism by powerful ironmasters, the Bessemer process was only a first step, and one that quickly gained a reputation for unreliability. Apart from the difficulties posed by phosphorous and other impurities, the real problem was inherent in the speed of the process itself, each 'blow' lasting just 20–30 minutes. While this was a merit insofar as rapidity of production was concerned, it gave inadequate time for close quality control, especially in judging exactly when the correct carbon content had been achieved in the deoxidising stage. The ingots produced were of variable quality – often hard and brittle, though this did not disqualify the metal from a small number of applications. The railways were the chief consumers of Bessemer steel in the early years, using it for rails and locomotive tyres, axles and boilers. The London and North Western Railway (L&NWR) Company even established its own converters in 1863 – soon becoming one of the largest integrated steelmaking and engineering plants in the country – and in March of that year laid steel rails at Crewe Station. Wrought-iron rails at the same station had to be turned every 9 months; the steel substitutes were still in place 18 years later, showing wear to one head only.[4] Despite showing itself to be longer lasting and

technically superior, Bessemer steel was fundamentally a 'moderate-quality, general-purpose steel',[5] and consequently found few other early applications. It has even been estimated that through most of the 1860s there was surplus Bessemer capacity in Britain.[6] In 1866 Lloyd's authorised construction of a 1,550-ton 'experimental' steel ship, with a 25 per cent reduction in the sectional area of the deck beams. This trial was not repeated until the general introduction of open-hearth steel in the 1870s because it was considered impossible 'to produce a material of a uniform and trustworthy character'.[7]

The development in the 1860s of an alternative method of tonnage steelmaking, the open-hearth process, held greater assurance of high-quality steel for all uses, including construction (Fig 1.2). Patented by (Sir) Charles William Siemens in 1866, following work by Emile and Pierre Martin, licensees of his original regenerative furnace, this process was destined to transform British shipbuilding and boilermaking, and, in turn, constructional engineering. Because it was far slower in operation, each 'melt' lasting anywhere from 7 to 15 hours, it was possible to take samples at leisure, ensuring production of a bath of molten metal whose exact chemical composition was known. Through precise control of temperature and the

Fig 1.2
Ranks of open-hearth furnaces for the production of high-quality, structural-grade steel at Dorman Long's impressive Britannia works, Middlesbrough c 1910.
[Pocket Companion … of steel manufactured by Dorman Long & Co. Ltd, Middlesbrough, England (1913)]

content of carbon and other key constituents, mild steel of a far higher quality and more uniform nature could be produced. Until 1875 open-hearth steel production went almost exclusively into rails; the first were laid at Paddington in 1867, and the following year the L&NWR installed open-hearth furnaces alongside their Bessemer converters, becoming the first British steelmakers to take up the process on a manufacturing scale.[8] A growing reputation for reliability saw the process make inroads into other markets, notably shipbuilding. Plates, angles and other parts for use in ships' hulls had to be of the highest strength, quality and trustworthiness, and the ability of the open-hearth process to meet these requirements encouraged shipbuilders to make the transition from iron to steel.[9] In 1867, under its chief naval architect E J Reed (1830–1906), the Admiralty began building in steel, but steel shipbuilding for the merchant navy was delayed until 1877 when Lloyd's, content that mild steel of the highest quality could be produced by the open-hearth process, agreed to a reduction of 20 per cent in the weight of plates and scantlings relative to iron. Weight for weight, steel was still more expensive than iron at this time, but the sanctioning of thinner sections widened its economic reach.

It was in the 1870s, with mounting appreciation of the quality and dependability of the steel being produced for shipbuilding, that a number of engineers and steelmakers began turning their attention to bridges and other structural applications. The Board of Trade was the main obstacle to the transfer of the technology, Bessemer steel having set a tenor of mistrust. As early as 1859, (Sir) John Hawkshaw had proposed to use Bessemer steel in constructing the Charing Cross railway bridge across the Thames, but he was prevented from doing so by Board of Trade restrictions.[10] These placed the material properties of steel on the same footing as those of wrought iron, limiting the permissible stresses to 5 tons per square inch in tension, and 4 tons per square inch in compression. By refusing to sanction the higher stresses that steel could safely accommodate – which would have made it, weight for weight, competitive in price with wrought iron – a powerful economic disincentive was set in place. Hawkshaw had to resort to wrought iron for his lattice-girder bridge, and the Board stubbornly refused to relax its strictures for almost two decades despite the use of steel in Continental bridge

building. Most notable was a group of three long-span bridges built in Holland in 1862,[11] with steel supplied from Sheffield by John Brown's Bessemer works;[12] lesser known examples included a railway bridge of 1866 across the Gota River in Sweden,[13] and an 82ft-long shallow arch bridge built for the 1867 Paris Exhibition.[14] In England, *The Mechanics' Magazine* commented in 1868: 'In our own country … there are but one or two isolated examples existing of the employment of the latter material [steel] in the designs of bridge work.'[15] One documented example was a railway swing bridge built over the Sankey Canal in 1865 to designs by Samuel Barton Worthington. Bessemer-steel plates were used in the fabrication of four 56ft-long girders; these weighed about five-sixths of the weight of wrought-iron equivalents.[16] But the great majority of engineers were sufficiently deterred. Rowland Mason Ordish originally proposed that his Albert Bridge (1871–3) across the Thames should make extensive use of steel, *The Mechanics' Magazine* noting in 1866 that 'In the construction of every proportion of the roadway, girders, and tension rods of this bridge, steel will, we believe, be used, to the exclusion of wrought iron.'[17] Although the material was tested at David Kirkaldy's recently established Testing Works in Southwark and 'the results of these experiments … proved in every way satisfactory',[18] wrought iron won the day. It was elsewhere, in a bridge over the River Moldau in Prague of 1867–8, that Ordish was able to exploit the superior tensile properties of steel.[19] *The Mechanics' Magazine* lamented the obstacles to the take up of steel, stating, in 1868:

One insuperable drawback against its general adoption is the apathy displayed by the Board of Trade, and their entire unrecognition of its merits. So long as that great and inappealable authority continues to deny its superior qualities, and refuses to make any alterations to its arbitrary rules with respect to metallic structures for railway purposes, so long will it be impossible for the engineer having regard to economy, to adopt it in his designs.[20]

Mindful of the increasing quality of steel for shipbuilding, in 1874 the British Association for the Advancement of Science recommended a permissive stress in both compression and tension of 8 tons per square inch. It took another

3 years, with mounting agitation from engineers and steelmakers, coupled with the examples set by the Admiralty and other bodies, for the Board of Trade to appoint a specialist committee to review the conservative strength coefficients it had previously laid down.[21] In 1878 this committee, made up of Hawkshaw, William Henry Barlow and Colonel W Yolland, endorsed the British Association's recommendation of 8 tons per square inch. Despite this commendation, the Board would only sanction stresses of 6½ tons, 'still leaving British practice far behind continental practice where stresses up to 10 tons were not uncommon'.[22] This was enough to permit the use of appreciably thinner steel sections, making the material more competitive in relation to wrought iron. Almost immediately use of the newer metal began to grow. Arthur Thomas Walmisley (1849–1923), a respected authority on long-span iron roofs, who had already used steel for the tie-rods of the 160ft-span roof of Central Station, Liverpool (1874), used it for both the tie- and suspension-rods of the 170ft-span roof of Queen Street Station, Glasgow (1879).[23] In the same year, at least two large railway viaducts

were completed using steel, including the Llandular Viaduct on the L&NWR line to Holyhead,[24] and another at Meldon on the London & South Western line from Exeter to Plymouth, which used Warren girders and trestles fabricated from steel supplied from J & R Moor's Stockton-on-Tees works.[25] Doubtless other similar structures were built at this time.

In the 1880s, steel, and specifically Siemens-Martin open-hearth steel, showed itself to be the engineering material of the future. Sir John Fowler's and Sir Benjamin Baker's mighty Forth Bridge, Scotland, spectacularly symbolised the indomitable ascent of structural steel over iron for bridges and other long-span structures (Fig 1.3). Erected between 1882 and 1890 by Tancred, Arrol and Company, it used some 58,000 tons of the metal supplied from steelworks in Wales and Scotland.[26] In the wake of the Tay Bridge disaster of 1879, it had to be designed to withstand wind pressures as high as 56 pounds per square foot.[27] The bridge's designers convinced the Board of Trade to increase the maximum stress from 6½ to 7½ tons per square inch by using steel of the best quality, the ultimate strength of which was at least 30 tons per

Fig 1.3
The mighty Forth Bridge takes shape, July 1889. Some 58,000 tons of Siemens-Martin open-hearth steel was employed, announcing the eclipse of wrought iron for engineering structures. Engineered to withstand wind pressures as high as 56 pounds per square foot (3.6 tons per square inch), Sir John Fowler and Sir Benjamin Baker believed their design would not have been possible without steel.
[Courtesy of Institution of Civil Engineers]

square inch.[28] In a metropolitan context, both Tower Bridge, begun in 1890, and Walmisley's re-fit of Hawkshaw's Charing Cross Bridge in 1888 were also of importance in signalling the eclipse of iron. The structural engineering of Tower Bridge, however, was literally masked by its architectural cladding, and its significance, at least in the context of this story, is to do with how it re-ignited old controversies within architectural philosophy (*see* p 56). Charing Cross Bridge was a more 'functional' structure, one that demonstrated clearly and unambiguously the advantages of steel. Walmisley used wrought iron for the main and cross-girders, but, to economise on weight, utilised steel for the main platform girders and those resting upon the cylinders of the old structure. *The Building News* remarked (of both the Scottish and London bridges) that 'These instances of the employment of steel are significant, and indicate the increasing trust which engineers place in that material, which is now superseding both wrought and cast iron wherever strength and lightness are necessary.'[29] Further afield, another end-of-the-decade structure, historiographically overshadowed in Britain by the Forth Bridge, dramatically demonstrated the potential expediency of steel. This was the breathtaking Galerie des Machines, erected for the Paris International Exposition of 1889 to designs by Ferdinand Dutert and Victor Contamin (Fig 1.4). Two years before it was constructed, *The Building News* thought this exhibition building, with a clear span of 364ft, would be 'the first great exemplification of the value of steel for structural purposes of this kind ... [and] will confirm the opinion of engineers that steel can be used in roofing with economic results that cannot be obtained in iron'.[30] However, in the event it seems that at the last minute wrought iron was substituted, 'on the two-fold ground of expense and the necessity of hastening the execution of the work'.[31]

If the use of structural steel in engineering structures was postponed because of official mistrust, then its delayed use in buildings was perhaps less due to the conservatism of architects – as is conventionally thought – and more to the unavailability of useful structural sections. Unaffected by the Board of Trade regulations (which governed merchant shipping, transport and other interests relating to internal and foreign trade, but not buildings) steel girders made their world debut in a London building, the London & County Bank,

Lombard Street, built in 1860–1. These puddled-steel members, however, were almost certainly manufactured to order (*see* pp 19–20), as probably were those used in Siemens's country house, 'Sherwood', near Tunbridge Wells, Kent (Fig 1.5), perhaps the earliest documented example of open-hearth beams used in a 'polite' British building.[32] Not surprisingly, the first buildings to make consistent use of steel beams were those built for the railways – the first large-scale users of steel – and specifically those built for the L&NWR, who early on became producers in their own right. That company's Heaton Norris Goods Warehouse (1878–80) and the Huddersfield Goods & Wool Warehouse (1881–4) both used joists rolled by the Horseley Company, Staffordshire, which research has shown to be steel.[33] Nonetheless, the great majority of architects and patrons did not have such close links to the nascent steel industry. Indeed, to most architects away from the vanguard of metallurgical and structural developments, steel must have seemed a strange and alien material. Yet their professional journals were eager to

Fig 1.4
The Galerie des Machines of the Paris International Exposition (1889; Dutert and Contamin, designers). Like the Forth Bridge, to contemporaries this was a monumental harbinger of the 'Steel Age'.
[Bridgeman Art Library 205724]

educate and familiarise. As early as 1869, the year the Iron and Steel Institute was formed and began running its recurring discussions on 'steel versus iron', *The Builder* assured its readers that 'Steel is, in fact, no more a different material from iron, than are wrought-iron or cast iron from each other.' The same periodical, at this juncture, thought the 'wider use of Bessemer steel will afford architects an extended range for carrying out constructive and ornamental details'.[34] Three years earlier, it had reported on some French experiments into the compressive strength of steel columns, concluding 'if … confirmed by future trials, steel will prove a valuable material for resisting compressive strains'.[35]

It was in the early 1880s, with the confluence of a number of factors, technological, economic and productive, that the decisive changes began. Intense discussion in the engineering world, stimulated by the reality of structural-grade steel, the use of which by this date was less hampered by officialdom, began flowing into the architectural world, with architectural magazines and journals the chief conduit. Perhaps most influential were the summaries given by *The Builder* of two papers presented to the Royal Institute of British Architects (RIBA) by Professor Alexander W B Kennedy and J Allanson Picton, entitled, respectively, '"Mild steel" and its application to building purposes', and 'On iron as a material for architectural construction'.[36] Both papers,

given in 1880, were of course printed in full, with lengthy discussions, by RIBA's own journal.[37] Later in the year, a paper read before the Society of Engineers entitled 'Modern steels as structural materials' was reported,[38] and in 1882 so too was Ewing Matheson's paper 'Steel for structures', presented before the Institution of Civil Engineers.[39] Through the rest of the decade, the topic continued to receive coverage, and steel attracted increasing numbers of advocates as accurate testing demonstrated its superior physical properties and structural utility.[40] To all intents and purposes steel became a technically proven material. Discussion continued with fervour through the 1890s, by which time practical examples of steelwork in buildings were standard.[41]

Steel had one other major hurdle to overcome from a constructional standpoint. Before it could start being regularly used in buildings, it had to be produced in useful and efficient forms. Steel angles, bulbs, plates and deck beams for shipbuilding were being rolled before 1875,[42] but it was not until the mid- to late 1880s that a number of foresighted firms began bulk-rolling structural sections. The forms these took were based on those of wrought iron, and it was the flourishing and lucrative trade in such sections – met mostly by Continental producers – that induced many producers to diversify. How this came about, and the constructional revolution it engendered, is examined in the next chapter.

Fig 1.5
'Sherwood', Tunbridge Wells, Kent. Built in 1867 to designs by Frederick Johnstone, architect, this Italianate mansion was purchased by Siemens in 1874, who lived here and in his London home in Kensington until his death in 1883.
[Reproduced by courtesy of Savills (L&P) Limited]

2

Constructional steelwork and its iron inheritance

Iron and steel are every year taking a more important place in architectural designs, large hotels, theatres, banks, and warehouses being constructed with a metallic framework, which economises space. The low price of steel has encouraged such methods. (1893)[1]

So in January 1893 said *The Builder*, perhaps the most influential mouthpiece of the British architectural and building world. This world was in the process of a dynamic transformation, in the midst of a changeover from iron to steel as its leading-edge structural material. Driving this change was a fundamental shift in the structural trade, in which British manufacturers had finally woken to the fact that a vast market existed for rolled structural steel shapes for use in buildings of all kinds, a market that hitherto had been met predominantly by Continental suppliers and merchants of rolled (wrought) iron. By investing heavily in the basic open-hearth and in powerful rolling plant capable of working the hot metal, a handful of English and Scottish companies were able to wrestle control from Continental producers by mass producing a stronger, lighter and more versatile product. Mass production of structural sections in wrought iron had never been feasible for British producers, due to limitations in the rolling technology and the expense of producing the metal. In steel it was a natural corollary, and the 1890s saw an explosion in the range of sizes and shapes of rolled shapes manufactured for use in building and engineering, so much so that in 1901 a special committee was created to implement standardisation. By this date, open-hearth steel had eclipsed wrought iron in virtually all structural applications, the combined result of its superior physical properties and its mass availability at competitive prices. Cast iron still clung on, but only in compressive situations, in the form of columns and stanchions, but even there its use

was steadily declining. Metal, in the form of mild steel, now routinely entered into the construction of virtually all major metropolitan buildings to a degree that would have been both implausible and untenable in the eras of both cast and wrought iron. Architects and builders, who a generation earlier might have shied away from exploiting metal structure on economic or philosophical grounds, now saw it as an indispensable aid to creating large, safe, well-lit, non-combustible spaces.

Joists, beams and girders

> The introduction of the I beam was an important event in the history of building techniques, because with it some completely new forms of construction emerged, and most importantly the principles of building took another direction. The iron period in building truly began with this material.[2]

Eduard Mäurer, a German industrialist, wrote this perceptive statement in 1865, less than 20 years after the world's first wrought-iron beams had been rolled in France. It could equally have been written by a French or Belgian commentator, given the rapidity with which wrought iron was supplanting cast iron in tensile applications in Continental building practice. It is less likely that such words would have come from a British observer. Even at this date, the conservatism of some British builders and architects, aware of the manifold advantages of wrought iron, but constrained by matters of availability and cost, was ensuring the continued use of cast iron, alongside timber and masonry. A year later, *The Builder* could still note 'although every architect knows that cast-iron girders are really dearer than wrought-iron girders, inasmuch as they require more than double the weight to furnish equal strength, and even then that cast-iron girders cannot be relied on through defects inherent in their manufacture, yet at this very moment we may behold them placed in more than one costly erection progressing in the metropolis'.[3] Even so, the early stages of a revolution in construction technique were unequivocally afoot, with London leading the rest of the country in the application of wrought-iron beams into buildings of all types. Too discreet in its architectural manifestations to attract more than occasional comment by contemporaries, it ultimately effected far more widespread and pervasive changes than were ever possible in the era of cast iron.

The dangers and limitations of cast-iron beams

About 20 minutes to 11 o'clock a sudden snap, something similar to the report of a gun, was heard, and the next moment the men, to their horror, found the building tumbling beneath them … Some of the poor fellows, by an extraordinary and almost superhuman effort, bounded, as the floor gradually gave way, on to the roof of the church, while others, in as astounding a manner, contrived to leap on to the adjoining houses. The great majority of the workmen, however, fell with the building, and the scene that ensued may possibly be imagined … From some cause or other the uppermost girder suddenly snapped in two; its excessive weight dragged the wall out of its perpendicular, and the girder, getting loose, fell and broke the girder below. This instantly brought the walls down with the floors. (1851)[4]

By the 1850s there was a growing appreciation among architects that wrought-iron beams were lighter, stronger, safer and more reliable than their cast-iron equivalents. Cast iron was unsuited to use as beams, since it had little tensile strength and gave no warning of failure. Already, engineering confidence in cast iron had been badly shaken by the collapse of Robert Stephenson's Dee Bridge, Chester, in 1846. A series of well-publicised and devastating collapses of multi-storeyed textile mills, internally framed in iron, including Radcliffe's Mill, Oldham, in 1844, and Gray's Mill, Manchester, in 1847, brought the material's inability to absorb tensile stresses alarmingly to the fore within the building world.[5] A particularly horrifying accident in Gracechurch Street in London, described in the epigraph above, propelled the issue firmly into the mindset of metropolitan builders. Not only did it highlight the unpredictability of cast iron, with its potential for hidden flaws and defects, it underscored the absolute necessity of rigorously proof-loading every beam, not just a sample of material or actual members.[6] Thereafter the approach to testing adopted at the Albert Dock warehouses in Liverpool (1843–5), where strips of iron that were cast integrally with the beam were removed and tested in bending, would never offer complete peace of mind.[7] For the Great Exhibition, Hyde Park, in 1851, every cast-iron beam was independently tested, and the entire structure, once erected, was further tested by marching men and rolling canon balls across the galleries.[8] Such precautionary measures were costly and time consuming. They also continually underscored the treacherousness of the material to those architects and builders who chose to use it. In contrast to the brittle, crystalline nature of cast iron, wrought iron was a ductile, plastic material that deformed under excessive load, giving visual warning of impending disaster.

There were other inherent limitations surrounding the design and manufacture of cast-iron beams that had in any case held back their general use in the wider building world beyond the specialised field of industrial buildings. By 1831 the cast-iron beam had reached its theoretical apogee in terms of efficient form, that is, the maximum strength for cross-sectional area. This was the result of Eaton Hodgkinson's experimental analysis of beam action, in which he improved upon Thomas Tredgold's symmetrical I-section by widening the bottom flange to compensate for the metal's lower strength in tension. Whereas Tredgold had deduced incorrectly that the material was equally strong in tension and compression, Hodgkinson's 'ideal section' correctly accounted for the fact that cast iron was some five or six times weaker in tension than in compression. But Hodgkinson's theoretical ideal was rarely if ever achieved in practice, for a casting with such marked differences in flange and web dimensions would have cooled and set unevenly, generating internal stresses within the material. Consequently, the ideal had to be adapted to the requirements of foundry practice, with maximum compressive and tensile stresses typically nearer three than the ideal ratio of five or six.[9] Cast-iron beams were thus always inherently inefficient in cross-sectional shape, and, as a result, remained heavy, wasteful and, compared to timber, expensive. This was of less concern in a multi-storey iron-framed mill, where standardised bay lengths meant that, through economies of scale, numerous castings to the same pattern could be advantageously applied. In the context of public and commercial buildings, however, where varying room dimensions militated against such standardisation, the use of cast-iron beams tended to be restricted to the upper echelon of building projects, where costs were less assiduously cut. In London, what evidence there is for the use of cast-iron beams in 'polite' buildings suggests that it was the leading, most technologically adventurous architects who championed their employment, albeit using the older, less efficient but more readily cast inverted T-section form. (Sir) Charles Barry or his pupils used inverted T-section beams at the Travellers Club (1829–32), the Reform Club (1837–41), the new Houses of Parliament (1838–43), and No. 12 Kensington Palace Gardens (1845–6). William Wilkins employed similar forms at University College, London (1827–8), and in the National Gallery (1833).[10]

In all these examples, cast iron had been employed for its perceived 'fireproof' qualities, supporting non-combustible brick arch or ceramic floors from the beams' wide flanges, or for its ability to provide large-span floors that did not sag or bounce. This second attribute was developed independently within the domain of fashionable architecture, not in textile mills, where bay spans rarely exceeded 20–25ft. It was the need to create generous, rigid floors in grand buildings that stimulated innovation towards longer-span cast-iron beams. John Nash used mostly solid inverted T-beams of up to 37ft in length at Buckingham Palace (1825–30), but the acme in span-length was reached by Robert Smirke (1781–1867) in the King's Library at the British Museum, of 1824–5. Smirke, who, like Nash, had already trialled iron beams in the floors and roofs of his country houses, was persuaded by John Urpeth Rastrick of the iron-founding firm Foster, Rastrick & Co., that girders as long as 44ft could be cast in one piece. The resulting members were of advanced design, incorporating oval voids in the web to save metal and weight, but their huge length necessitated an enormous depth at mid-span – over 3ft, illustrating one of the drawbacks of using cast iron in flexure (Fig 2.1).[11] In profile, when simply supported at either end, cast-iron beams typically increased in depth towards the centre, to accommodate the

Fig 2.1
Long-span iron floor construction at the King's Library in the British Museum (1824–5; Robert Smirke and John Urpeth Rastrick, designers).

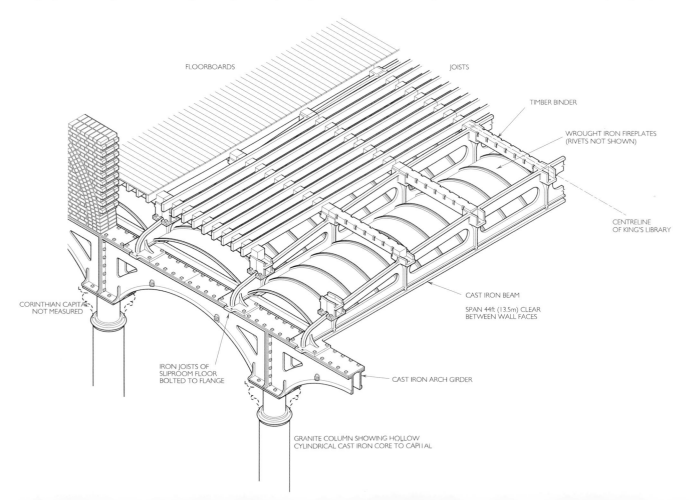

FLOORBOARDS

JOISTS

TIMBER BINDER

WROUGHT IRON FIREPLATES (RIVETS NOT SHOWN)

CENTRELINE OF KING'S LIBRARY

CAST IRON BEAM

SPAN 44ft (13.5m) CLEAR BETWEEN WALL FACES

CORINTHIAN CAPITAL NOT MEASURED

IRON JOISTS OF SLIPROOM FLOOR BOLTED TO FLANGE

CAST IRON ARCH GIRDER

GRANITE COLUMN SHOWING HOLLOW CYLINDRICAL CAST IRON CORE TO CAPITAL

increased bending stresses there. This typically took the form of a curved 'hog-backed' or 'fish-bellied' profile, which became more pronounced the longer the span. While structurally efficient in its distribution of metal, it was this maximum depth at mid-span that dictated the depth of the floor where it was decided not to coffer the ceiling or, rarely, leave the metal exposed. Architecturally and economically, this was an extravagant waste of available space. Such beams were also extremely heavy, requiring thick brick or masonry load-bearing walls, and awkward to manoeuvre, involving difficult lifting procedures. Long-span cast-iron beams, trussed with wrought-iron tie-rods designed to remedy the brittle material's tensile deficiencies were also extremely deep, as an example from Goldsmith's Hall in the City of London shows (Fig 2.2). Such composite solutions to the problem of providing safer, longer and incombustible spans were highly ingenious, but were employed less often than cast-iron beams, partly because of the prohibitive design and fabrication costs.

Despite their unreliability and unwieldiness, cast-iron beams would almost certainly have continued to proliferate – mainly on account of their incombustibility – had a superior material not emerged. From the mid-1840s, cast-iron beams saw widening application in London buildings, largely because of a provision within the Metropolitan Building Act 1844 (*see* p 42). The emergence of wrought iron as a structural material for beams was principally a result of groundbreaking research by William Fairbairn and Hodgkinson in connection with the design of the Conway and Britannia Railway bridges. Their experimental work, carried out in 1845–7, established the box-beam, and subsequently the I-beam, built up from riveted plates and angles, as the most efficient structural forms for wrought-iron beams. The latter were destined to become almost the automatic choice for long spans, but the development of the all-purpose rolled-iron joist had increasing impact in the building world over time. Both brought about constructional change of a kind that was notionally unattainable in cast iron.

The rolled-iron joist and the built-up beam

Wrought iron is as strong in tension as it is in compression. In terms of production and use, this gave it a major advantage over cast iron, for it meant that there was no need to differentiate between the size or shape of the top and bottom flanges of a beam. At its most simple, the joist or beam could be placed with either flange uppermost, preventing 'all chance of mistakes which occur with other girders put up by ignorant men'.[12] Equal flanged I-sections (or H-sections

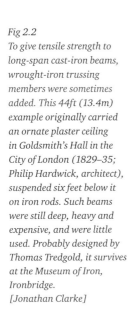

Fig 2.2
To give tensile strength to long-span cast-iron beams, wrought-iron trussing members were sometimes added. This 44ft (13.4m) example originally carried an ornate plaster ceiling in Goldsmith's Hall in the City of London (1829–35; Philip Hardwick, architect), suspended six feet below it on iron rods. Such beams were still deep, heavy and expensive, and were little used. Probably designed by Thomas Tredgold, it survives at the Museum of Iron, Ironbridge.
[Jonathan Clarke]

as they are sometimes called) were structurally efficient, and were more convenient and wieldy than cast-iron beams. Once cut to the requisite profile, the grooved rolls of the rolling mill could turn out infinite repetitions of the same section, in lengths that did not have to be prescribed in advance. Indeed, as the demand increased later in the 19th century, it became possible to cut beams to length from premanufactured supplies in the stockpile yards, or indeed on site and, in some cases, such was the speed of supply that the ironwork contractor could simply measure the dimensions from the brickwork of the building as it progressed, thus dispensing with the need to refer to an architect's plans.[13] More fundamentally, the shape of I-sections enabled a simplification and standardisation of connection technology, for they could be directly bolted or riveted together or to column brackets through the web or flange, the ductility of the metal absorbing the localised stresses. Because the flanges were equally strong, they could be safely fixed to their supports, which in turn greatly added to their strength, and facilitated experimentation in framing members rigidly together. By contrast, cast-iron beams were usually simply supported at their ends, either on walls or storey-height columns, and an ordinarily designed beam (with the lower flange larger than the upper flange) could not be rigidly connected at either end because of the change from compression to tension in the top flange towards either end. Accordingly, cast-iron girders had to be specially designed where they were to be fixed at either end, with upper flanges enlarging at the extremities to absorb the acute stresses there. Even then, with such a brittle material, rigid connections were never fully satisfactory, and the difficulties inherent in detailing and casting continuous beams, coupled with their unwieldiness, helped restrict their application. It was this versatility of the I-section that put it at the heart of constructional developments through the second half of the 19th century. Nevertheless, before the arrival of steel, the full impact of the technology was held back by the essentially craft methods of the puddling process.

The Palm House at Kew (1844–8) seems to have marked the first architectural use of rolled wrought-iron I-beams; the first arched rib, rolled at Millwall under licence from Kennedy & Vernon of Liverpool, was erected on 15 October 1845.[14] Commercial production of straight, equal-flanged joist sections aimed specifically

at the building trade (as opposed to the bulbed T-deck-beam section used in shipbuilding)[15] began the following year at La Villette near Paris, on the instructions of Eugène Flachat for the Saint-Lazare railway station. A carpenters' strike in Paris in 1845, lasting several months, had provided a powerful incentive to devise alternative solutions for strong, economical flooring systems, and by 1849 at least four firms, including La Providence in Belgium, were competing in the manufacture of I-section beams.[16] While other structural sections of utility to the building industry had already been rolled in England by 1821, including angles and tees, which could be used in glass houses, roof structures and 'fireproof' flooring,[17] this was a significant breakthrough involving a great deal of mechanical power and manual dexterity owing to the geometrical complexity of the shape. With an unvarying depth of about 6–7in. these joist sections were ideally suited to their intended purpose of forming the backbone of the first generation of proprietary 'fireproof' floor systems, being completely encased within the floor depth, their regular shape facilitating the standardisation and connection of other components. Two such systems were brought to the attention of British architects in February 1854 by Henry Hockey Burnell, one accredited to a Monsieur Thuasne, and a more easily constructed, unpatented and hence popular system (Fig 2.3). Subsequent discussion on

Fig 2.3
French systems of fire-resisting floors built around rolled-iron joists were quickly brought to the attention of British architects and builders. From the 1850s, British firms began devising their own in increasing number.
[The Builder, 25 March 1854]

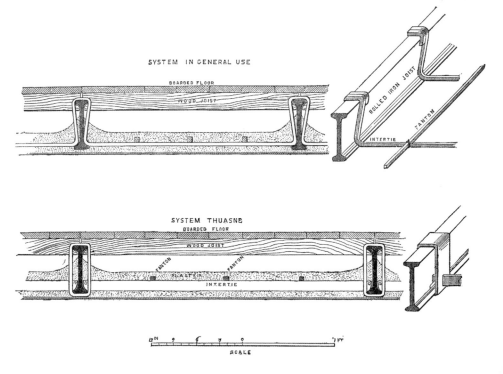

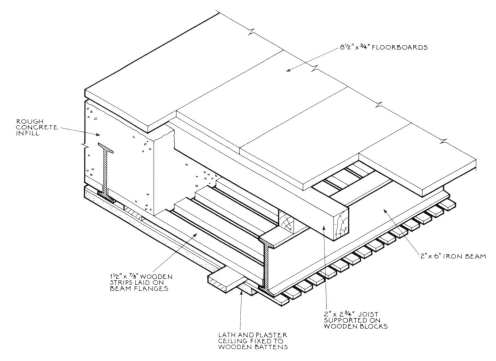

ROUGH
CONCRETE
INFILL

8½" x ¾" FLOORBOARDS

2" x 6" IRON BEAM

1½" x ⅞" WOODEN
STRIPS LAID ON
BEAM FLANGES

2" x 2¾" JOIST
SUPPORTED ON
WOODEN BLOCKS

LATH AND PLASTER
CEILING FIXED TO
WOODEN BATTENS

Fig 2.4

*The Fox and Barrett floor,
an English 'fireproof' floor
system devised in the 1830s
and initially reliant on
inverted T-shaped cast-iron
joists incorporated rolled-
iron joists from c 1851.
This drawing shows the
system as used at Finsbury
Barracks, City Road,
Islington (1853–4).*

Burnell's paper 'The French method of con-structing fireproof floors'[18] continued for a further three monthly meetings of RIBA, with contributions by, among others, Thomas Leverton Donaldson and (Sir) William Tite.[19] Although the systems were considered too expensive to be adopted in England, owing not to the metal, but to the cost of obtaining French plaster, they were an early verification of the expediency of rolled joists.

By 1854 a number of English 'fireproof' floor systems based on rolled joists had in any case been patented. First among these was the Fox and Barrett floor, a system initially reliant on inverted T-shaped cast-iron joists, quickly sup-planted from 1850 or 1851 with rolled joists, owing to a widening prejudice against cast iron in general, and possibly in direct response to the Gracechurch Street disaster.[20] This substitu-tion was effected by the more entrepreneurial partner, James Barrett (1808–59), who con-sulted a number of manufacturers before one, probably the Butterley Iron Works of Alfreton, Derbyshire, was persuaded to prepare rolls for producing I-section joists. Rid of the treacher-ies of cast iron, these stout members, measuring no more than 7 or 8in. deep with narrow flanges, helped spread the Fox and Barrett system, which saw widespread application in both met-ropolitan and provincial buildings, particularly after the expiration of its patent in 1859.[21] Early or notable examples included Finsbury Barracks

(1856–7)[22] (Fig 2.4), Brompton Hospital, East Wing (1854)[23] and the Royal Albert Hall (1867–71).[24] It continued in use until the 1880s, by which time an array of other patented systems, some derivative, had crowded the market. Some of these used concrete or corrugated-iron jack arches to span between the lower flanges; others, including the lightweight Doulton-Peto system introduced in 1885 had specially shaped hollow fireclay blocks to protect the lower flanges of joists and girders from flame (Fig 2.5). Nevertheless, in the 1850s even modestly sized rolled I-section joists were not readily available, and were commonly built up from other struc-tural shapes. For the fire-resistant tile-arched upper floors and stone-flagged stair-landings of the Union Bank, Nos 13 and 14 Fleet Street (1856–7), George Aitchison Senior used joists made up of an inverted T-section with two angles riveted to them, forming a 5 by 4¾in. I-section (*see* Fig 11.15).

Small-section joists, with their restricted span and load-bearing potential, had limited practicality outside the increasingly specialised field of fire-resistant flooring. This was recog-nised as early as 1854 by Fairbairn, who in his classic work *On the Application of Cast and Wrought Iron to Building Purposes*, urged iron manufacturers to roll larger sections:

it is more than probable that all those [beams] under 12 cwt. might be delivered at once, of the required form, from the rolling-mill; and it would be premature to assume, that even larger sizes, such as we have just described, could not be manufactured in the same way. The skill and intelligence of the iron manufacturers of this country have sur-mounted greater difficulties; and I have no doubt that a demand has only to be created in order to insure perfect success in all the manipulations connected with that impor-tant process. If this could be accomplished, a very important saving of the mineral treas-ures of the country would be effected; nearly two-thirds of the metal would be saved, and the price (supposing the beams to be taken from the rolls), reduced to nearly one-half, or from 16l. to 8l. or 10l. per ton.[25]

In the prevailing climate of guardedness towards cast iron *The Builder* agreed entirely, stating in its review of Fairbairn's book 'We have before now recommended that Govern-ment should offer a sufficiently large premium, say 1,000l. for the production of the best machinery to roll large girders.'[26] British manufacturers remained unconvinced by Fairbairn's suggestion, despite it being made in flattering terms, and no financial carrot was ever offered by government. The ironmasters confined their interests to the traditional markets of sheet, bar, strip and rod, moving only to rails, angles and plate for the shipbuilding and boiler-making industries. The production of large I-sections would have required technical acumen and considerable investment in more powerful rolling plant, all for a product the market for which was new and untested. Across the Atlantic, Peter Cooper's and Abram Hewitt's Trenton Iron Works, New Jersey, laboured for over 5 years before successfully producing what was effectively America's first rolled joist in spring 1854, a 7in. high T-rail.[27] Even in the 1860s, a typical ironworks in South Staffordshire – the centre of heavy wrought-iron products – was too small and ill-equipped to meet the challenge, assuming it even possessed the incentive:

> The capital cost was small – about £15,000 – and the mechanism was primitive in the extreme. A dozen or so puddling furnaces made 'loops' weighing some five cwt. which were shingled under a helve and then rolled out in the forge train or roughing mill into puddle bars; these were cut into one-foot lengths, piled together in bundles, reheated in a coal-fired furnace and rolled off the finishing mills into bars, rails or plates. Helve, forge train and finishing mills were all driven by beam engine and speeds were regulated by trains of gear wheels, the whole being steadied by a huge and heavy flywheel. The rolls and gears were usually mounted on timber foundations to absorb shock. The gearing consumed a great deal of power and was constantly getting out of alignment; serious breakages were frequent … mills for rails and the heavier plates, requiring more power, were usually separately driven. Even so the heaviest products weighed only about three cwt.[28]

Three hundredweight (336lb) was considerably less than Fairbairn's optimistic 12 hundredweight (1344lb) and the practical limits of production by the labour-intensive, highly skilled puddling process typically resulted in a mass of hot iron ready for rolling of only some 2cwt (224lb). Such a mass of metal, formed from two or more blooms 'piled' or welded

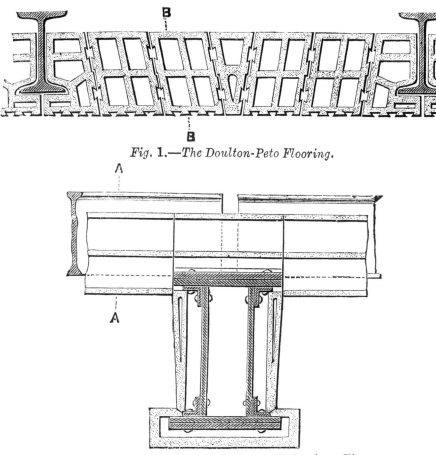

Fig. 1.—The Doulton-Peto Flooring.

Fig. 2.—Method of Protecting a Girder from Fire.

together under the hammer, could produce a 12 by 6in. rolled-iron joist (which typically weighed 43.1lb/ft),[29] but only in lengths of 4.6 to 5.2ft, which were obviously of little use. Theoretically, more puddle bars could be stacked together into a larger pile, reheated in a furnace and re-rolled, but the size of the resulting piece would be too large for the manual handling techniques typical of British works of the time. William Menelaus (1818–82), 'probably the most adventurous British ironworks engineer of that time',[30] succeeded in rolling a 50ft-long 12in.-deep beam in 1859 at his Dowlais works in South Wales, and another (presumably shorter) beam that was 18in. deep. However, both appear to have been rolled for the 1862 Exhibition, and were thus brazen demonstrations of potential rather than practice. Menelaus's main market was in rails, and he had scaled up the number of puddling and balling furnaces to match the capacity of his powerful 'Goat Mill', which had rolled the beams. Most British puddling and rolling works were considerably smaller. In 1859 (Sir) John G N Alleyne (1820–1912),

Fig 2.5
Numerous proprietary flooring systems designed around rolled-iron joists had evolved and flourished by the 1880s. The Doulton-Peto system used flat arches of interlocking ceramic blocks that, unlike most rival systems, protected the underside of the metal. [The Builder, 19 December 1885]

Fig 2.6
John Alleyne's Patent of 1859 for producing deeper I-section joists by welding together two Ts.
[© The British Library Board (F60110–03)]

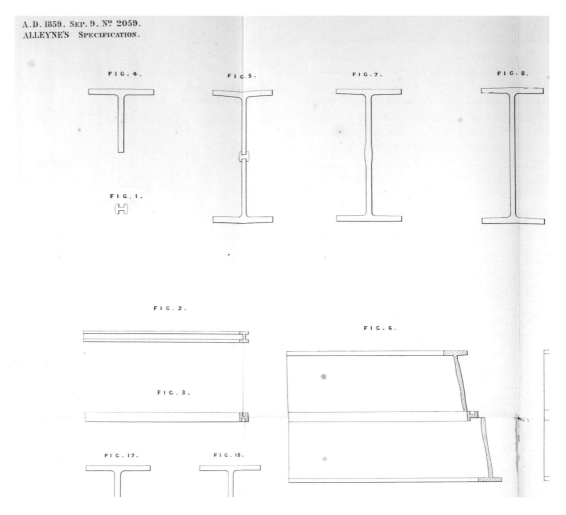

Fig 2.7
One of the few identified applications of Alleyne's 'Double T's' was in a huge beer trans-shipment warehouse alongside the Regent's Canal at St Pancras Goods Yard, erected by the Midland Railway for Bass & Co. in 1864–5 (demolished). The join between the two Ts was all but invisible, and Butterley's conspicuous rolling mark in the centre of the web may have been placed to demonstrate this.
[Malcolm Tucker (528/18)]

manager of the Butterley Iron Works, succeeded in producing a deeper I-section joist by joining two Ts using an H-section glut of readily weldable iron (Fig 2.6).[31] Although patented in that year, and commercially produced from 1863, few structures incorporating these 'Double T's'[32] have come to light, and hence the technique's wider success is doubtful. More effectual was the 'Butterley Bulb', a rolled T-section with a bulbed foot that was essentially a deeper version of Kennedy & Vernon's deck beam of 1844. Large quantities of these were produced in the 1860s and 1870s, finding ready application as floor joists in industrial buildings, particularly those connected with the Midland Railway, such as in the former Bass Ale Warehouse, St Pancras Goods Yard (1865, demolished) (Fig 2.7). The name Butterley Bulb came to be used for all sections of this type, regardless of manufacturer and material, since they were later rolled in steel.

Alleyne, who consulted with the engineer W H Barlow on the fabrication of the wrought-iron members for the St Pancras train shed of 1863–7, was almost alone among British iron-works managers in being responsive to the needs of the construction market. He was concerned with heavy structural sections, and his most important contribution in this area was the introduction in 1861 of mechanical travers-ers that facilitated the manipulation of heavier pieces. He also designed and patented a new type of rolling mill that enabled the piece to be rolled in both directions, but which (unlike the standard two-high mill) did not require the piece to be lifted high on the return pass.[33] Yet it was not until c 1876 that Butterley began commercial production of large I-beams, rolled in one piece.[34] Apart from the Dowlais and Butter-ley works, both of which were at the forefront of technical improvements in puddling and roll-ing technology, until the 1880s there was little interest among British ironworks in consis-tently rolling large structural sections suited for the building industry.[35] Even then, British man-ufacturers seem to have been incapable of producing sound joists exceeding 12in. in depth, due to the stresses caused by the webs cooling and contracting before the flanges.[36]

The only practical alternative to using cast iron for sections large enough for heavy loading or greater than usual span was to build up a heavier I- or box-section from rolled plates and angles riveted together. In 1847 Henry Fielder took out a patent for composite beams fabri-cated from both cast and wrought iron, but which also included an I-section plate girder made from plates and angle irons.[37] Possibly these were being manufactured before that date, but Matthew T Shaw was among the first producers of these for the building trade, com-mencing fabrication in the early to mid-1850s at works in Queen's Wharf, Bankside.[38] One of the first examples of their use in a 'polite' con-text was at Windsor Castle in 1853, where plate girders were used to span the full width of the Prince of Wales Tower,[39] and by the end of the decade they were witnessing more general application in the metropolis. For the rebuild-ing of the Rainbow Tavern, No. 15 Fleet Street, in 1859–60, the architect Rawlinson-Parkinson used 40ft-long I-section girders on two floors to assume the load-bearing role formerly taken by a dividing wall, enabling grander, more coher-ently planned spaces.[40] The form of these girders, possibly designed by the ironwork sup-plier (a little-recorded City-based iron founder/ merchant named Frasi) is shown in Fig 2.8.

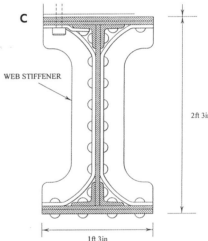

WEB STIFFENER

2ft 3in

1ft 3in

Fig 2.8 a (left), b (below left) and c (below) In the rebuilding of the Rainbow Tavern, No. 15 Fleet Street, City of London, in 1859–60, 40ft-long wrought-iron girders helped carry the upper floors. Built up from plates and angles, and well-stiffened along their length, they were concealed in timber boxing with no fire protection. [a BB009753; b BB009752]

Over the next two decades many other compa-nies that would have lasting significance in London either came into being, or built upon earlier origins. These included Glaswegian companies such as P & W MacLellan which opened offices in Cannon Street in the City, but mostly comprised those of more local prove-nance such as the St Pancras Iron-Work Co.; Macnaught, Robertson & Co. of Bankside;

Fig 2.9
*Julius Homan's system,
patented 1865, of riveted
together rolled-iron joists,
with or without plates,
revolutionised constructional
ironwork in the later
19th century.*

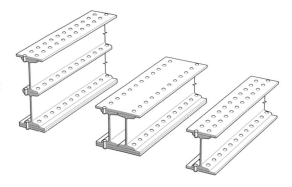

Richard Moreland & Son of Old Street; Rownson, Drew & Co. of Upper Thames Street, and Yates, Haywood & Co., Dyers' Hall Wharf, Upper Thames Street. By 1870, when the first edition of the *Kelly's Directory of the Building Trades* was published, some 50 'Iron girder manufacturers' were listed for the London area, showing the growing importance of fabricated wrought-iron members within the metropolitan building world. Some of these companies, such as P & W MacLellan, Richard Moreland & Son, and Matthew T Shaw were also fabricating entire structures such as iron roofs and bridges, and constitute the first generation of constructional engineering contractors. Leading firms, with facilities for the rapid delivery of orders in bulk, such as Moreland, which boasted 'all kinds of iron girders [made] by the most improved steam machinery – supplied at a day or two's notice',[41] had markets well beyond London, and even competed for international contracts. The emergence of such companies ensured that built-up wrought-iron members, especially the I-section (which, unlike the box form, permitted inspection of virtually all the metal for signs of corrosion), saw steadily increasing application until the 1890s, when the form began to be reproduced in steel. Nevertheless, such built-up sections, while structurally efficient, were inherently more inefficient in their use of material than rolled beams of equivalent sectional area, chiefly because of the diminished area of the lower (tension) flange caused by the rivet holes. Their fabrication was also a costly, time-consuming and highly skilled process.[42] As *Gwilt's Encyclopaedia of Architecture* noted in 1867, 'A riveted plate girder is not always adaptable for general purposes.'[43] Some of the companies involved in fabricating these members were clearly aware of the need and potential market for rolled sections. Matthew T Shaw and his business partner Thomas Howard Head, for example, were

granted a patent in 1868 for an early universal mill for rolling I-beams,[44] and by the following year were turning out sections ranging in size from 7 by 3½in. to 12 by 6in. They even erected a testing machine at Bankside 'for the purpose of enabling architects, engineers, and others to witness personally the testing of various sections of wrought iron before purchased'.[45] Yet M T Shaw, Head & Co. was very much the exception at this time: most firms imported rolled iron from the Continent, in the form of plates and angles as well as joists and beams, which they riveted together into larger, built-up beams and other sections.

The innovation of the 'compound' beam or girder, fabricated by riveting together rolled joists and plate sections, helped unleash the latent utility of structural wrought iron within British buildings. Patented on 7 October 1865 by London-based Julius Homan, formerly an inventor of sewing machines and later a partner within a leading constructional engineering firm, the compound beam was a simple but highly successful response to the problem of limited joist sizes and the expense of fabricating members from plates and angles.[46] In its simplest form, I-sections were given additional resistance to sideways buckling by plates riveted to their upper or lower flanges, effecting a claimed 'strength double that afforded by riveted plate girders of equal sectional area'.[47] For more demanding situations, two or more rolled joists were riveted together on top of each other directly through the flanges (Fig 2.9). Homan's compound beams were tested at Kirkaldy's Testing Works, and the results showed that an 8in.-deep joist, probably of Continental origin, was strengthened by some 40 per cent by the addition of a flange plate, and that this compound beam was some 2.4 to 3.7 times stronger than rolled girders supplied by the Butterley Company. Two 8in.-deep joists riveted on top of one another with a flange plate were some 3.4 to 5.3 times stronger than the Butterley rolled girders.[48] The 1867 edition of *Gwilt's Encyclopaedia* emphasised the importance of the innovation by pointing out that 'the present sections of rolled irons are so limited in depth that they have hitherto only [been] valuable where light loads and limited spans occur'.[49] Although the patent was in Homan's name, it would seem that the true inventor was William or Thomas Phillips, who sold him the rights.[50] The circumstances and outcome of this transferral of rights is somewhat obscure, for the

DRAPERS' HALL DOME.

firm W & T Phillips which had fabricating yards in Borough (South London), Manchester and Belgium, soon set about promoting the 'Phillips's Patent Girder System'.[51] Early known applications of the technique were in the New University Club, St James's Street (1865–9, demolished, *see* p 146), and in a Clerkenwell warehouse erected in 1868–9 (*see* p 286). Further proof of its efficacy came with the remodelling of Drapers' Hall, Throgmorton Street, City of London, in 1866–70, where the architect, Herbert Williams, used curved slender compound ribs to frame a double-dome (Fig 2.10).[52] Each rib was formed from two curved H-section iron beams riveted together along the flanges. Springing from a continuous iron curb, cast in sections

with shoes to fit the ends of the ribs, the ribs were secured at the top with a wrought-iron ring, from which sprang an upper dome formed in the same manner. Prior to this, the principal techniques at hand for iron-framed domes had been the individual casting of heavy ribs, as in J B Bunning's celebrated Coal Exchange (1847–9), although for larger structures, such as the dome of the Royal Albert Hall (1867–71), the laborious practice of riveting bars and angles into lattice ribs continued unabated. By the mid-1870s, Homan & Rodgers could claim that 'Homan's Patent Girders' had been used in nearly 700 buildings in Britain, the Continent and America,[53] and that these, together with its fire-resisting floors were 'superseding every other kind'[54] (Fig 2.11). Both Homan's and Phillips's systems enjoyed huge success in the 1860s and 1870s before the rights lapsed and other companies capitalised on the technique. The basic principle of compounding smaller structural components to form larger beams, girders and stanchions was to revolutionise the British application of structural iron, greatly adding to the utility of simple rolled-iron sections, and later crossed over into the age of steel. The technique enjoyed immense popularity until the 1950s, when universal sections, available in a

Fig 2.10 (left)
One of the first uses of compound girders was in framing the dome of the Drapers' Hall, Throgmorton Street, City of London (1866–70; Herbert Williams, architect). Timbers spanning between the ribs formed the framework for an exterior covering of lead and an interior sheathing of plastered canvas, enriched with carton-pierre ornaments. [The Architect, 2 January 1869; RIBA69812, RIBA Library Photographs Collections]

Fig 2.11
Homan's technique was exploited by Homan & Rodgers and W & T Phillips in the 1870s, and thereafter compound beams became standard, crossing over into steel from the 1880s. [Kelly's Directory of the Building Trades, 1874. © The British Library Board (F60105–64)]

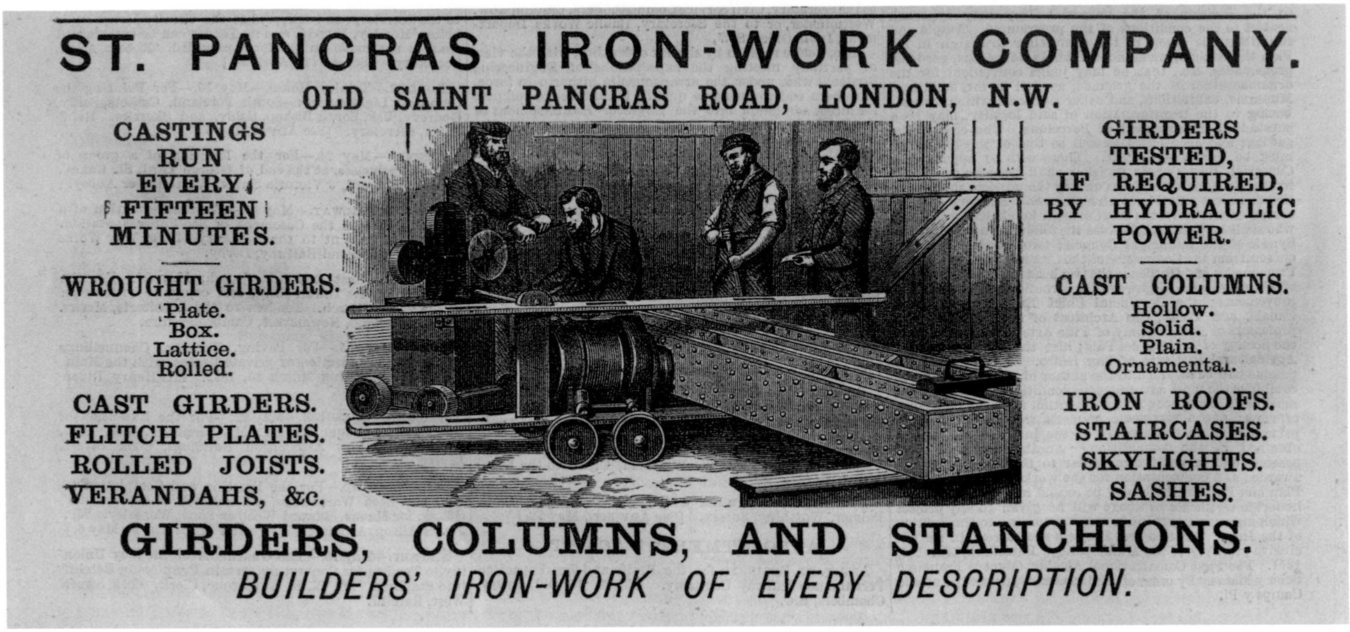

Fig 2.12
Advertisement in an 1871 issue of The Architect, *showing the range of products, including rolled-iron girders, offered by one London company.*
[The Architect, 13 May 1871; RIBA46495, RIBA Library Photographs Collection]

wide range of sizes, began replacing compound beams.

In addition to the boost given to the application of structural ironwork by the use of compound beams, a small number of British manufacturers besides Butterley's were rolling heavy structural sections by the 1880s. The St Pancras Iron-Work Company was supplying rolled girders and joists in addition to 'Wrought-Iron Plate, Box, and Lattice Girders' (Fig 2.12) and the Shelton Bar Iron Company, Stoke-on-Trent, was rolling girders, angles, Ts and channels.[55] Other producers, many in the Black Country, including the Earl of Dudley's Round Oak Works at Brierley Hill, the Patent Shaft and Axletree Company at Wednesbury, and Thorneycroft's at Wolverhampton,[56] and the Lowmoor Company in Bradford,[57] were turning out rolled beams, at least intermittently. In 1881 *The Building News* could observe:

the rolled iron joist or I-shaped girder is one of the most useful sections in construction, and is manufactured in a variety of forms, proportions and sizes. It is rolled in depths of from 3 in. to 14 in., but every manufacturer publishes his own sections, with the weight per foot run and the distributed weight each section will bear. The uses of this favourite section are too well-known to need mention; it may be remarked that its employment for fireproof floors and large roofs on the Continent increases yearly. Small and narrow sections are the cheapest. Beyond 9in. deep

and 4in. wide the price increases from £1 to £2 a ton. If wider flanges than 6in. are required, the price is greater. For constructive purposes joists of more than 12in. deep are better in riveted iron, though sections up to 20in. deep are used on the Continent, and by riveting two sections together, either side by side or vertically, rolled joists of this section can be applied to almost any ordinary purpose.[58]

By this date, a mature, thriving trade in structural sections for the building trades had evolved, one in which engineering contractors, not architects, typically took control of the design of constructional ironwork. Nevertheless, despite the increasing range and sizes of rolled members offered by a small handful of British manufacturers, it was overwhelmingly the Continental ironworks, and especially those in Belgium, which supplied the London – and wider British markets – during the period *c* 1865–1900. In 1875 one South Staffordshire ironworks manager demanded 'How is it that Belgium supplies England with two-thirds of the large rolled iron girders which are used there for building purposes?'[59] Continental manufacturers, with their more advanced puddling and rolling technologies, and better organisation, were able to produce larger sections in bulk at economically viable prices. In this country, mass production was chiefly held back by the volumetric limitations of the technology of actually producing the metal. Given its status and close historical and geographical links with the Continent, London inevitably became the chief entrepôt for imported ironwork, and scores of middlemen and merchants arose, satisfying and driving the market in structural sections. It was this Continental dimension (*see* Chapter 3), first apparent in the late 1860s, that really stimulated the structural trade, both directly, in supplying a demand not met by British manufacturers, but subsequently indirectly, by fostering that demand to the point where home interests were sufficiently threatened to respond. The British response only became effective in the late 1880s, once a number of firms put down rolls specifically for the manufacture of heavy structural sections in steel. Only then did continuous, bulk production of large structural sections become economically feasible in this country, and the age of building in steel began.

Rolled steel beams

In 1860 the first steel girders in the world were manufactured in Sheffield, the hub of traditional crucible steel production and trade. At this date, the city still 'wrapped itself in absolute security and believed it could afford to laugh at the absurd notion of making 5 tons of cast steel from pig iron in 20 to 30 minutes when, by its own system, 14 or 15 days and nights were required to obtain a 40 to 50 pound crucible of cast steel from pig iron'.[60] Absurd or not, Messrs Brown, Bragge & Co. (later John Brown & Co.), the first Sheffield firm to adopt both the manufacture of 'puddled' steel, and Bessemer's pneumatic steelmaking process,[61] achieved this impressive feat, producing steel in large enough quantities to roll the steel plate and angles that were used to fabricate at least two 32½ft-long girders. They were made not for an exhibition of technical virtuosity, but to order, for the headquarters of the London and County Bank, Lombard Street, completed in 1861 to designs by Charles Octavius Parnell (Fig 2.13).[62] Although the metal used in this instance was 'puddled steel', made by decarburising cast iron in a refractory puddling furnace, the Sheffield firm stated that 'being the largest manufacturer of steel for such purposes', it could 'furnish … information and data for girders made from puddle or cast steel'.[63] The term cast steel was traditionally applied to crucible-steel, but in the 1860s, it came to be applied to steel cast directly from the Bessemer converter

into ingot moulds for subsequent remelting and casting or rolling.[64] The bank was demolished in 1965 and the exact form of the plate-girders is unknown, but in the 1880s the engineer John Richard Ravenhill (1824–1894) stated that H-section steel members 'had been made in Sheffield and used in London twenty years ago'.[65] Possibly they conformed to those patented by John Brown in 1858, which were 'built up of specially-rolled sections, put together hot or cold, and finished by rolling or pressing'.[66]

John Brown's Atlas Works in Sheffield rolled 'the first Bessemer rail made in the ordinary course of business' in May 1861. The firm,

Fig 2.13
London and County Bank, Lombard Street (1860–1, demolished c 1965), perhaps the location for the first use of steel beams in building construction anywhere.
[*The Builder, 23 August 1862*]

LONDON AND COUNTY BANK.—MR. C. O. PARNELL, ARCHITECT.

relaunched as John Brown & Co. on 1 April 1864, pioneered the mass production of rolled-steel armour plate and became the world leader in this field.[67] Brown was knighted in 1867, and 3 years later his enterprise (now Sir John Brown & Co.) was listed under the headings for both 'Iron girder manufacturers' and 'Iron roof manufacturers' in *Kelly's Directory of the Building Trades*. As this firm had London offices, at No. 10 John Street, Adelphi, under the management of the enterprising John Clowes Bayley (1834–1909),[68] it seems unlikely that the London and County Bank was the only metropolitan building of the third quarter of the 19th century to make use of either puddled or Bessemer steel. No others have come to light, perhaps because structural sections of this period encountered through the course of renovation have been assumed to be iron, and perhaps also because, during this period, the architectural press tended to use 'iron' as a somewhat catch-all term that included mild steel. Nevertheless, in the absence of more direct and unequivocal evidence, Parnell's bank remains a constructional anomaly. Even were evidence of

comparable buildings to become known, it would not alter the fact that neither puddled nor Bessemer steel failed to bring about any constructional revolution in this country.

Almost two decades elapsed before specific discussion of steel in building construction reoccurred. Most significant was the publication of (Sir) Alexander Kennedy's landmark paper 'Mild steel and its applications to building purposes', read before RIBA on 19 April 1880. Kennedy, a consulting structural engineer and Professor of Engineering at University College, London, observed:

> There has hitherto been very little demand for rolled joists in steel, and there is some difference of opinion on the part of those most competent to judge as to the relative difficulty of rolling them in steel and in iron. *At the present moment, however, one if not more of our largest firms are rolling steel joists at very much the same price as that at which firms of equal standing are rolling them in iron,* a price perhaps 30 per cent in excess of what is paid for Belgian joists. The difference of strength, in favour of the steel over the latter, would probably be about 80 per cent., and over English (iron) joists perhaps 60 per cent. In either case there would be a large balance of economy in favour of the stronger material.[69]

Kennedy was referring to Bolckow, Vaughan & Co., a long-established Middlesbrough firm, whose Eston Steel Works, operating on the new Basic (Bessemer) process, he had described earlier in detail, possibly following a site visit.[70] It is unclear precisely when Bolckow Vaughan began rolling joists, but by November 1880 *The Times* could report that 'every mechanical appliance which can be devised is in operation [for rolling] steel girders and angles'.[71] Kennedy had been sent 12 by 6in. steel joists for testing by the firm's general manager, E Windsor Richards, and these compared auspiciously with iron joists of equivalent dimension, proving to be some three times the strength, 12 times the ductility and, unlike the iron, having an equal tenacity both along and across the metal's fibre.[72] Nevertheless, Bolckow Vaughan stood virtually alone among British manufacturers in the production of steel joists during the early 1880s.[73] In reply to Kennedy's paper, constructional engineer Ewing Matheson stated that steel joists are being made 'only by one or two

Fig 2.14
The two-high reversing mill was the typical apparatus used by most early British steel rollers

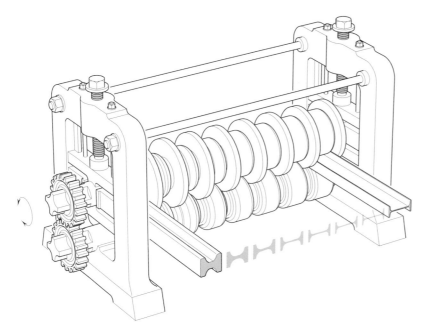

TWO-HIGH REVERSING MILL

PROFILE DEVELOPMENT FROM BLANK TO FINAL I-SECTION
AS THE BAR PROGRESSES THROUGH THE ROLLS

manufacturers, and that there are certain diffi-culties in making them which have hitherto shut them out of ordinary commerce'.[74] Two years later he could still say they 'have not yet become an ordinary article of commerce', despite being 'extremely useful, and of which large quantities could be employed in build-ings'.[75] The technology for rolling steel joists at this time was entirely based on experience in iron rolling, and experience soon showed that more powerful rolling machinery was needed to reduce a rectangular shaped steel ingot or bloom to an I-section than that required to form an equivalent section from piled wrought iron.[76] Bolckow Vaughan's Eston works was among the most advanced in the country, fitted with reversing cogging mills (in which white-hot steel ingots were roughly shaped into billets or slabs before being shaped for girder and other sections in finishing rolls), powerful two-high reversing mills, and hydraulic machinery for moving the hot metal about the mills quickly, fostering the kind of continuous uni-form production seen in the most advanced American mills (Figs 2.14 and 2.15).[77] Even so, Bolckow Vaughan, which by 1882 was produc-ing some 100,000 tons per annum of basic

(Bessemer) steel and an equivalent amount of acid (Bessemer) steel, never specialised in structural sections. At this time, rails, their tra-ditional product, still formed around 80 to 90 per cent of their total output.[78] The manufac-ture of structural sections, like the firm's subsequent attempts to diversify into steel-plate for shipbuilding, was simply a strategy to avoid over-dependence on rails, the future market of which was uncertain. It was not a strategy with any long-term viability or profitability, because switching between rails, plates and structural sections required frequent roll changes and kept the mill inoperative for hours at a time. And although the firm could produce bulk quantities of metal that could be poured into large ingots for direct rolling into beams, there were still technical problems in correctly work-ing the metal through the rolls at this stage: the flanges of I-sections were appreciably thinner and hence more difficult to form than the much thicker heads and bases of rails. However, the fundamental problem at this stage lay in the quality of metal for structural purposes.

It was Dorman Long & Co., also of Middles-brough, who effectively launched the steel structural trade in Britain. Formed in 1876 by

Fig 2.15
A typical late 19th-century steel rolling mill, with grooved rolls lying idle in the foreground, lines of two-high mills in the middle-ground, and the flywheels of the steam engines that powered them in the distance.
[© The British Library Board (F60105–65)]

for rolling bigger pieces. This machine was either based on a Belgian design, or imported directly from that country.[80] It could roll 60ft-long plates up to 50in. wide and 2½in. thick.[81] In 1883, with the lucrative structural trade almost wholly driven by Belgium and Germany, Dorman Long turned to the production of iron girders, which became its hallmark (Fig 2.16). Clearly aware that the future lay in high-quality open-hearth steel, in 1886 the firm replaced one-half of its 120-strong battery of puddling furnaces with seven Siemens-Martin open-hearth steel furnaces to enable the production of steel joists and girders and the fabrication of built-up members (*see* Fig 1.2).[82] By sidestepping the increasingly discredited Bessemer process, Dorman Long took the first step in out-manoeuvring its Continental rivals.

Arthur Dorman and Albert de Lande Long, and based at the former West Marsh Iron Works site, this company specialised from the outset in rolling iron bars and angles for the fast-developing local shipbuilding industry. By 1879 the partners had expanded onto an adjoining site, the Britannia Ironworks, supplementing their original 20 puddling furnaces with a further 120, together with furnaces, forges, hammers and rolling mills.[79] In that year they put down one of Britain's first universal mills at Britannia, incorporating vertical as well as horizontal rolls

In 1887 Dorman Long published its first section book of steel and wrought-iron members. It included an extensive range of channel and angle shapes, and 33 different joist and beam sizes, the largest of which measured 20 by 8in.[83] This was the first British handbook to include tables explaining the structural properties of the sections, enabling designers to analyse and determine safe, efficient built-up members and framed structures. Prior to this, manufacturers' catalogues only included the sizes and weights of joists and the safe distributed loads each could bear for various spans, meaning that

Fig 2.18
Dorman Long erected new bridge and construction shops at their Britannia Works, Middlesbrough, in c 1900. These state-of-the-art sheds were fitted with electrically driven overhead travelling cranes.
[© The British Library Board (x52913857)]

structural properties had to be painstakingly calculated from first principles.[84] The company had clearly taken on board engineering expertise, and it was in this direction that it moved, establishing a constructional and bridge-building department in 1890.[85] By vertically integrating itself with engineering, the company actively sought rather than waited for demand for constructional steel. In the same year it began circulating weekly stock lists to merchants and architects to promote its products, and made arrangements with engineering contractors Peter & Walter MacLellan to keep a stock of their girders in Glasgow. By 1894, in a bid to capitalise on the burgeoning London trade in joists, Dorman Long established the largest stockyard in the metropolis, complete with riveting and drilling shops, on a riverside wharf at Nine Elms Lane, Vauxhall (Fig 2.17).[86] The same year the firm sent a representative to visit Andrew Carnegie's Pittsburgh works, to obtain the technical expertise for rolling a 24in.-deep girder, the largest then in existence.[87] This was not practically achieved until 1902,[88] but Dorman Long did manage to attract something else of more immediate consequence from Carnegie, structural engineer Charles Vivian Childs. In 1894 Childs took charge of Dorman Long's engineering and drawing office, under H B Powell, London manager,[89] and from

here he would design the constructional steelwork for a number of major buildings in the capital (*see* Chapter 7).

The growth of Dorman Long during the 1890s was phenomenal, its structural steel output increasing yearly to the detriment of rolled iron, production of the latter ceasing entirely by mid-decade. In 1897 Arthur Dorman could report that the Constructive Engineering Department formed a key, profitable part of the business, and the following year the company was generating 11,000 tons of fabricated work yearly, including compound girders, stanchions and bridges.[90] By 1900 the demand for structural steelwork having considerably outgrown the original fabricating shop, Dorman Long erected new, state-of-the-art bridge and construction shops in Middlesbrough (Fig 2.18). In that year, nearly 30,000 tons of structural steelwork was produced at this works alone, which easily made the company the leading British producer.[91]

Other steel producers quickly followed Dorman Long's triumphant lead. The Frodingham Iron and Steel Company in Scunthorpe put down open-hearth steelmaking plant in 1888, and from *c* 1892 began adding further stands of rolls for producing sections. By 1899 weekly production of sections was averaging 1,144 tons.[92] The Glengarnock Iron and Steel Company, Scotland, formed in 1890 as an offshoot of an established Bessemer-steel-producing company with, among others, E Windsor Richards on the board of directors, moved rapidly into the market for rolled joists, fitting acid-lined Siemens-Martin furnaces and new rolling plant in 1892.[93] David Colville and Sons, from their Dalzell Works, Motherwell, also began diversifying from shipbuilding into structural sections during the 1890s, as did a number of other Scottish steel companies, including the Lanarkshire Steel Company, and the Steel Company of Scotland. In England, the Earl of Dudley at Round Oak, and the Leeds Steel Works were among the first to capitalise on the burgeoning demand for rolled sections.[94]

Although Dorman Long was perhaps the only steel producer to move into constructional engineering during the 1880s, two Scottish constructional firms quickly embraced the new material, turning out riveted members and structures from home-produced open-hearth steel sections. Redpath, Brown & Co. of Edinburgh began manufacturing built-up steel beams in 1886, and in that year opened a constructional department.[95] It published its first catalogue in 1892, containing safe load tables for both iron and steel-rolled joists, and a limited range of compound girders.[96] Having erected new structural workshops in Edinburgh, by 1900 this firm had a 'branch office' at No. 25 Victoria Street, London, and, by June 1903 had begun

production in earnest at a London works, occupying a 24-acre site in East Greenwich fronting the Thames (Fig 2.19).[97] Prior to this, from *c* 1878, the firm had been acting as merchants and fabricators, supplying Belgian rolled-iron joists and built-up sections for the building and engineering trades.[98] Redpath Brown was to be responsible for Mathias Robinson's furniture warehouse in Stockton-on-Tees (1899–1901), one of the first fully steel-framed buildings in Britain and, indeed, a great number of the first generation of such structures. The other leading constructional firm, Sir William Arrol and Co., had already risen to national prominence as the main contractor for the Forth Bridge, the world's largest all-steel bridge until 1917. In 1887–8 the firm laid down its 'upper girder works' at its Dalmarnock Ironworks,[99] for the fabrication of steel sections for the engineering and building market. Like Dorman Long and Redpath Brown, this company expanded massively in the 1890s, extending from fabricating steelwork to designing and erecting complete structures to order, such as engineering sheds and bridges.

Of equal significance to the London building world was the profusion of smaller contractors and merchants that had come into being to supply built-up wrought-iron beams and girders in the 1860s and 1870s, but who increasingly began supplying rolled-steel sections during

Fig 2.19
Redpath, Brown & Co. Ltd established a 24-acre works at East Greenwich In the early 1900s. In full production by 1903, it boasted huge fabricating shops, each crane-served bay designed for easy manoeuvring and assembly of heavy steelwork. In the foreground here, a workman assembles components for a steel roof truss.
[Redpath, Brown & Co., Ltd, Contructional Steel Work, 1905 edn]

the 1880s and 1890s. In 1881, Macnaught, Robertson and Co., of Bankend, Southwark, published a 34-page section book detailing its extensive rolled-iron and steel sections, the heaviest of which measured 20in. deep by 10in. wide.[100] By 1882 Measures Brothers & Co. of Southwark were advertising a similar range as part of its 'Town Stock' of 5,000 tons of 'Builders' Ironwork', all ready for 'prompt delivery at lowest market prices'.[101] It also produced pocket books detailing safe load tables, which *The Building News* decreed as 'fully reliable, being verified by actual tests, so that an architect will be quite safe in taking them as those to which the girders may be subjected without deflection'.[102] By 1895 Measures Brothers Ltd could boast that their 'Town Stock' comprised some 6,500 tons of Siemens-Martin and Bessemer steel joists, 3 to 20in. deep. Their stock of rolled-iron sections had dwindled to some 1,000 tons by this date.[103] Section books of iron and, increasingly, steel members, with attendant tables, proliferated through this period, published for companies such as H J Skelton, of Finsbury Pavement (established 1883);[104] Homan & Rodgers, Gracechurch Street and Vauxhall Steel Wharf;[105] Walter Jones, Magnet Wharf, Bow;[106] Alexander Penney & Co., Fenchurch Street and Glasgow;[107] and P & W MacLellan, Ltd., Glasgow and London.[108] Much of the earlier steel was imported from the Continent, but as English and Scottish manufacturers rose to prominence in the 1890s, these metropolitan firms came to rely more exclusively on British steel.

By the start of the 1890s steel rather than iron sections were the norm, one handbook noting that 'the great competition of Belgian manufacturers has, during the last 10 years, practically withdrawn this trade from England, and iron joists are only now made in England by a few works'.[109] Production of rolled-iron joists had all but ceased in this country by the end of the century. Because the quality of rolled iron was directly related to the amount of re-working, which advantageously redistributed scale and other impurities, only the most prohibitively expensive brands could compete with the homogeneity of steel produced by the open-hearth process. Siemens's often-repeated castigation of rolled-iron sections – a 'sandwich of iron and dirt'[110] – had by this time undoubtedly diffused out of metallurgical circles into the mindset of ordinary engineers and architects. Open-hearth steel, by contrast, was a fully fused, homogenous metal, some 30–50

per cent stronger, equally ductile and capable of withstanding rougher treatment during construction.

The advent of flexural members in steel, produced in bulk at competitive prices, revolutionised the British building industry of the 1890s. By the start of the decade, steel suppliers and fabricators were competing for advertising space in periodicals such as *The Builder* (Fig 2.20). I-sections, in all sizes, truly became a commodity, more so than the wrought-iron members of the 1870s and 1880s, which as often as not required further post-rolling fabrication to enable them to be used in strenuous, demanding situations. The common rolled joist or beam

Fig 2.20
Engineering contractors jostle for space on the advertisement pages of The Builder *in 1892. By this date, steel was making steady inroads into a market formerly the preserve of iron.* [The Builder, 2 July 1892]

became the most versatile, standard structural member available to the builders and engineers, supplanting the built-up girder for most applications.[111] It shared basic geometric characteristics with its wrought-iron forebear (indeed, one of the reasons for the ease of transition was that it did not require significant modification of the rolling plant, and no conceptual change was required on the part of the builder), but could be produced, and used safely, in greater lengths. It could also be used advantageously in compressive situations, enabling connections that were more simple yet rigid compared to those between cast-iron columns and rolled beams. But for various reasons, cast-iron pillars and columns remained enduringly popular.

Columns and stanchions

Although cast iron had shown itself to be a potentially treacherous material in tensile or bending situations, its use in compressive applications actually continued beyond the era of wrought iron into that of steel. Indeed, as the use of cast-iron columns had preceded that of cast-iron beams by 30 years, so too it endured longer, by some 50 years.[112] Cast-iron columns remained popular through the 1890s, and continued in use, albeit rarely, even into the 1920s.[113] Cast iron was ideally suited for use as columns because of its great compressive strength, typically some 2.5 times greater than that of wrought iron and 1.5 that of mild steel. Used in this way, it was only likely to fail following a fire, when the heated metal was apt to shatter if cooled suddenly by water. The earliest forms employed were narrow, solid cylindrical or cruciform sections, but from about 1805, hollow cylindrical sections increasingly came into use. This form remained far and away the most popular until the 1890s, not just in industrial buildings, but also in commercial and institutional architecture, though perhaps for a number of reasons, cruciform columns retained great popularity in London, especially in warehouse construction (see pp 284–5). Relative cost aside, the main reason for the endurance of the hollow cylindrical section was that, for the same cross-sectional area or volume of metal, it offered the greatest stiffness and hence could carry more load without buckling. This was learned empirically from the outset, but was subsequently confirmed by the experimental and theoretical work of Hodgkinson, Lewis D B Gordon and W J Macquorn Rankine in the mid-19th century.[114] Unlike cast-iron beams, hollow cast-iron columns were inherently efficient in cross-sectional shape, and as a result were economical. The only real difficulty was ensuring there were no variations in thickness of the column wall, caused by misplacement of the core during casting, but as foundry techniques improved throughout the century, this was less of a problem. Another reason for their enduring popularity was the ease with which decorative shapes and patterns could be cast integrally with the member. Classical bases, fluted shafts and elaborate foliated capitals were easily incorporated into the moulds of round sections, and once the mould had been made, many identical castings could be reproduced. Riveted wrought iron or steel rarely attained the ornamental qualities of cast iron, and this alone ensured that decorative cast-iron columns enjoyed continued high-profile application in hotels, banks and other grand commercial or public spaces.

Despite the unrivalled strength of the cylindrical form, variations in cross-sectional shape did appear in the second half of the 19th century. There were two main reasons for this. One was the increasing use of wrought-iron (and later steel) beams, which encouraged new beam-to-column connections; another was the growing concern for 'fireproofing', which demanded new forms of column that could be more easily embedded in masonry piers and perimeter walls. The most important shape to evolve was the H-section stanchion, which anticipated developments in steel, but of much significance also were the C- and E-section pillars, which enjoyed great popularity in commercial buildings because of the ease with which they could be built into brick walls and pillars. The earliest recorded use of the H-type of stanchion was in some iron buildings manufactured by H & M D Grissell in 1844–5 for export to Mauritius (see p 353, note 18), and it seems likely that it was Henry Grissell who invented it. In the following decade H-section stanchions of cast iron found use in other sophisticated works produced by this company, including Godfrey Greene's No. 7 Slip, Chatham (1851), and the Royal Navy Boat Store, Sheerness (1858–60). The principal advantage of this form was that it enabled beams to be directly bolted to the flange or through the web of the stanchion, effecting a more positive, stiffer connection than possible with a circular column.[115]

It was also easier to cast H-sections with integral seatings or brackets for resting the beams; casting flaws were more readily spotted, and all surfaces could be protected from rust by paint.[116] They were rapidly adopted by the American building industry following their introduction as late as 1894,[117] but in Britain, surprisingly few examples have come to light outside of the Royal Dockyards. Significant among these are the machine halls of a number of mid-Victorian Glaswegian engineering works, where enormous open-webbed H-section stanchions were employed to support both the roof structure and overhead travelling cranes.[118] H-sections were ideally suited to eccentric (unbalanced) loading – inevitable on external columns and those supporting crane gantries. Engineering sheds, including some built in London, were one of the first types of building to employ steel H-section stanchions. Within 'polite' buildings, H-sections had the advantage of concealment within the walls of the building, which protected them from direct exposure to fire. Because of their open, flat-sided geometry, brickwork could be built into and around the member. This facility was exploited by architect (Sir) Robert William Edis for his office rebuilding of No. 10 Fleet Street (1885–6), City of London, where cast-iron H-sections, embedded in an interior wall, carried the end of a long compound beam on two floors (Fig 2.21).[119] That this only came to light during refurbishment shows the form's capacity for 'invisibility' within a finished building, and may indicate more widespread use than is generally appreciated. Both C- and E-section columns were specifically produced for embedding within walls, although some engineers cautioned against the use of such asymmetrical sections, 'on account of the very unequal distribution of compressive stress over its section, and … its liability to become bowed on cooling'.[120]

The unpopularity of wrought-iron stanchions

If the wrought-iron beam set in train a constructional revolution that was fully realised in the age of steel, then the wrought-iron stanchion was a poor and unsuccessful addendum, at least in Britain. In America, prefabricated wrought-iron columns saw enormous and wide-ranging application from the 1860s until the 1890s, a condition that owed much to the widespread success of one product that assured the fortune of the company that produced it.

Fig 2.21
Cast-iron stanchion in first floor of No. 10 Fleet Street, City of London, an office rebuilt in 1885–6 by Sir Robert William Edis (1839–1927).
[AA025290]

America and Japan.[124] Among the first London buildings to use these distinctive wrought-iron members may have been at the western extension of Smithfield Market, where an eight-segmented version, incorporating vertical filler pieces for enhanced rigidity, was employed to eye-catching effect (1879–83) (Fig 2.22). Conceptually similar prefabricated cylindrical columns, presumably manufactured under a British patent, were used in this country in pier construction, but outside such specialised applications this type saw little use in wrought iron. Its unpopularity probably owed less to the wrought iron's inferior compressive strength or resistance to rust, and more to the reluctance – or inability – of British rolling mills to produce such particularised sections, and the diffidence of consumers and designers. Plates, channels and angles had multiple uses, and indeed were produced primarily for shipbuilders, but flanged, segmental sections were fit only for column assembly. Even columns built out of the more common sections never achieved any real popularity, largely because of the time and expense of riveting together the individual pieces, and because many architects disliked their brutal appearance. Smithfield Meat Market (1866–7) was built using box columns, built up from channels, angles and plates, but the architect, Sir Horace Jones, was an anomalous enthusiast of iron, and this utilitarian building was not 'polite' in any conventional sense. The Royal English Opera House (1889–91) was designedly polite and, unusually, employed long cruciform-section columns made from four angles to support the stage. However, like all the other iron and steelwork, this was hidden from view.

Steel stanchions and columns

As the number of steel sections produced by rolling mills continued to increase from the late 1880s, British constructional engineering firms began to experiment with new forms of built-up steel stanchions (Fig 2.23). The basic I-section, whether used alone in short lengths for lighter loads or compounded with plates for more demanding situations was the basis for many, but other sections also found favour, including channels and angles. Their design was dictated by fabrication costs, which rose with the number of rivets and individual pieces used, and an appreciation of the fact that as much of the metal should be located as far

Fig 2.22
Detail of Phoenix Column at Smithfield Market (1879–83), used to support the wrought-iron roof girders. For additional strength, filler pieces were riveted between the arc segments, and for adornment, wrought-iron foliate capitals screened the cast-iron junction.
[DP001754]

Patented in 1862 by Samuel J Reeves of the Phoenix Iron Company of Phoenixville, Pennsylvania, the Phoenix Column was a hollow cylindrical assemblage, made up of four, six or eight wrought-iron flanged segments riveted together.[121] It was considerably lighter than a cast-iron column, but its real advantage was that it could be riveted to other structural components for strong connections. Of immediate application in bridges, viaducts and elevated railways throughout the continent, Phoenix columns also played a key role in the construction of tall urban buildings in the late 19th century because they permitted experimentation in wind-bracing,[122] and, as one contemporary advocate noted, were stronger than rival built-up forms, and better equipped for eccentric loads.[123] They were exported for use in structures in Europe and as far away as South

Fig 2.23
The later 19th century saw
great experimentation and
diversity in forms of built-up
stanchions, although few
were used in practice.

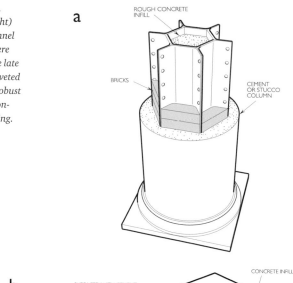

*Fig 2.24 a (right),
b and c (below right)
Splayed steel-channel
sections, which were
produced from the late
1880s, could be riveted
together to form robust
stanchions and non-
combustible flooring.*

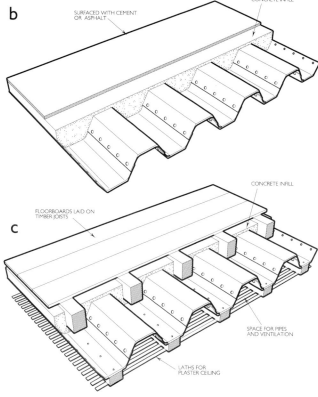

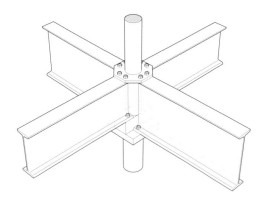

*Fig 2.25
Method of connecting rolled
steel beams to patent solid
steel columns, using
wrought-steel pieces shrunk
onto the column shaft.
Richard Moreland and Son,
who began manufacture in
the 1890s, claimed that
'though the joint is an
exceedingly simple one, it
is nevertheless absolutely
rigid when completed'.*

from the central axis as possible to prevent buckling.[125] Consistently popular with British designers were built-up box columns, since their simple geometry made them relatively easy to fabricate and connect to beams. Latticed box sections, formed from angles or channels riveted together by zigzagging narrow flat sections were the most decorative form 'naked' steel had to offer, and hence were particularly popular in railway stations of the 1890s and early 1900s. Because these were lighter than solid-sided stanchions, they were occasionally employed in the upper storeys or roofs of buildings, to reduce the overall weight on the foundations. Especially popular in America, they were also favoured by those engineers who had worked over there, such as Sven Bylander, who used them in the framework of The Ritz Hotel (*see* Fig 10.44). They were also used by British designers who looked across the Atlantic for inspiration, such as Edward Graham Wood (1854–1930; *see* Fig 8.18).

Phoenix columns were essentially circular sections, and therefore were the strongest possible form of built-up section for their weight.[126] The Phoenix Iron Company began manufacturing them in open-hearth steel in 1889,[127] but in this country, Dorman Long was already producing them by 1887,[128] and by 1900, most large British rolling mills were manufacturing them.[129] Clearly, they were being used in quantity, but the lack of direct evidence in buildings of this period suggests in a concealed manner. Importantly, the generic form of the Phoenix column influenced British steelwork designers. As early as 1887, constructional engineering firm W H Lindsay and Co. of Paddington, London, began fabricating hexagonal steel columns riveted together from splayed channels manufactured by Dorman Long.[130] For increased loading, the columns incorporated bulb-headed flats running the entire height between the channels' flanges, which added rigidity because of the greater cross-sectional area. This technique was seemingly appropriated from the 'Improved Phoenix Column', which employed vertical filler pieces between the flanges.[131] Like the Phoenix column, Lindsay's columns could be rendered 'fireproof' through encasement in non-combustible materials (Fig 2.24a). Lindsay's hexagonal columns were a particularly adroit utilisation of an existing section, originally used by the company (and others, such as Westwood & Baillie) to make steel trough decking for bridges, and as non-combustible flooring in buildings (Fig 2. 24b and c).[132] Lindsay's steel columns and flooring found much application in London buildings, and mark some of the earliest uses of structural steel in the country, with Alfred Waterhouse among its advocates.

Innovation in steel column design that was particular to England stemmed from another London firm of constructional engineers, Richard Moreland & Son, which had emerged as one of the leading firms during the wrought-iron era. This company began supplying 'solid steel columns' from the 1890s (Fig 2.25).[133] Fitted with shrunk-on caps and bases of 'solid wrought steel' to spread the concentrated loads to the foundations, and permit the connection of beams,[134] this unusual structural member saw great application in metropolitan buildings into the 1920s and beyond, with other manufacturers turning them out.[135] Its unique advantage was its slenderness, which meant it intruded less on sightlines than a conventional column or stanchions. Moreland claimed 'a Solid Steel Column, 9in. diameter, 14ft high, will carry 213 tons, whereas it would be necessary to use a Cast Iron Column, 16in. diameter, or a Steel Box Stanchion, 15in. × 18in., to carry the same load'.[136] This combination of strength and slenderness ensured this form's popularity in theatres, music halls and shops, where it was usually concealed within 'fireproof' or decorative sheaths. Besides an inferior load capacity, hollow cast-iron columns of similar diameter would have been more difficult to connect to steel beams, and would have been more prone to impact damage.

Such idiosyncratic innovations in stanchion design contributed to the often eclectic character of London's formative constructional steelwork. From 1901, however, standardisation of structural sections by the Engineering Standards Committee and increasing rationality in design began narrowing the range of forms, while also ensuring the supremacy of mild steel as a building material.

Standardisation of British steel sections

Despite the unprecedented degree of efficiency offered by the industrialised, integrated methods of steel production and rolling, the large number of firms involved meant that there was a bewildering array of sections available, albeit conforming to certain geometric proportions.

The size of the basic I-section was designated by its depth, width in inches and weight in pounds per running foot. Manufacturers increased the weight and section for a given depth by moving the rollers fractionally further apart, which both thickened the web and increased the flange width by the same amount. On top of this scope for variation was the slope of the interior flange surfaces, which varied from company to company, and occasionally from mill to mill within the same company, depending on the contour cut into the roller (Fig 2.26). There was also a variety of other

Fig 2.26
With steel sections being rolled by dozens of companies, British and overseas, there was considerable dissimilarity in their shapes and proportions. This diagram shows the variations in the flange and web area of commonly stocked I-sections produced by four leading companies in 1900.
[The Builder, 1 December 1900]

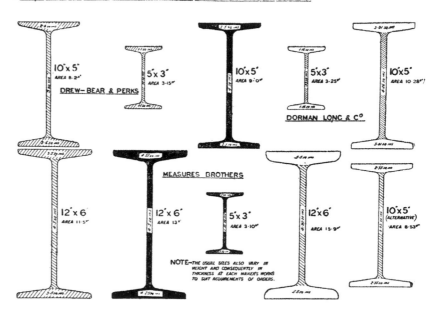

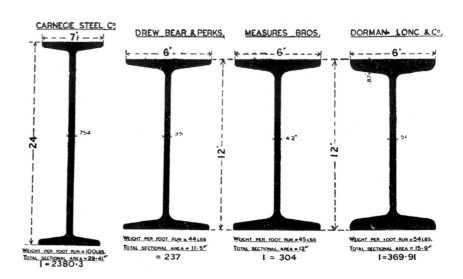

shapes – angles, channels, tees and zeds – in numerous sizes, which were variously produced with unequal flanges or legs, or with 'bulb' ends, where one leg or flange was teardrop-shaped.[137] Many of these sections, however, were used primarily by the makers of ships, railway wagons and bridges, and their use in building construction – in fabricating built-up sections – was being increasingly supplanted by the I-beam.[138]

By the turn of the century, British manufacturers were rolling over 170 different kinds of section.[139] This was a source of confusion to the designer, and, arguably, a hindrance to rational design. On a more practical level, it was a cause of much inconvenience and delay in the building process, since all too frequently, the architect, engineer or contractor would find that the particular sizes required were not kept in stock, or that the quantity he required was insufficient to induce the manufacturer to change rolls and make them. Variability in form also had a bearing on the safety of the member, and, ultimately, on the confidence of the designer. Such difficulties had, by this date, already led both the USA and Germany to standardise their structural sections, reducing the number to 49 and 67 respectively.[140] In America (where throughout the 1880s and 1890s there had already been increasing standardisation), the adoption in January 1896[141] of the Association of American Steel Manufacturers' (AASM) 'American standard beams' brought economic benefits to the steel producers, since, sure of a market or saleable surplus, they gained economies of scale in rolling huge quantities of any given section. Consumers also benefited because of quicker deliveries and keener competition.[142]

In Britain, the issue had been anticipated as early as 1885, when a writer in the *Iron and Coal Trades Review*, in referring to Belgian domination of the girder trade, proposed that English ironmasters 'learn from the enemy, and agree to roll certain standard sections of joists, each one, heavy or light, for which his mill is best adapted, and let him keep these few sections always in stock, so that the merchant may be able to make up a small but varied order at short notice'.[143] Similar sentiments were expressed in 1891 by the well-known iron and steel merchant Harry John Skelton (1858–1934) in his *Economics of Iron and Steel*, in which he decried the 'needless and wasteful elaboration of variety in sections'.[144] In January

1895, in response to a letter to *The Times* quoting the case of a contractor who was dismayed to find that, after many transfers of his order from one manufacturer to another, he was ultimately supplied from Belgium, Skelton penned an influential reply:

> If the members of the engineering professions had any real interest in the English working man or showed anything like scientific method in the practice of their profession, such an instance as the one he related would be impossible. Rolled steel girders are imported into Britain from Belgium and Germany because we have too much individualism in this country, where collective action would be economically advantageous. As a result, architects and engineers specify such unnecessary diverse types of sectional material for given work that anything like economical and continuous manufacture becomes impossible ... no two professional men are agreed upon the size and weight of girder to employ for given work and the British manufacturer is everlastingly changing his rolls or appliances, at greatly increased cost, to meet the irregular unscientific requirements of professional architects and engineers.[145]

No action was taken until June 1900, when Skelton was invited to read a paper before the British Iron Trades Association on 'The competitive outlook for the British iron trade'. Urging standardisation of structural steel members, it was particularly well received by leading consulting engineer Sir John Wolfe-Barry, a recent president of the Institution of Civil Engineers, and a serving member of its Council. With the salutary German and American precedents already set, and with the nation's economic advancement in mind, Wolfe-Barry resolved to standardise steel sections for various applications, including construction. He achieved this by forming a new body, the Engineering Standards Committee, made up of representatives of the Institution of Civil Engineers, the Institution of Mechanical Engineers, the Institution of Naval Architects and the Iron and Steel Institute. The forerunner of the British Engineering Standards Association and ultimately the British Standards Institution, the body was officially launched on 26 April 1901.[146] Soon comprising four specialised committees, including a Bridges and Buildings Sectional Committee, this body's first achievement was BS No. 1 – a pared-down list of rolled iron and steel sections for structural purposes, including just 30 beam sizes ranging from 3 by 1½in. to 24 by 7½in.[147] In the same year, 1903, it published BS No. 4, a handbook defining the structural properties of beams, and in 1904, BS No. 6, which provided the same data across the full range of standard shapes.[148] The standards were defined in terms of weight per foot, thickness of web and flange, sectional area and, for those designers with a scientific understanding, the centre of gravity, the moment of inertia, the radius of gyration and the moment of resistance.[149] In essence, this pamphlet enabled the architect or contractor to order a 'BSB 19' (or '10in by 8in/70lb') without specifying a particular company, since a safe, uniform product free from accidental variation would, in theory, arrive promptly. These sections remained standard until the introduction of New British Standard Beams in 1921, which reproduced virtually the same range of sizes, albeit in more efficient forms with fractionally reduced web and flange thickness and sectional area.

From the start, leading British manufacturers took a keen interest in standardisation. Representatives from firms such as Dorman Long & Co., Andrew Handyside & Co., Sir William Arrol & Company and P & W McClellan attended the Bridges and Buildings Sectional Committee meetings, chaired initially by Sir Benjamin Baker. Some of the largest producers of structural shapes even began rolling standard sections in advance of the published standards, motivated by the economic benefits they would bring. Dorman Long's fourth handbook, published in 1900, actually listed BSBs 1 to 30, in addition to a (diminished) range of traditional sections (Fig 2.27). Although unattributed in contemporary accounts, it would seem that it was this leading company that actually designed the new standards that were contained in BS Nos 1, 4 and 6. Some forward-looking designers were also seemingly keen to use them as soon as they became available. Aston Webb, for example, made extensive use of 'BSB 19' from Dorman Long's handbook in the interior framing of his headquarters offices in Tooley Street, London, for Board & Son distillers (1899–1901).[150]

Standardisation had real and growing impact from 1903. In the year to 1904 it was estimated that the reduced range of sections had already saved the iron and steel industry some £200,000 to £300,000.[151] In that year, both the Admiralty

and the Public Works Department of the Government of India expressed their intent to use, wherever possible, standard structural sections.[152] By 1914 numerous other bodies, including Lloyd's Register of Shipping, had adopted British Standard steel sections, and returns from the largest steelworks showed that 95 per cent of their output of rolled sections in the year 1913–1914 conformed to British Standards.[153] Within the space of a few years, Britain had become more standardised than Germany and America across a whole range of manufactured products, including tram rails, locomotives, electrical cables, Portland cement and structural sections. For architects and engineers in Britain and the colonies, a truly versatile, structurally safe and efficient building component of assured quality was a full reality.

Fig 2.27
Extract from Dorman Long & Co. Ltd's handbook of 1910, showing British Standard Beam 19, which had a slightly reduced web and flange thickness compared to equivalent-sized beams manufactured by this and other companies. Dorman Long began rolling BSBs in 1900, and other British firms soon followed suit. [Courtesy of Institution of Civil Engineers]

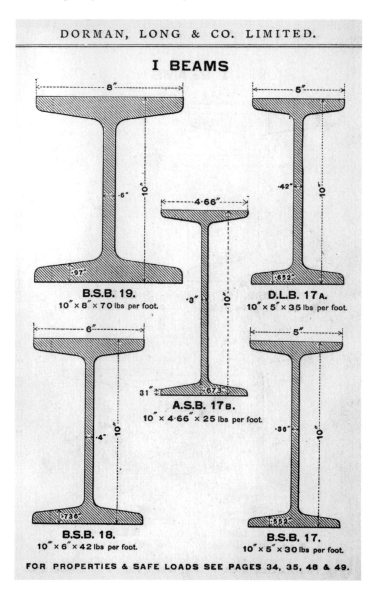

33

3

The Continental dimension: iron and steel girder imports

Let us look around in our own city, in London and elsewhere, and note the vast amount of steel and iron in the form of girders which now enters into the construction of all modern buildings. Where do all these girders come from? Why, from Belgium.   (1894)[1]

The Paris Exhibition of 1867 brought before the public the complacent state of the English iron industry, and the great strides that had been made by foreign competitors. Nowhere was this disparity more apparent than in the displays of Continental rolled products, such as girders and armour plate with their conspicuous 'appeal to the eye'.[2] French manufacturers Pétin Gaudet & Co., flaunted a rolled beam almost 40in. deep and over 32ft long, and another that was 12in. deep and 106ft long. Châttilon and Commentry, also French, boasted a 100ft-long specimen, 9in. deep, and Burbach of Prussia

displayed one that was 47ft long and 15in. high.[3] Next to these brawny demonstrations of technical skill, those English producers that had bothered to make a show simply 'intruded slovenly heaps of raw material mingled with pieces of rusty iron'.[4] English iron manufacturers were blatantly outclassed, and this was widely taken as symptomatic of a more general decline in British industry. *The Times* duly took the English manufacturers to task, and two public inquiries – one into the state of scientific instruction, the other into the impact of trades unions – were started in 1867.[5] Two years later, in response to the adverse criticism encountered at the Paris Exhibition, the Iron and Steel Institute was founded to increase scientific progress in British manufacture.[6]

What really caught attention was the growing importation of rolled iron from Belgium, to values of £103,000 in 1866, £135,000 in 1867,

Fig 3.1
St Thomas's Hospital, Lambeth (1868–71), as photographed by Henry Taunt in 1878. The 1,200 tons of wrought iron used to support the floors came from Belgium.
[CC56/01077]

Fig 3.2
Iron-rolling spread from Belgium in the 1820s to France and Germany. This engraving depicts an early French rolling mill, the skilled workers using tongs to feed red-hot iron back and forth through revolving rolls to make flat bars.
[Paul Poiré, Simples lectures sur les principales *3rd edn, Paris: Librairie Hachette, 1878. (Library of Congress, General Collections, T47.P6 1878)]*

£146,000 in 1868 and £124,000 in 1869.[7] Through the findings of the two inquiries, people were dismayed to learn that the iron-work contractor for Alfred Waterhouse's South Kensington Museum was buying girders from Belgium.[8] Equally, St Thomas's Hospital, then in course of erection, was using some 1,200 tons of structural wrought iron to support its floors, all 'framed and fitted in Belgium, after designs prepared by the architect' (Fig 3.1).[9] Other public edifices, including the Gaiety Theatre (1867–8), were also using Belgian ironwork,[10] and beyond the metropolis, it was revealed that buildings in Sheffield, Middlesbrough and Glasgow – all iron or steel producing localities – incorporated Belgian girders.[11] To one critic, this state of affairs was clearly the result of one of two causes:

> 'either the British workman has, until lately, been requiring a very extravagant rate of wages for converting pig into manufactured iron, or our British rolling mills are inferior in construction and our ironmasters deficient in skill (as compared with their Belgian and French competitors) in the manipulation and production of the forms and sizes of rolled iron required in the present day'.[12]

Both inferences were broadly accurate, and would be reiterated repeatedly by British commentators over the next three decades as Continental girders continued to pour in, meeting an ever-growing market among British architects and builders.[13] Consequently, the great majority of structural iron and an appreciable amount of structural steel in London buildings in the last third of the century came from Continental rolling mills, with Belgium by far the single biggest source.

Belgium had emerged as Britain's chief rival in the production of iron by the 1850s, its twin advantages of abundant raw materials and cheap labour having been given an early boost in the demand for rails prompted by the Belgian government's decision to build a system of state railways in 1833.[14] English expertise, capital and labour had played a decisive role in laying the groundwork for this; the Lancashire mechanic William Cockerill started an engineering works at Liège in 1798, and his sons, John and James, instituted the famous foundries at Seraing in 1814, with English workmen and technological innovations.[15] In 1821 Huart & Henrard of Couillet, in the province of Hainault, erected the first puddling furnace in Belgium[16] and by 1823 Joseph-Michel Orban had introduced puddling furnaces and grooved rolls at his works at Grivegnée, Liège.[17] From then on, wrought-iron production spread rapidly throughout the country and into neighbouring France and Germany, the trade secrets of the 'mysterious art' of puddling, and the technical complexities of rolling being widely divulged following the lapse, in 1824, of an injunction prohibiting the emigration of British skilled workers (Fig 3.2).[18] One source wrongly claimed that 'Rolled iron joists were first made in 1849 by the Providence Works in Belgium;

the invention, which was patented by them, was wholly theirs.'[19] That the French had already produced rolled-iron joists for construction 3 years earlier invalidates this claim (*see* p 11); nonetheless it is symptomatic of the advanced technological climate in Belgium at this date. In the 1860s, Belgian ironmasters began looking abroad for markets, exporting cheap manufactured iron products to Britain, which, as the champion of international free trade, was unprotected by trade tariffs.[20] In 1866, when Belgian iron imports appeared for the first time in British market statistics, rolled girders made their debut.[21] In that year *The Builder* noted 'It appears that the importation of iron girders and joists from Belgium is now a regular branch of London Trade, a business circular which we have seen, offering to "undertake delivery as early as for girders made in this country," and presenting an illustrated sheet of sections, for which orders will be received.'[22] Exploiting a yawning niche in the British market, Belgian manufacturers quickly established a firm foothold, one given additional purchase in 1874 when a series of strikes among English puddlers saw 'very considerable orders for girders' placed by English customers.[23] In the mid- to late 1870s a steady rise in imports of heavy 'section iron' began, comprising principally beams and girders and penetrating even into the Midlands, the traditional centre of heavy finished products: from 24,994 tons in 1876,[24] rising to 32,896 tons for the first 9 months of 1879.[25] Operating through agents and merchants with offices and stockyards in the metropolis, Belgian producers were able to offer rolled-iron sections at highly competitive, if not unassailable prices and delivery times, and in unrivalled sizes – up to 20in. deep – for spanning greater distances.[26]

A number of British manufacturers, including the Derbyshire pioneers, the Butterley Iron Works, responded to the demand for rolled girders but struggled to compete with the Continental competition. Benjamin Hingley (1830–1905) admitted in 1881 that he and fellow Staffordshire ironmasters had failed in their attempts to contend with Belgian girder iron.[27] Five years later, a Sheffield Corporation architect at a conference arranged by the Society of Architects 'thought it a pity English firms did not turn out girders at a price within the reach of English architects and their clients'. Furthermore, 'he could not say that he had ever yet come across an English Girder in an English building, except, of course, in engineering'.[28]

The success of the Belgians in penetrating and dominating the British market was partly attributable to the lower wages and considerably longer hours accepted by Belgian puddlers and rollers, enabling them to undercut competitors' prices.[29] Puddlers were the 'aristocrats' of the industry,[30] and in Britain they were fully aware of this: highly skilled, independent and, from the 1860s, unionised. Continental plants also had better rolling technology and organisation of practice. They were open to new techniques of fuel economy and other cost-saving advances, including 'rolling in one heat', which eliminated the reheating process between cogging and finishing.[31] Belgian producers quickly embraced the universal mill[32] and three-high mill[33] (both of which originated in Britain) in preference to the simpler two-high mill, which was the type preferred by British works. The former had the advantage of producing beams and H-sections of great width and depth with minimal taper on the flanges,[34] while the latter bypassed the 'dead pass' associated with the two-high mill, whereby the piece being rolled had to be manually lifted over the top of the mill between 'live' passes, thereby allowing it to cool and slowing production.[35] The three-high mill was also better suited to continuous operation without roll changing, which suited the relatively few sizes and types of sections favoured by builders on the Continent. In this respect, Continental rollers had a huge advantage in a well-established, flourishing home market for structural sections born of a more progressive architectural and engineering climate with stronger links to the ironmasters. This fact was acknowledged by Cleveland ironmaster (Sir) Isaac Lowthian Bell, who pointed out, in his apologia in 1884 to the Royal Commission on Depression of Trade and Industry:

It happens that abroad both architects and engineers for many years past have devoted their attention to the introduction of iron girders for structural purposes. In the new streets of Paris and in Germany iron girders have been largely introduced, and the consequence was that mills were erected in Germany and in France suitable for rolling girders. The demand for this form of iron became so large that it paid the masters to keep a large stock of which orders could be supplied without delay.[36]

Similar sentiments had been expressed almost 20 years earlier by *The Builder*, which had noted 'Several years have elapsed since the Continental architects and builders began the adoption of rolled iron beams in place of ordinary wooden girders and joists, and found advantages in the substitution, not only in respect of cost, but also in the facility they obtained for working out conceptions which never could have had an actual existence under the old mode of construction.'[37] While English producers had been 'resting somewhat on their oars … their rivals had been hard at work',[38] perfecting useful and economical joists and beams for the building market. By 1886, Belgium was producing some 120,000 tons per annum, the typical rolling mill turning out an average of 2,700 tons per month.[39]

The ability to roll sections up to 20in. deep in great lengths owed much to ingenious techniques of 'piling'. The bloom of wrought iron was first passed through roughing rolls and then through finishing rolls where it emerged as a 'puddle-bar'. Cut into lengths by powerful mechanical shears, the 'puddle-bars' were then stacked together to form a 'pile' of a form approaching that of the desired section. The pile was then reheated in a reverberatory furnace and rolled, the pieces welded together as they passed through the rolls. Successive passes through further finishing rolls defined the final section. To ensure equal diffusion of heat and pressure throughout the metal, the precise arrangement of bars within the pile was important (Fig 3.3). Much care had to be exercised in manipulating the heated pile through the rolls to prevent buckling to the web or flanges. The flanges were particularly prone to 'work-hardening' and defective workmanship, and possibly for this reason some Belgian firms placed better quality 'No. 2' bars at the corners of the pile.[40] The plastic nature of Belgian iron lent itself to the extrusion of structural shapes, and was peculiarly suited to wide and difficult sections.[41] The superiority of Belgian techniques was such that American industrialist Andrew Carnegie appropriated one aspect for the production of deeper, more efficient steel beams. In 1889, he wrote 'We should experiment on making a Beam for building purposes with narrow heavy flanges such as the Belgian in which we could do as they do – use all raw rails with a small piece of muck bar at each end of pile to act as flux.'[42]

The chief, and perhaps, only weapon employed in Britain's defence, and one exploited by manufacturers and commentators alike, was an

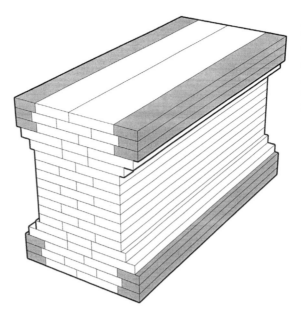

Fig 3.3
The Belgian method piled wrought-iron bars for rolling into an I-section. More ductile bars at the corners resulted in keener flanges.

insistence that Belgian iron was of inferior quality. Almost from the outset, Belgian ironwork gained a poor reputation among British engineers on account of its perceived low strength and the fact that it was not routinely subjected to testing. From the 1860s until the 1890s this tarnished image diffused through many sectors of the British building world,[43] and the term 'Belgian iron' or 'Belgian joist', a somewhat generic expression that usually meant ironwork of Continental origin,[44] was a notorious byword for a cheap, second-rate product. One contemporary remarked 'such inferior metal has commonly been used for [joists] that Belgian iron has a decidedly bad name in England, and is often specially excluded in specifications for bridge work'.[45] Although it was certainly true that some manufacturers placed their emphasis on quantity over quality, this was by no means universal, and some Belgian metal exhibited equal or even superior characteristics to British counterparts.[46] One of Britain's largest importers of Belgian joists, a long-time supplier of Sheffield, Birmingham, Newcastle, Glasgow and other iron centres, doubtless irritated readers of *The Building News* in 1874 when he stated 'the iron used by the Belgian makers is more suitable for the joist section, and that these sections of Belgian iron are stronger than those of English iron of same weight, per foot'.[47] Essentially, it was 'a question of price', noted one Engineer, 'for there is as good iron to be got in Belgium as in England'.[48] Belgium simply supplied what British consumers were prepared to pay. Later commentators saw through the jingoistic hype. In 1901, architect-turned-constructional engineer Archibald Dawnay stated, 'Architects have been frequently misled by grossly exaggerated and inaccurate reports as to the foreign joists being so very inferior to English as to be untrustworthy.'[49] By this date British manufacturers were finally dominating their own structural trade, mass-producing steel sections at highly competitive prices. With a reduced threat to home interests there was less need for propaganda.

In architectural terms, one of the most important consequences of the influx of rolled sections from the Continent was that it helped foster two very different approaches to the design and specification of ironwork. On the one hand, at the upper end of the building market, where quality was allowed to override cost, ironwork was designed and specified by specialist consulting engineers who valued safety and durability over the demands of economy.[50] They tended to endorse and specify only ironwork manufactured and fabricated in Britain, frequently incorporating in their specifications provision for testing. On the other hand, in lesser building projects, comprising the vast majority, architects or builders specified the ironwork, and economy was generally their overriding concern. Typically it was left to the building contractor to obtain the ironwork, and he invariably turned to the numerous iron merchants acting as intermediaries between Continental manufacturers and the British consumer. By bulk-buying from different sources, these merchants were able to execute small orders, often for a wide variety of sizes or sections, something that would have been financially unfeasible direct from the mills, given the loss of time and output involved in changing rolls.[51] The problem was, at least to begin with, that these merchants, 'the sole distributors of rolled joists to the building trades … neither possessed nor professed any technical knowledge'.[52] Nor did the majority of architects, who had next to no training in constructional ironwork, and thus did not prepare detailed structural drawings, let alone inspect the work as it progressed. On this last note *The Builder* provides some insight into the ramshackle nature of the procurement process:

> Those architects employing rolled beams, joists, girders, and other constructional ironworks from Belgium, are in the habit of leaving all inspections to a foreman, or to the

clerks of the works. The iron is brought to the buildings and cast off trolleys into the road, and labourers at once are at work in hoisting it to its intended position.[53]

Such lack of control in the design of iron saw the term 'builders' ironwork' enter the lexicon – almost to the point of becoming a trade term – alongside 'Belgian iron'. Both were used to 'designate an inferior description of work which is considered good enough for ordinary buildings'.[54] Nevertheless, the distinction between the engineered, high-end approach to building, and that of run of the mill, low-end, was often blurred and many architects took an interest in both the quality of the metal used, irrespective of source, and how the structural members were joined together. Alfred Waterhouse, for example, used Continental ironwork contractor John Simeon Bergheim (1844–1912) time and again throughout the 1860s and 1870s, in both major and smaller works.

The deluge of Continental ironwork had another, more significant and lasting effect, both economically and materially. It ultimately compelled British manufacturers to roll structural sections in steel. British ironworks, alerted to the huge market that existed for rolled joists and girders by the Continental monopoly had tried, unsuccessfully, to compete in iron. However, by rolling in steel, they could offer a superior product in unlimited quantities that would be used by both architects and engineers. The first stirrings of this British backlash were felt in the mid-1880s, when the Middlesbrough manufacturers really began applying themselves to the challenge, initially by rolling larger sections in iron, later switching to steel. Dorman Long & Co., which began rolling in iron in 1883, turning to steel in 1886, precipitated the reversal of the status quo. In February 1886, both *Iron* and *The Builder* could report: 'Finding that Middlesbrough makers are getting their products more and more into the market, the foreigners are accepting lower prices than before.' Both notices ended with the low prices offered by Dorman Long.[55] By 1892, the reversal in British fortunes was becoming manifestly clear. Skelton could write:

> No attempt has been made to show British manufacturing capabilities in rolled girders or pieces of H section in puddle iron. For cheap employment in buildings, the British makers of rolled joists have long abandoned

any attempt to equal the low prices of foreign competitors. A new industry in Great Britain is, however, rapidly springing up in the manufacture of the H shape in mild steel (ingot iron). Engineers who objected to use such a shape in wrought iron, find little or no objection to it when rolled from an ingot.[56]

Not surprisingly, given their responsiveness to the needs of the construction industry, Continental producers had already stolen a lead in structural steel. By 1865, several European ironworks had installed Bessemer plant, including Krupps of Essen, Germany, John Cockerill of Seraing, Belgium, and, in France, James Jackson and Co. of St Severin, near Bordeaux, and Pétin, Gaudet et Cie, at Rive-de-Gier, between Lyon and Saint-Étienne.[57] Pétin, Gaudet et Cie were rolling heavy structural sections including I-beams in iron from 1860, when it brought a universal mill into use,[58] and were in all likelihood rolling in steel by the following decade. In the mid-1870s, following earlier experimentation with the Bessemer process, two of Germany's leading joist-rolling mills, the Burbach Iron and Steelworks in the Saar basin, and the Gutehoffnungshütte (GHH

by abbreviation) at Oberhausen in the Ruhr district began producing structural steel sections for use in bridge- and shipbuilding, and for export, by way of the Dortmunder Union canal.[59] The former company, which had exhibited large structural sections at the Paris Universal Exposition, exhibited rolled-steel girders at the 1876 World Fair in Philadelphia, earning the grade 'Premium Quality' by the exhibition judges. Employing some 1,550 workmen by 1872 and producing over 21,000 tons of rolled structural shapes annually, it was a significant presence in the late 19th-century girder trade (Fig 3.4). In 1885, Burbach supplied 6.4m-long (21ft) steel beams, comprising 260mm-deep (10.2in.) channel sections, for use in the Chilworth Gunpowder Company's Incorporating Mills in Surrey; these are among the earliest known surviving *in situ* rolled-steel beams in Britain.[60] Documentary evidence suggests that significant quantities of Burbach rolled steel were imported during the period *c* 1884–1908 through the company's London agents, Otto Gössell of Cannon Street, and Zeitz & Co. of Lime Street (Fig 3.5), although only rarely is this steel visible within surviving buildings.[61] One notable exception is a two-storey

Fig 3.4
The mighty German Burbach Iron and Steelworks in full night-time production, c 1876. By this date its annual production of rolled structural shapes exceeded 21,000 tons and the quality of its rolled-steel girders had been prized at the Philadelphia World Fair. [Landesarchiv Saarbrücken (B 1715/3)]

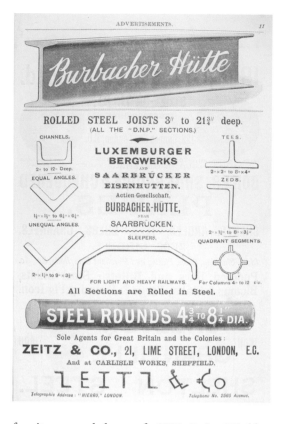

furniture workshop of 1901–2 in Winkley Street, Bethnal Green (Fig 3.6). More significantly, Burbach supplied the thousands of tons of steel framing The Ritz Hotel, Piccadilly (1903–5) – London's most important and influential early steel-framed building.

Technologically adventurous Continental architects and engineers were probably using steel beams by the early1880s, before William

LeBaron Jenney's feted use of Bessemer steel beams in the upper storeys of the Home Insurance Building, Chicago (1883–5).[62] From a British perspective, some of the beams imported during the 1880s would have been of Bessemer steel, although their greater cost relative to iron would have confined their use to more prestigious works. J S Bergheim, Waterhouse's contractor, may have substituted steel for iron beams in some of the architect's works of the early 1880s, or even 1870s, but uncertainty and inconsistency in the technical nomenclature precludes clarity about this.[63] Belgium's niche was unequivocally in rolled iron, and even in the mid-1880s this metal still accounted for the majority of structural sections.[64] Nevertheless, some Belgian producers could probably cater for special orders in steel around this time. The London constructional engineering company Rownson, Drew and Co., for example, produced a section book in 1885 illustrating rolled joists and compound girders, 'of Belgian manufacture chiefly, though most of the sections can be rolled in steel'.[65] Earlier catalogues incorporating both rolled iron and steel sections, such as those by Macnaught, Robertson and Co. (1881),[66] and Measures Brothers Ltd (1882), do not detail the source of the steel, but in all likelihood it was of Continental origin.

Precise dates for when Continental producers began rolling predominantly, and then exclusively, in steel are lacking, but the ever more formidable presence of British steel in the girder trade of the early to mid-1890s probably induced many to switch to the superior metal. Competition, especially from Germany and Belgium, remained extremely fierce, so much so that it was reported in 1894 that 'a Scotch merchant of long standing' (probably Redpath, Brown & Co.) placed a repeat order of 5,000 tons of fabricated steel with a German manufacturer because neither English or Scottish prices could come close.[67] Continental imports continued to exercise a strong, though diminishing grip on the market through the 1900s, maintaining the pressure on home producers. One specialised quarter exploited by a firm from Luxembourg was the broad-flange beam, a particularly advanced form of structural section that British producers were disinclined to roll, presumably because of the cost associated with reinvesting in the plant necessary to do this. Broad-flange, or H-beams, were the product of the 'Grey' reversing mill, a specialised form of universal mill developed by Henry Grey in Duluth, America in 1897, but perfected by Max Meyer at the Differdingen Steel Works of Luxembourg in 1902.[68] The 'Grey' mill had four rollers instead of the usual two, enabling the production of beams with unusually thin, wide flanges with considerably reduced interior slope (some 9 per cent compared to the 14 per cent of ordinary joists)[69] – precursors of the present-day universal columns and universal beams that have parallel flanges with no inward inclination. Inherently the most efficient rolled structural section in their use of material, the wider flanges provided additional strength, enabling greater unbraced lengths when used as beams. Because the thickness of the flanges could be increased or decreased as well as that of the web, manufacturers were able to supply a great range of sizes, meaning there was less need for fabricators to add riveted cover plates when used as girders. They also made for better stanchions; unlike I-sections, broad-flange beams have high moments of inertia around both axes, providing utmost resistance against buckling without the need for additional flange plates (Fig 3.7). Their most potent advantage was that afforded by the reduced interior slope of the flanges, enabling faster, easier splicing of stanchions and beams in framed construction. One specialist British periodical announced in 1906 that 'broad-flange beams offer many advantages in steel frame construction, and, consequently, they will no doubt come into much greater use in modern buildings in the next few years'.[70] Until 1959, however, when Dorman Long put down a universal mill for the production of true 'universal' sections, the principal British source for large, broad-flange beams remained H J Skelton & Co., which purchased them from the Differdingen Steel Works.[71] These distinctive sections did find use in London buildings (Fig 3.8), but compared to British standard sections, this was marginal. Surviving examples are particularly rare.

Continental competition fostered increased British production through economies of scale, ensuring that steel beams and girders soon came into the price range of virtually every British architect and builder. Even allowing for broad-flange beams, the comparatively limited range of shapes and sizes rolled by Continental works also set a powerful example, one that contributed to the adoption of standard sections in 1903. By simplifying and rationalising the design process, this too helped broaden the use of steel.

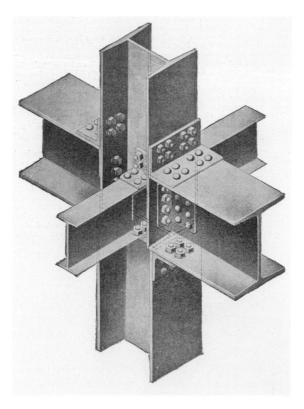

Fig 3.7
Broad-flange beams were exceptionally strong when used as stanchions, and generally did not require additional riveting of plates onto their flanges. This illustration of 1906 shows the manner in which they could be connected to beams using simple angle brackets.
[Builders' Journal and Architectural Engineer (C&S Supplement), 5 December 1906]

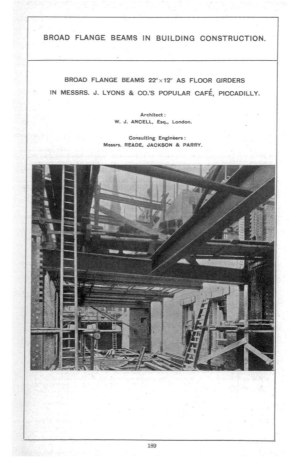

BROAD FLANGE BEAMS IN BUILDING CONSTRUCTION.

BROAD FLANGE BEAMS 22″×12″ AS FLOOR GIRDERS IN MESSRS. J. LYONS & CO.'S POPULAR CAFÉ, PICCADILLY.

Architect:
W. J. ANCELL, Esq., London.

Consulting Engineers:
Messrs. READE, JACKSON & PARRY.

189

Fig 3.8
Steel from the Differdingen Steel Works of Luxembourg frames an archetypal London café (1906). Imported, broad-flange beams were an uncommon choice for London architects and engineers, despite their superior characteristics.
[Lawrance Hurst, H J Skelton & Co., Structural Steel Handbook No. 10]

4

The London building regulations

Probably no city in the world has so carefully devised a code of building regulations as has London, or so complete and efficient machinery for seeing that they are complied with.

(1898)[1]

The destruction of almost four-fifths of the city in the Great Fire of 1666 saw the passing of a series of buildings acts that placed London at the vanguard of building control and regulation until the middle of the 19th century. Drafted primarily to thwart or restrict the spread of fire, rather than improve the health and comfort of the populace, the London Building Acts of the pre-Victorian era introduced a range of controls over the construction of new buildings that were more comprehensive than anything else in Britain, and indeed served as models for a number of provincial towns and cities.[2] The most important of these were the London Building Acts 1667 and 1774, the latter more so.[3] Although the initial Act of 1667 introduced requirements that set the mould for subsequent London Acts of the 18th and 19th centuries, including external walls of incombustible brick or stone, projecting party walls, rules governing wall thickness and timber scantling, and a (largely ineffectual) provision to appoint surveyors to enforce its regulations, that of 1774 was perhaps inevitably more considered and wide-ranging. Its guiding tenets, 'the further and better regulation of buildings, and party-walls', and 'more effectually preventing mischiefs by fire'[4] echoed former acts, but its jurisdiction was the whole built-up area of London, rigorously administered and enforced by a body of district surveyors. The 1774 Act classified buildings for the first time, rating them according to their size, location and resistance to fire; it also introduced nascent principles of compartmentalisation and isolation.[5] Like its predecessors, it did not, however, make any mention of structural iron (although

curiously, iron and steel were permitted for use in crane houses on the banks of the Thames).[6] Except for the exclusion of combustible materials in particularly risky situations, it was not the purpose of the Act to stipulate the materials of various types of buildings, and there was nothing in the 1774 Act proscribing or encouraging the use of iron components. The legislation was formulated at a time when timber-framed or masonry-arched floor construction was universal, and although cramps, ties and straps were in quite widespread use before the Act, it was almost certainly not anticipated that iron would become a primary element in building structures. But if structural iron was not enough of an issue to inform, and be included within, London's Georgian building regulations, the incremental use of iron beams, posts and ties in a growing number of building types through the subsequent decades brought eventual legislative recognition.

It took another 70 years before iron construction was formally acknowledged in London's building legislation. Prompted by concerns about fire safety, and no doubt informed by widening use of iron columns and beams, the Metropolitan Building Act 1844,[7] the first of London's three major 19th-century building acts, actively encouraged iron construction in theatres, concert halls, hotels and other public buildings. It demanded that the floors of staircases, landings, lobbies, corridors and all other means of ingress to public buildings be constructed of, and carried by, 'fireproof' materials. At this date it was widely believed that iron was 'fireproof', whether protected by incombustible material or not, and thus iron joists, beams and girders saw extensive use in these contexts. Similar but lesser strictures were demanded of buildings of the 'dwelling houses class',[8] although iron enjoyed no such fillip in warehouses and factories, the other principal class of building categorised by the 1844 Act. For the first time, however, iron bressummers in addition to storey posts could be used in exterior walls of any buildings.[9] This effectively sanctioned the use of long iron girders, resting on iron or masonry supports, in carrying entire façades at first-floor level, enabling extensively glazed shop-fronts on the ground floor. Tacitly, it also enabled iron-framed interiors composed of columns and beams. One of the first major industrial buildings to be erected under this Act was the City Flour Mills, Upper Thames Street (1850–1),[10] an enormous, constructionally sophisticated building that was subsequently destroyed by fire (*see* p 286). More generally, the most remarkable and far-reaching aspect of the new legislation was the establishment of a coterie of district surveyors and official referees under the jurisdiction of the Metropolitan Building Office (from 1855, the Metropolitan Board of Works).[11] This 'system of building police'[12] was able to monitor the bulk of London's built area to an unprecedented degree, although the district surveyors' discretionary powers in interpreting or extending the provisions of the Act led, inevitably, to unevenness between and even within the districts.[13]

The second of London's principal mid-century acts, the Metropolitan Building Act 1855,[14] consolidated and improved many of the technical regulations governing the construction of buildings, but made few fundamental changes. The chief purposes of the Act, like those preceding it, were 'unquestionably security against fire and stability of structure',[15] but contrary to what some historians have maintained, there was no prohibition on the use of exposed iron in exterior walls or in building interiors.[16] In fact, for certain classes of buildings exempt from the Act, including warehouses under 216,000 cu ft (or those in excess of that, internally subdivided by party walls) the 1855 Act actually permitted wholly iron construction, provided the design and particulars were approved by the Metropolitan Board of Works.[17] The overwhelming majority of architects and builders, however, gave primacy to the traditional load-bearing and stabilising role of thick brick or masonry envelopes in their designs, and this was reflected in the provisions of the Act. The 1855 Act reiterated and expanded upon earlier controls specifying minimum wall thicknesses in relation to their height, length and number of storeys. It also limited the proportion of openings or recesses in relation to that of the surrounding wall area. It was the combination

of these two factors that mitigated against iron framing extending to the exterior of ordinary buildings.[18] The stability of exterior walls was held to be crucial, and this was reflected in the 'Rules as to bressummers', which stated:

> Every bressummer must have a bearing in the direction of its length of 4 inches at the least at each end, upon a sufficient pier of brick or stone, or upon a timber or iron storey post fixed on a solid foundation, in addition to its bearing upon any party wall.[19]

This meagre stipulation would form the sole official guideline governing the design of structural iron and steelwork, internal or external, until 1909.[20] Because the rules regarding bressummers were under the jurisdiction of the district surveyors, there was considerable variation across London in how they were interpreted, a factor that helps explain the marked preponderance of iron and steel lintels, sills and columns in mid- to late 19th-century commercial buildings in Clerkenwell, Shoreditch and other districts fringing the City (Fig 4.1). In more fashionable districts, district surveyors almost certainly discouraged such overt displays of ironwork, at least on the street frontages.[21]

If the material's frank exposure in building façades was an aesthetically controversial issue mediated by a body of professionals whose examining body was the RIBA, more problematic was the mounting evidence that unprotected iron quickly loses strength when subjected to high temperatures. Worse, it could expand and literally push the walls out, or in the case of cast iron, shatter when doused with cold water, causing the collapse of walls, floors and even entire buildings. All this was brought sharply into focus when a devastating fire laid waste to dozens of iron-framed warehouses in Southwark, the notorious Tooley Street fire of 1861. James Braidwood, the head of the Metropolitan Fire Brigade, and one of the principal opponents of iron 'fireproof' construction, died fighting the conflagration, buried under a collapsed wall.[22] Insurance companies were quick to advocate a return to traditional timber construction, but officialdom was slower to react. Not until 1873 did the Metropolitan Board of Works bring in its own Bill to look into this and other issues. The following year a Parliamentary Select Committee sought evidence from experts. One of these, Captain E M Shaw – Braidwood's successor – provided alarming

testimony of drooping tie beams in jack-arch floors and failed columns supporting the corners of buildings and bressummers, all of which had potentially devastating consequences for life or property. Shaw had fewer objections to the use of iron in floors than as vertical supports, but considered 'the ordinary wooden floor, supported on beams, and well plastered on the underside, would be more fire-resisting than a floor constructed of arches supported on the flanges of iron girders'.[23] The use of the term 'fire resisting' in preference to 'fireproof' instanced a more mature understanding of the comparative behaviours of different materials under prolonged, intense heat, an appreciation summed up at the inquiry by Walter Newall, Principal Clerk in the Department of the Superintending Architect of the Metropolitan Board of Works, in his belief that 'no material is absolutely fireproof'.[24] No action regarding fire-resisting construction was taken, the Bill was dropped, and the use of exposed ironwork continued unabated. However, its findings did seemingly inform the Metropolis Management and Building Acts Amendment Act 1878, part of which was specifically drafted with theatres and music halls in mind – historically among the most conflagratory of all public buildings. This demanded that all floors, including the auditorium, be made of 'fire-resisting materials', and that all structural ironwork be protected from the action of fire (see p 113).

The London Building Act 1894 was arguably the most significant and comprehensive piece of building legislation in the metropolis since the Act of 1774.[25] It attempted to regulate 'widths of streets, lines of frontages, open spaces to dwellings, heights of buildings and projections there from, ventilation and height of habitable rooms and the control and prevention of the spread of fire'.[26] 'Fire-resisting' materials were now defined, and included iron and steel, as well as any new material that might be proved to be fire-resisting to the satisfaction of the London County Council (LCC). Although it had a greater range of technical clauses than any previous legislation, putting London in many respects ahead of the rest of the country,[27] none of these addressed iron or steel construction despite its ever-growing and increasingly ambitious use. Like earlier Acts, it failed to give any specification or guidance on the type or quality of metal to be used, the allowable stresses and loads for beams, girders, columns or stanchions, how these should be connected or indeed any basic requirement for safe structural design. It merely added a further clause to the (re-enacted) bressummer rules to account for thermal expansion of beams or girders, requiring that a space be left at each end of every 'metallic bressummer' calculated at a ¼in. for every 10ft. If anything, this stipulation actually made buildings that did exploit framed metal elements more unstable, for it effectively called for bolted rather than riveted connections, the bolts passing through slotted holes at the ends of the beams or girders. Bolted connections were necessary anyway where cast-iron columns or stanchions were employed, yet, by 1895, steel stanchions, either rolled or built up of separate elements, were coming into increasing use, potentially permitting the employment of stronger, riveted frameworks. The failure of the 1894 Act to recognise this, coupled with the absence of any guidelines and an unwavering insistence on thick walls, had two major, interrelated, consequences for constructional steelwork up to 1909: highly diverse forms of interior framing persisted, much reliant on the bracing action of exterior walls, and the introduction of the all-steel skeleton frame in commercial buildings was held back. Like earlier acts, the 1894 Act was formulated around the exigencies of traditional bearing-wall construction, unaware or unrecognising of the potential of framed structural members of iron or steel to create large, well-lit, structurally stable spaces. According to the technical author Walter Noble Twelvetrees (1853–1941), it was

'at least fifty years behind the most advanced architectural practice, and a century behind engineering practice'.[28]

The 1894 Act required that 'Every building shall be enclosed with walls constructed of brick, stone or other hard and incombustible substances', though this did except certain 'special' buildings (*see* p 48). The height for all buildings except churches or chapels was limited to 80ft above pavement level, exclusive of two storeys in the roof and ornamental towers, turrets or other architectural features. (In exceptional cases, buildings of greater height could be erected, but this required the consent of the LCC, which was most unlikely to be granted).[29] As demanded by the 1855 Act, the thickness of walls varied according to their length and height, increasing from the top downwards in a series of steps with each storey (Fig 4.2). The support and stability offered by these obligatory monolithic walls to interior framing was such that there was little reason to make internal iron or steelwork self-sufficient. As Twelvetrees put it:

> 'If the law requires the architect to think that massive brick walls will cover all sins against the canons of steel construction, there is no encouragement for him to connect up iron and steel members, employed for convenience, into a rigid framework, and there is a direct incentive to the disjointed insertion of stanchions and beams in such a way as to constitute a source of possible danger.'[30]

Obedience to the requirement for bolted connections using cast-iron stanchions compounded the issue, giving rise, in many cases, to unsteady, partial frameworks that were wholly reliant on the bracing action of the brick or masonry envelope. The ambitious framing employed in the Hotel Cecil (1889–95), designed in compliance with the 1855 Act, was illustrative of this. There the ends of the steel girders spanning the great halls were 'laid loose' in cast-iron wall-boxes resting on cast-iron stanchions implanted in the brickwork of the walls (*see* p 164). Professor Alexander Kennedy, a consulting engineer and authority on steel, designed it, so the construction was sound. Other buildings, perhaps lacking expert engineering input, were less fortunate. Some even collapsed during construction.[31] A district surveyor for Holborn later noted that, prior to the 1909 Act, 'builders, in using steelwork in

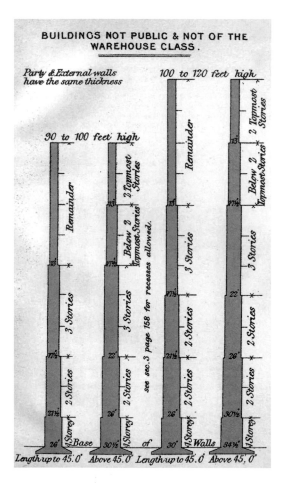

BUILDINGS NOT PUBLIC & NOT OF THE WAREHOUSE CLASS.

Fig 4.2
Wall thicknesses stipulated for one of the three classes of buildings categorised by the London Building Act 1894. [Fletcher, The London Building Acts 5th edn, 1915]

45

building simply piled one piece on top of another, stuck a few bolts in and called it constructional steelwork'.[32] This was, of course, a partial view, for an increasing number of commercial and industrial buildings of the 1890s and 1900s made use of extremely robust interior frameworks, including the Piccadilly branch of the National Provincial Bank, the Bovril Building and the Hotel Russell. The last two buildings, indicative of a trend towards all-steel framing by the turn of the century, may have benefited from a relaxation, reinterpretation or complete waiving of the requirement for bolted column–beam connections: it seems doubtful that Sven Bylander was the first to persuade the LCC of the need for riveted connections when lobbying for his skeleton for The Ritz Hotel in 1904 (*see* pp 182–3).

The second major consequence of the 1894 Act for constructional steelwork was that it created an economic impediment for designers who had the motivation and technical wherewithal to erect a skeleton-framed building. While the LCC was prepared to sanction riveted connections, it remained dogmatic about its wall thickness minima: Selfridges (1908–9) was the only major building where this was waived before legislative change (*see* p 247). The Act did not make it unlawful to erect a multi-storey frame, but this had to be enclosed or carry, storey-by-storey, walls of load-bearing thickness in a misguided belief that the masonry envelope was still doing the work.[33] Structurally illogical, this stipulation nullified or diminished many of the key advantages of skeleton framing. Compared to curtain walls, thick walls were costly and time-consuming to build, devoured floor space, diminished natural lighting, and added to the expense of the foundations and framework, which had to be over-engineered to accommodate excessive, eccentric loads. In addition, there was a requirement that the window-to-wall ratio should not exceed 50 per cent, another hangover from the 1855 Act, which had hardly ever been exceeded in earlier buildings, and only then by cunning interpretation of the legislation.[34] This quashed the potential for thinly clad frameworks to support large, virtually continuous windows on each floor.[35] Yet it was not until about 1905 that London's architects and engineers began railing against this 'vexatious and oppressive' restriction, Twelvetrees remarking that 'at the time the Act was passed it was not very far from representing the practice of average architects,

and that the profession cannot be entirely absolved from blame in permitting the enactment of so unsatisfactory a measure'.[36] Certainly in the period immediately before and after the passing of the Act there was much faultfinding of its provisions in the architectural press, but none of it directed at its failure to accommodate or foresee a metal-frame building system.[37] *The Builder*'s comment of 1892, in discussing the American system 'known as steel or iron frame construction', was symptomatic both of the architectural profession's bearing-wall mentality, and of just how alien a concept skeleton-framing was at that date: 'In some cases the walls of the upper stories are carried by the steel framework independently of the walls below, though for what possible object, except to the danger of the stability of the building, an English architect would find difficult to understand.'[38]

According to Twelvetrees, at least a decade elapsed from the passing of the Act before London architects 'fully awakened to the merits of skeleton building construction', finding 'to their annoyance that, during the period of somnolence, building regulations have been enacted effectually barring the way of progress'.[39] The Savoy Hotel extension (1902–4) and, more potently, The Ritz Hotel (1903–5) were the chief exemplars, showing that even under the existing wall-thickness regulations, skeleton construction still offered advantages in terms of reduced building time (giving savings in ground rent during the time saved in erection), greater stability, and for prestige buildings, a modern, cosmopolitan identity (*see* p 186). The success of these buildings forced the issue on the LCC, which approached the Council of the RIBA to make recommendations for an Amendment Bill to the 1894 Act. As early as February 1904, a committee headed by John Macvicar Anderson (1835–1915), Joseph Douglas Mathews (1838–1923) and Lewis Solomon (1848–1928), together with two members of the Council, submitted recommendations and amendments to the Act. Using the most recent American Building Acts as models, including those of New York and Philadelphia, it recommended the legal sanction of 'iron or steel skeleton construction' in combination with enclosing walls at least 8½in. thick for the uppermost 20ft, and not less than 13in. thick for the remainder of a building's height. A raft of provisions prescribed *inter alia* allowable stresses for iron and steel; demanded that wrought-iron or steel stanchions be spliced together using riveted cover plates; established loading factors for specific building types; required that framing members be completely enclosed by brickwork or other fire-resisting materials; forbade cast iron for beams and or girders; and stipulated that detailed construction drawings and calculations be deposited with the district surveyor 1 month before commencement of the building.[40] Technically and bureaucratically detailed, these recommendations proved highly contentious. Subsequent disagreement between the Council and the Building Act Committee hindered legislative reform. In particular, the Superintendent Architect of the LCC, W E Riley, insisted that 'internal control' over steel-framed buildings should be set out in the legislation, and not left to the discretion of the district surveyor. This factionalised the issues, exacerbating delay.[41] When passed, the London Building Acts (Amendment) Act 1905,[42] which required more stringent fire-safety provision in new buildings, did not acknowledge skeleton construction. Nor did the Building Act reforms of 1908, which sanctioned greater interior cubic footage allowance between party walls, restrictions which, less decisively and directly, also hindered the adoption of full steel framing in certain types of building including department stores and factories (see pp 246 and 294). Not until 1909, after more than 5 years of mounting agitation by architects, engineers and developers, as well as pressure from professional bodies, including the RIBA, the Institution of Civil Engineers, the Surveyors' Institution and the recently formed Concrete Institute (1908), did the Council accept the Building Act Committee's proposals, and push a bill through Parliament. The London County Council (General Powers) Act 1909 decreed that 'it shall be lawful to erect … buildings wherein the loads and stresses are transmitted through each storey to the foundations by a skeleton framework of metal, or partly by a skeleton framework of metal, and partly by a party wall or party walls …'.[43] With only some modifications, the Act endorsed the original recommendations, including reduced wall thicknesses ('the great reform long awaited by the advocates of skeleton building construction in this country') saving floor space, brickwork and the load on foundations (Fig 4.3)[44] but augmented these with a host of other provisions, all detailed under 35 subsections. It was not the constructive panacea that some had hoped for, yet the date of its passing, 17 years after

equivalent legislation was passed in New York,[45] was of great significance to London and Britain, marking the beginnings of the widespread adoption of the multi-storey steel skeleton frame. A sizeable number of skeleton-framed buildings had been erected in the capital before 1909, the cumulative, yearly construction of which served to keep the issue very much alive, but such was the throttlehold of the 1894 Act that this system of building was barred from widespread use. As one (Canadian) engineer noted in 1907, in some buildings 'the design is

Fig 4.3
Sections and plans showing the reduction in wall thickness of a steel skeleton building of the Warehouse Class under the 1909 Act compared to conventional building under the 1894 Act.
[Rivington's Notes on Building Construction (1912)]

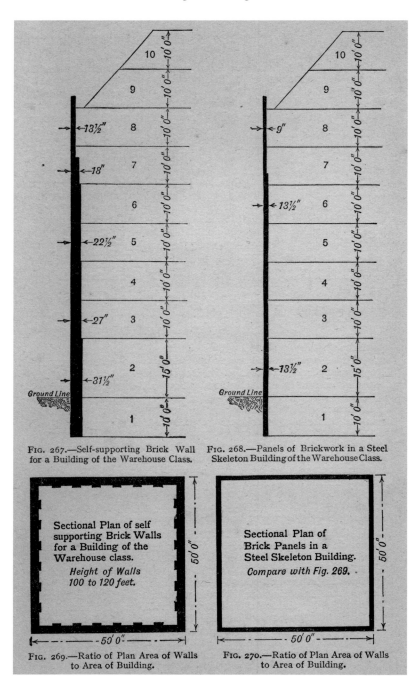

FIG. 267.—Self-supporting Brick Wall for a Building of the Warehouse Class.

FIG. 268.—Panels of Brickwork in a Steel Skeleton Building of the Warehouse Class.

FIG. 269.—Ratio of Plan Area of Walls to Area of Building.

FIG. 270.—Ratio of Plan Area of Walls to Area of Building.

modified in the direction of confining the steel frame to interior work and using it in conjunction with ordinary outside walls, but this does not secure the full advantage of steel-frame construction, either in economy or stability'.[46] With the benefit of hindsight, this was a huge underestimate.

Like the legislation that preceded it, the 1894 Act was not an inhibitor to fully framed metal construction in all buildings. Certain 'special' buildings were exempt from all or some of their provisions, including government buildings, courts of law, inns of court and police buildings in London and Middlesex, the County Council buildings of London and Middlesex, the Bank of England, exhibition buildings sanctioned by the Commissioners of the 1851 Exhibition, buildings erected by railway (including underground), canal, dock and gas companies, and electricity generating and marine engineering sheds. Subject to the Council's approval, in such cases an 'iron building or structure, or any other building or structure' was permitted, so long as it did not exceed 216,000 cu ft under the 1855 Act, or 250,000 cu ft under the 1894 Act, unless divided by party walls into compartments each no larger than these volumes.[47] For a fortunate few, comprising (from 1860) 'any building which,

being at a greater distance than 2 miles from St. Paul's Cathedral, is used wholly for the manufacture of the machinery and boilers of steam vessels, or for a retort house',[48] and additionally, from 1894, those for 'the manufacture of gas, or for generating electricity', these volumetric restrictions could be considerably raised, provided 'that such building consist of one floor only, and be constructed of brick, stone, iron, or other incombustible material throughout'.[49] Unfettered by regulations governing wall thicknesses or cubical capacity, some of these buildings became key vehicles for the development of specialised frames that were increasingly independent of their masonry enclosures. In particular, engineering sheds and power stations swiftly advanced in structural sophistication, their stability dependent on the careful engineering of column–beam connections, while dock companies occasionally employed lengthy single- or double-storey metal roofing systems for their transit sheds, capitalising on the freedom to omit exterior walls, and the ability of metal framing to confer strength and stability. Yet these were particular responses to highly specialised functional requirements. The all-purpose, multi-storey skeleton frame was, in many respects, a creation of a different order,

Fig 4.4
Fire damage, Old Bailey (1908). As well as illustrating the devastating effects that heat could have on unprotected steel girders, this shows traditional load-bearing techniques used even in a building exempt from the London Building Act 1894.
[Builders' Journal and Architectural Engineer, 26 August 1908]

and even in those exempt buildings that might have benefited from it, was rarely, if ever used in London before the early years of the 20th century. Until then, designers continued to rely on, or exploit, thick exterior walls. Both the Great Northern Railway (GNR)'s Western Goods Shed, King's Cross Goods Yard, (1897–9) and the Great Central Railway (GCR)'s Goods Warehouse, Marylebone (1896–9, demolished) utilised monumental walls to help stabilise their internal frameworks (*see* pp 300–1). In this type of structure there was little to be gained from fully framed construction, which would have been an unnecessary and expensive innovation. Government buildings and courts of law could have exploited steel frames, but for prestige architecture the monolithic dignity long linked with the masonry tradition discouraged change before Sir Henry Tanner demonstrated the grand possibilities of classically dressed full-framing, albeit in reinforced concrete, for the new General Post Office building in the City of London (1907–11, demolished 1998).[50] Leeming & Leeming's (revised) Admiralty Extension (1888–1905, in stages) was described by *The Builders' Journal* in 1895 as being 'constructed of steel framing cased externally in red brick and Portland stone',[51] yet this building,

and E W Mountford's competition-winning designs for the Central Criminal Court (Old Bailey) (1902–7), one of the most celebrated examples of Baroque Revival architecture, relied heavily on the massiveness of their brickwork for support and stability. A little more than a year after the latter building's opening, a fire swept through part of the site fronting Fleet Lane, revealing both the devastating effects of flame upon unprotected rolled-steel girders and iron columns, and the traditional construction techniques used (Fig 4.4).[52] One unmistakable – though small and unassuming – example of the advantages of multi-storey full-framing being realised in the absence of any legislative restriction was the construction of two lavatory blocks for the Mount Pleasant Parcel Office, the Post Office's central sorting depot in London (Fig 4.5) (demolished). Forming part of the largest and final phase to the quadrangular Parcel Office, of 1898–1900, these 30ft-long projecting bays were rigidly framed with rolled-steel joists and stanchions, with the outermost 10ft cantilevered from cast-iron columns set about two-thirds the length of the joists. Only two iron columns thus encroached into the yard at ground floor level; above that the five-storey bays were enclosed

Fig 4.5
Mount Pleasant Sorting Office (1897–9), following bomb damage in 1943. The interior had breeze-concrete filler-joist floors on long-span steel plate-girders. More radical were two skeleton-framed lavatory blocks of c 1900 (one visible to the right).
[London Metropolitan Archives, City of London (SC/PHL/02/061/F1636)]

by walls just 9in. thick, built directly on the filler-joist concrete floor slabs. They were designed by the young W J H Leverton, then Sir Henry Tanner's principal architectural assistant,[53] and were in keeping with Tanner's dour red brick Parcel Office, which was internally framed with steel beams and iron columns, and derided by *The Times* as 'appallingly ugly'.[54] This was hardly monumental government architecture; where steel was put to unusually ambitious use in more stately buildings such as the Land Registry Building, Lincolns Inn Fields (1903–5) (Fig 4.6), James Osborne Smith's massive Central Offices for the Royal Arsenal, Woolwich (1905–11)[55] or Sir Aston Webb's Admiralty Arch off Trafalgar Square (designed 1905–7, built 1908–12),[56] there was little desire to reduce wall thicknesses, at least in external walls. It was not until the second decade of the 20th century that skeleton framing in steel was taken up with commitment, in buildings such as the Public Trustee Office, Kingsway (1912–16).[57]

The 19th-century Acts were thus broadly agents of delay as regards innovation in constructional iron and steel, but the legislative limitations were mediated and complicated by circumstance and context. The 1894 Act can not be said to have held back the appearance of the multi-storeyed skeleton frame in the years immediately following its enactment because there was little if any awareness or desire for it among architects and builders. On the other hand, the Act did not encourage its natural development either. Not until the early years of the new century, with the appearance of the steel frame in fully developed form, did the 'tyranny of [London's] masonry tradition',[58] codified in centuries of legislation, become apparent. This was also true in most other British cities, where compulsory thick walls did not become a problem until after 1900, when skeleton framing was introduced. Maximum wall thicknesses specified by building regulations in Glasgow, Liverpool and Manchester were all in the same order as in London; respectively 3ft 9in., 2ft 4in. and 2ft 3in. for Manchester.[59]

In Manchester at least, Americanised techniques of skeleton framing effectively managed to absorb the constraints incumbent in the need for supporting heavy exterior walls. At Trafford Park, the British Westinghouse Electric & Manufacturing Co.'s vast works (1900–2) comprised 'the most remarkable example of steel frame construction hitherto completed in this country'. Each of the skeleton-framed buildings, including the 10 engineering sheds, had to accommodate 'brick-enclosing walls 18 in. and 23in. thick in the workshops, and somewhat thicker in the office building'.[60] Charles Trubshaw's Midland Hotel (1899–1903), the steel skeleton frame of which, like that of the Westinghouse complex, was erected under the supervision of James Stewart & Co., likewise had to carry thick walls (*see* pp 75–6). Twelvetrees noted in 1906 'Thus it will be seen that the Manchester building regulations limit the scope

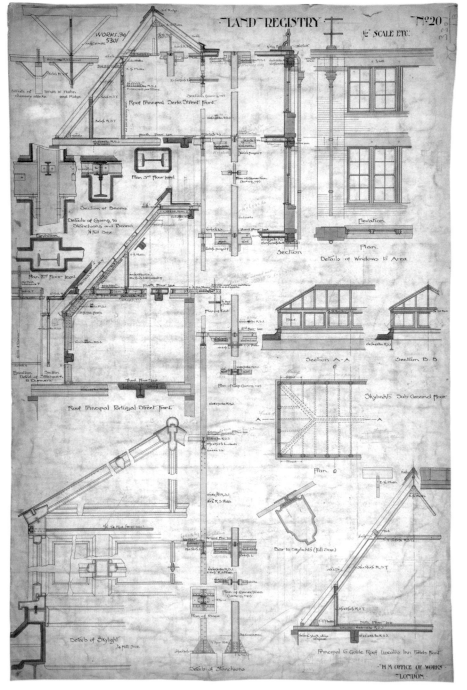

of structural engineers as well as the universally condemned Act applying to the metropolis.' [61]

If by this date an insistence on walls of load-bearing thickness was proving a surmountable obstacle in other British cities, legislative leniency in other areas gave added incentive to skeleton framing. Compared to London's rigorous and effective legislative system, enforced and policed by an army of district surveyors, provincial cities may have been more open to – or unmindful of – new constructional methods, enabling them by turning a blind eye, or conceivably, in lacking the bureaucratic machinery to see that regulations were complied with, failing to look in the first place. Hoyles Warehouse, Manchester, rapidly erected to designs by Charles Heathcote & Sons,[62] benefited from a seeming absence, or waiver, of any clause governing window-to-wall quotient (Fig 4.7). The rear elevation of this six-storey building sported an astonishing amount of glazing, arranged in near-continuous bands between

Figs 4.7 (above) and 4.8 (left) Joshua Hoyle's warehouse, Nos 39–50 Piccadilly, Manchester (1904, now the Malmaison Hotel), showing the contrast between the façade and rear elevations. The back did not have to impress, so the daylight-admitting qualities of the minimally clad skeleton were exploited in a way not possible in London at this time.
[AA022354; Builders' Journal and Architectural Engineer, taken 20 November 1904]

horizontal brick panels (Fig 4.8). Unlike the prestige façade, heavily overloaded with ornament, this rationalised exterior was made to exploit fully the daylight-admitting possibilities of the steel skeleton. Sheffield too may have been more indulgent in this regard. A factory on Moore Street for cardboard-box manufacturers Joseph Pickering & Sons Ltd (1906–7, C & C M Hadfield, architects) also capitalised on licence for extensive glazing, the terracotta-clad grid framing outsized windows on each floor (Fig 4.9).[63] Edwardian London offered a less legislatively fertile ground for the full steel frame, but with land values and commercial competition unequalled anywhere in Europe, it always provided the economic conditions for it to take root and thrive. When London did legislate in favour of the steel skeleton frame, it did so in a far-reaching manner, conscious that the rest of the country would follow.

Fig 4.9
Joseph Pickering & Sons Ltd's cardboard box factory, Moore Street, Sheffield (1906–7; C & C M E Hadfield, architects). It was once skeleton-framed, but only the generously glazed, Renaissance-style terracotta façade was retained when the rest of the building was demolished in the 1990s. [DP143354]

5

Philosophical concerns with iron, steel and framed construction

Architects as a body have neglected and slighted this universally useful metal, either neglecting it altogether, or employing it as it were under protest, and as if they were ashamed of it.

(1874–5)[1]

Iron in construction [was] the most difficult question in the whole field of architectural philosophy … It was easy to say iron should be used as a new material; but it was practically impossible for the designer to forget all the forms in brick, wood, and stone, in which he had been educated all his lifetime, and it would probably require several generations to pass away before the most effective mode of using the material was devised. (1882)[2]

From the 1850s there was a change in attitude to building with structural iron. What had been seen as an innovative, intrepid material became the subject of grave concern among influential voices in the architectural world. A W N Pugin and John Ruskin (Fig 5.1) were the most potent forces, driving the 'iron problem' that flared up in the 1850s,[3] the repercussions of which were felt even into the 20th century. These highly influential theorists imbued architectural practice with deep-seated anxieties about constructional ironwork in all its forms, and in whatever stylistic medium. Pugin was more abstract in his disapproval, itself part of a wider nostalgic tenet that saw past practice rather than innovation as desirable, and which deplored industrialisation, mass production and the interchangeableness of structural and decorative elements. Cast-iron columns that were as massive as stone equivalents, as employed by John Nash at Carlton House Terrace (1827–33), or imitative of other forms, most fancifully in the guise of palm-tree trunks in the Banqueting Room Gallery of the Royal Pavilion, Brighton, also by Nash, were, for Pugin, insincere. More importantly, structural members concealed

Fig 5.1
John Ruskin. By 1885, when John Edwin Mayall took this photograph, Ruskin's predictions about 'metallic construction' were manifest. [Getty Images (3435734)]

behind veneers of wood or plaster, or within floors, roofs or walls were likewise, by extension, moral deceits. Pugin was willing to accept iron, but only within his hierarchical notions of propriety, 'as a cement but not as a support, for binding but not for propping'.[4] Ruskin too was disdainful of mass production, which separated design from manufacture, but his notions regarding iron were more explicit and frequently more damning. Taking one of Pugin's cornerstones of the Gothic Revival, an insistence on honesty of construction and truth to materials, he singled out iron for condemnation, stating in 'The lamp of truth' (1849) that 'true architecture does not admit iron as a constructive material', and 'the entire or principal employment of metallic framework would, therefore, be generally felt as a departure from the first principles of the art'.[5] Like Pugin, he believed that iron should be subservient to traditional materials such as stone, brick and timber. It could abet, but not usurp, their structural role. Yet his musings on iron were often mystifying and complex. In the same book, he declared: 'Abstractedly there appears be no reason why iron should not be used as well as wood; and the time is probably near when a new system of architectural laws will be developed, adapted entirely to metallic construction'.[6] In practice, however, Ruskin remained sceptical, and seemingly unwilling to try. Despite his well-known involvement with Deane and Woodward's Oxford Museum project (1855–60), a building that attempted 'to try how Gothic art could deal with those railway materials, iron and glass',[7] Ruskin seems to have pushed for an alternative roof of stone and timber even before he saw the architects' vague designs in metal. The richly ornamental design in wrought and cast iron that finally materialised by the Coventry metalworker Francis Skidmore, with engineering input by William Fairbairn, evidently failed to win the approval of Ruskin, who later lectured an audience seated in the museum 'never build them [roofs] of iron, but only of wood or stone'.[8] Not all pronouncements were so harsh, but many echoed Ruskin's disdain of industrial production and Crystal Palace-influenced utilitarianism, including *The Building News* which remained unconvinced that the Oxford Museum roof would 'convert the world to a belief in the universal applicability of Crystal Palace architecture Gothicised'.[9]

The opinions of intellectual theorists such as Pugin and Ruskin, often bowdlerised and misunderstood, held great influence over a broad spectrum of mid-century architectural thinking, and not just that of the most ardent disciples. Ecclesiastical buildings echoed their sentiments most clearly, the great majority of Gothic churches reviving the medieval practice of open timber roofs with no plaster ceilings to hide the construction. Iron found little admissibility beyond bolts, ties, straps and cramps, and where it was used for key structural members, it was often accompanied by guilt. St Paul's, Hammersmith (1882–7), designed by Gough and Seddon was, by the architects' reckoning, the first British church to employ wrought-iron lattice principals. But by innovating in this way, they felt duty-bound to issue an apologia (*see* p 260). This embarrassment – shame even – extended beyond the stern canons of Puginian ecclesiology, and into buildings of all types. George Edmund Street, one-time president of the RIBA, was doubtless not alone in feeling that 'there was something almost sinful about an architect employing iron girders'.[10] He himself used exposed ironwork, as in the great library at Dunecht, Aberdeenshire (1861–71), yet whenever he did so he 'he felt conscious of having done something which he ought not to have done.'[11] More immoral was the use of concealed ironwork, and there were few apologists for the universally derided phenomenon of concealed shop-front bressummers.[12] Despite the dogma, unassailable convenience on practical and economic grounds ensured the universal use of iron girders in this context. Some architects got around the polemic by abstracting, synthesising, or reformulating the basic tenets, others by ignoring them completely. For his exquisite office building at Nos 59–61 Mark Lane (1864), George Aitchison junior combined the stylistic elasticity of Ruskinian Gothic with the structural logic of Viollet-le-Duc, appropriating Ruskin's wall-veil premise – 'the façade as expressive skin', to articulate 'truthfully' the (ostensibly) independent iron-cage behind (*see* p 200).[13] George Gilbert Scott considered Gothic to be sufficiently broad and robust that it might absorb technical innovation without compromising its credibility.[14] Confident in his opinion that iron construction offered 'a perfectly new field for architectural development', he made unconcealed use of it in many of his major works, including Kelham Hall, Nottinghamshire (1858–62), the War and Foreign Office, Whitehall (1862–73), and the Midland Grand Hotel, St Pancras (1868–74) (*see* pp 146–7). In these buildings the riveted girders were ornamented with paint, their use ostensibly justified within Puginian philosophy, which held that ornamentation should be decorated structure, a tenet that William Butterfield had exploited as early as 1850 in his use of Gothicised iron beams in the presbytery of All Saints, Margaret Street (1849–59).[15] Yet for every girder that Scott chose to show, there was at least one that remained hidden, and his use of a giant cruciform wrought-iron structure in framing the Albert Memorial (1862–72), neither visible nor articulated,[16] brought in accusations of illogicality and immorality as late as 1894.[17] Such contradictions between theory and practice were symptomatic of a great deal of mid-century architectural use of structural iron.

To many architects, exposed riveted ironwork may have been honest, but despite the best attempts to apply painted ornament, or perforate the web-plate with patterns, it still looked ugly and prosaic. It could never appeal to higher sensibilities, on which architecture prided itself. Cast iron was seen by some as 'more tractable', lending itself 'more to design',[18] so much so that structural logic was entirely sacrificed to aesthetic fitness in some buildings. About 1884, when the tensile properties of cast iron had long been discredited, George Gordon Hoskins used enormous cast-iron arched beams to carry a patent Dennett & Ingle 'fireproof' floor over the public hall of Middlesbrough Town Hall (1883–9) (Fig 5.2). Their spandrels cast with decorative cutout motifs and the council's coat of arms, these members were specifically chosen over wrought-iron equivalents because they appealed to the Gothic eye and intellect, even if they were technologically anachronistic and the floor they supported was framed with concealed rolled-iron joists.[19] The

concealing of structural ironwork within fire-resisting materials, increasingly widespread from the 1870s, may have given technological justification to some architects, but for others it doubtless exacerbated their fretful dilemmas. T Hayter Lewis, a leading promoter of fire-resisting construction placed his own technological common sense behind that of moral conviction when he left iron columns exposed rather than protected with plaster jackets.[20] Constructional ironwork, whether exposed or concealed, stylistically articulated or taciturn, was to some architects a bad habit, but like any bad habit, difficult to abandon. To others, it was probably something that was an inescapable part of architectural practice, the philosophical quandaries of which were best left unexplored. Such was the irrefutable utility and convenience of structural iron that, by the late 19th century, it was fully absorbed within the mainstream of architectural practice. Even architectural theory, riddled with contradictions and characterised by eclectic historicism, was forced to find it an uncertain place, a niche that was soon occupied by steel.

The architectural response to steel was in some respects analogous to that accorded iron.

Fig 5.2
Structural logic sacrificed to aesthetic strictures in Middlesbrough Town Hall (1883–9). Although long discredited for its tensile weakness, cast iron was chosen for the decorative long-span arched girders that carried the floor over the public hall for appearance's sake. Ironically, the floor itself was 'dishonestly' framed with concealed rolled-iron joists.
[DP143350]

Visually indistinguishable, and riveted together in the same structural forms as wrought iron, it took a long while for architects to differentiate steelwork from ironwork. It inherited from iron much of the baggage of architectural and technical debate, although by the 1880s, some of the more abstract hostility to iron had quietened as it became clear that eclecticism in materials, as in styles, was inescapable. Steel also brought with it some new concerns, notably a radically new system of building – multi-storey skeleton framing – that iron had only mildly threatened. This did not happen until after 1900; in the two decades preceding, a new generation of architectural luminaries, already conversant with iron, were setting strong examples to the rest of the profession. They gestured that steel, clothed or unclothed, was a perfectly admissible material where convenience dictated it, even in ecclesiological contexts. Richard Norman Shaw used wood cladding to hide the steel principals and girders of his Holy Trinity, Latimer Road (Harrow Mission Church) (1887–9), but for his mission hall in Church Street (now New Church Road), Camberwell (1894–5), he elected to expose the riveted steelwork (*see* p 267). Even a traditionalist with Ruskinian leanings, John Dando Sedding seems to have made extensive use of concealed steel framing at the Holy Redeemer, Exmouth Market, Clerkenwell (1887–8). Briggs's All Saints' Mission Church, Pentonville (1901–2, demolished early 1960s) even assayed a brash metal aesthetic, tempered only by sub-Art Nouveau wrought-iron ribbon-work (*see* p 271). This building was on the ecclesiological fringes, but it instanced the degree to which some architects were prepared to make open show of constructional steelwork, even for permanent places of worship. Outside of that doctrine-ridden field, eminent architects, such as Alfred Waterhouse, Aston Webb and John Belcher, helped to legitimise the material in large institutional, public and commercial buildings, unashamedly making the most of the structural freedom and forms the material could provide.

By the 1890s constructional steelwork may have become an ordinary part of architectural practice, but the old dilemmas were not entirely extinct. The construction of Tower Bridge (Sir Horace Jones and Sir John Wolfe Barry, 1886–94) caused them to resurface in spectacular fashion (Fig 5.3). At the time the bridge was, in constructional terms, London's closest answer to a Chicago skyscraper. Two giant towers were each formed from four octagonal steel columns connected by horizontal girders and braced together through their full height. For aesthetic reasons, and in deference to the neighbouring Tower of London, these multi-storey skeletons of steel, connected together by a high-level footbridge and supporting the massive suspension chains at either end, were clad in a Gothic dress of granite and brick. To Jones, like most Victorian architects, 'a structure without decorative expression was no more beautiful than a skeleton without flesh and blood'.[21] In appeasing this dictum, he contravened another, truthfulness in construction. The most vociferous critic of all, H H Statham, editor of *The Builder* thought it was

the most terrible and monstrous piece of sham ever erected. The masonry towers, considered per se, are very poor concerns, – tawdry Gothic, totally deficient in breadth of style or power of effect, – but they become simply contemptible when we know that they are only masonry skins carrying nothing, and, in fact, supported by the steel work which they conceal. To make the farce complete, the heavy suspension chains of the side spans of the bridge are led up through 'hawse-holes' into the masonry, so as to keep up the pretence that the towers are supporting the chains, which would, in fact, pull them down immediately. This is one of the cases in which one feels that the naked exhibition of the steel structure, with no pretence at architectural treatment of any kind, would have been infinitely preferable to this foolish piece of elaborately-constructed scenery.[22]

Statham was so incensed by it that he stated in a *Builder* editorial of 1894 that the magazine would not carry plates of the masonry towers since they did not constitute architecture.[23] He even produced his own design, in which 'genuine' masonry towers carried the high-level girders, the projecting ends of which were shown clearly to carry the suspension chains.[24] He was not alone in his condemnation. Many other architects including Peter Waterhouse (son of Alfred), W D Caroe, Beresford Pite and Horace Cubitt joined the chorus, which continued as late as 1909, and probably beyond.[25] Statham would go on to condemn John Belcher's Electra House, 84 Moorgate, City (1901–3), once he realised that the façade was carried on a skeleton frame of steel. It was only

the structural engineer, Thomas C Cunnington's allusion to this that alerted him, and Statham's shocked response was almost as extreme as that meted out to the reviled bridge (*see* p 223).

Multi-storey skeleton construction presented the most acute threat of all to traditionalists like Statham. In iron, this system of building had failed to achieve architectural recognition or currency. As the culmination of a long line of prefabricated iron-and-glass structures, the Crystal Palace in many ways marked the closing stage rather than the beginning of a particular line of development. Its prefabricated progeny, and those buildings that absorbed its structural details and techniques – exhibition halls, market halls, arcades, railway stations, and buildings and structures for export – remained marginal to everyday practice.[26] Architects were able to neutralise the threat these posed to established aesthetic criteria and structural understand-

ings by insisting that they lay outside the world of architecture. By the 1870s what search there had been for a 'ferro-vitreous' architecture, or indeed for a universally applicable system of iron architecture, had been abandoned as a 'wild goose chase'.[27] After 1900, with the reintroduction of skeleton framing in a more potent and all-embracing form, the philosophical issues returned in force. Here was a truly versatile system of building that went to the core of mainstream practice, impinging on commercial, institutional and public architecture. To many architects 'the concept of full load-bearing steel construction confronted the most basic aspects of Victorian and Edwardian architectural logic: truth in construction and stability of form'.[28] The steel skeleton entirely usurped the structural role of the masonry or brick wall (although visually this was not the case), assuming total responsibility for the stability and safety of the

Fig 5.3
Tower Bridge under construction in 1892; one contemporary grumbled that as with American skyscrapers 'the masonry is simply a facing for the gaunt anatomy beneath' (The Building News, *15 April 1898, 518*).
[Getty Images (2673453)]

building. It was the quintessence of constructional deceit, a betrayal that, in combination with the limitations of the London Building Act 1894, actually dissuaded at least one architect from using skeleton construction in 1890s London.[29]

For the majority of English architects full-steel framing would also have been intrinsically related to the phenomenon of the American skyscraper. Skyscrapers were routinely derided for being excessively tall (blocking sunlight, creating windy canyons, being unhygienic, and rising above the height to which fire hoses could reach), for being disproportioned and lacking in surface treatment or monumentality, and for being erected at an uncivilised pace. On this last point, Howard Colls, of the London building firm Trollope & Colls, was of the opinion that 'No building that was constructed beyond a certain speed could possibly be as good as one which was more leisurely, but still speedily and properly done.'[30] Even the American-trained engineer C V Childs, who had himself designed skyscrapers before relocating back to London, declared 'Who would wish to see the streets of London turned into cañons, with precipitous cliffs of buildings on either side, showing only a narrow streak of sky just visible overhead?'[31] Such widely held negative associations with tall buildings were embodied in the 1894 Act's height restriction, a check that

seems never to have been criticised in the pages of the architectural press either in the 1890s or 1900s. Nevertheless, in the early to mid-1900s there was a change in thinking that conceptually dislocated the skeleton frame from the skyscraper. In London, the Savoy Hotel extension and The Ritz Hotel physically demonstrated not only that skeleton framing did not have to equate to buildings of monstrous height, but also that it could be intrinsic to architecture of worth and distinction. These works, and those that followed, did embody concepts unpalatable to traditionalists such as Statham, but such diehards represented a fast-diminishing if vocal minority. More troublesome was the fact that such buildings could only result from full architect–engineer collaboration, for this was at the core of architectural unease with iron, steel and framed construction. Underlying the Puginian rhetoric, the Ruskinian dogma, and the ensuing angst and dilemmas, was the architectural profession's crisis of confidence about its role and professional status. It was engineers, and not architects, who truly grappled with and understood iron and steel construction, and this put them in an envied position of authority, enabling them to intrude ever more into architecture. More than any other matter thus far, iron, and above all steel, changed the roles of the architect and engineer, altering the dynamics of their relationship.

6

Professional conflicts: architect–engineer dynamics

Many of the constructions which formerly belonged to architecture have already passed into the province of the civil engineer; and it can scarcely be doubted that all those which require the exertion of science rather than taste for their execution will speedily follow. No one now thinks of applying to architects to build bridges, or piers, or lighthouses, or anything indeed which is not to be embellished. (1842)[1]

The relation between engineers and architects has been rather one of antagonism. The former has regarded the architect's ideas as chimerical or impracticable, as based on imperfect knowledge and experience of iron and steel; while the architect has looked upon the engineer's work as crude and inartistic, based on mathematical theories and formulae, which leave much unsatisfied, and which seldom give any sense of pleasure. (1904)[2]

In Britain, as in Europe, the notion of architects and engineers as separate 'professions' emerged in the later middle ages, their roles differentiated more 'by the hierarchies they belonged to and the tasks they carried out than by the technologies they used or the design skills they deployed'.[3] From 1400 until about 1750, British engineers were identified as either military fortification specialists or authorities on machinery and water infrastructural projects. Designers of non-military buildings, secular and non-secular, tended to be thought of as architects. Nevertheless there was substantial fluidity, interchange and overlap, a state of affairs that persisted through the second half of the 18th century and into the 19th century, despite the emergence of professional institutions, and increasing territorial demarcation.

The Society of Civil Engineers betokened the emergence of civil engineering as a professional endeavour distinct from military engineering. Founded in 1771, and owing much to the enter-prise of John Smeaton, the first individual to publicly adopt the term 'civil engineer', it was effectively a London-based social club that brought together engineers and fellow professionals keen to promote and garner parliamentary support for the new projects of the age – canal and river navigation works.[4] Demand for improved inland navigation, stimulated by the phenomenal growth in industrial manufacturing and attendant transport requirements created a new constructional niche, and with it a new class of specialist engineers: in late 18th-century England 'in effect, "civil" meant water'.[5] Early members, including William Jessop and John Rennie, were emblematic of this new breed, but the core of 'real engineers, employed in public works or private undertakings of great magnitude'[6] also included the likes of Matthew Boulton and James Watt, whose expertise lay in mechanical engineering. Membership in the early years, and following its reconstitution in 1793, also included allied craftsmen and professionals, including printers, geographers, instrument makers and architects; Robert Mylne, who styled himself 'Architect and Engineer' was a founding member and twice the Society's Vice President, and Samuel Wyatt was another member whose expertise was as much in architecture as engineering. However, despite such overlaps, some of the engineer members doubtless saw their province, and value, as quite distinct from that of their fellow architects: in the overblown words of the preface to Smeaton's engineering reports, civil engineering was concerned with 'those wonderful works, not of pompous and useless magnificence, but of real utility'.[7]

The Society of Civil Engineers fostered credibility and status for the profession, but with few members, limited technical exchange, and an exclusive constitution it failed 'to meet the wants of so large and mixed a body as soon became involved in engineering'.[8] Nurtured by the expanding opportunities presented by the Industrial Revolution, the identity of the growing array of fledgling 'civilian engineers' found further professional distinction with the formation of the Institution of Civil Engineers in 1818, especially with the appointment of the highly respected and well-connected Thomas Telford as its first president in 1820. Telford, who began his career as a mason-architect, was instrumental in drumming up membership for the world's first professional engineering body, and obtaining a Royal Charter for it in 1828 that famously defined the tasks of its members as 'the art of directing the great sources of power in nature for the use and convenience of man'. Worded by Thomas Tredgold, it once again underscored the utility of the civil engineer's endeavours, recognising also his extended range by that date, which encompassed 'the construction of roads, bridges, aqueducts, canals, river navigation, and docks … ports, harbours, moles, breakwaters, and light-houses, and … the construction and adaptation of machinery, and … the drainage of cities and towns.'[9] Few architects would have bemoaned this remit, although some might have felt threatened by the growing cohesiveness and professionalisation of engineers. Georgian architects tended to view themselves as individual practitioners rather than as a collective group, and with less incentive to form associations, none of the early initiatives, including the Architects Club (formed in 1791), and the London Architectural Society (formed in 1806) proved long-lasting or successful.[10]

Probably the area of greatest overlap between the two professions in the late Georgian period was bridge building, a sphere in which both might have claimed equal expertise, at least to begin with. This held even with the introduction of cast iron as a credible alternative to masonry arches, with both architects and engineers working closely with ironfounders to realise the first generation of cast-iron arch bridges. The Iron Bridge at Coalbrookdale (1777–9), designed by the architect Thomas Farnolls Pritchard (1723–77)[11] in collaboration with ironmaster John Wilkinson, is perhaps the most celebrated example, but others testify to architects' early experimentation and capability in this embryonic field. John Nash's second bridge over the River Thames, cast at the Coalbrookdale foundry and based on a patent he filed in 1797, instanced an early use of hollow-box voussoirs and spandrels, and a local

architect, Thomas Wilson supervised the construction of the 236ft-span Sunderland bridge across the River Wear (1793–6). Yet neither were unmitigated cases of architectural originality or aptitude. Nash's bridge, which replaced an earlier design of his that failed, was possibly based on ideas by the American engineer Robert Fulton,[12] and was strongly influenced by Sunderland Bridge for which Wilson seems to have had little design responsibility, that being mostly the work of the patron, Rowland Burdon MP 'engineer, architect and paymaster'.[13] Wilson went on to register a patent for an improved method of connecting cast-iron 'keystones' of the type used in Sunderland Bridge, employing the technique for his bridges at Staines (1803) and at Yarn (1805). Both, however, were unsuccessful.[14]

Failures in early cast-iron bridges were not entirely restricted to those designed by architects – one by William Jessop, built about 1805 at Bristol, collapsed – but already by that date, leading engineers and ironfounders were moving from empirical methods towards a more scientific consideration of iron structures that made safe design more certain. Thomas Telford, who switched from architecture to civil engineering in 1793 (finding it 'a much wider field, and … a much more agreeable pursuit')[15] made experiments on the ribs of the Buildwas Bridge and the columns of the Longdon aqueduct, each completed to his designs in 1796. The findings of the latter were utilised by Charles Bage in the design of his iron-framed Ditherington flax mill.[16] Telford spent much time in 1800–1 calculating the form of the ribs and spandrel frames of his and James Douglas's famous design for the new London bridge. Although the project was abandoned (not because the 600ft span was deemed theoretically unfeasible by the parliamentary select committee, or the experts it consulted, which included scientists, eminent engineers and ironmasters but not architects), Telford's calculations informed the design of his subsequent iron bridges.[17] Following additional experimental work, including that undertaken by John Banks, employee of the Yorkshire ironfounders Aydon & Elwell, major bridge building in iron began slipping out of the hands of architects and into those of engineers. Understanding, based on calculations and proof-testing of components, grew slowly, but from the 1810s, engineer–ironfounder affiliations increasingly dominated the field, above all those of Telford and William Hazledine and Rennie and the Butterley Company, whose partners included William Jessop and the civil engineer Benjamin Outram. The structures produced by these collaborations, notably the Craigellachie, Banffshire (1812–15), Waterloo Bridge Betws-y-Coed, Caernarvonshire (1815–6), and Southwark Bridge (1811–18) were elegant, daring and efficient, and were generally accepted as such. Unsurprisingly, the first stirrings of an architectural estrangement – backlash even – to this state of affairs began emerging, evinced in one architect's observation from 1819 that 'a new class of men, calling themselves Civil Engineers, but who are in fact little better than millwrights, now usurp the abused title of Architects; and our bridges now erecting, and lately erected, however ingenious in construction, are abominable as works of art'.[18] Not long after, in 1824, the architect Joseph Gwilt made similar play of the architectural profession's artistic aptitude, also upholding its authority in masonry bridge building: 'as a substitute for stone in the arches of bridges, it [cast iron] is so unarchitectural [sic] in appearance that one may incline to think the architect, if left to his own choice, will never adopt it; and in many instances, experience has taught us that a few years, and partial settlements, make it much more inexpedient than stone'.[19] Yet by that date, rare indeed was the architect who could use the material more expediently, or efficiently, in a mathematical or scientific sense, than an engineer.

The field of endeavour in which the architectural profession could really claim expertise, and precedent, was the design of buildings. As with bridges, Georgian architects were quick to experiment with new materials and techniques for the buildings they designed. Samuel Wyatt, for example, employed remarkably bold timber framing and inverted-arch foundations for his Albion Mill, Southwark (1783–86);[20] John Nash, and the carpenter-constructor William Nixon designed and built an unprecedented timber-framed self-supporting tented enclosure in Carlton House gardens (1814), the span of which equalled the dome of St Paul's,[21] and Robert Smirke, acting on John Rennie's suggestion, used a raft foundation of lime and gravel concrete to support the Millbank Penitentiary (1816–17).[22] In the use of iron, some architects were equally enthusiastic. Nash used cast iron for the Gothick balustrades and brackets in the Grand Hall at Corsham, Wiltshire (1797–1800) and Longer Hall, Shropshire (c 1805), and

S P Cockerell seems to have taken similar delight in its novelty, employing guilloche patterned segmental iron girders to carry the main stair at Sezincote, Gloucestershire (1805).[23] Thomas Hopper made still more extensive use and greater visual play in his Gothic cast-iron fan-vaulted conservatory at Carlton House (1807), and Nash used it to fanciful effect in the Brighton Pavilion (1815–22), particularly in the 'bamboo' staircases and palm-tree columned great kitchen (Fig 6.1).[24] Cast iron also found more sober, covert, structural applications by leading architects. John Nash and Robert Smirke were especially given to using concealed iron beams in many of their country houses and public buildings of the early 19th century, to give wide-span floors that did not sag or bounce, and Nash employed great skeletal frameworks of cast iron for the two onion-domes at the Brighton Pavilion.[25] For reasons of incombustibility, and preserving sightlines, architects made extensive use of

Fig 6.1
The great kitchen, Royal Pavilion, Brighton, photographed by Eric de Maré in the 1960s. [FF98/00155]

iron in theatres; Henry Holland, at the Theatre Royal, Drury Lane (1791–4); John Foulston at the Theatre Royal, Plymouth (1811–13); and Samuel Beazley and Albinus Martin at the New English Opera-house, Strand (1815–16) (*see* p 107).[26]

Architects responsible for dock and dockyard buildings embraced, where circumstance encouraged, iron's strength and/or 'fireproof' qualities, and it was in this field that they made most progressive use of the material. Daniel Asher Alexander, who had already designed warehouses at Bankside and Mark Lane before his surveyorship with the London Dock Company, 1796–1831, during which he designed all the buildings,[27] made innovative use of iron for the Tobacco Warehouse at the London Docks (1811–14; *see* Fig 13.22). For the 'Skin Floor', Alexander used widely spaced cruciform-section cast-iron columns with raking struts to support wide-span timber roof trusses, an elegant solution that might have been achieved in timber, albeit less efficiently, elegantly and expediently, given that Napoleon's blockade of the Baltic saw timber in short supply. Telford had already used cruciform-section raking struts to help support his London Aqueduct, Shropshire (1896–7), but Alexander took his cue from the previous building, which had roof trusses of identical span, carried on timber columns with raking struts or knee braces.[28] His tree-like columns seem to have had few successors, at least that have survived or been documented, although it embodied structural principles expressed more maturely in the arched beam, a form early on employed in the Rum Quay Sheds, West India Docks (1813), probably to Rennie's design.[29] The branching column form reappeared at Sheerness Naval Dockyard in the following decade, where it was used to achieve wide uninterrupted floor spans in the Working Mast House (1823–6) and the North Saw Pits (*c* 1828). Both buildings were designed by the Navy Board's Surveyor, Edward Holl, arguably the most committed, and innovative Georgian architect working with iron. Between 1812 and his death in 1824, Holl was directly responsible for at least 14 dockyard buildings with significant quantities of structural iron, including four large multi-storey iron-framed warehouses, and, in conjunction with Rennie, the celebrated Smithery building at Woolwich dockyard (1814– 16), with a complete cast iron frame of columns, arched girders and cast- and wrought-iron roof trusses.[30]

By training or title, Alexander and Holl were unquestionably architects, and would have considered utilitarian buildings, and iron structure, part of their professional territory. Yet the demand for new or improved types of buildings engendered by the industrial revolution, including multi-storey 'fireproof' mills and warehouses, helped bring practicing engineers into the field of building construction, and in due course civil architecture. Rennie's involvement with iron-framed dockside buildings has been mentioned, but textile mills were the medium for the most widespread application and systematic investigation of cast iron. Although their designers were mostly 'engineers only by courtesy',[31] rather than of the emergent professional kind, the knowledge and lessons they accrued, some of it theoretically rationalised, seemingly informed the emergent professional engineering community far more than its architectural counterpart. Charles Bage, the initiator of cast-iron beams in buildings, for example provided Telford with the results of full-scale tests on a form of beam intended for mills he designed in 1802–4 at Meadow Lane, Leeds, and the Canal Terminus, Shrewsbury,[32] reciprocating the latter's earlier experimental input at Ditherington. Methodical, rational experimentation with cast-iron beams and columns within these buildings fostered a growing corpus of practical and theoretical knowledge that while initially largely confined within certain millwright–ironfounder circles, soon permeated to their closest professional peers, the civil and mechanical engineers. The conduit for this knowledge transfer, besides pamphlets, technical books, professional society meetings and papers, were the ironfounders. Working closely with both millwrights and engineers, these specialists strove to produce structural elements that gave the maximum strength for a minimum weight (and hence cost) of iron. Many emerged as engineers in their own right, specialising in the design and erection of structural ironwork. William Fairbairn, who by the early 1830s ran perhaps the largest ironworks in the world, and who collaborated with Eaton Hodgkinson (1789–1861) to design the most geometrically rational cast-iron beam possible (using 20 to 30 per cent less iron),[33] designed iron-framed buildings for export, and John Urpeth Rastrick (1780–1856) designed the huge beams that spanned the column-free King's Library of the British Museum (1823–7). Robert Smirke had originally wanted to use a cast-iron girder trussed with wrought-iron ties, but Rastrick, who in 1811 had become 'engineer' at the Hazledine's Bridgnorth Foundry, persuaded him otherwise.[34]

By the 1820s, expertise regarding the most structurally efficient ironwork was increasingly vested in engineers and specialist ironfounders. Architects such as Nash, Smirke, Charles Barry and William Wilkins were likely able to 'size' the cast-iron beams in their private and institutional buildings, perhaps using tables included in Tredgold's *A Practical Essay on the Strength of Cast Iron* (1822),[35] but probably none could have had much grounding in scientific knowledge and understanding of structural behaviour. Certainly there is no evidence to suggest that architects collaborated with scientists or ironfounders to produce beams of more efficient shape or longer span, or structures using the minimum quantity of iron. This want of structural understanding was rarely a problem per se, for much of the iron that went into public buildings in the 1820s and 1830s was in the form of beams of unexceptional span. But occasionally it was. The failure, three days after it opened, of the iron roof at the Brunswick Theatre, Whitechapel, in February 1828, killing 13 people, showed the limitations of architects' structural proficiency in iron (Fig 6.2). Although the building's architect, James T S Whitwell (formerly an assistant of Daniel Alexander) did not design the triangulated wrought-iron truss (this being the work of the supplier, Bristol ironmaster, James Wellington), his uncritical confidence in its ability to support the additional load imposed by suspended equipment – a habit in theatres of the time – were equally blameworthy.[36] *The Mechanics Magazine*, fearful that the accident might exacerbate 'a prejudice that is beginning to prevail against the use of iron for roofs' laid the blame at outmoded empirical design rules based on proof-loading, urging instead engineering rationale: 'Architects would do well … to keep in mind the example which Mr. Telford has set them in the erection of that bridge [The Menai Bridge, 1818–26]; he

Fig 6.2
The collapse in 1828 of Thomas Stedman Whitwell's Brunswick Theatre, Whitechapel. Aquatint by Robert Cruikshank.
[London Metropolitan Archives, City of London (SC/GL/PR/S3/WEL/squ)]

ascertained before hand, by a numerous series of experiments, varied in every possible way, what such and such combinations of parts would perform: he did not erect the bridge first, and then try whether it could bear a thousand head of cattle or not: he proved that it would do so before he proceeded a single step.'[37] Some architects, often responding to the exigencies of particular building types, did show masterly understanding of cast iron. Charles Fowler, rather than the engineer-contractor Joseph Bramah, designed the acclaimed Hungerford Fish Market, near Charing Cross (1831–3), perhaps the first freestanding building wholly reliant for stability on its rigid column–beam connections, and one whose double butterfly roof form set the mould for thousands of smaller railway station platform canopies.[38] Despite instances such as this, and John Nash's oft-quoted opinion 'the Architect ought to be the most competent judge of the form of the castings he requires, and their application and strength',[39] structural iron of any complexity came to be seen increasingly as the preserve of engineers.

If the emergence of the civil engineering profession, structural cast iron and multi-storey textile mills were interweaving strands that helped bring engineers into the time-honoured territory of the architect in the late Georgian period, then wrought-iron and the railways accelerated the territorial overlap in the early Victorian era. The demand for railway structures, including bridges and long-span roofs, the introduction of structural wrought-iron, and the rapid strides in design methods based on model-testing and knowledge of materials, were key interrelated developments of the 1830s and 1840s with engineers and engineering at the fore. The development of the triangulated truss, including the first entirely of wrought iron in 1838 at Euston by (Sir) Charles Fox, and the increasing use in the following decades of statics to calculate the forces in such structures were further instances of engineering authority, but the key labours that saw structural engineering emerge as a truly scientifically based profession followed the collapse of the cast-iron Dee Bridge at Chester in 1847. Subsequent rigorous analytical work undertaken by Fairbairn, Hodgkinson, William Pole, Charles Wild and others for the Britannia and Conway tubular bridges served to launch wrought iron as the structural material of the future, but also widen the epistemological gulf between engineering and architecture: 'It was at this point, more than at any earlier stage in the use of iron or other materials, that engineering slipped beyond the grasp of most practising architects'.[40]

The yielding of structural expertise to civil engineers in the mid-19th century was nonetheless a drawn-out, somewhat guarded affair. In 1834, when architects at last found institutional voice with the establishment of the Institute of British Architects, it was the encroachment of builders, surveyors and others connected with an increasingly complex and fragmenting building world that posed more immediate threat than engineers. One of its chief self-styled aims was to promote 'uniformity and respectability of practice in the profession', and another, in spite of the emerging and excluding notion of the 'art-architect', was 'for the promotion of the different branches of science connected with it'.[41] The Institute did not, however, admit the consulting engineer William Tierney Clark, one of the most successful British designers of suspension bridges, within its list of fellows.[42] Bridge design had by this date been largely conceded to the engineer, but the centrality of structure within architectural theory, and architects' longstanding association with iron made similar concessions vis-à-vis architecture less acceptable. To the majority of practitioners, perhaps, Joseph Gwilt's definition of the architect's role in 1842 as 'A person competent to design and superintend the execution of any building', was entirely credible, railway stations included.[43] Until 1850, architects designed most of the rural and town stations, including the timber, iron or timber-and-iron canopies, and occasionally they were responsible for wider-span iron train-sheds, such as John Dobson's triple-span roof at Newcastle (1849–50), which employed curved wrought-iron ribs formed on special rollers at the architect's behest.[44] (Sir) William Tite seemingly designed the iron sheds at his larger stations, although was prevented from doing so by the L&NWR at Lime Street, which told him to keep to architecture and engaged the iron manufacturer-engineer Richard Turner, designer of the Royal Botanic Gardens, Kew (1844–8) instead.[45] As an architect, Tite was particularly enthusiastic, and adept at iron construction, having seemingly used riveted T-section, rolled-iron joists as early as the mid-1820s at Mill-Hill School, Middlesex (1825–7),[46] also using cast iron extensively and efficiently in his greatest work, The Royal Exchange (1842–4). In many respects Tite, a member of the Institution of Civil

Engineers from 1829, represented a continuation of the type of constructionally engaged architects and architect-engineers of the Georgian era. Others, such as John Whichcord and Edward I'Anson – both pupils of Douglas Asher Alexander, and George Aitchison senior (1792–1861) and junior (1825–1910), succeeding architects for the St Katharine's Docks Company, used iron extensively in their purpose-built offices of the mid-19th century, seemingly without specialist help. Aitchison junior, whose most celebrated speculative office at Nos 59–61 Mark Lane, was built on a cast-and-wrought-iron frame structurally independent of the Venetian-Byzantine façade (*see* p 200), insisted that 'Architects are before constructors; architects without any constructive knowledge are even worse than sculptors without anatomy.'[47]

Despite such sentiments, and areas of expertise, engineers and engineering contractors came to monopolise certain building types, including train sheds, exhibition and market halls, conservatories and other wide-span roof structures. The Crystal Palace, perhaps more than any other building, showed that architects could be virtually dispensed with in designing and erecting the most high-profile buildings of the age (Fig 6.3).[48] This, and many other

Fig 6.3
Hand-coloured view, probably by Philip Henry Delamotte, of the vast nave of the Crystal Palace. The engineer was R M Ordish, who had a hand in its original design for Hyde Park, and a lead role in its modification for the Sydenham site in 1854. [FF91/00334]

projects with a high engineering or iron component, including the 'Brompton Boilers', provoked considerable scorn from an increasingly protective and endangered architectural community, and provided a subtext to the 'iron problem' of the 1850s (*see* p 53). Architectural theorising could accommodate or neutralise the threat posed by such works by claiming that they lay beyond or beneath true architecture, but for reasons of fire-resistance, spatial openness, flexibility planning and so forth, iron was routinely entering the fabric of large buildings, and with it came the expertise of the specialist contractor, and sometimes the consulting engineer. On a day-to-day level, this overlapping of professional roles on shared territory demanded cooperation, rather than competition. This still left the question of who had ultimate design responsibility. In practice, it seems that this was ceded to the architect in almost all cases: 'In order to facilitate this cooperation effectively, and play a creative role in the making of architecture, the engineer developed a series of highly specialized services, as consultant structural, mechanical and [later] electrical engineers, providing expert, indeed essential, professional input into a collaborative design process.'[49] The role of the consultant engineer thus became highly focused and specialised, working to the wider schematic requirements of the architect. To many architects, even this circumscribed responsibility must have been an ominous source of irritation, a reminder that a critical aspect of the building was not, and could not be, of their making. One tactic architects used to maintain an illusion of complete control was stressing the aesthetic or stylistic component at the expense, even to complete omission, of the constructive element and its designer and fabricator. Most contemporary accounts of buildings in *The Builder* and *The Building News and Engineering Journal* (to give it its full title) tended to focus on stylistic aspects, partly because the architect typically supplied the textual and graphic information on which these were based. In this sense, the full contribution of engineers, including those working for ironwork companies, was often written out of the building world, and, as a result, of history.

Partnership and parity in design contribution were not always unacknowledged. Sir George Gilbert Scott made public the structural engineering that went into the Albert Memorial, stating 'For the arrangement and construction of the iron work forming the interior of the fleche, as well as of the great girders by which it was supported, I am indebted to my friend Mr F W Shields, the eminent engineer.'[50] Scott, who trained with the ironwork contractors Grissell and Peto during their Hungerford Market commission, maintained strong and productive collaborations with some of the most accomplished structural engineers of the day, including Rowland Mason Ordish. So impressed was Scott with Ordish's advanced design for the winter garden of Leeds Infirmary (1869) that thereafter, until his death, he 'consulted Mr. Ordish on all important questions of structure'.[51] Consulting engineers, such as Ordish, were happy to work on buildings where the architect's input was minor, as in the dome of the Royal Albert Hall (1867–71, H Y D Scott, engineer), or more fundamental, such as the Pandora Theatre, Leicester Square (1882–4, Thomas Verity, architect). In this latter building, 'under Mr. Verity's direction' he brought his considerable skills to bear in realising advanced cantilever construction and long-span, wrought-iron roof trusses for the auditorium (*see* pp 113–14).

As a building type, theatres were especially dependent on increasingly ambitious metal construction, and for this reason architects tended to see the role of engineers as indispensable and legitimate. There was a certain licence given to theatre design that relinquished it from the prevailing moral values applied to other public architecture. At any rate, the often awe-inspiring structural engineering was difficult to disguise. Augustus W Tanner was doubtless not alone in thinking:

> Good building, and therefore in a wide sense good architecture, is the result of the united efforts of the engineer and the architect. It is not given to one man to be a successful follower alike of both callings, and few better subjects than a theatre could be chosen for members of both professions to meet on friendly footing and together design the building, the one to exhibit 'the beauty of fitness' to its utmost degree, and the other to complete the edifice and enchant the senses, making that 'music to the eye,' which has been aptly applied to noble architecture.[52]

Edwin O Sachs (1870–1919), an architect with specialist knowledge in theatre construction, also acknowledged the need for full collaboration given the complexities of this building type, outlining the division of responsibility:

> Speaking of the engineering problem, it would not be out of place here to point out the necessity, especially with respect to large theatres, for the architect to work in consultation with a competent civil engineer. The many and varying interests that have to be considered by the architect in the design of a playhouse, if given due attention, will tax the highest professional ability which need not, however, include an intimate knowledge of metal construction, a subject in which, speaking generally, even the cleverest architects may be, at best, amateurs. It is, perhaps, hard to define the distinct point in the designing and supervision of a building of this class where the architect's duty ends and that of the civil engineer commences, but there is not the slightest doubt that some line can be drawn, and that whilst the general conception and direction remains in the hands of the architect, the detail and execution can become specialised in certain directions.[53]

Such was the recurrence of the 'engineering problem' in theatre design that one of the largest and most prolific specialist architectural practices, Frank Matcham & Co., brought a structural engineer, Robert Alexander Briggs, within its fold in the 1880s (*see* p 356, note 93). The establishment of inter-disciplinary teams, incorporating engineering expertise, was unusual but not unprecedented in the Victorian era; for example, in 1842 the engineer Alfred Meeson joined Sir Charles Barry's design team as 'superintendent of constructional and engineering details' for the Houses of Parliament[54] before becoming an independent consultant, designing the structural ironwork of E M Barry's Theatre Royal, Covent Garden (1858–60) and the 1862 International Exhibition among other projects.[55] Matcham's and Brigg's alliance seems especially long-lasting and fruitful, resulting in a geographically diverse range of theatres that made progressive use of steel-cantilevered balconies, the largest of which were built on their own patented technique (*see* pp 135–6). Whether such instances are emblematic of a shared enthusiasm for structure, or an attempt to avoid over-reliance on consulting engineers, is difficult to determine. Certainly, despite a differentiation of roles through the 19th century, the boundaries between engineering and architecture were never entirely fixed. Engineers frequently lectured about their projects and the design of constructional iron and steelwork to the RIBA and other architectural bodies, and occasionally architects brought their subjects before an engineering audience. In some instances there were full professional 'border crossings'. Archibald Dawnay, for example, who began his career with a foot in both camps, permanently switched to constructional engineering in 1872, his letter of resignation to the RIBA stating that his interest in 'a certain iron construction' necessitated his being engaged as an engineering contractor.[56] And even an architect of such distinction as Richard Norman Shaw was prepared to admit that he was 'intended by nature to be an engineer'.[57] Yet because of the specialisation in roles, and design skills required, no Victorian practitioner could truly style himself 'architect and engineer', unlike Mylne and some of his Georgian contemporaries.

Theatres, and buildings requiring unusually complex or generous framing, were the exception not the rule, at least in Victorian London. For most run-of-the-mill projects, architects were mostly able to eschew the expensive – even galling – services of consulting engineers while still reaping the benefits of structural iron and steel. The multiplicity of engineering contractors, some of which had arisen in the railway era but which subsequently diversified into other fields, made this possible. In many senses these firms posed less of a threat: they would get on and do the specified job discreetly and inexpensively, and they were regarded as lower down the professional and social scale than independent consulting engineers. Certainly they were less likely to adopt a 'we-can-do-without-you' stance, or refuse to 'put themselves into architectural harness', which it was claimed some engineers did.[58] Specialised firms such as Matthew T Shaw, Richard Moreland & Son, Archibald Dawnay & Sons, Homan & Rodgers, and Dennett and Ingle evolved and flourished in the era of wrought iron, supplying and fitting their own fire-resisting flooring systems and standardised or custom-built framing components. Like other subcontractors, such as specialised heating, lighting or ventilation firms, their role was clear-cut, circumscribed and unlikely to challenge the architect's programmatic conception of the whole building.

At its simplest, the iron- or steel-joisted floor did not pose a conceptual threat, and could be

readily incorporated within the traditional way of doing things (Fig 6.4). In City office buildings, it was common practice for the architect to design the building and then let the floors out for competitive tender to contractors with their own floor systems. The architect did not choose what system would be used until the tenders were received.[59] Because this input was low down and chronologically late in the hierarchy of decision making, it remained marginal within the design process, and did not significantly affect the form of the building or the architect's command. In this way the ever-changing technology of construction was appropriated expeditiously in realising the architect's spatial and formal intentions, rather than presenting a deep-seated challenge to them. Yet even in engineering contractors' boundaries and roles were not always clear-cut. Sometimes the contract for framed iron- or steelwork was let to one company and that for the fire-resisting floor to another. And while little is known about the history and make-up of these firms, it is recognised that many had highly qualified and expert engineers, either in overall management or in their employ. Men such as Richard Moreland junior (1834–1927) and senior (1805–1891), Archibald Dawnay (1842–1919) – all associates of the Institution of Civil Engineers – and Thomas Charles Cunnington (1852–1917), who worked for William Lindsay & Co.,[60] forged very close working relationships with architects, and for more complex buildings they must have had input at the earlier design stages. Near the end of his career, Ordish, who had started off working for engineering contractors, joined Dennett and Ingle,[61] and in this more anonymous role he continued to design bespoke solutions to complex structural problems. The precise contributions of such individuals were often subsumed under the company's name, which might itself be omitted from building accounts in the architectural periodicals.

By subcontracting constructional iron and steelwork to specialist firms, and in cases of great size or complexity, to consulting engineers, most architects avoided really grappling with the materials' technical complexities. Some architects might have disliked the unavoidable input of these designers, but there was probably comfort in the fact that it was still beholden to the architect's overall drawings and specifications. C V Childs, who was used to working in a more collaborative manner with American architects, was struck by this difference in practice when he moved to London, noting 'Certain of the American features might be introduced into the English buildings with advantage; the difficulty of doing so lies in the fact that the architect's plans are practically complete before the engineer's work begins'.[62] At a more fundamental level, some architects found such collaboration troublingly symptomatic of the separation between architecture as service and architecture as art. Architects might claim authorship of the aesthetic content of a building, but the fact that they had conceded intellectual command of its structure meant that this could not truly be the case, because the two were inseparable. By the late 1890s, with steel construction abounding, some architects were aware that the days of technical avoidance or complacency on their part were numbered. Following a discussion of Thomas Cunnington's paper on 'Constructional Steelwork', in 1898 the Chairman of the Architectural Association thought:

architects ought to understand the principles of engineering work, even though they might not have the opportunity of working out strains and stresses, and he thought it would be well if less dependence were placed upon the engineer. Quite recently the Discussion Section of the Association held in that room a joint meeting with the Institution of Junior Engineers, when a discussion took place upon the desirability of a closer union between architects and engineers. The general opinion of that meeting seemed to be that architects could not do without engineers, while engineers could not do without architects. At the rate that iron and steel construction had been going on during the last ten or twenty years, architects would soon be left behind by the engineers.[63]

The years after 1900 saw mounting pressure on the architect to acquire technical expertise in steelwork, to 'restore … his long-lost status of head-master of technical knowledge and work'.[64] Architectural periodicals began carrying adverts promoting correspondence courses in constructional steelwork, one asking 'Can you Design your own Steelwork? Do you know how to properly proportion your columns, stanchions, and girders?'[65] The number of articles and textbooks designed to present the subject in a demystified light to architects rose accordingly. In its review of Walter Twelvetrees's

Fig 6.4 (facing page) 'Fireproof' floors reliant on iron or steel members, and the specialist firms that fitted them, did not pose a fundamental threat to most architects. [RIBA81729, RIBA Library Photographs Collection]

Fig 6.5
The Franco-British
Exhibition of 1908,
'White City', Shepherd's
Bush (demolished). Spread
over a 140-acre site, the
exhibition buildings were
almost all of steel-skeleton
construction, with
architects' responsibilities
limited to applying
decorative plaster dresses.
[London Metropolitan
Archives, City of London
(SC/PHL/02/946/05443)]

Structural Iron and Steel (1900), the *Journal of the RIBA* pointed out that 'Many things taught in Mr. Twelvetrees' book are what architects too generally neither know nor think of, and yet things of which every architect should feel ashamed to be ignorant.'[66] Not all architects were badly informed: John Burnet for example had already adapted himself to the challenge, his Beaux-Arts training an obvious bonus. 'He often designs forms which, from his engineering training, he knows can be constructed, but he engages engineers to work out the stresses and exact sizes.'[67] T E Collcutt was another architect with uncommon interest, and understanding, of structural steelwork, and one who eagerly engaged with his engineering colleagues, such as with the American architect-engineer L C Mullgardt for the Savoy Hotel extension. Although Mullgardt, working for contractors James Stewart & Co., had prime responsibility for the skeleton framing, Collcutt and his partner Hamp made significant contributions to it (*see* p 179). Such genuine collaborations were rare. Indeed, the skeleton-framed building presented the greatest challenge of all, for it inverted established architectural practices and theories, and presented a conundrum:

Unless the architect himself has some considerable knowledge of the subject, the engineer is presented with a drawing which takes little account of the important steel frame, and he consequently has to make his steel frame fit the design. Obviously this is contrary to the elementary rules of architecture. And if the position be reversed, and the architect follows the engineer's skeleton plan in designing his elevations, the elevations are bound to suffer.[68]

Elite Gardens from Flip-Flap, The Great White City, Lond

Of course the determining influence of the skeleton was more than skin deep. The repetition and regularity of the structural grid demanded standardised floor spaces, thus limiting the architect's freedom in internal planning and spatial organisation. It also required that the engineer and the architect get together from the start to ensure that the different aspects of the building meshed. Yet from an aesthetic point of view, it was the elevational form and character that was most immediately palpable, and thus troublesome. Architects were seemingly required to provide an appropriate expression or stylistic dress to the skeleton, in essence to decorate structural engineering. To many architects this was unpalatable, but it was a state of affairs soon taken to its logical conclusion in the Franco-British Exhibition of 1908 at White City (Fig 6.5). There the role of architects was reduced to simply designing façades and inner screen walls for great halls, palaces and other steel-framed exhibition structures, using regulation non-combustible fibrous plaster. Architectural critic J Horsfield Nixon was aghast, noting:

> I am informed that the block plan was devised by Imre Kiralfy, the Commissioner-General to the exhibition, and that he has not only allocated the site of the buildings, but actually ordered their steel framework, this fixing their dimensions and general form. Then, and not until then, were the architects consulted. It is due to the architects who

have been engaged to point out this unfortunate example of the vulgar error of putting the wrong end of the stick before the horse.[69]

As in the Great Exhibition over half a century earlier, architects were sidelined, and the nature of architecture was thrown into question. This was an extreme example, but one emblematic of a changed and changing milieu, with steel construction the prime mover. As many architects feared, the London County Council (General Powers) Act of 1909 – 'essentially an Engineer's Act'[70] – effectively formalised the interdependency of the two professions, but also placed engineering in a position of greater authority than ever before. In discussing the Act a few years later, one engineer mused:

> it seems within the possibilities of the near future that, provided the engineer takes advantage of his opportunities, he might assume the more important position – and in such case the Architect would confine his attention solely to the architectural treatment – unless of course, the Architect more fully appreciates the rapid change in modern building conditions and construction.[71]

Structural engineering became an inseparable aspect of architecture through the 20th century, and like the previous century, territorial blurring, collaboration and conflict continues to characterise the relationship of the two professions.

7

American influence

Many persons think that in the use of steel for buildings we in England are a little behind our American cousins ... We are in full sympathy with the method adopted across the water, and are glad that English architects and engineers are now falling into line, and the American system is being largely adopted. (1898)[1]

To an engineer the steel construction is, of course, of great interest. The Savoy Hotel, the Westinghouse works at Trafford Park, and the Midland Hotel at Manchester are about the only buildings which we have built on the American plan, and the general feeling is that once we get the building by-laws into line, steel construction will become almost as common with us as it is with you. I think we shall draw the line at skyscrapers which look like pieces of cake set on edge, however. (1904)[2]

England provided technological inspiration to the development of American prefabricated iron building systems and fireproof technologies in the first half of the 19th century, but over the next 50 years the direction of technology transfer was steadily reversed.[3] During the 1860s and 1870s, the British architectural press began reporting on iron framing in American commercial buildings.[4] In the next decade British architects began making transatlantic visits to witness at first hand the developments in construction techniques, presenting their findings to audiences back home.[5] One such was John Bradshaw Gass, who talked to the RIBA on William Le Baron Jenney's Home Insurance Building (1884–5), noting that 'iron columns run up the full height in the centre of brick piers, [with] iron window lintels resting on the columns'.[6] In a subsequent discussion at the same organisation, John Slater (1847–1924) exclaimed how he thought 'the constructional methods of America seemed to show a boldness and breadth of aim, and a lack of conventionalism, which was really refreshing'.[7] His own paper to the RIBA, entitled 'New materials and inventions', made extensive reference to American techniques, including fire-resisting flooring systems and methods of encasing wrought-iron columns with 'fireclay' blocks, and was evidently based on first-hand experience, for he declared 'Our cousins across the water are far more inventive than we are and I have had to go to them for several of the novelties which I shall bring before you.'[8] In 1888–9 a large house named 'Thriplands' incorporating steel framing was built to his designs in Kensington (*see* p 250) and before this, Richard D'Oyly Carte was inspired enough by one building he saw on his own American travels to use it as a model for the Savoy Hotel, Strand (1884–9) (*see* p 160).

From intermittent coverage in the 1880s, English reportage of American building progress mushroomed in the 1890s as the remarkable phenomenon of steel-framed skyscrapers gathered momentum.[9] Many of these reports were tinged with both awe and unease, one commenting, 'Such structures may be practical and economical, but the building has ceased to be architecture, in any sense in which that expression has hitherto been understood in the world.'[10] Others were kinder. *The Builder* in 1892 thought it was 'to the credit of Chicago

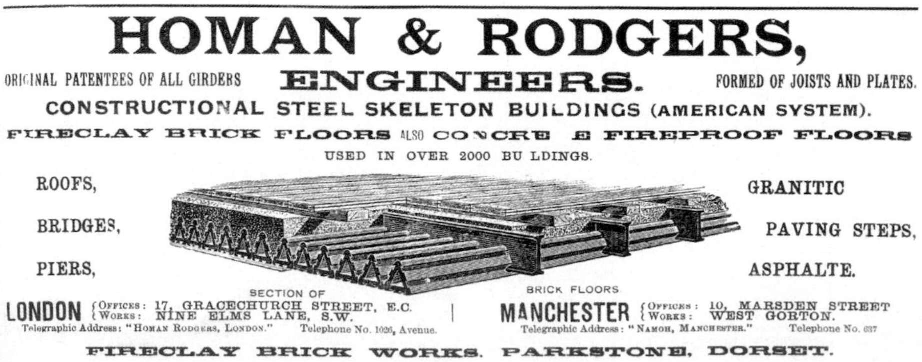

Fig 7.1
Advertisement in Builders' Journal and Architectural Review *7 (22 June 1898).*

architects that they should have produced not only a characteristic group of buildings, but buildings well worth studying as examples of the most modern tendencies in architecture and their expression in architectural form'.[11] The same periodical's review of William Birkmire's textbook *Skeleton Construction in Buildings* (1893) at least recognised its potential usefulness to British readers:

> Whether this method of building will ever be tolerated or practised in by English architects it is not easy to prophesy; in London at all events the legal limitations as to height will take away its chief or only advantage. If we ever do come to that, however, the architects implicated will find useful information in Mr. Birkmire's book; but we counsel them in that case to be content with the honest steel building, and not make the matter worse by erecting a sham masonry building outside it.[12]

Similar reports continued through the decade in the architectural and engineering periodicals.[13] Significantly, British structural engineers were beginning to take note. Edward Graham Wood (1854–1930), a Manchester-based engineer and author of the first epigraph to this section, seems to have quickly adopted what he termed the 'American system', in which buildings were 'constructed almost entirely of steel, with the walls and floors filled in afterwards'.[14] In 1898 he gave a talk to the Manchester Association of Engineers, in which he tried

> to show why steel is the best material to be used in the constructional parts of warehouses, manufactories, engineering works, etc., and to illustrate how it may afterwards be covered over or clothed with masonry, marble, terracotta brickwork or woodwork; thus preserving all the architectural features of a highly artistic building, while at the same time the building itself is strong, reliable, and of a permanent nature.[15]

Among the illustrations was one showing a steel girder spanning between exterior and interior stanchions, 'suitable for carrying a Manchester warehouse floor in general conditions', and another of a built-up steel stanchion 'suggested for use in the seven-storey building in our illustration' (*see* Figs 8.17 and 8.18).[16] Edward Wood and Co. Ltd would subsequently

erect the skeleton frames of a number of London office buildings, notably Mappin House on Oxford Street (1906–8) (*see* p 231). Also by 1898, Manchester and London-based constructional engineers Homan & Rodgers were advertising 'Constructional Steel Skeleton Buildings (American System) in addition to their usual structural sections and fire-resistant floor systems' (Fig 7.1).[17] According to the *The Builders' Journal and Architectural Engineer*, this firm 'were the first to advocate the advisability of introducing … the complete steel skeleton-frame construction … in order to satisfy certain conditions as a result of their study of American buildings'.[18] Beyond studying and adapting techniques from widely accessible published examples, it seems likely that representatives of both of these companies visited America in the 1890s.

In terms of direct intervention, Charles Vivian Childs (1864–1941) was one of the first structural engineers with American experience to arrive in England. Prior to his move to London in 1894 to take up a post in Dorman Long & Co.'s London office, he designed the skeleton framing of a number of high-rise buildings, including the Carnegie Building, Pittsburgh, erected in 1892 (Fig 7.2).[19] For economic and

Fig 7.2
*The steel skeleton of the Carnegie Building, Pittsburgh (1893–5, Longfellow, Alden & Harlow, architects; demolished 1952), engineered by Charles Vivian Childs. Temporary timber struts were used to brace the framework during erection, to counteract thermally induced distortions caused by sunlight striking one side. [*The Engineering Magazine*, vol 15 (May 1898). Library of Congress, General Collections]*

practical reasons, he adapted his advanced methods to conventional English practice, restricting the all-steel framework to the interior, or supporting the building's weight on huge girders and stanchions placed at first-floor level. He explained this restraint in an engineering periodical article of 1898,[20] which was summarised in *The Building News* the same year.[21] Childs designed the steelwork for a number of major buildings of high architectural merit, including the Bovril Building, the Hotel Russell, Harrods and Waring & Gillow's, but resisted full steel framing as unsuitable to British circumstance. His influence was thus overtaken by events after 1900. In 1899 advanced fabrication methods and skilled labourers were directly imported from Pittsburgh to realise a steel-framed power station for the Bristol Tramways & Carriage Company, erected in 1899–1900 (*see* p 318), but the wider influence of this seems again to have been limited.

Decisive change came in the first years of the 20th century, beginning with the arrival of James Stewart & Co., drafted in to expedite the construction of the British Westinghouse Electric and Manufacturing Co.'s works at

Trafford Park. The largest manufacturing plant in Europe when completed in 1902, these colossal works, comprising 10 gigantic engineering sheds, all skeleton-framed in steel, and a seven-storey office building, were a virtual replica of parallel works in East Pittsburgh designed by Westinghouse's engineer-cum-architect, Thomas Rodd (1849–1929). For the Manchester site, Charles H Heathcote adapted Rodd's plans and specifications to local conditions, and a local firm was engaged to lay the foundations, with another Manchester firm, Mayoh and Haley,[22] contracted to erect the 17,000 tons of steelwork supplied by Dorman Long & Co. Work began in 1900, but aghast at the sluggish rate of progress, George Westinghouse brought in James Stewart & Co. as building managers, their instructions being 'to expedite its completion in every way possible' (Fig 7.3).[23] By April the following year James Christian Stewart and his team of American assistants were on the 130-acre site, instituting highly streamlined and efficient working practices to the existing contractors. Bricklayers were shown how to lay 1,400–1,800 bricks per day instead of their usual 400–500, steam

Fig 7.3
American contractors James Stewart & Co., outside their site offices in Trafford Park, c 1901. James Stewart himself is on the right, front row.
[© Trafford Local Studies Centre (T2232)]

Fig 7.4 (left)
The British Westinghouse
Electric and Manufacturing
Co.'s works at Trafford
Park, Manchester, during
construction in 1901.
Dorman Long & Co. supplied
17,000 tons of steelwork,
erected under the
supervision of James
Stewart & Co.
[Some Stewart Structures
(1909), 91–2]

hoists replaced human hod-carriers, and Dorman Long's riveters were given automatic tools, thus quadrupling their output.[24] The Middlesbrough firm's steelwork, 'the most beautiful [Westinghouse] had seen', rose at the rate of more than 100 tons daily.[25] Eight of the sheds were up and operational within 10 months, and within 1 year the works were essentially complete, with installation of plant well under way (Figs 7.4 and 7.5). Westinghouse and Rodd had estimated that it would have taken the existing contractors 5 years to bring the plant into operation.[26]

One of the largest and most progressive engineering enterprises in the world, the construction project attracted enormous attention:

'Not only did the prominent English and Continental papers frequently comment on its progress, but the engineering papers, monthly magazines, weeklies and dailies for months after its completion contained articles fully describing the work and highly commending the builders'.[27] Such efficient construction management, and levels of productivity, had little if any precedent in Britain, and were attributable to Stewart's fresh management approach, his ability to motivate the British workforce and a sensible approach to logistics and materials handling'.[28]

After Trafford Park, Stewart moved to central Manchester to expedite the construction of Charles Trubshaw's Midland Hotel (1899–1903)

Fig 7.5 (below)
Completed in 1902, and
comprising gargantuan
engineering sheds and an
imposing office building,
these long-demolished
works were a virtual replica
of others in Pittsburgh,
and were designed by
Westinghouse's engineer-
cum-architect Thomas
Rodd, with assistance
from Manchester architect
Charles H Heathcote.
[Some Stewart Structures
(1909), 91–2]

(see p 179)

(Figs 7.6 and 7.7), that city's first hotel built on a skeleton frame, and designed following an American tour by the architect.[29] Brought in by the Midland Railway Co. in 1902 when the work was hardly one-fourth complete, the American company – by then joined by Louis Christian Mullgardt (*see* p 179) – hustled the contractors along, the steelwork being erected by Edward Wood and Co. Ltd, using steel supplied by Dorman Long. As 'managers of construction', the American firm was under agreement to complete the building within 1 year; in the event, it secured a bonus for earlier completion.[30] On the strength of that contract Stewart and his entourage moved to London to erect the Savoy Hotel extension. After that, 'numerous power houses, sections of the London underground, the reconstruction of the tunnel under the Mersey, all followed'.[31]

By working closely with local engineering contractors, and garnering huge publicity, Stewart & Co. undoubtedly helped advance English working practices. The American firm showed just how fast skeleton-framed construction could proceed, pronouncing speediness as one of its chief merits. The building press closely followed the working methods of the company, stimulating considerable interest among British architectural and construction firms. Such instances of hard-nosed efficiency almost certainly provided a model for British contractors, and the architect Charles Heathcote drew attention to the reasons for this, possibly as a result of an earlier transatlantic visit.[32] *The Builders' Journal and Architectural Engineer* having already published coverage of the rapid construction of the Westinghouse works, the Midland Hotel, and the Savoy Hotel extension, resolved to give detailed coverage of the skeleton-framed Hoyle's Warehouse (1904) designed by Charles Heathcote & Sons.

Fig 7.6
James Stewart & Co. expedited construction of Manchester's Midland Railway Hotel, seen here on 22 July 1902 from the south west corner: 'For six months the laying of strong blue bricks ... went on out of sight from the public, and then was seen the erection of a framework of girders – a skeleton of steel – going up apparently to a sixth story.'
[The Builder, 24 January 1903]

Each week we shall publish a photograph showing the regular progress of the building, which is to be a complete steel frame structure of six storeys, including basement. We think the illustrations will show clearly that the ability to erect a building speedily is not confined to the much-talked about American contractor [James Stewart].[33]

Heathcote was conversant with skeleton framing even before the Westinghouse works (*see* p 96). But he was spurred on by the example set by Stewart, demanding similar levels of efficiency and speed from the local contractors, Messrs Neil & Sons. Hoyle's Warehouse was erected and fitted out in less than 10 months.[34] A year earlier, the same architects and contractors had erected Peak's Warehouse, Portland Street (1902–3) in 20 weeks from signing the contract.[35] The arrival of Stewart effectively 'upped the ante', a situation irreversibly reinforced with the arrival of the Waring White Building Co. in the heart of London with Sven Bylander at the helm (*see* p 181). This marked the real onset of 'The American Invasion', a development predicted by skyscraper contractors and engineers.[36] Had the plans of an Anglo-American syndicate prepared in 1901 come to fruition, for a proposed 7,000-room office block to be erected on the Strand on a

Fig 7.7
The Midland Railway Hotel (1903).
[The Builder, 24 January 1903]

77

skeleton framework that was partly 10-storeys in height, then the entrée would have begun earlier, with perhaps much opposition.[37]

America also provided inspiration to some British architects for how they might approach the architectural treatment of multi-storey steel frames. Most preferred dresses of heavy Classical or Baroque ornament, but others sought more exotic inspiration. John Belcher (1841–1913) and his younger partner John James Joass (1868–1952) were among the foremost designers to exploit the Mannerist possibilities of rectilinear frameworks in later Edwardian years, and the historian Alastair Service attributes great significance to the American influence behind this.[38]

Another British architect to chart a similarly progressive course in steel-framed architecture in the 1900s was John James Burnet (1857–1938).[39] Burnet, unusually for a British architect, had been trained as an engineer as well as an

architect in 1874–7 at the École des Beaux-Arts in Paris. He made his first visit to America in 1896, where he used his French connections to make contact with his American contemporaries at the École des Beaux-Arts, including Louis Sullivan in Chicago. Besides seeing examples of the work of Sullivan's renowned partnership with the German-born engineer Dankmar Adler, including the Auditorium Building (1886–90), Burnet cannot have missed the pioneering frame buildings in Chicago of Sullivan's former associate, William Le Baron Jenney, including the First Leiter Building (1879) and the Home Insurance Building (1883–5), nor more recent edifices, such as Daniel Burnham's Reliance Building (1894–95).[40] In Britain Burnet moved increasingly away from a residual Baroque to an ever more stripped Classicism, 'the Chicagoan influence was constantly present ... based on a Sullivanesque grid'.[41] *The Builders' Journal and Architectural Engineer*

Fig 7.8
John Burnet & Son's warehouse-cum-department store for R W Forsyth Ltd, Princess Street, Edinburgh, during erection in 1907. Burnet was one of a number of leading Edwardian architects to have visited Chicago.
[RIBA46488, RIBA Library Photographs Collection]

credited his warehouse for R W Forsyth, Ltd, on Princes Street, Edinburgh, as the 'First steel-frame building in Edinburgh' (Figs 7.8 and 7.9).[42] The structural engineer William Basil Scott (1876–1933) believed it to be the 'first steel-framed building in Scotland'.[43] Burnet's London buildings perhaps represent the more extreme end of this stylistic journey: certainly his highly influential Kodak Building, Kingsway (1910–11; with Thomas S Tait) for the New Yorker George Eastman, feted as one of the first examples of modern architecture in Britain, was the most 'rationalist' in terms of exploiting the vertical power of the frame (*see* p 335).

Fig 7.9
Burnet's building for Forsyth betokened a move away from Baroque and towards stripped classicism as more fitting for the British steel skeleton.
[Jonathan Clarke]

8

The evolution of the fully framed building

[The skeleton system of construction] may be said to have been incubated, rather than invented, and the simple, triumphant method of constructing the most marvellous of modern buildings is found upon examination to be but an enlarged use of preceding methods. (1898)[1]

The step that completed the most radical transformation in the structural art since the development of the Gothic system of construction in the twelfth century was the invention of complete iron framing or skeletal construction. (1964)[2]

The American perspective

The most important technological advance in 19th-century building construction was the development of the multi-storey full metal frame. Overturning conventional practice in which walls provided primary support and stability to buildings with multiple floors, this was a structural system with a versatility, precision, speed and, in certain conditions, clear gain in functional performance that saw its reach spread across the world in the early 20th century. An industrialised, forward-looking system born of an industrialised world, it emerged most vigorously and systematically in steel in late 19th-century America, where the conditions favouring its use were inescapably compelling. Chicago and, a little later, New York, were hot-houses for the rapid development of the full steel frame, with the skyscraper being both a spectacular outcome and a vehicle.

Although not reliant on fully metal-framed technologies to begin with, this approach to construction soon established itself as a prerequisite for buildings above 10–12 storeys. Above that height, conventional building technologies became both uneconomic and impractical, for the enormous thickness of load-bearing walls consumed rentable floor space, diminished natural lighting and, in the case of Chicago's compressible soil, demanded expensive foundations because of the excessive loads. Skeleton frames could also be erected with astonishing rapidity, bringing prompt returns on site investment. In Chicago, the all-steel skeleton frame, ardently promoted by Pittsburgh steelmakers, emerged by 1890 as the pre-eminent structural system for high-rise buildings (Fig 8.1).

In New York, a more restrictive building code nullified the major advantages of skeleton construction and contributed to the persistence of a hybrid 'cage' construction reliant on self-supporting walls and iron framing elements.[3] In cage construction the frame supported only the floors and interior partitions, unlike skeleton construction, in which all the weight of the building, inner and outer walls and floors, was carried to the foundations by the frame. The masonry or brick envelopes of cages were structurally independent, bearing their own weight, although they might offer stability to the interior framework. New York's first crop of skeleton-framed tall buildings, including the 11-storey Tower Building (1888–9), the 10-storey Lancashire Insurance Building (1889–90) and the 12-storey Columbia Building (1890–1), which all used iron framing elements alongside those of steel. The first of more substantial size, the 350ft-tall Manhattan Life Insurance Company Building (1893–5), used cast-iron columns in the lower six floors and the front wall was not carried by the frame.[4] In fact, many of New York's pioneer skeleton skyscrapers were not as materially or structurally 'pure' as most contemporaries suggested: in addition to the use of iron, many were a mixture of structural systems, using load-bearing party, interior or rear walls.[5] Not until the second half of the 1890s, following the well-publicised collapses of two cage-type buildings incorporating cast-iron columns, and the reduction of building-code-proscribed minimum wall thicknesses,[6] was a mature high-rise building technology, using rolled-steel members riveted or bolted together, diagonal or portal wind-bracing, fire-resistant cladding and flooring, and caisson foundations, in standard use in New York and other American cities.

American skyscrapers were the most spectacular expression of steel-skeleton building technology. Designed to maximise rentable space in the central business districts of cities where land values were soaring, they rose hastily in an unusually fertile commercial environment in which engineers, architects, steelmakers, building developers, agents and contractors all played a part, and where building regulations were supportive or indulgent. Advances in metal framing aside, their extraordinarily rapid advance also resulted from the coalescence of a number of technologies, including power-operated construction equipment and elevators, forced-draught ventilation, fire protection, steam heating and so forth. In late 19th-century London, as in other British and European cities, some of these essential preconditions were present. London in the 1890s featured a financial district whose land

values nearly equalled those across the Atlantic; the greatest concentration of engineers, architects, engineering contractors and builders in the country; and huge steel stockyards and fabricating workshops. The British capital had also seen the emergence of advanced iron-framed building technologies, power-operated passenger lifts, non-combustible flooring systems and electric lighting. Yet it did not develop or import the skyscraper in response to the intersection of such technologies and circumstances, and building developments involving constructional steelwork remained closer to the ground. The reasons for this are complex and varied, but the 100ft-height limit imposed by the LCC in 1894 seems to have reflected the reservations and fears of much of the architectural profession, as well as those of engineers and ordinary Londoners over the impact that tall buildings would have on their historic city. Had taller buildings been both wanted and sanctioned, multi-storey framing technologies might have evolved at a faster rate than they did. London did not see its first documented multi-storey commercial building built on a complete all-steel skeleton frame until 1902–4,

Fig 8.1
The steel skeleton frame of the Fair Store, Chicago (1890–1) during construction. Designed by William Le Baron Jenney, it would soon rise to 11 storeys. [Industrial Chicago v2 after p 186 (Goodspeed Publishing, 1891). By courtesy of the University of Illinois at Urbana-Champaign Library]

with the construction of the eight-storey Savoy Hotel extension, quickly followed by the more renowned Ritz Hotel (1904–5). Anglo-American collaboration produced the structural design of the earlier building, while a Swedish-born structural engineer who had formerly worked on skeleton skyscrapers in America devised that of the second. American example and expertise played a key role in the pioneer phase of the multi-storey steel skeleton frame's development and acceptance in British architecture.

This is only part of the story, and not the first part. By the third quarter of the 19th century British and French engineers had progressed iron framing to the full skeleton form for a small number of multi-storey buildings. Collectively and retrospectively, these milestones of iron construction, some well known, others less so, form part of an intermittent line of structural development that, classically, originates with late 18th-century textile mills and culminates with the Chicago skyscrapers. This technological progress, most ardently explored in the mid-20th century through Modernist historiography, remains broadly accepted by historians on both sides of the Atlantic, though American scholars have placed differing degrees of emphasis on certain British and French contributions, and, justifiably, more on their own indigenous pre-skyscraper fully framed examples. Yet no matter the perspective, or the contestable step forward offered by each structure, the line of development was disjointed and discontinuous.

Recent additions to the stock of examples that help flesh out the story only serve to make this clearer. Despite the technical capability there was no orchestrated movement towards full framing prior to the late 19th century in any country, with the possible exception of France. In America, at least, the story has unity, but in Britain there was somewhat of a hiatus from the 1870s until the mid-1890s, when fully framed construction was taken up with conviction using mass-produced steel. A reprise of pre-existing technologies, this phase marked, in the words of this chapter's first epigraph, an 'enlarged use of preceding methods' – methods that stemmed in more dormant, embryonic form from Britain, and more boldly and determinedly from across the Atlantic. The question as to which was Britain's first fully steel-framed building remains almost as debatable as that of which was the first fully framed American skyscraper. While this may never be satisfactorily answered because of insufficient documentary or physical evidence, it seems clear that it was not in London.

For the purposes of this book, it is useful to present a sequential record of British, and, where pertinent, overseas examples of fully framed multi-storey construction in the period 1850–1905. Significant or influential precursors before this are briefly considered. Many are familiar, others less so, but part of the purpose of this account is to show why such buildings were built the way they were, an aspect often neglected at the expense of how they were built. Among other factors, it also shows that in Britain at least, the expense of fully framed construction helped restrict its application until the coming of steel. However, the main intention is to present a more balanced, less polarised view that neither attributes the advent of steel-frame architecture in Britain solely to the importation of American expertise,[7] nor over-stresses Britain's technological self-reliance, which has more recently emerged as a revisionist interpretation.[8] The circumstances were far from simple or monocausal; independent invention spurred by suitable conditions was just as much a feature of the story as was the diffusion and transmission of knowledge across land or sea.

The iron skeleton frame: developments up to 1850

Multi-storey iron framing might have originated in textile mills in the 1790s, but this type of building was not destined to be the vehicle in which the skeleton frame evolved. Complete internal framing using cast-iron beams and columns was first used in the six-storey Ditherington Flax Mill (1796–7) to give wide expanses of open 'fireproof' floor, but as in every subsequent iron-framed mill, lateral stability was provided by the enclosing box of load-bearing walls (Fig 8.2).[9] As one structural historian put it, structures like these, 'while marking a first step towards a fully framed form, nevertheless remained primarily a bearing-wall one … [their] internal framing was still as far as from being self-sufficient as it had been in the Egyptian house and in the great audience halls at Persepolis'.[10] Mills played a cardinal role in driving the structural investigation and application of cast iron until the mid-1840s,[11] but they did not progress full framing beyond this initial stage. It was not until the Edwardian era that the fully framed textile mill appeared in Britain.[12]

Any move away from reliance on the bracing action of the exterior envelope required that lateral stability be provided either by means of diagonal braces, or, with more sophistication, by the inherent stiffness of the orthogonal frame itself, through rigid column–beam connections. Both techniques were employed to some extent in the first half of the 19th century to enable independent, freestanding structures. With few exceptions these were all single-storey roofing systems, or unassuming structures that did not exceed two or possibly three storeys. Particularly notable were Charles Fowler's Hungerford Fish Market (1831–3) (Fig 8.3), which used spandrel brackets to connect the girders carrying the cantilevered roof rigidly to the columns, creating what has been said to be the first iron frame with wind bracing,[13] and a

Fig 8.2 (left)
Ditherington Flax Mill, Shrewsbury (1796–7), the world's first multi-storey building to use iron columns and beams.
[AA022794]

Fig 8.3 (below)
Charles Fowler's innovative Hungerford Market, a freestanding structure of 1835 framed entirely in cast iron.
[RIBA12250, RIBA Library Photographs Collection]

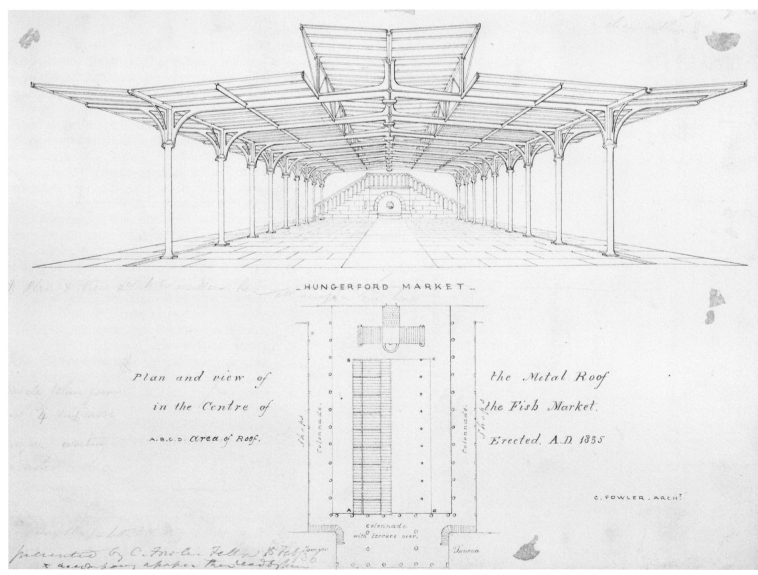

HUNGERFORD MARKET

Plan and view of in the Centre of A.B.C.D area of Roof.

the Metal Roof the Fish Market. Erected. A.D. 1835

C. FOWLER. ARCH.

had to install diagonal bracing, the column–beam connection proving unsatisfactory.[15] Other iron-framed buildings were prefabricated for export and re-erection overseas in this period, most influentially a three-storey corn mill designed by William Fairbairn Sons & Co and sent to Istanbul in 1840.[16] Proudly exhibited on completion at the company's Millwall works before being dismantled and shipped, it accelerated a trend in prefabricated iron houses, churches and warehouses for overseas export, and formed the inspiration for James Bogardus's own prefabricated building system in New York City.[17] However, these portable buildings were invariably of modest dimensions,[18] and it would seem many were not entirely free from reliance on brickwork or in some cases timber framing.[19]

The iron skeleton frame: 1850–90

With the onset of wrought-iron beams, substantial, multi-storey, fully framed buildings were erected in the 1850s. The first of these was the Crystal Palace (1850–1), a vast 1,848ft-long exhibition hall comprising a lofty single-storey nave and transepts flanked by three-storey aisles and galleries. Designed as a modular, prefabricated and demountable framework of cast and wrought iron, this was the most remarkable feat of true skeleton construction the world had seen. For sheer scale, speed and rationality of design, fabrication and assembly, it remained unmatched throughout the 19th century. Utilising the concept of the portal frame, that is, rigidly jointed columns and beams, repeated over and over on two different scales and on both axes, its overall stability was derived principally from the stiffness of its connections, achieved by driving cast-iron and oak wedges between the column connectors and the beam ends (Fig 8.5). This joint detailing, designed by (Sir) Charles Fox, was highly ingenious, giving the frames considerable fixity, but it was insufficient for a building of this scale. For additional stability full diagonal bracing was added at either end of the building and in the bays flanking the central transept. Dismantled and re-erected for permanence in larger and more robust form at Sydenham in 1852–4, the Crystal Palace was critical in the development of framed structures.[20] It stimulated the construction of a whole line of similar iron-and-glass exhibition buildings, initially in

Fig 8.4 (above)
Cast-iron-framed water tower at Portsmouth Royal Naval Dockyard. When it was erected in 1843, it carried a huge water tank and was only subsequently clad, to serve as a fire station.
[AA034962]

two-tier water tower at the Royal Naval Dockyard in Portsmouth (1843) (Fig 8.4).[14] The latter, designed by Captain Roger Steward Beatson, had columns set on a 12ft grid joined by arched beams, the stiffness of the jointing intended to give the necessary stability. In the event, however, the contractors, Fox Henderson,

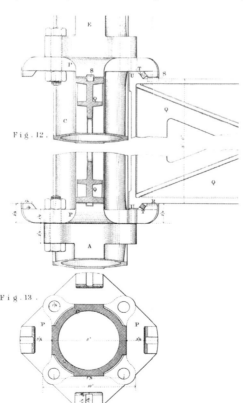

Fig 8.5
Detail of the column–beam connection at the Crystal Palace (1850–1). Wedges at S and T fixed the girders to the column collars.
[Charles Downes, The Building erected in Hyde Park … (1852), plate 8. Courtesy of Institution of Civil Engineers]

New York and Dublin (both completed in 1853), as well as railway and dockside structures, and, most notoriously, the Museum of Science and Art on Brompton Road (1855–6) (Fig 8.6). This building, essentially a triple-roofed shed with mezzanine galleries clad in corrugated iron, fully framed in cast and wrought iron with no diagonal bracing, quickly earned the disparaging moniker of 'the Brompton Boilers' because of its stark no-frills look. Fabricated by C D Young and Company to designs by the Civil Engineer William Dredge,[21] it contributed enormously to the denigration of iron architecture, but it nonetheless signalled a move towards the use of full framing in buildings more suited for daily occupation than the ferro-vitreous glasshouses and exhibition buildings that had preceded it[22] – a move that was being paralled in America by this date.[23] Its structural arrangement, comprising a central open nave flanked by floored aisles (Fig 8.7), the perimeter of which were framed using

Fig 8.6 (above)
Hugely disparaged by contemporaries for its utilitarian appearance, the Museum of Science & Art, South Kensington (1856), was soon dismantled and re-erected in Bethnal Green, where it survives as the Museum of Childhood. There brick replaced corrugated iron for the façades.
[2006BD5277]

Fig 8.7
The Museum of Science & Art, South Kensington, 1856.
[Illustrated London News, 12 April 1856, Mary Evans Picture Library (10514449)]

H-section cast-iron wall columns (one of the earliest documented uses of this shape) formed the basic template for numerous subsequent industrial sheds.

In the wake of the Brompton Boilers was the historiographically celebrated Boat Store, designed in 1858 by Col. Godfrey Greene and erected in 1859–60 for the Royal Navy at Sheerness Dockyard (Fig 8.8).[24] This was a larger, more robust and innovative structure, designed to serve as a foundry, a fitting-out shop and boat store. Framed almost entirely in massive, H-section cast-iron stanchions (the corner columns were cast in square section) and I-section wrought-iron girders (the transverse I-section beams were of cast iron), it consisted of a lofty 210ft-long open nave flanked by two four-storey aisles. The extensive use of H-section stanchions, each cast with brackets to carry the beams and girders, enabled exceptionally rigid connections that ensured full longitudinal and transverse stability without recourse to any diagonal bracing, which would have hindered the free manoeuvre of boats. Such was the strength and stability of the framework that it could easily absorb the forces generated by the three travelling girders that spanned the nave, serving the timber-floored galleries to either side. The moment-resisting rigidity of the connections, achieved by bolting the girder ends throughout their entire depth to the webs of the stanchions, was of an order of sophistication greater than anything hitherto, and was prophetic of techniques used well into the next century for skeleton structures. Even the exterior cladding, with its long continuous bands of windows alternating between lightweight corrugated-iron infill panels, gave an appearance of refined simplicity, seeming to foretell the thinly panelled portal-framed industrial sheds of the 20th century.[25] Yet despite its 'forwardness', the wider influence of the Sheerness Boat Store was marginal. Secreted away in a remote naval dockyard, it remained 'undiscovered' and unpublicised until the mid-20th century. This does not mean that engaged contemporaries were unaware of it.[26]

Fig 8.8
The Boat Store, Sheerness (1858–60; G T Greene, engineer). Sophisticated connection details, combined with H-section columns set this structural design decades ahead of its time.
[Ed Swale]

In the next evolutionary stage, the multi-storey skeleton was re-invented in France, this time to support brick exterior walls and fire-resisting floors in a massive building more suited to daily habitation and more aspirant of representational form. It was in the Paris suburb of Saint-Ouen, outside the influence of the capital's building code and aesthetic mores, that the first cast- and wrought-iron skeleton frame supporting both brick walls and 'fireproof' floors appeared.[27] Erected in 1864–5 after the plans of the architect Hippolyte Fontaine and the structural detailing of the architect-engineer Préfontaine, the six-storey warehouse for the Saint-Ouen Railway and Dock Company was a vast structure.[28] Six hundred and eighty feet long, and enveloping three sides of a dock basin, its structural frame of iron with brick infill panels was left frankly exposed on all but one elevation (Fig 8.9). Carried storey by storey on cast-iron spandrel beams spanning between the tiers of superimposed columns, the brick-panel walls measured just 15in. thick through the full height of the building. Only the formal Louis XIII-style façade facing the Route de la Revolte was executed as traditional masonry construction, and it remains unclear whether even this might have been carried by the frame at each floor level, rather than simply supporting its own weight. Inside, wrought iron was used for the long-span primary girders that spanned between the two rows of interior columns and framed into the external columns. These built-up I-sections carried cast-iron beams of half their length, which in turn helped carry the 'fireproof' jack-arch floors. Formed from cement-mortared hollow bricks, haunched with concrete and covered with asphalt, the floors were immensely strong.[29] In essence this structure synthesised, in one dramatic gesture, over 70 years of English textile mill development with the newly emerging fully framed form. According to one American historian 'the only element in the warehouse that belongs to a day that was rapidly passing is the use of cast rather than wrought iron for the spandrel beams'.[30] In fact, although these arched members (similar in design to those used in the Portsmouth water tower) gave the construction a *retardataire* appearance, they were probably consciously used in replacement of rolled beams (which had been first manufactured in France a decade earlier) for decorative reasons.[31]

Even with its dockyard setting, the structural system of the Saint-Ouen warehouse excited some measure of overseas publicity and interest. It was illustrated in *The Builder* in April 1865,[32] and almost certainly did not escape the attention of William LeBaron Jenney,[33] designer of the 10-storey Home Insurance Building in Chicago (1883–5), regarded by many as the first instance of skyscraper construction. France's first documented instance of multi-storey skeleton building had not emerged from a vacuum. Both theoretically and practically, 1860s France was an extraordinarily fertile setting for the development of rationalised iron architecture,

Fig 8.9
The Saint-Ouen Docks Warehouse, north of Paris, erected in 1864–5 to designs by Préfontaine and Hippolyte Fontaine. This was the first use of the multi-storey skeleton in combination with fire-resisting floors. [The Builder, 29 April 1865]

Elevation, Rue de Ceinture.   *Elevation, Place des Docks.*

THE SAINT OUEN DOCKS, PARIS.——M. PREFONTAINE, ARCHITECT AND ENGINEER.

Fig 8.10
*Imaginary Pans-de-Fer
apartment building for
Paris (Viollet-le-Duc, 1863).
Both influential and
prescient, it was illustrative
of the great interest in
1860s France in skeleton-
framed architecture.*
*[Entretiens sur
l'architecture, Atlas,
pl. xxxvi (Courtesy of
Institution of Civil
Engineers)]*

witnessing the publication of the first volume of Viollet-le-Duc's *Entretiens sur l'architecture*, which depicted an imaginary apartment building that 'honestly' revealed its diagonally braced skeleton (Fig 8.10), as well as buildings actually constructed using skeleton-framed walls. *Pan de fer*, as it was known, was used for the front and rear façades of a two-storey worker's cottage designed by Stanislas Ferrand and Maurice Grand for the 1867 Paris Exposition; and in 1869, another influential proponent of the system, C A Opermann, claimed that skeleton framing had been in limited use in France for at least 10 years.[34] Also in 1869, Louis-Charles Boileau proposed that the new Les Magasins-Réunis department store be erected on a full

iron frame,[35] but more importantly and productively, plans commenced on a building that would be one of the last French contributions to the evolution of the multi-storey skeleton.

The Menier Chocolate Factory at Noisiel-sur-Marne (1871–2) represents another approach to full framing. This was the first building to use wrought iron exclusively in forming its structural skeleton (Fig 8.11). Like the Saint-Ouen warehouse, this landmark of architectural engineering appeared in the outskirts of Paris, where the prohibitive building code did not reach. Erected to designs by Jules Saulnier, with the assistance of the Parisian engineer Moisant, structurally and visually the Chocolaterie was like a huge lattice-truss bridge. Indeed it was designed to span between and cantilever over three pre-existing masonry piers in the riverbed that had carried a medieval mill and a half-timbered replacement of 1840. Functionally, it followed its timber-framed predecessors in deriving power from the fast-flowing River Marne (in this case using turbines). There was also structural continuity. Rising from a deck structure of deep box girders, a three-storey framework of I-section stanchions and I- and box-section girders carried the walls and floors of the superstructure. For added stability, I-section diagonal members were used to triangulate the orthogonal framework, and riveted knee-braces were introduced at major column-girder junctions. It was largely the lozenge-shaped secondary system of bracing, overlaid on the non-load-bearing polychrome walls, which gave the building great distinction. Reminiscent of late-medieval French timber framing, and thus of Viollet-le-Duc's similarly historicist apartment building project of 1863, it earned the praise of the great theorist, who saw it as the realisation of some of his theoretical aspirations for contemporary architecture. Indeed his illustration might actually have inspired it.[36] In turn the Chocolaterie was highly influential. It was published in the second volume of *Entretiens* (1872), enjoyed coverage in the *Encyclopedie d'architecture* (1874), and continued to feature in works of a much later date, including Sturgis's *Dictionary of Architecture and Building* (1901–2). A climax in the progression of structural and aesthetic principles for French multi-storey construction, it marked the beginning of a new style, its technology inspiring and informing a whole line of commercial buildings that openly and rationally expressed structural frames of iron. From 1878, with the

liberalising of the Paris building code to permit exterior walls facing streets to be less than 500mm-thick, skeleton-framed iron construction blossomed in the capital.[37] The Chocolaterie was doubtless known to Jenney,[38] and in England was surely known to Richard Norman Shaw, who, in 1857, spent a short period in Viollet-le-Duc's atelier as part of a lengthy French tour.[39] Shaw later exploited a technique used in the Chocolaterie, that is, the suspension of the uppermost floor from the roof structure so that the floor below is left completely free of columns.[40] He did this most assuredly in the Royal Insurance Building, Liverpool (1897–1902), an unorthodox skeleton-framed building that was also one of Britain's first in steel (*see* pp 98–9).

Accounts of the origins of full multi-storey framing in Europe invariably end with the Saint-Ouen Dock warehouse or the Menier Chocolate Factory, whereupon the story shifts to America. In a purely technical sense the Sheerness Boat Store might have been 'one of the last English contributions to the evolution of the iron frame',[41] but in the interval between the construction of the two French mileposts, an English-designed fully framed building erected in a British colonial city took the genre in a new and prescient functional direction. Designed in London in 1865–6, prefabricated in Derby in *c* 1866 and shipped to and erected in India's fast-emerging commercial capital in 1867–9, Watson's Esplanade Hotel, Bombay (Mumbai)

Fig 8.11
The wrought-iron skeleton of the chocolate factory at Noisiel-sur-Marne, east of Paris (1871–2), depicted without its polychrome brick skin for structural clarity. This striking building, which survives, drew inspiration from medieval timber framing as well as from Viollet-le-Duc (see Fig 8.10). [Sturgis, Dictionary of Architecture and Building, *Vol. II (1901)]*

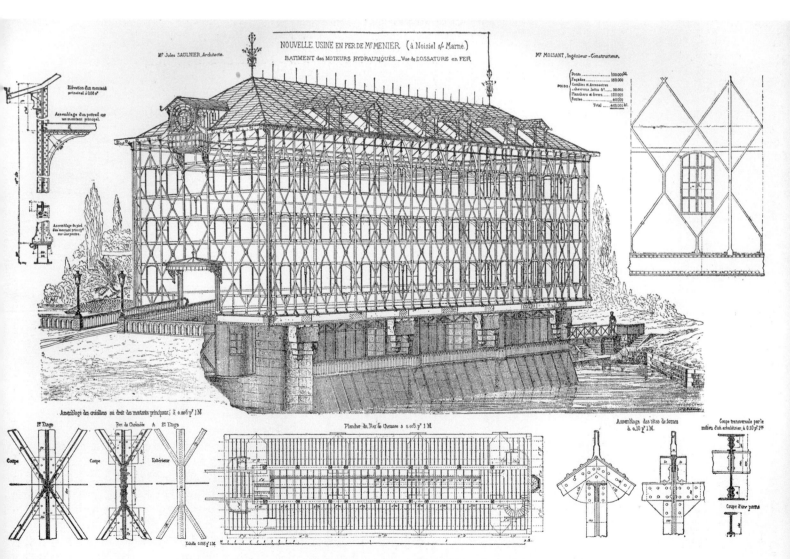

NOUVELLE USINE EN FER DE Mr MENIER. (à Noisiel s/ Marne.)
Mr Jules SAULNIER, Architecte.   BATIMENT des MOTEURS HYDRAULIQUES.—Vue de L'OSSATURE en FER   Mr MOISANT, Ingénieur-Constructeur.

IRON CONSTRUCTION: FACTORY AT NOISIEL, SEINE ET MARNE, FRANCE.

Decorative treatment of iron framework.   The spaces between the members of the skeleton shown were filled with brick and tile in ornamental patterns, the construction thus being a modern fireproof elaboration of the half-timbered method.

marked the next stage in the maturity of the framed building, the application of the fully framed system to permanently habitable, high-profile commercial architecture (Fig 8.12). Still surviving in dilapidated condition, this five-storey edifice was designed by Rowland Mason

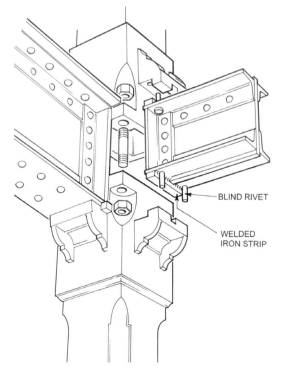

BLIND RIVET

WELDED
IRON STRIP

Fig 8.13
Technique of fixing the
structural joints in Watson's
Hotel, conjectural
reconstruction.

Ordish (1824–86), an extraordinarily accomplished and versatile engineer who had helped detail the structural design of the Crystal Palace and that of its modified successor at Sydenham, and who later served under Greene. 'Without a doubt the finest hotel in Bombay',[42] and the most ambitious and technically accomplished of all the exported buildings of the 19th century, Watson's was built entirely on a huge cast- and wrought-iron skeleton. The framework supporting the thin, hollow-brick panel walls was everywhere left exposed. As at Saint-Ouen, cast-iron spandrel beams and girders were used for ornamental reasons on the elevations, but within, built-up wrought-iron girders, rigidly connected to the columns by means of an ingenious column-to-column detail, were used exclusively as the flexural members (Fig 8.13). This form of connection, with the columns spliced midway through the depth of the girders, was an entirely novel, logical technique, representing an advance on the Crystal Palace, eliminating the need for interconnecting collars between the columns and diagonal bracing. The hotel was not as sophisticated in its joint detailing as the Boat Store, and the teak-boarded floors suggest that Ordish, or the patron, John Hudson Watson, did not aspire to the fire-resisting techniques used at Saint-Ouen. These 'demerits' possibly stemmed from the functional and aesthetic considerations of portability, ease of assembly, expediency (teak was locally available) and ornamentation (the exposed, chamfered, square-section columns, had at least a degree of enrichment one step above H-section equivalents). Like the Crystal Palace, Watson's Hotel was conceived as a modular assembly, based on an 11ft 9in. square structural unit that spanned in both directions and was stackable in multi-storey heights, foreshadowing the non-directional, addable grids of the next century.

The chief significance of Watson's Hotel has to do with service not structure. Until this building took its place on downtown Bombay's principal thoroughfare, all known fully framed buildings were built to serve the needs of exhibition, industry or storage. With a sumptuous, top-lit ground-floor restaurant and first-floor saloon, three upper storeys given to 130 bedrooms, provision for bathing and internal transportation (reputedly housing India's first power-operated elevator), Watson's was of course designedly habitable, and for public view. Maximisation of space and provision for ventilation were two key reasons that skeleton-framing was invoked and speed of erection was another major concern. With such a novel and controversial constructional form, the hotel was delayed in part by architects within the Bombay government, who, like some of the city's inhabitants, were aghast at its unabashed metal frame. Yet it was sanctioned by the people that mattered, the Bombay Public Works Department, which, composed largely of Royal and civil engineers, was arguably less bound by the 'anti-iron' philosophical dictates of the architectural profession. In a sense, the Watson's project arose by a kind of fluke, enabled by the progressive attitude of the Public Works Department, which saw it as a 'valuable experiment' conducted through a London-based engineer. By contrast, in 19th-century London the dominance of building regulations and the time-honoured bearing-wall mindset of architects and clients meant that no permanently habitable commercial building flaunting an iron skeleton was built.[43]

Watson's Esplanade Hotel was not an influential building, neither in Bombay nor further afield. *The Architect*, the only organ of the architectural press to feature the building, thought it 'deserve[d] the attention of architects',[44] but there is no evidence to suggest that they took note. Yet it did produce one offspring in the

town of its manufacture. In 1871–2 the Phoenix Foundry Company, Derby, which prefabricated Watson's Hotel, erected a four-storey, cast-iron-fronted shop in the town centre (Fig 8.14). Unlike most iron-fronted buildings in America, England and Scotland, this was actually carried on a full skeleton frame of cast and wrought iron. As the retail showpiece of the company's ironmongery branch, no expense was spared: Owen Jones designed the ornamental cast-iron front, and the framework was designed by Ordish in collaboration with James Haywood and another notable engineer, Perry Fairfax Nursey (1830–1907).[45] This was Britain's first fully framed shop, and it may also have been the first such building to have its framework dressed in protective cladding, thus anticipating a challenge

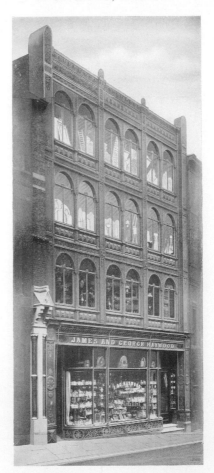

Messrs. J. & G. HAYWOOD'S (Ltd.) PREMISES, IRONGATE, DERBY.

THE framework of this building is entirely constructed of cast and wrought iron. The cast-iron facia was designed by the celebrated Architect, Owen Jones, the author of the "Grammar of Ornament," and is supposed to be the first iron structure of this class erected in Great Britain.

that Viollet-le-Duc issued in the prophetic second volume of *Entretiens sur l'architecture*. In this he stated 'A practical architect might not unnaturally conceive the idea of erecting a vast edifice whose frame should entirely be of iron, and clothing that frame, preserving it by means of a casing of stone.'[46] Purely on the evidence of the photograph (*see* Fig 8.14), it would seem that corner stanchions flanking the cast-iron fascia were clad in ornamental terracotta, and it seems plausible that the interior framework may have been similarly clad in ceramic or brick sleeves, for decorative, or possibly even for 'fireproof' reasons. It should be remembered that Jones, a noted expert in the use of colourful decoration, was much in demand for his encaustic tile and terracotta designs, and it would be logical to expect him to have contributed to the interior design of the shop. Notwithstanding the Ruskinian philosophical concerns of 'deceit' and 'sham', the fully clothed frame would mark one of the final stages in the technological maturity of the multi-storey skeleton, both in terms of its protection from fire, and of its aesthetic admissibility into mainstream commercial and institutional architecture. Chicago in the 1880s is held to be the setting for this innovatory, and concluding step, but it seems possible that it was to some extent foretold in Derby in 1871.[47]

Soon after J & G Haywoods' premises opened, *The Architect* declared the building a 'decided success, and Messrs Haywood deserve credit for thus effectively developing the system of iron architecture, which is gradually growing up in our midst'.[48] Yet Watson's Hotel and the Haywoods' shop seem to be among the last major British offerings to the development of the iron skeleton. After 1870 the full frame in Britain appears to have withdrawn to its industrial origins rather than building on the achievements of the 1850s and 1860s. In 1871 a remote paper mill near Aberdeen was built on a multi-storey skeleton, apparently escaping architectural attention until 1911, when a local architect proffered it as a Scottish precursor to skyscraper technology. Speaking before the Aberdeen Architectural Association on 'Structural ironwork and steelwork for architects', Harbourne McLennan thought:

The beginning of this construction in America may be said to have been about 1883, and it is sometimes supposed to have been an American invention; but this is hardly the case, as examples of skeleton construction, on a much less scale, certainly, but having all the requisite features, are to be found in this country long before the above date. At Stoneywood Paper Works, near Aberdeen, for instance, iron-framed buildings of several stories in height and thin panel walls are to be found bearing the date 1871, and at Union Works, in Aberdeen, there is a similar building which appears to be even older.[49]

Central Manchester is the setting for the next stage of the story, some 7 years later (Fig 8.15). In 1877–8 a six-storey textile warehouse was erected for Messrs John Rylands and Reuben Spencer in Tib Street (demolished 1985). It was fabricated by Messrs William and Sons of Stalybridge and erected by local contractors Messrs Robert Nield and Sons to the designs of engineer John Henry Lynde (1843–1919). *The Engineer* proclaimed 'this class of building' as 'altogether a new feature in Manchester'.[50] Apart from the brick rear wall, the building, covering some 600 square yards,[51] was fully framed in cast iron, using columns and square E-section stanchions tied 'firmly together' by cast-iron girders and spandrel beams that carried the timber floors and ornamental iron fronts.[52] Such a mode of construction reduced wall space, and permitted the iron-fronted elevations to be glazed with large sheets of glass to admit maximum light.[53] Technologically the use of cast iron for all the framing elements was retrograde at this date, but economically prudent, reflecting the low price to which the material had fallen. Combined with the economies of scale of using repetitive structural elements, this made skeleton construction economically viable.

It was cost, perhaps more than any other factor, that discouraged the spread of multi-storeyed fully iron-framed building technology in 19th-century Britain and her colonies. Innovation tended to win through only when there was a clear gain in functional performance over load-bearing, stabilising walls and interior framing. At Sheerness, where the ground was poor, smaller foundation loads helped offset the cost of the skeleton.[54] In Bombay it became economically feasible to use iron as the principal structural material only because the cost of traditional materials rose markedly there in the early to mid-1860s. Of the Menier Chocolate Factory, Giedion noted that 'Saulnier's type of construction was considerably more expensive

MANCHESTER WAREHOUSES

Fig 8.15
Rylands & Sons Ltd warehouses, Tib Street, Manchester, as illustrated in c 1900. The block to the far right was erected in 1878 on a skeleton of cast iron, affording a far greater degree of glazing than that on its bearing-wall neighbours. This structurally exceptional building was demolished in 1985.
[© The British Library Board, Port of Manchester, 1901 (PP 1786d6)]

than any other'.[55] In Derby, where patron and contractor were one and the same, the costs were borne in-house, and it seems that the same iron members were used in an associated warehouse development, giving economies of scale. Long before this, in the wake of the interest stirred by the Crystal Palace, cost alone had been enough to deter at least one architect, William White, from using iron framing, in this case for a private school chapel. 'The gentleman who advocated iron' informed him that 'the lowest estimate for producing the very simplest effect in iron was between twice and three times its actual cost in brick and stone'.[56] With aesthetic strictures and building code requirements often hostile it is hardly surprising that the iron skeleton frame was only used sporadically in Britain. Not until the era of bulk-produced structural steel sections did this form of construction become economically viable in an extended range of circumstances, a real alternative to traditional techniques. Philosophical and legislative hindrances soon vanished.

The steel skeleton frame in Britain: 1890–1905

Industry rather than commerce seems to have been the driving force behind Britain's first steel skeleton frames and, bridges aside, the Royal Arsenal at Woolwich was perhaps the preliminary test bed. Long the setting for tech-nological innovation, including advancements in iron construction, it was here that steel was early on pressed into service for creating large-volume freestanding enclosures. Documented early examples included Buildings D81 and D73, erected in 1891 and 1892, respectively, for the Royal Gun Foundry department (*see* p 311). Neither of these can be regarded as true build-ings, nor wholly differentiated from other examples of framed structures using steel, such as George Trewby's Beckton No. 9 and Kensall Green No. 6 gasholders (both 1890–92).[57] Despite some evidence to suppose that 'Thrip-lands', a maverick London town-house of 1888–9, may have continued what Watson's and Hay-wood's premises began (*see* p 250), it would seem that the exigencies of industry gave rise to Britain's first floored steel-skeleton buildings.

The Manchester Ship Canal was an indisput-ably vital conduit for the reintroduction to Britain of the multi-storey skeleton frame. Built to connect Manchester with the Mersey estuary upriver of Liverpool, and by so doing circum-vent the high rates charged by that rival city, this long seaway permanently reduced the price of transport into Manchester. 'The Big Ditch' opened on 1 January 1894, having taken over 6 years to construct. Within 2 years of opening, Manchester rose to become the country's second largest importer of cotton, soon becoming the country's principal banking, commercial and transport centre outside London.[58] From the quaysides of its terminus the canal helped usher steel-framed construction into Manchester and beyond. The Manchester docks, comprising four basins upstream of the Trafford Road Bridge (Pomona Docks) and three downstream (Salford Docks), were given dedicated terminal facilities tailored to the nature of the intended import and export transit trade (Fig 8.16). This called for the con-

struction of specialised transit sheds and open-sided warehouses, which would make effective use of the restricted dockland available, hasten the turnover of goods and avoid traffic congestion.[59] The first two sheds were built in 1893, but the real spate of construction began in the middle of the decade, the first set of 15 buildings opening in June 1896.[60] Local

Fig 8.16
The terminus of the Manchester Ship Canal as it looked around 1900. This view, looking east, shows the Pomona Docks in the distance, with the Salford Docks in the foreground. In 1890, the engineering contractors Edward Wood & Co. relocated to the spur of land between these two docks before erecting many of the transit sheds and warehouses depicted.
[© The British Library Board (F60105–66)]

constructional engineers Edward Wood & Co. secured the lion's share of orders, fabricating both single-storeyed transit sheds and multi-storeyed warehouses of advanced design.[61] Many were built to the plans of Sir E Leader Williams, Chief Engineer of the Canal, who had previously designed the Anderton Boat Lift on the Weaver Navigation (1875) and the Barton Swing Aqueduct (1882), both mechanically and structurally sophisticated works.[62] The multi-storey warehouses he designed for the Salford and Pomona Docks were no less refined, and bore similarities to the Sheerness Boat Store, incorporating such advances as the era conferred: all-steel construction for strength, concrete-filled lattice stanchions for economy

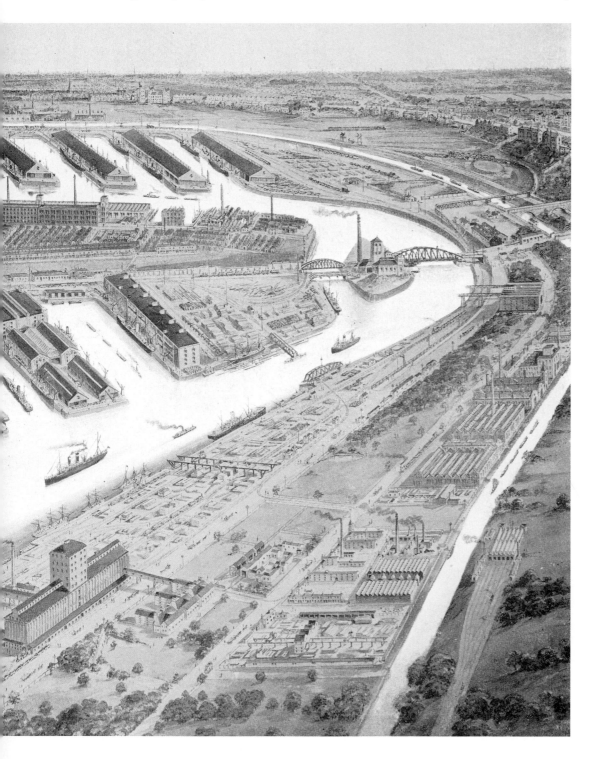

loads on the compressible alluvial deposits on which the quays were built. As at Sheerness, lightness of construction helped make full framing more cost effective. The Manchester sheds and warehouses proved financially rewarding from the outset. From 1897 further warehouses were built,[63] in all likelihood by Edward Wood & Co. By 1900 the Port of Manchester could boast almost 50 roofed structures, mostly for transit, including 6 three-storey and 7 five-storey transit sheds, 1 four-storey transit warehouse and 13 seven-storey warehouses. It is unlikely that all of these were steel framed and certainly Hennebique reinforced concrete was being extensively used by 1905.[64] Either way, the terminus of the ship canal formed a decisive, intense testing ground for advanced building technologies that caught the attention of architects.

Wider influence aside, the designers of some of the Ship Canal buildings had a direct connection with the architecture of central Manchester. Charles H Heathcote (1850–1938), credited as having built 15 warehouses for the Manchester Ship Canal by 1899,[65] and Edward Graham Wood (1854–1930) seem to have played important roles in introducing the skeleton frame into the streets of central Manchester. In November 1898, Wood presented a paper to the Manchester

Fig 8.17 (above)
Contemporary view of one of the steel-skeleton-framed multi- storey transit sheds erected by Edward Wood & Co. in the 1890s, in this case used to trans-ship cotton. Its constructional similarity to the Sheerness Boat Store is palpable.
[© The British Library Board (04A01217)]

Fig 8.18 (right)
Illustration of a steel-skeleton-framed building for central Manchester, drawn for a paper read by Edward Wood in November 1898.
[© The British Library Board (F60105–68)]

Fig 8.19 (facing page)
Stanchions fabricated by Edward Wood & Co. Ltd, used to illustrate the same talk. Charles Trubshaw designed that on the left, Edward Wood that on the right, which was intended for the building in Fig 8.18.
[© The British Library Board (F60105–67)]

of weight and materials, and concrete filler-joist floors for fire-resistance (Fig 8.17). Besides the obvious advantages of rapidity of construction and the maximisation of storage space, these skeletal structures exerted smaller foundation

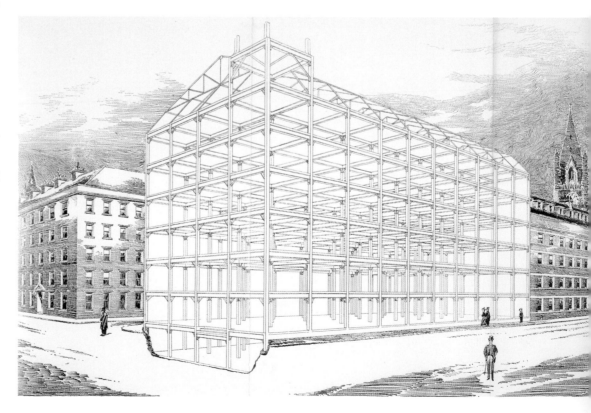

Association of Engineers entitled 'The use of steel in buildings'.[66] He illustrated this with drawings for a seven-storey steel-skeleton-framed commercial building in central Manchester (Figs 8.18 and 8.19), possibly never built. Wood was eager to apply his firm's constructional proficiency to the challenge of commercial architecture. Manchester was already seeing the construction of the Great Northern Railway Warehouse at Deansgate (1896–9), built on an interior framework of steel stanchions and wrought-iron beams (*see* p 301), and more pertinently, the eponymous 'Manchester warehouse' was undergoing rapid constructional development. Stimulated by the increased textile trade brought by the ship canal, ever-grander Manchester edifices, such as the warehouses for Tootal, Broadhurst and Lee on Oxford Street (1896, J Gibbons Sankey, architect) and Horrockses and Crewdson at 107 Piccadilly (1897–9, Charles Heathcote, architect)[67] were already using steel for their interior skeletons. To designers with the experience of Wood and Heathcote it must have seemed a natural step to make the framework carry the outer envelope as well, as they had done at the Manchester Docks. Only the patrons and the district surveyors needed convincing. Heathcote was a hugely prolific architect, and although his first documented fully skeleton-framed commercial building, a warehouse for Joshua Hoyle, was not erected until 1904 (by the same contractors who built the 1870s warehouse for Rylands and Spencer), it is likely that others, still awaiting discovery, preceded it. For his part, Wood soon realised his ambitions, in fabricating the giant skeleton of the Midland Railway Hotel (1899–1903), which was designed by the company's architect, Charles Trubshaw (1841–1917), following a tour of American hotels.[68] Trubshaw was another architect with structural expertise, and Wood attributed the design of one of this building's giant stanchions to him. The Midland Hotel would serve as one of the main springboards of the technology to London, largely through the publicity-grabbing methods of James Stewart and Co., the American contractors brought in to expedite its construction. On the strength of this works, that firm secured the contract for the Savoy Hotel extension (1902–4) (*see* p 176).

By 1900 skeleton framing in steel was beginning to take root across Britain, especially in those towns or cities where building regulations, or the authorities, permitted the use of

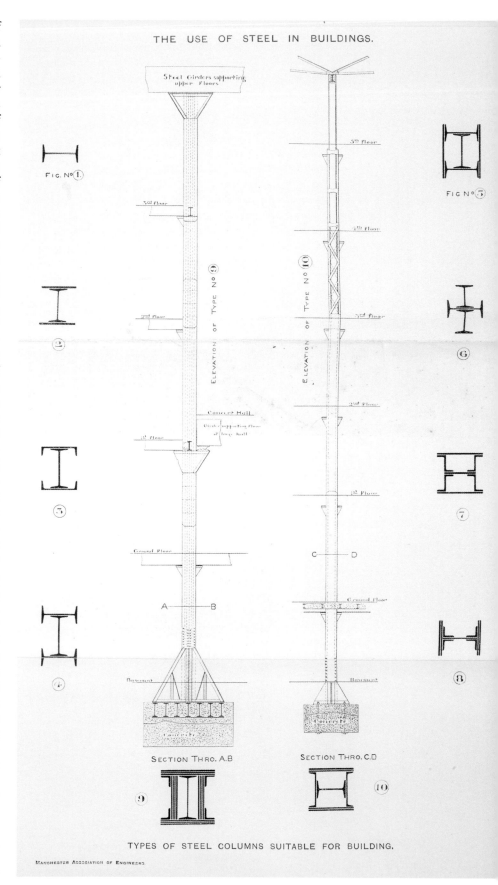

THE USE OF STEEL IN BUILDINGS.

TYPES OF STEEL COLUMNS SUITABLE FOR BUILDING.

MANCHESTER ASSOCIATION OF ENGINEERS.

Fig 8.20 (above)
Bacon factory under construction for the Co-operative Wholesale Society, Trafford Wharf, 1900. The CWS had been involved with the Manchester Ship Canal since its inception, and would pioneer the use of both steel and reinforced concrete.
[Salford City Archive, The Frank Mullineux Collection (T1103)]

Fig 8.21 (right)
The column-free ground-floor office, Royal Insurance Building, Liverpool (1897–1902; Richard Norman Shaw, architect), which was suspended from giant arch trusses above.
[© The British Library Board (G70017–50)]

Fig 8.22 (far right)
Cross-section of the unorthodox steel skeleton.
[© The British Library Board (G70017–51)]

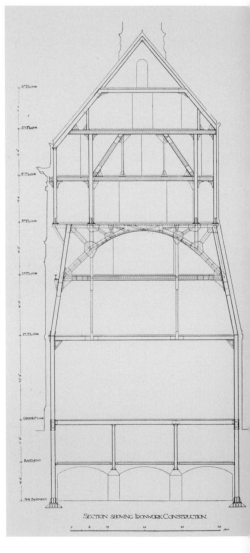

SECTION SHOWING IRONWORK CONSTRUCTION.

panel walls or extensive glazing. In this sense, its early application paralleled that of reinforced concrete. In 1900 the Co-operative Wholesale Society began erecting a steel-skeleton bacon factory at Trafford Park, next to the canal (Fig 8.20, demolished), and, nearby, work began on the British Westinghouse Electric and Manufacturing Co.'s engineering works and office, designed partly by Heathcote (demolished, *see* p 74). In Liverpool, construction of the Royal Insurance Building, Liverpool (1897–1902) was well under way. Designed by J F Doyle and Richard Norman Shaw, its eccentric fully clothed skeleton incorporated what amounted to an arched bridge structure at second floor level, upon which was hung the floor below, thus skilfully meeting the directors' wish for a truly unobstructed ground-floor office (Fig 8.21 and 8.22).[69] In Stockton-on-Tees, work was progressing on the skeleton of Mathias Robinson's furniture warehouse (1899–1901), designed by the structural engineer W Basil Scott, of Redpath, Brown & Co., in conjunction with the Sunderland architects Barnes & Coates. This warehouse replaced a short-lived, burnt-out building erected in 1896, long considered (incorrectly) to be Britain's first steel-framed building.[70] For the replacement Scott designed a full steel skeleton that was encased in fire-resisting plaster and clearly expressed in the glass-filled façade.[71]

Further afield, construction had commenced on the Scotsman Building on Edinburgh's North Bridge (1899–1903) (Fig 8.23). The most celebrated work of James B Dunn and James Leslie Findlay, this massive newspaper printing office was the largest building then erected in that city by private enterprise, and seems to have been fully framed.[72] Across the Irish Sea, work began on what would be Ireland's first substantially steel-framed building, the Market Street Storehouse in Dublin (1902–4) (Fig 8.24). Rapidly erected for the Guinness Brewery to overall designs by the company's Engineer-in-Chief, A H Hignett, this eight-storey block exploited a rigid, 3,600-ton cage of steel and iron to create a coherent amalgam of open floors and galleries, top-lit central bays and extensive glazing on all four sides. This structural grid, which made minor use of cast-iron columns besides massive, widely spaced stanchions, was fabricated and erected by Sir William Arrol & Co., another company that had pioneered the manufacture of steel-framed sheds as well as bridges. Architecturally, if not structurally, the building drew

inspiration from Chicago, the three main façades executed in brick and rock-faced granite, their semicircular arches and corbelled windows reminiscent of early H H Richardson or Adler and Sullivan (Fig 8.25). These 125ft-high monolithic walls supported only their own weight, enabling generous amounts of glazing. The plain west elevation, however, originally intended as a temporary wall for future expansion, was a series of 15in.-thick panels carried by the frame.[73]

Fig 8.23
The Scotsman Building on Edinburgh's North Bridge (1899–1903; Dunn & Findlay, architects). [Jonathan Clarke]

By about 1905, with architects taking increasing note of its spatial, structural and timesaving qualities, the multi-storeyed steel-framed building had become firmly established in Britain. In 1903–4 Cambridge saw the construction of the Laurie & McConnell shop on Fitzroy Street.[74] R Frank Atkinson, who would later design the stylistically comparable Waring and Gillow department store in London (*see* pp 242–3), concealed the concrete-encased framework in distinctive 'Wrenaissance' dress (Fig 8.26). One year later, Newcastle-upon-Tyne saw the completion of a skeleton-framed office block, Milburn House, on Dean Street (1902–5) (Fig 8.27). Where not enclosed by masonry, the giant frame, fabricated by Swinney Brothers of Morpeth, was encased in expanded metal and fire-resisting plaster. The architects, Oliver, Leeson & Wood, gave the interior a rich Art Nouveau décor throughout, themed like an ocean liner to reinforce the Milburn family's connection with shipping.[75] Parallel with this, on the other side of the country, another huge office was arising in the port of Liverpool using advanced technology and the profits of mercantile trade. The Mersey Docks and Harbour

Board Offices, erected in 1903–7 at Pier Head, were, however, built around an independent floor-supporting steel cage rather than on a skeleton, such was the monumental thickness of the enclosing walls (Fig 8.28). This high-profile Baroque-inspired work was designed by the local architect Arnold Thornely (with Briggs and Wolstenholme). Its 1,651-ton frame was fabricated by Dorman Long, and possibly erected by Manchester contractors William Brown & Sons.[76] Both companies had worked alongside Edward Wood & Co. on the erection of Trubshaw's Midland Hotel, an illustration of the city's importance in nurturing and showcasing the technology. Indeed, it was Manchester that arguably witnessed the most remarkable, concentrated outburst of fully skeleton-framed buildings in Edwardian Britain. Perhaps originating with Wood's prescient design of 1898, or with the construction of Joshua Hoyle's warehouse at Piccadilly in 1904, the technology of central Manchester's warehouse construction underwent a rapid transformation. Full framing became de rigueur as the city's wealthy merchants and industrialists wholeheartedly embraced it, demanding the maximum space

Fig 8.26
The former Laurie &
McConnell shop, Fitzroy
Street, Cambridge.
Stylistically and structurally,
this was an early exercise in
steel-frame 'Wrenaissance'
architecture for R Frank
Atkinson, comparable to his
subsequent store for Waring
& Gillow in London.
[DP110812]

and light from cramped or awkward sites. Local architects, such as Birkett, Fairhurst, Heathcote and Higginbottom, were happy to oblige, enlisting the specialist services of engineering firms like Edward Wood & Co., and Richard Moreland & Son of London.[77] A productive combination of factors, including established links between architects, engineers and clients, less restrictive by-laws, cheaper steel and a tight geographical focus, saw the appearance of imposing brick- and terracotta-clad skeletons that vied for status along the city's foremost streets, most sited cheek-by-jowl or within sight of each other. Despite their exuberant attire, the façades of these towering monuments of commerce betrayed, or tacitly acknowledged, the geometric order of their

skeletal underpinnings to a degree unmatched anywhere else in the country at this date (Figs 8.29–8.37). More than central London, which (though wealthier) was more constrained by architectural tradition and legislative stricture, central Manchester in this period offered a fertile ground for fully developed steel-frame architecture, albeit largely mediated through a highly specific building type. Following the construction of the influential Ritz Hotel, often believed to have been Britain's first steel-framed building,[78] the technology spread rapidly in the capital, where demand and finance for new buildings were virtually continuous, and where a new generation of converts within its huge architectural community were keen to emulate and progress.

Fig 8.27
Milburn House, Newcastle-
upon-Tyne, 1904, showing
the skeleton of what was to
be a giant office block.
[Builders' Journal and
Architectural Review, *12*
April 1905]

Fig 8.28
Mersey Docks and Harbour
Board Building, Pier Head,
Liverpool, showing the steel
cage arising within the
masonry shell in 1904.
[Builders' Journal and
Architectural Review, *28*
December 1904]

Fig 8.29

Fig 8.31

Fig 8.33

Fig 8.35

Fig 8.37

Fig 8.36

Figs 8.29–8.37
Edwardian steel-framed splendour, Manchester style.

Facing page: Figs 8.29 and 8.30 National Buildings (formerly National Boiler Generator Insurance Company), St Mary's Parsonage (1905–9; steelwork by Edward Wood & Co., Harry S Fairhurst, architect). [Edward Wood & Co., Ltd 1916; Jonathan Clarke]; Figs 8.31 and 8.32 The distinctive skeleton of Manchester House by Edward Wood & Co., with its angled sides elevation and light well setback, built as a pair with Asia House, Princess Street by I R E Birkett in 1906–9. [Edward Wood & Co., Ltd 1916; AA022663]; Figs 8.33 and 8.34 Langley Buildings, Dale Street (1908–9; steelwork by Edward Wood & Co., R Argile, architect). [Edward Wood & Co., Ltd 1916; AA003170]

This page: Fig 8.35 The rear elevations of Asia House, barely concealing their skeletons. [AA022303]; Fig 8.36 Canada House, Chepstow Road (1905–9; W and G Higginsbottom, architects). [AA022681]; Fig 8.37 India House, Whitworth Street (1905–9; Harry S Fairhurst, architect). [AA022605]

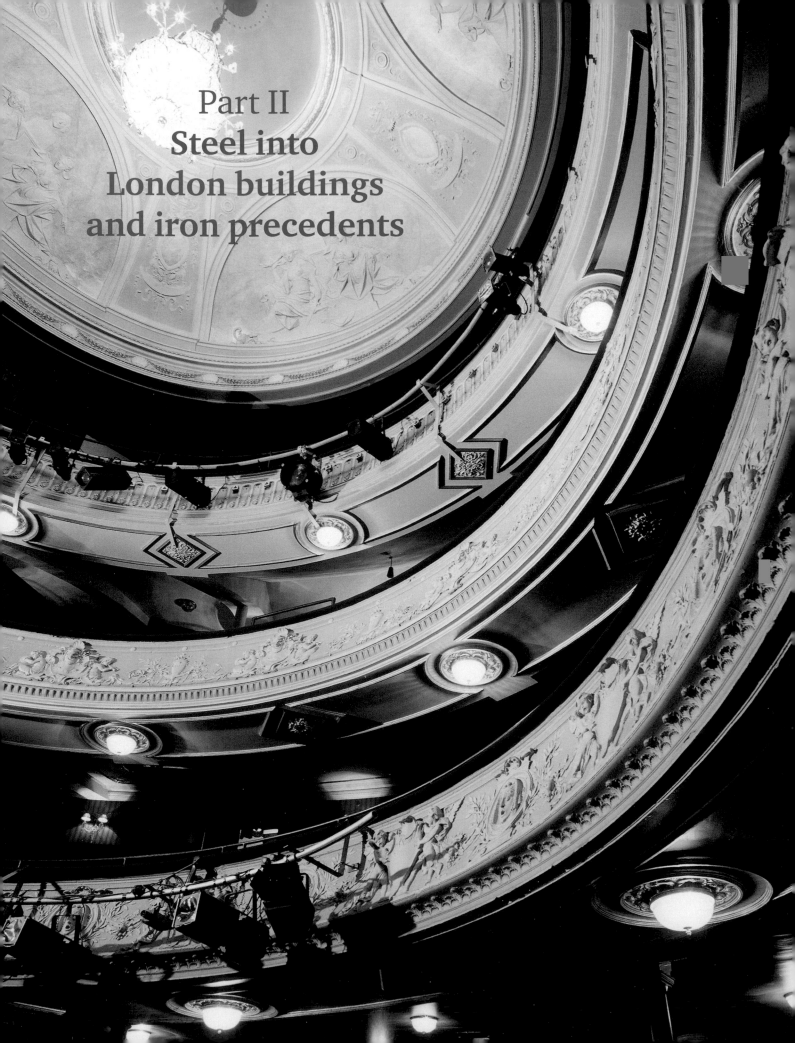

Part II
Steel into London buildings and iron precedents

9

Theatres and music halls: advances in safety and sighting

... but in ten or twelve minutes it ran up the front boxes, and spread like kindled flax. This may be accounted for from the body of air which so large a hollow afforded, and also the circumstances of the whole being a wooden case. For our readers will recollect, that the immense pile was constructed of timber. (1809)[1]

Iron forerunners

The spectre of fire long exerted a commanding influence on the structural character of London theatres. When Henry Holland was called in to rebuild the Theatre Royal, Drury Lane, in 1791, he was acutely aware of this, the greatest of all theatre hazards. Opened in 1794, Holland's new edifice was the last word in British 'fireproof' theatre construction. It was also unprecedented in size, being twice as large as London's existing theatre, the Haymarket Opera House; in terms of seating capacity, it was the biggest in Europe. Fitted with four large reservoirs of water on top of the building, an iron curtain that would separate the auditorium from the stage in the event of fire, a panoply of devices to prevent or resist fire, including Hartley's fireplaces[2] covering the timberwork, and commodious stair-halls to enable rapid egress, it was considered quite safe. Unfortunately, when fire did break out only 15 years later, in 1809, the reservoirs were virtually empty and the iron safety curtain had rusted and been removed. The mammoth theatre burned to the ground, leaving little more than a brick shell standing.[3] The tragedy was compounded by the destruction a few months earlier of the Covent Garden Theatre, a building that Holland had refitted in 1792, employing similar protective measures.[4]

Holland's Drury Lane was the last major London theatre to be built at a time when there was no real alternative to structural timber or masonry in polite architecture. Cast iron, in the form of slight columns, was used sporadically, and had been used by Holland to support tiers of boxes around the horseshoe auditorium, but limited availability, cost and constructional conservatism meant that it was not a common choice. When the fourth Drury Lane theatre opened in 1812, the constructional scene was changing. Benjamin Dean Wyatt's competition-winning scheme, smaller than its predecessor, made significant use of cast iron, the non-combustible properties of which were by this time being widely exploited in industrial buildings. In the auditorium, slender and generously spaced cast-iron columns were used to support the fronts of the five horseshoe tiers of boxes and galleries, whose raked floors were framed with joists and beams of the same metal. Elsewhere in the building, however, timber was used extensively, framing the galleries, the grid and flies above the stage and all the roofs.[5] At this date, more extensive use of both cast and wrought iron in theatre building was only being tried outside the metropolis, in the construction of John Foulston's extraordinary Theatre Royal, Plymouth (1811–13). Foulston claimed it was 'the only fire-proof Theatre in the country',[6] and, while that proved illfounded (the building suffered two fires later in the 19th century, the second, in 1878, catastrophic), its abundance of cast-iron and wrought-iron framing was probably unexampled in any other public building in Britain.[7]

Both these theatres were early forerunners of a basic constructional form that would flourish well into the Victorian era, having relatively shallow and vertically aligned balconies projecting from the auditorium walls, supported at their fronts by successive tiers of interconnecting cast-iron columns. Two important mid-19th-century developments – one legislative, the other material – saw the widening and increasingly ambitious application of structural iron to London theatre buildings. The first, the

Metropolitan Building Acts 1844 and 1855,[8] required that all circulatory spaces in public buildings – corridors, staircases and so forth – be made incombustible, thus effectively prohibiting timber. The second, the growing availability of an array of rolled sections, including joists, greatly extended the utility of wrought iron, enabling new structural forms that foreshadowed developments in steel. Two technologically and architecturally advanced theatres, both completed in 1858, began this trend.

The first, the Royal Opera House, Covent Garden, was built to designs by Edward M Barry to replace Robert Smirke's Theatre Royal of 1808–9, destroyed by fire in 1856. In 1846–7 the interior of Smirke's theatre had been completely rebuilt by an engineer, Benedict Albano, who had used cast-iron columns and girders to frame six tiers of boxes, and introduced stone-flagged corridors and staircases in compliance with the 1844 Act. The inquest after the 1856 fire pointed to the carpenters' room in the timber-framed roof-space over the stage as the likely origin of the fire. Explaining his own designs, Barry stated that he 'sought to avoid this source of danger by making the roof and its covering entirely fireproof', conceding that 'though it is hardly possible at present to render a theatre actually fireproof, the new building was intended to be an advance in that direction'.[9] Barry, who was assisted in the building's structural design by the consulting engineer Alfred Meeson,[10]

introduced giant wrought-iron trellis girders to support a ridge-and-furrow slate roof and carry the paint room above the stage, alongside many other mechanical fire precautions.[11] By using partially cantilevered balconies, Barry also took the structural character of auditoria in a new direction, as *The Builder* recognised.[12] The banishment of view-obstructing columns and the traditional concerns of fireproofing together became the order of the day.

Within the void defined by the building's thick brick carcase, Barry placed a largely independent auditorium, explaining that 'the boxes and corridors are, in fact, a separate structure of iron, stone and wood, erected inside the chamber formed by the main walls of the building'.[13] Three tiers of boxes formed the back of this horseshoe-shaped auditorium (Fig 9.1). Each tier was supported by vertically connected hollow-cylindrical iron columns, yet these were not placed on the line of the box fronts, but 6ft 6in. rearwards, close to the back of the boxes. The boxes rested on wrought-iron cantilever girders that were anchored in the brickwork of the corridor walls to the rear (Fig 9.2). These I-section members slotted through cast-iron junction boxes that connected one column to the next, and which also served to transmit the loads around the cantilevers, not through them. A line of principal cross girders or 'binders' also slotted into the column heads at right angles to the cantilevers, providing the main lateral sup-

Fig 9.1

Longitudinal section of the Royal Opera House, Covent Garden (E M Barry, architect), as built in 1858 with boxes on modestly cantilevered wrought-iron girders.

[The Builder, *16 April 1859*]

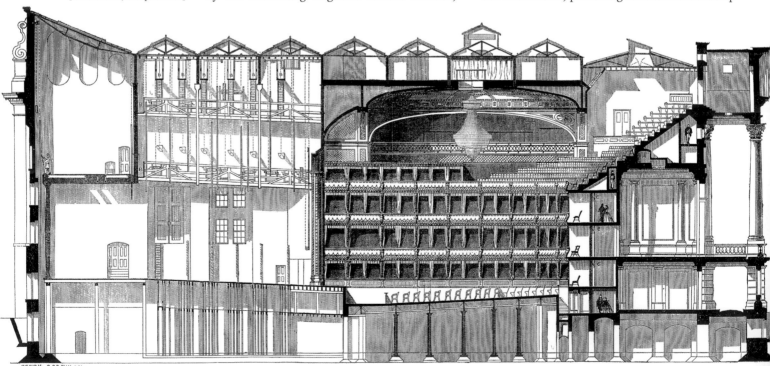

RENRY. GODWIN.DEL.

port to the framework. Another line of rolled-iron cross girders, spanning between the cantilevers, carried the stone floor of the corridor behind. Timber joists, fixed between the cantilevers and parallel with the cross-girders, supported the boarded floors of the tiers. This iron frame, probably designed in collaboration with Henry Grissell of the Regent's Canal Ironworks,[14] enabled exceptional levels of visibility for an audience who wanted both to see and to be seen. It also facilitated a new concept in auditorium design – adaptability. Each structural bay, occupying the space between the columns, was divided into two boxes of a standard size for opera houses, but the non-structural timber divisions were designedly removable, 'so that the whole of what will be usually private boxes can be thrown together, should the theatre be

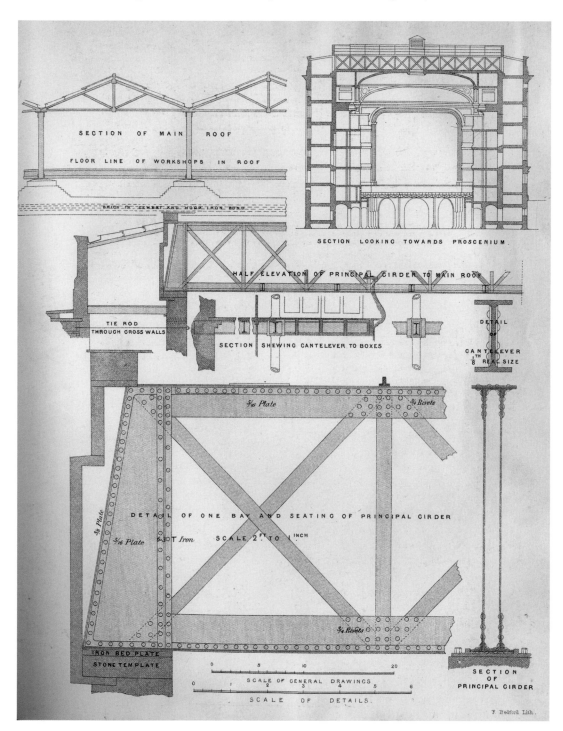

Fig 9.2
Structural ironwork in the Royal Opera House (Alfred Meeson, consulting engineer), showing that each of the eight wrought-iron trellis-girders spanning the full width of the auditorium and stage was 96ft-long and 9ft-high, dimensions unprecedented in any British theatre. [RIBA54983, RIBA Library Photographs Collection]

109

Fig 9.3

The unobstructed view from the partially cantilevered grand-tier balcony of the Theatre Royal, Adelphi (1858; T H Wyatt, and Stephen Salter junior, architects). This, and the curvilinear character of the auditorium, were enabled by the use of 'riveted iron'. [The Builder, 25 December 1858]

devoted to ordinary dramatic entertainments'.[15] Additionally, the rolled cross girders of the tier over the central grand state-box, the most privileged audience space, were bolted and not riveted to the cantilever girders. This enabled their removal, to give the box additional height 'on special and extraordinary occasions'.[16] Grissell made eight colossal main trellis girders support both a continuous ridge-and-furrow roof, and the floor of the workshops, high above the stage and auditorium. With a relatively shallow depth of 9ft – dictated by the minimum effective height that would not inconvenience workmen who needed to pass through them – they were less wasteful in their use of space than the typical timber queen-post-truss arrangement. They also enabled better lighting, concomitant with a ridge-and-furrow roof, and, tested to a loading of 300 tons, the girders above the stage could safely carry heavy machinery and scenery.[17]

The other pioneering theatre of 1858 was the Theatre Royal, Adelphi (Fig 9.3). Executed to designs by Thomas Henry Wyatt, with assistance from Stephen Salter junior, the theatre was considered 'remarkable' by *The Builder* 'for the application of iron in modes which are

altering the structural character of this class of buildings'.[18] The ceiling, with a low central dome, was hung from wrought-iron roof trusses carried on tiered iron stanchions rising from the concrete foundations and set within or beside, but independent of, the exterior walls. This arrangement was adopted to avoid the removal and reinstatement of a party wall, though the wall was found to be seriously defective and was removed anyway. The use of the stanchions permitted an external wall thickness of only 1ft 6in. all the way up, and also precluded the use of columns in the top gallery, traditionally used to help support the auditorium ceiling.[19] Vertically connected, widely spaced columns were used to support the three tiers. Decisively, the lower tier or dress circle – the most privileged seating level – projected 6ft in advance of the column lines or fulcrums. Contemporary accounts do not detail the specific structural techniques, but it seems likely that wrought-iron cantilever girders were used, projecting from the pit-level column heads. The three tiers were all framed in wrought iron, and it was the ease with which this metal could be shaped into curved structural members that lent the interior its distinctive flowing lines:

The use of riveted iron has readily allowed of that sinuosity of character, so to speak, which we have described in the plan, and which in section is equally shown in the girders bent to the forms of the boxes, sometimes in the reverse direction of that of strength, as it would have appeared on some data not of this age.[20]

The plasticity of wrought iron was also exploited in a time-honoured and obviously decorative way, in forming the open-work front of the dress circle, 'so that the ladies' dresses may show through'.[21]

Through the 1860s for reasons of economy and conservatism, most theatres continued to be built in the traditional manner, without partially cantilevered balconies. Balconies, framed in wrought iron, grew steadily deeper to accommodate more people, but these were typically supported at their outer edges by tiers of columns. The Royal Surrey Theatre, Lambeth, even used two concentric ranks of columns to extend the open balconies over the floor of the auditorium or 'pit'. Built in 1865 to designs by John Ellis, this theatre incorporated much structural iron, including wrought-iron roof trusses, flies and trussed girders to support the deep gallery.[22] However, partially cantilevered balconies, usually restricted to the most expensive dress circle, did increasingly enter the structural vocabulary of central London theatres. The diminutive Holborn Theatre Royal, High Holborn, built in 1866 to designs by Finch, Hill and Paraire,[23] essentially repeated Ellis's ploy in extending the first tier over the pit, but with far fewer columns, the majority being set well back from the edge. The auditorium of the Old Vic, Lambeth, remodelled in 1871 by J T Robinson and consisting of two lyre-shaped balconies supported on iron columns, was perhaps typical of the era, with all of the upper-tier seats, but only the front rows of the lower tier, benefiting from uninterrupted views of the stage (Fig 9.4).

Fig 9.4
The Old Vic, Lambeth, substantially unaltered since 1871 when remodelled by J T Robinson, and typical of the era, its partial application of cantilevering. [© Alberto Azoz /Axiom]

In the National Standard Theatre, Shoreditch, built in 1867, the two shallow upper tiers were carried on ornamental cast-iron figurines that projected from columns set at the rear of the auditorium.[24] The use of such exposed cast-iron cantilever brackets, with their limited reach, prefigure most notably in Bunning's Coal Exchange (1847–9), was, however, a technological step backward. Perhaps the most progressive aspect of this theatre was the arrangement of the tiers, each one being set back from the one below it.

By 1870 new directions in cantilevered and fire-resisting construction were becoming evident in London's foremost theatre district, the West End. Her Majesty's Theatre, Haymarket (1868–9, demolished 1896), was probably the first British theatre to abandon visible iron columns in the auditorium. To achieve this, Charles Lee, the architect behind this rebuilding, assisted by his sons and partner as Lee, Sons & Pain, skilfully positioned the interlock-ing columns within an outer fireproof corridor that wrapped around the horseshoe-shaped auditorium (Fig 9.5). Extending from the basement to the roof, these stacks of columns provided mid-length support to specially shaped Phillips's patent girders[25] that radiated inwards into the auditorium, their outer portions embedded in the two concentric walls forming the corridor. The cantilevered sections of these compound girders projected some 12ft beyond the columns, carrying the weight of the boxes, an arrangement that caused *The Builder* to exclaim 'All the tiers of boxes are so built that there is no need of columns for the support of any part.'[26] As in the Royal Opera House, the partitions of the boxes were designed to be removable, so that the whole of the first tier could be opened out into a dress circle for winter performances.

Her Majesty's Theatre was also built using the latest fire-resisting techniques, its predeces-

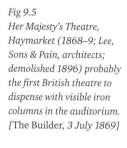

Fig 9.5
Her Majesty's Theatre, Haymarket (1868–9; Lee, Sons & Pain, architects; demolished 1896) probably the first British theatre to dispense with visible iron columns in the auditorium. [The Builder, 3 July 1869]

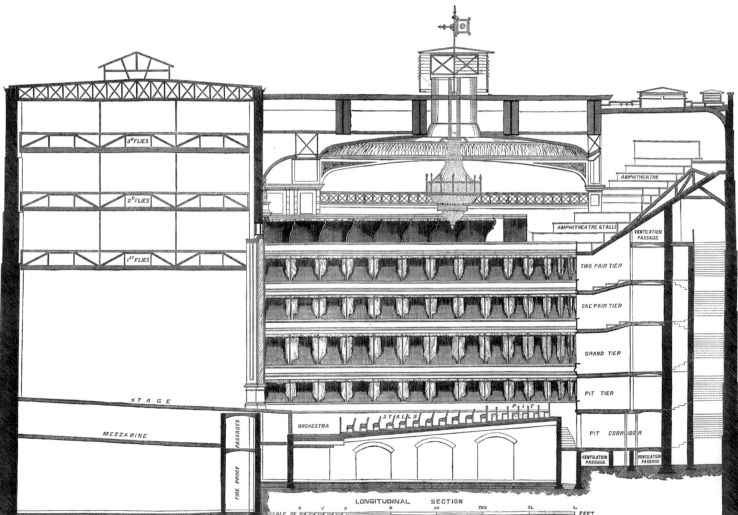

sor having been destroyed by fire. It used nearly an acre of Dennett's 'fireproof' concrete to form the floors, and the stage was separated from the auditorium by a full-height wall.[27] The auditorium roof was carried by four 90ft wrought-iron trellis girders, and that of the stage by five 56ft girders.[28] The Gaiety Theatre, on the Strand, completed in 1868 to designs by C J Phipps, also exploited the greater strength and spanning potential of Phillips's compound girders to frame its innovatively shaped dress circle, 'the front forming a semicircle, opening out by arms of a contrary flexure to the proscenium column', and the gallery above, the front of which was implausibly described as forming 'a complete circle'.[29] Although the dress circle projected over the pit to a remarkable degree ('only a limited portion of the first four benches is clear of the balcony, and the remainder is covered in by the boxes so closely that a tall man with a hat on would find it somewhat difficult to stand upright'),[30] cantilever construction was apparently not used.[31] Yet, in its attention to non-combustibility, the Gaiety was perhaps unrivalled, at least in London:

> In fact, the whole construction of the building is as nearly as possible fireproof, for not only are all staircases, passages, and corridors of stone or cement, and separated in every case by brick walls, but the several tiers – balcony, upper boxes, and gallery – have no wood in their construction, except the flooring boards; they are entirely built of an iron framework, embedded in and filled between with a solid mass of cement concrete, much on the principle adopted at the Grand Opera and the New Vaudeville Theatre at Paris, which system was adopted there as being the most perfect that could be devised, as by diminishing the amount of inflammable material in a building the risk of its even taking fire is rendered almost impossible, while the prevention of a fire spreading is insured. With the exception of the two Theatres at Paris before mentioned, the 'GAIETY' will be the only Theatre in Europe so constructed.[32]

Both Her Majesty's and the Gaiety were extraordinary in their progressive use of iron structure protected by concrete, but further legislation by the Metropolitan Board of Works in the 1870s ensured that this became almost standard practice. The Metropolis Management

and Building Acts Amendment Act 1878[33] dealt with, among other things, the design and construction of theatres and music halls whose area exceeded 500 sq ft.[34] It went far further than the mid-century Acts in requiring that all floors, including those of auditoria, be made of fire-resisting materials, and that all ironwork be protected against the action of fire. It also regulated facilities of egress, scaling the staircases and corridors to the size of the audience – in effect, promoting 'roomy places of amusement'.[35] Although it did not address cantilever construction, it did, arguably, encourage its widening uptake through two stipulations: one prescribing a minimum of 2 sq ft to be allotted to each person in the galleries, the other requiring that each tier be provided with its own direct means of exit to the street.[36] In effect, the former encouraged deeper balconies if a reasonable number of rows were to be accommodated, the latter promulgated the type of arrangement pioneered at Her Majesty's, where each tier of the auditorium was encircled by its own fireproof corridor. Deeper balconies did not, of course, preclude columnar support, but cantilevering could keep these to a minimum. The concentric walls of fireproof corridors behind the auditorium provided the necessary anchorage – and sometimes the fulcrum support – for cantilevered girders that simultaneously carried the passage floors above.

The new regulations of 1878, according to one contemporary architect, 'caused a complete revolution in the construction of theatres'.[37] Specialist consulting engineers were increasingly brought in to provide innovative and economical solutions to structural problems that were beyond the capabilities of most architects and ironwork contractors. For the design of the constructional ironwork of the Pandora (later the Empire) Theatre, Leicester Square (1882–4, demolished 1928), the architect Thomas Verity enlisted R M Ordish, arguably the most experienced and proficient structural engineer then working in iron.[38] The Pandora was not an entirely new structure, but a partial conversion by Verity of an existing oval building, the Royal London Panorama (1880–1), a short-lived venture designed by a M L Dumoulin of Paris for a French company.[39] By simply cutting away one end of the elliptical envelope for the insertion of the orchestra and stage, Verity created one of the largest auditorium spaces in the metropolis. Here, Ordish, 'under Mr. Verity's direction', implemented 'the cantilever principle ... to the greatest perfection' in respect of the dress circle or balcony and the two tiers above that.[40] A series of inwardly radiating wrought-iron cantilever girders, supported at mid-length by cast-iron columns set some 13ft forward from the auditorium wall (in which the cantilevers were embedded) permitted unrestricted views of the stage from virtually all parts of these crescent-shaped balconies, each of which receded from the one below. The girders were raked, the cantilevered portions gradually diminishing from 12ft 4½in. in the centre of the auditorium to 4ft 2in. at the front, next to the proscenium wall, allowing the supporting columns to be placed 'as many as five rows back from the balcony fronts'.[41] In earlier theatres, such as the Royal Opera House, cantilevered beams were connected to the cast-iron columns through a boxlike casting bolted between the lower column capital and upper column base. Ordish redesigned this junction for a more rigid connection, strengthening the upright portion of the cantilever between the columns 'in the most scientific way',[42] so that it transmitted the compressive forces, effectively becoming a continuous portion of the columns. This reinforced fulcrum was also the best point at which to connect the principal transverse girders that helped support the fireproof floor structure.

Under Ordish's guidance, the Pandora's inspired iron construction was not limited to the gallery. Four extremely heavy 50ft-tall cast-iron stanchions, imbedded in brick piers, rose from the foundations and supported three built-up I-section girders that formed three sides of a rectangle high above the stage. This framework of iron supported the combined weight of the scenery battens, borders and flies, and the central wrought-iron roof trusses. The two larger, solid-web, side girders, 52ft long, 6ft deep and 'stronger than any railway girders of like span'[43] also carried much of the walls, floors and roofs of the rooms to either side of the stage. Because these girders were so heavy – the larger two weighing some 15 tons apiece – the ironwork contractors Dennett & Ingle assembled them *in situ* on temporary staging, the plates and angles having rivet-holes pre-prepared. This work and that of hoisting and fabricating the wrought-iron roof trusses comprised 'perhaps the largest and heaviest piece of constructive ironwork ever before placed in any one building in London'.[44]

The introduction of steel

The theatre designer of a generation ago would doubtless see in the construction of the modern playhouse many things that would surprise him, and, perhaps, in no respect more so than in the way in which steel is now used throughout these structures ... The wonderful adaptability of steel and iron has encouraged many forms of construction that would, without it, have been practically impossible, or, at any rate, would have offered little temptation to the engineer or architect to depart from stereotyped forms.   (1906)[45]

While the Pandora Theatre was being built, another large theatre was nearing completion in Leicester Square, its structural design also dependent on a leading consulting engineer. Although contemporary accounts in the building press mention only cast and wrought iron in this building, the Alhambra Theatre (1883, demolished 1936) was almost certainly the first British theatre in which structural steel was used extensively.[46] Built to replace a predecessor largely destroyed by fire, this theatre was probably the first essay in structural steel by (Sir) Alexander Kennedy, the first Professor of Engineering at University College London. Kennedy had been asked by the theatre directors to work alongside the architect, T Hayter Lewis, and builders Perry & Reed, for structural alterations to the earlier building.[47] These alterations – which included the widening of the proscenium by replacing two of the iron columns supporting the main roof with a sturdier pair of cast-iron stanchions and a huge compound girder – had proved inadequate in the face of fire, despite being designed in accordance with Metropolitan Board of Works regulations.[48] *The Builder* noted that 'one of the most striking features in the ruins after the fire was the great iron girder which was, at the time of these alterations, built across the proscenium, measuring some 60ft by 5ft, lying in festoons across some unbroken cast-iron stanchions like a piece of tripe – showing well the relative effect of the heat on cast and wrought iron'.[49] This spectacle may well have shaken the proprietors' and architects faith in wrought iron, and Kennedy, an expert on materials, was presumably recalled to provide a sturdier, less vulnerable structural solution, this time in steel and concrete.

Fig 9.6 (facing page, top) The iron- and steel-framed rotunda of the Alhambra Theatre, Leicester Square, 1883, demolished 1936. Engineered by Professor Alexander Kennedy, this was a largely self-supporting structure.
[The Builder, 16 June 1883]

Fig 9.7 (facing page, bottom) Underside of the Alhambra's grand tier, showing faience-clad steel cantilevers and columns.
[Sachs and Woodrow, Modern Opera Houses and Theatres, 1896, vol. 1]

The framework of the new theatre, built to the designs of Kennedy and his young assistant, Thomas Hudson Beare, was, by all available accounts, a triumph of what might best be described as proto-cage construction. The original façade to Leicester Square was retained, carried on a continuous girder, but behind that 'a perfect internal structure of iron'[50] of 'unique construction'[51] was built (Fig 9.6). In essence this consisted of a largely self-supporting iron- and steel-framed rotunda that rose four storeys and lent mid-span support to cantilever girders, which radiated inwards from the enclosing front and party walls. Beyond the columns of the rotunda, which occupied the same position as within the previous structure, the cantilevered portions of these steel members carried the balconies four rows into the interior, enabling unobstructed views not shared by the previous building. Each cantilever girder was made up of two plates stiffened and joined together by angles, and was designed to resist the concentrated stresses at the fulcrum, where it passed between the columns above and below. At this point, as with Ordish's cantilevers, the girders were at their deepest, diminishing in size towards the inner edges of the balconies. Like the supporting columns, the girders were clad in a riot of orientalising faience (Fig 9.7). Rolled I-section steel joists, gently curved to conform to the rounded plan of the tiers, spanned transversely between the cantilevers, fixed in place by angle brackets riveted to the web-plates. These joists – forming the risers or steps of each tier – varied in size according to the length of span between the cantilevers, from 16in. by 6in. at the back to 8in. by 3in. on the front edge of the gallery. Flat plate sections were used to brace the joists at intervals, making the whole structure more rigid. The building impressed. Less than a decade after its completion, the architect Ernest Woodrow was of the (mistaken) opinion that it was the first theatre to adopt cantilever construction.[52]

The 73ft-diameter, flat-edged conical roof over the auditorium was framed in steel and rested on the uppermost tier of iron columns. Steel ribs sprung from the column capitals and abutted against a 12ft-diameter ring beam, which was itself crowned by a cupola fitted with what was claimed to be the largest sun-burner in the world.[53] Rolled-steel joists, measuring 4in. by 3in., rested on the upper flanges of these meridionally arranged principals, and supported a concrete and asphalt roof

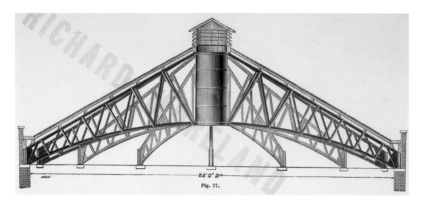

Fig 9.8
The steel trussed-roof of
Hengler's Circus, Argyle
Street (1883, demolished).
[© The British Library
Board (F60077–63)]

covering and, below that, a flat inner ceiling of fibrous plaster.

The stage area was also partly framed in steel. H-section stanchions, probably of cast iron, were embedded in the piers of the stage opening and carried a 6ft-deep box girder, used to support the solid wall of the proscenium. This wall continued from the foundations to above the roof level, forming a fireproof separation between stage and auditorium in accordance with Metropolitan Board of Works regulations. Above the stage, a large steel truss supported the gridiron from which the scenery could be safely manoeuvred. To make the public parts of the interior as fireproof as possible, all the iron and steelwork was covered by incombustible materials.[54] The projecting cantilevers, joists and iron columns forming the rotunda were embedded in fibrous plaster, while the floors and ceilings of the landings and steps were covered by a 7in. layer of concrete. To reduce the weight borne by the cantilevers, the floors of the galleries and dress circle were formed in concrete made with coke breeze aggregate.[55] Even the 2½in.-thick curving walls dividing the boxes were of concrete.

One of the most remarkable aspects of the rebuilding was the rapidity with which it was executed: building work started on 23 April 1883, and the Alhambra re-opened on 3 December the same year. Such a feat was testament to the design efficiency and collaboration of the architect-engineer team as well as to the capability of the contractor, William Brass, in supplying, fabricating and erecting the iron and steelwork. Brass (1813–88)[56] was unusual at this time insofar as he supplied his own metal (tested by Kennedy as it arrived); *The Building News,* in applauding the 'unexampled progress' of the works, noted 'there are few contractors who supply and fix their own ironwork, the consequence of which is very often serious

delay'.[57] The source of the metal is not known, but its supply may have been subcontracted to Richard Moreland & Son, who like Brass had premises on Old Street.[58]

It is astonishing that so much steel seems to have entered the Alhambra's structure alongside iron at so early a date. Only one other comparable building seems to have made significant use of the metal at this time. The roof of Hengler's Circus, Argyle Street, was 'constructed wholly of steel in the year 1883'.[59] This distinctive, polygonal structure was designed by C J Phipps in consultation with Moreland (Fig 9.8). Most theatre designers of the 1880s probably felt happier using wrought iron, and it was only the most technologically adventurous consulting engineers who began experimenting with the new, expensive material, albeit in cost-effective ways. For the diminutive Terry's Theatre on the Strand (1885–7, demolished 1923), its architect Walter Emden entrusted the design of the structural metalwork to Max Am Ende (1839–c 1916), a former pupil of Ordish who would go on to design the Olympia Exhibition Hall, Kensington (1886–7), with Arthur T Walmisley (1849–1923).[60] Ende made use of the by then unexceptional cantilever principle, 'characterised, as in most such cases, by galleries projecting far beyond the points of support – the columns – so that these may not obstruct the view of the stage to persons placed in the best seats'.[61] Nonetheless, his judicious use of solid steel columns coupled with a radically different approach to the structural design of the tiered floor, broke new ground (Fig 9.9). Rather than letting an array of cantilevered girders do all the work of supporting the tiers, Ende made the two stepped-tier floors rigid, virtually self-supporting frameworks in their own right. Curved wrought-iron lattice girders were used to form both the vertical (riser) and horizontal (tread) components of each step, the whole assemblage spanning between inclined girders (stringers) running parallel with the side walls of the auditorium. Diagonally braced in both horizontal and vertical planes, this stepped floor acted as a giant folded plate, and, like such plates, did not require supports at close intervals. Just two raked girders were needed to carry each floor near the centre of its span. Running parallel with the stringers, these cantilever girders formed part of a supporting framework that acted independently of the auditorium walls. Rather than being embedded in the masonry, the ends of the girders were riveted to two iron anchorage bars that rose from a huge wrought-iron bed-plate set in the pit floor. Two tiers of solid steel columns, set 11ft in from this stanchion and rising from the bed-plate, provided fulcrum support to the four girders, each of which cantilevered forwards some 13ft from these points. The great overhanging weight of the cantilevered portion of the balconies was thus not counterbalanced by the weight of the masonry – the typical solution, but one potentially susceptible to cracked brickwork under audience 'live loads'. Rather, the weight was firmly taken by the two anchorage bars, which formed an integral part of the rigidly connected framework. The anchorage bars, acting as vertical ties, were concealed within the back wall of the auditorium, but the steel columns, which also supported the stringers, were visible. With their slender dimensions

(the thickest not exceeding 5in.), they were considerably narrower and less visually intrusive than hollow cast-iron equivalents, but more importantly, they were safer, able to resist the considerable forces brought on them without risk of unexpected failure. Each column was fitted with ball pivots at either end to facilitate erection and evenly distribute the forces at the column/cantilever interfaces. Once the metal structure was erected, and the dead load had found its overall bearing on the columns, the sockets were wedged firm with steel pins to give them the strength of columns with fixed ends.

Ende's brilliant design enabled a column-free gallery 11 rows deep, and below that a first-floor tier that cantilevered forwards five rows from just two slight steel pillars. All the metal, fabricated and erected by Matthew T Shaw & Co., was encased in concrete to protect

Fig 9.9
Terry's Theatre, Strand (1885–7; Walter Emden, architect; demolished 1923). Max Am Ende designed this tier-supporting iron and steelwork with one of the first uses of solid steel columns in London, and perhaps Britain.
[The Engineer, 23 September 1887. Courtesy of Institution of Civil Engineers]

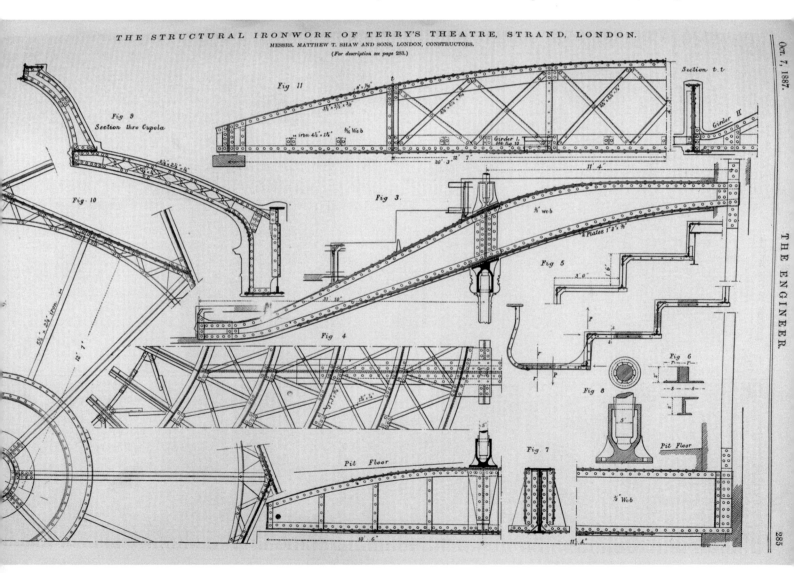

it from the effects of fire. In both its mode of action and its mathematical calculation, this structure was extremely sophisticated, the zigzag floor, restrained against flattening by the stiffening effect of the stringers, anticipating the folded-plate and other form-resistant surface structures of the next century. Ende considered this to be the first application of this technique to any theatre, and was aware of the importance of his innovation, stating

> If the zig-zags or corrugations are developed from the solid corrugated plate into light lattice work, with a corresponding enlargement of their depth and width, as in the present case, or in the case of the screens of the National Agricultural Hall at Kensington [Olympia], a new system of iron or steel construction is produced, which combines lightness with elegance with a fair degree of economy, and which might be applicable to other kinds of structures besides those here referred to.[62]

He would later apply the principle to two more of Emden's theatres, the Court Theatre, Sloane Square (with W R Crewe, 1887–8), and the Garrick Theatre, Charing Cross Road (with C J Phipps, 1888–9) (Figs 9.10 and 9.11). Like Terry's Theatre, both were small and narrow, accommodating under a thousand spectators apiece, which meant that the spanning capability of their zigzag floors was not over-reached. In the Court Theatre, Ende used neither columns nor cantilever girders in the auditorium proper, stating 'the whole circular structure carries itself free between the stringers, which are 31ft. 3in apart'.[63] This audacious stepped structure, with solid-plate rather than latticed steps and risers, survives intact. Such structural gymnastics were, however, ultimately reliant on steel columns, which supported the inclined plate-girders (stringers) on either side, and at the Garrick Theatre were used to provide intermediary support to the wider floor structure there.

Notwithstanding Ende's fresh take on the sightlines problem, his approach remained outside the main thrust of theatre engineering. Not only was it restricted to small, narrow auditoria, but its sheer mathematical complexity placed it outside the reach of all but the most skilful engineers, and beyond the financial grasp of most developers. Some principles and techniques – such as solid steel columns, curved girders spanning the width of the auditorium and U-shaped first treads, hung off the lowest treads of the tiers to extend the seating beyond the reach of the cantilever arms – did find life beyond these three theatres. But the main developments were in steel cantilevering and the facility this offered for deeper, stronger, column-free tiers in ever-larger auditoria.

The first theatre to exploit the technology fully and thus dispose of all view-obstructing pillars ('hitherto ... so objectionable to playgoers')[64] was the Royal English Opera House (now the Palace Theatre) in Cambridge Circus (1888–91). Built for Richard D'Oyly Carte, the technologically well-informed patron of the Savoy Theatre of 1881,[65] to the designs of architects T E Collcutt and Collings B Young, working with the contractor G H Holloway,[66] this building employed some 450 tons of constructional steel and ironwork behind its celebrated brick

Fig 9.10 (facing page) The (Royal) Court Theatre, Sloane Square (1887–8; Walter Emden and W R Crewe, architects), built by M T Shaw & Co. to designs by Ende. Late 1990s refurbishment exposed this daring zigzag iron structure. [© Alberto Azoz /Axiom]

Fig 9.11 (left) Tiers at the Garrick Theatre, Charing Cross (1888–9; Walter Emden and C J Phipps, architects), made possible by Ende's advanced structural design. [AA020204]

and terracotta 'Renaissance' exterior (Fig 9.12). Steel was ascendant, *The Builder* noting 'Steel has, with few exceptions, been used throughout', a circumstance perhaps owing to the progressive attitude of the consulting engineers, Messrs Reade, Reilly and Jackson of Bedford Row, Holborn, but probably also due to its diminishing cost.[67] The calculated efforts of this large team of architects and engineers, working with metal supplied and erected by the Horseley Ironwork Company of Tipton, Staffordshire, produced an auditorium with unrivalled sightlines, one whose daring, fully cantilevered balconies projected further forwards than had hitherto ever been tried (Fig 9.13). An architectural and technological *tour de force*, the Royal English Opera House was

Fig 9.12
The Royal English Opera House (1888–91; T E Collcutt and Collings B Young, architects); now the Palace Theatre.
[BB000928]

also one of the first theatres to profit from the expertise of a professional theatre consultant, J G Buckle (1852–1924), who advised on sightlines and safety.[68]

In the huge and opulent auditorium, all 1,697 seats were given the clearest views possible of the stage by the clever arrangement of the three tiers, the Royal Circle, the First Circle and the Amphitheatre-cum-Gallery. The front of the first and second tiers conformed almost to a semi-circle, but behind that the rows opened out, becoming shallow arcs at the rear so that the people seated there were still directly facing the stage. Both these tiers were extraordinarily deep, consisting respectively of nine and seven rows of seating. A semi-circular shape was also adopted for the four-row-deep amphitheatre,

and above that, the 10 rows of seats making up the gallery were arranged in shallow arcs. Each of these sculptured levels receded from the one below, each dipping gradually from the sides of the auditorium to the centre, and each completely bereft of columns. Steel building technology was being used to provide the greatest number of seats with the best possible sightlines, a response to an overtly commercial theatrical milieu. Traditionally cantilevering had been restricted largely to the dress circle, but the designers of the Royal English Opera House, in pursuit of profit, bestowed its benefits to a wider social spectrum.

Fig 9.13
The three-tiered auditorium of the Palace Theatre (formerly the Royal English Opera House).
[AA020778]

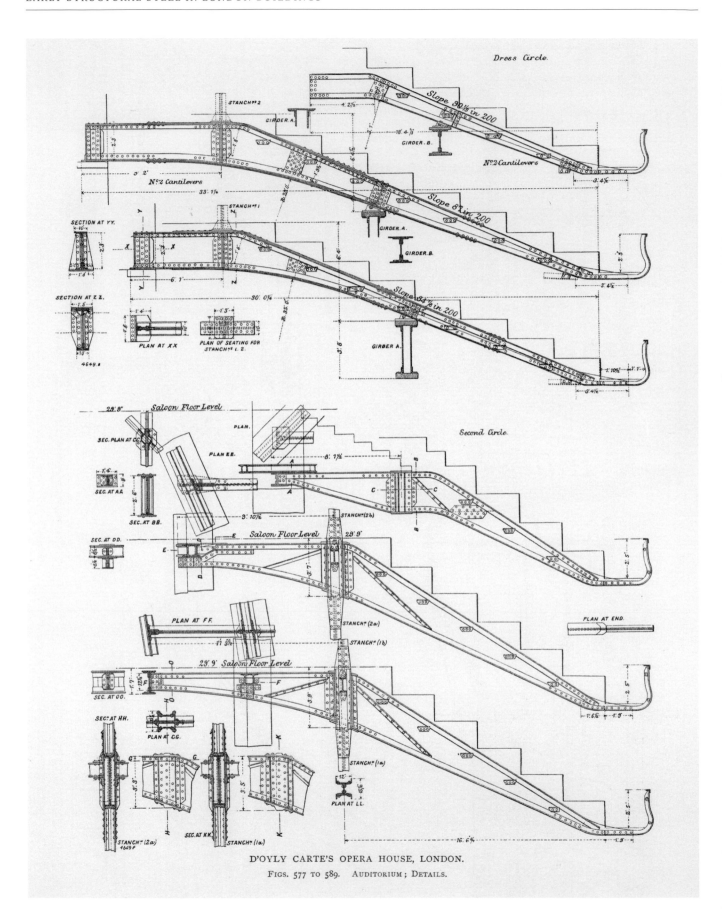

D'OYLY CARTE'S OPERA HOUSE, LONDON.

FIGS. 577 TO 589. AUDITORIUM; DETAILS.

The state-of-the-art steel-cantilever construction that enabled this new standard in audience satisfaction was necessarily complex, and differed in detail from level to level (Fig 9.14). Both the Royal and First Circles are carried on a series of inwardly radiating cantilever girders anchored into the side and rear walls of the auditorium, while those of the gallery are fixed to a deep box girder above the back wall. Measuring up to 33ft in length, the graceful cantilever girders of the Royal Circle are supported at approximately half their span on a deep steel box girder running transversely across the back of the theatre, spanning the side walls (Fig 9.15). At the corners of the auditorium, secondary girders branch off from this main girder at an angle, supporting the cantilever girders that carry the curved sides of the tier. Two huge box-section wrought-iron stanchions, set about 6ft from either wall and seated in the concrete foundations, take the main bearing for this huge girder. The shorter cantilevers of the second and third tiers are given fulcrum support either by stanchions standing on the cantilevers of the floors below, or from secondary beams carried by continuous cruciform-section stanchions rising from the ground floor. The cruciform stanchions, fabricated from wrought-iron angles, are built into the walls

separating the auditorium from the flanking corridors; they also partly support beams which carry the corridors.[69] This arrangement, while not eliminating reliance on intermediary support for the cantilevers, nevertheless enabled the uprights to be positioned at the sides and backs of the seating areas, permitting wholly unimpeded sightlines from all tiers.

The stepped floors of the tiers were of concrete, formed around iron angles supported on perforated firebrick walls forming the step faces. These light, non-combustible risers in turn rested on steel bearers spanning between, and bracing, the cantilever girders, the whole steel framework described as 'a perfect cobweb of steel-work'.[70] As in the Alhambra, incombustible concrete and Portland cement were used to encase the steelwork and form the floors, ceilings, stairs and partitions. Glazed brick was used for the interior walls of the corridors, dressing rooms, staircases and so forth, and the window frames were made of steel. Such extensive employment of 'fireproof' construction had never been tried on this scale, and, together with a completely isolated site and a massive tank of water over the stage, this precluded, uniquely, any fire insurance, as it was considered 'impossible for it to burn'.[71]

Fig 9.14 (facing page)
Steel cantilever-girders formed the principal elements in the 'perfect cobweb of steelwork' that framed each of the three tiers in the Royal English Opera House.
[Sachs and Woodrow, Modern Opera Houses and Theatres *(1898), vol 3]*

Fig 9.15 (below)
Royal English Opera House, Royal Circle structure.

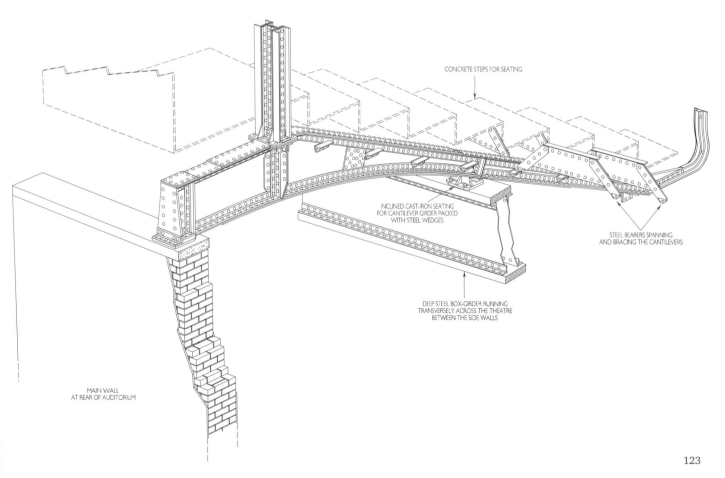

CONCRETE STEPS FOR SEATING

INCLINED CAST-IRON SEATING
FOR CANTILEVER GIRDER PACKED
WITH STEEL WEDGES

STEEL BEARERS SPANNING
AND BRACING THE CANTILEVERS

DEEP STEEL BOX-GIRDER RUNNING
TRANSVERSELY ACROSS THE THEATRE
BETWEEN THE SIDE WALLS

MAIN WALL
AT REAR OF AUDITORIUM

Steel was used extensively elsewhere in the building. The 34ft-wide roof over the auditorium, supported by girders spanning between the side walls, was framed in steel ribs encased in light coke-breeze concrete. The much larger roof and gridiron over the stage were hung from 69ft-long steel Warren trusses, and similarly configured deep lattice-girders supported the 'fly galleries' high above the stage. But it was in the deep, lavish auditorium tiers that steel broadcast its virtues most effectively, and it was this aspect of technical prowess that D'Oyly Carte's management proclaimed most fervently. Some 5 years after the theatre opened, E O Sachs, in his extensive treatise on Continental and British playhouses, *Modern Opera Houses*, could still note that 'the elaborate cantilever work … is unparalleled in any other theatre in Europe'.[72]

The Royal English Opera House was a potent advertisement for comprehensive cantilever construction in large auditoria, and, as Sachs noted, even 'second-rate' establishments strove to this end:

In Great Britain the ironwork is limited almost entirely to the principal constructional features of the tiers and their supports, whilst its application is particularly associated with the cantilever work in which we excel. In most of our modern playhouses there is a demand for an uninterrupted view of the stage from every seat in the auditorium, and the manager considers it essential for the prosperity of his house to be able to boldly advertise this fact. Hence, even in second-rate establishments, the cantilever is now frequently found, though it may be only in conjunction with the more elementary forms of wood bearers and flooring. But we may certainly pride ourselves on holding the first place in respect to the application of metalwork on the cantilever system, and as Vienna possesses the leading example of elaborate iron framing in the Court Theatre, so can London boast of the best example of cantilever work in D'Oyly Carte's Opera House.[73]

According to Ernest Woodrow, a key member of the LCC's department concerned with theatre safety, the Theatres and Music Halls Branch, and Sachs's collaborator on *Modern Opera Houses*, cantilever construction went hand-in-hand with 'fireproof' construction. A year after the opening of the Royal English Opera House he wrote:

It was the constant cry for the abolition of these obstacles [columns] that first gave birth to the cantilever construction now almost universally adopted in theatre building … Doing away with the supporting columns of the galleries has done far more than remove the annoyance from the audience of obstructing the view of the stage: it has also tended towards perfecting the fireproof qualities of the structure. Theatres that were built with supporting columns had the tiers and roof, as a rule, entirely constructed of wood. With the cantilevers the tiers are made of concrete or firebrick on steel girders imbedded in concrete, and the roofs are made fireproof … The knowledge that the building is fire-proof will do as much to reduce panic as good exits; in fact even more, as it will destroy the nervousness which generates a stampede.[74]

With the new benchmark set by D'Oyly Carte's Opera House, the 1890s saw steel and concrete become increasingly the standard structural and fire-resistant choices for new theatres and, more importantly, for music halls, including a growing array of 'second-rate' establishments. Indeed, the 1890s witnessed a fever of theatre and music-hall building activity in the capital and the provinces, with a distinct boom from 1896 that continued until curbed by the 1914–18 war.[75] The small handful of architects specialising in this type of building – Frank Matcham (1854–1920), Charles John Phipps (1835–97), W G R Sprague (1865–1933), Bertie Crewe (d1937), Ernest A E Woodrow (1860–1937) – satisfied the new-found public taste for cantilevered balconies in their lavish 'fireproof' designs, making use of the structural expertise of a growing array of specialist contractors and engineers. In Queen's Hall, Langham Place (1891–3; destroyed 1941), intermediary vertical support to the cantilever girders was completely eliminated (Fig 9.16). The steelwork for this impressive concert hall (measuring 125ft by 87ft and seating 3,000 people)[76] was designed as early as 1887 by engineering contractors Richard Moreland & Son, working in collaboration with the architects T E Knightley and C J Phipps.[77] The two gently raking tiers, which projected some 20ft into the auditorium, were entirely independent of one another, their component cantilevers simply embedded in an inner and outer wall that served both as a sound barrier and a stable anchorage (Fig 9.17). The steelwork was

Fig 9.16
Queen's Hall, Langham
Place (1891–3; T E Knightley
and C J Phipps, architects;
destroyed 1941) as
photographed by Bedford
Lemere and Company
in 1894.
[BL12577]

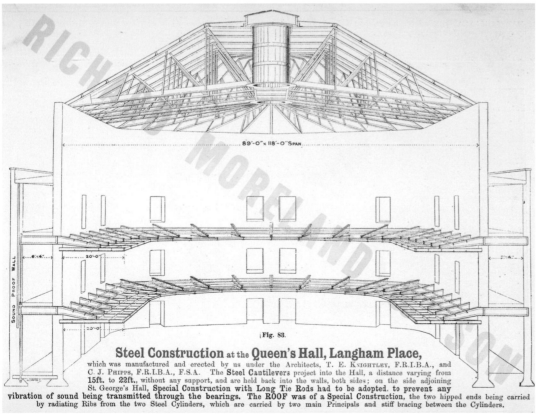

Steel Construction at the **Queen's Hall, Langham Place,**
which was manufactured and erected by us under the Architects, T. E. KNIGHTLEY, F.R.I.B.A., and
C. J. PHIPPS, F.R.I.B.A., F.S.A. The **Steel Cantilevers** project into the Hall, a distance varying from
15ft. to 22ft., without any support, and are held back into the walls, both sides; on the side adjoining
St. George's Hall, **Special Construction with Long Tie Rods** had to be adopted, to prevent any
vibration of sound being transmitted through the bearings. The **ROOF** was of a Special Construction, the two hipped ends being carried
by radiating Ribs from the two Steel Cylinders, which are carried by two main Principals and stiff bracing between the Cylinders.

Fig 9.17
The two tiers, projecting
20ft into the auditorium,
were carried round three
sides of the huge space on
cantilever girders.
[© The British Library
Board (F60077–64)]

noteworthy for the limited range of section sizes, the simplicity of the supported balcony floor-grid, and the angular form of the standardised cantilevers, all of which would have reduced fabrication and erection costs and time. The roof rested on light steel trusses, 'the principals showing outside appearing as flying buttresses'.[78]

Moreland & Son was one of the most prolific constructional engineering companies involved in the design and erection of steelwork for metropolitan theatres in the 1890s and 1900s, building close working relationships with the leading specialist architects of the era. Initially, its particular forte in this market was the use of

Fig 9.18
Solid steel columns offered an economical means of preserving sightlines in many London theatres of the 1880s and 1890s, before full cantilevering. The Lyric Theatre, Shaftesbury Avenue (1886–8; C J Phipps, architect), survives as an early example of this approach.
[AA020517]

small-diameter solid steel columns, an economical yet effective method of supporting the tiers while (largely) preserving sight lines, at least compared to more intrusive cast-iron counterparts. Another claimed advantage of these members was their resistance to buckling when heated by fire.[79] Theatres reliant on these in conjunction with cantilevering include the Lyric Theatre, Shaftesbury Avenue (Phipps, 1886–8) (Fig 9.18); the Shaftesbury Theatre, Shaftesbury Avenue (Phipps, 1888, demolished 1941); the Metropole Theatre, Denmark Hill (Crewe and Sprague, 1894, demolished 1937); the Shakespeare Theatre, Battersea (Sprague, 1896, demolished 1957);[80] Her Majesty's Theatre, Haymarket (Phipps, 1896–7); the Metropolitan Theatre, Edgware Road (Matcham, 1897–8, demolished 1963); Broadway Theatre, New Cross (Sprague, 1897, demolished in the 1960s); and the Camden Theatre, Camden Town (Sprague, 1901).[81]

In the late 1890s, Moreland's increasingly moved over to fully cantilevered construction, with steel columns or stanchions, where employed, concealed within the walls of the auditorium. In most cases, the cantilevers were probably simply anchored in the brickwork. Examples of 'unsupported' cantilever construction besides Queen's Hall include the Bedford Theatre, Camden High Street (Crewe, 1896–8, demolished 1969); the Grand Theatre, Fulham (Sprague, 1897, demolished 1958); the Empire Theatre, Holloway Road (Sprague, 1899, demolished 1920s); Wyndham's Theatre, Charing Cross Road (Sprague, 1899); the Camberwell Palace (Woodrow, 1899, demolished c 1955); the Grand Theatre and Opera House, Woolwich (Crewe, 1900, demolished);[82] the Apollo Theatre, Shaftesbury Avenue (Lewen Sharp, 1901); and the Imperial Theatre, Tothill Westminster (F T Verity, 1901, demolished).[83]

The specific techniques of the cantilevered construction in these buildings are not given in surviving records, but in most cases the standard approach will have been cantilever girders projecting from bearing walls or embedded stanchions at the rear of the auditorium, possibly with mid-span support provided by transverse girders. This is clearly seen in a Moreland publicity photograph of 1896 (Fig 9.19), showing the

Fig 9.19
Construction of the Shakespeare Theatre, Battersea, in 1896 (W G R Sprague, architect; demolished 1957), showing Richard Moreland and Son's steel truss for the auditorium and cantilever 'Crank Girders' (foreground, left and right).
[© The British Library Board (F60077–62)]

positioning of the 'Crank Girders' supporting the gallery above the semicircular dress circle of Sprague's Shakespeare Theatre in Battersea.[84] Simply anchored in the masonry, these cantilever girders, and the steel trussed girders carrying the auditorium roof, are indicative of the extent to which steel was entering the construction of smaller, suburban theatres by this date. *The Builder* noted 'the theatre will be constructed entirely of concrete and steel', adding 'As usual, in theatres of recent date, there will be no columns to impede sight.'[85]

The next logical step, and one the origins of which may lie in the technique used at the Royal English Opera House, was to cantilever raking girders from a deep girder spanning the side walls at the rear of the auditorium, thus bringing the whole assembly further forwards for greater projection. Ernest A E Woodrow, in alliance with Moreland & Son, were among the first designers to exploit this idea, in the Grand Palace of Varieties, Clapham (1899–1900) where a deep steel box girder was used to span the 65ft width between the side walls, forming a rigid anchorage for the 'Crank Girders' projecting deep into the auditorium (Fig 9.20). The tier above was similarly engineered, although with less projection and seating, required only a deep built-up I-section.[86] The only completely surviving theatre by this important theatre architect, it still preserves its excellent sightlines from every level, and much of its unusual Chinese-inflected interior decoration.[87] Similar constructional techniques were also being put to use in the West End, including the New Gaiety Theatre, Aldwych (1902–3, demolished 1957) (Fig 9.21). Occupying a prominent corner site between Aldwych and

the Strand, it replaced a predecessor whose demise was, for once, not brought about by fire, but rather by the Strand Improvement Scheme. It was built to the designs of the architects Ernest Runtz and George McLean Ford, and benefited both from £50,000 in compensation from the LCC, and from input by Richard Norman Shaw, who was brought in to provide an exterior worthy of the site. Shaw gave it an open colonnade of Ionic columns at the top of the building, but his crowning feature, set 90ft above the street, was a 40ft-diameter copper dome, framed in steel. Covering the crush room but also extending over part of the auditorium, this dome was carried on giant steel girders at roof level, whose loads were transmitted to the foundations via cast-iron stanchions embedded in the walls. But it was in the auditorium proper

that the spanning potential of steel girders was exploited more remarkably, namely in the circles, which were 'so designed as to carry from wall to wall of the auditorium without intermediate supports' (Fig 9.22).[88] The main structural member in each of the three gently curved tiers was a huge built-up girder, set forwards some 6ft from, and parallel with, the rear wall of the auditorium. Spanning the 60ft width between the side walls, each girder was seemingly built up from three lengths to give a polygonal arc, roughly approximating the curve of the tiers. The girders were set one above the other, so that their cumulative loads were transmitted to the foundations by stands of cast-iron columns and steel stanchions embedded in the side walls. An array of steel cantilevers projected from these main girders,

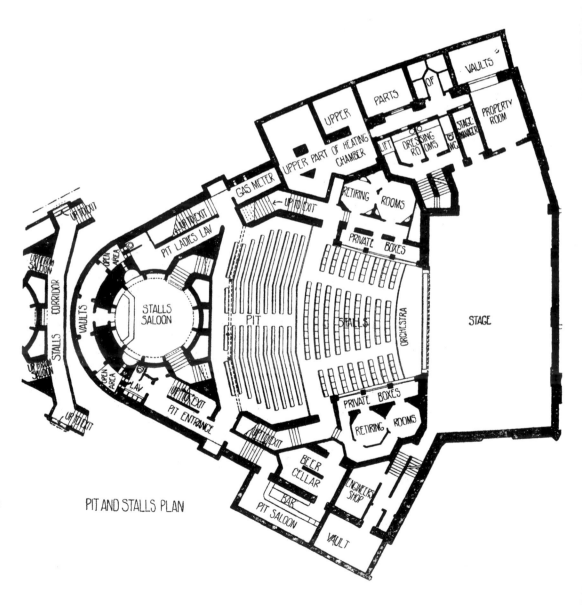

Fig 9.22
The New Gaiety Theatre, Aldwych (1902–3, demolished 1957). The position of the giant angled girder spanning the auditorium is denoted by dashed lines.
[The Building News, 26 September 1902]

given fulcrum support by lighter secondary girders parallel with the main girders. Measuring 2ft deep at the fulcrum, these cantilevers projected some 12ft to 15ft from the secondary transverse girders, giving an overall depth to the circles of almost 22ft. The stepped profile of the circles was formed in concrete on steel bearers spanning between the cantilevers, and the concrete ceilings of the tiers were formed around lathing suspended below the girders

Fig 9.23
Waldorf (Strand) Theatre, Aldwych (1904–5; W G R Sprague, architect). London's last three-tier theatre, entirely free of columns. [AA032969]

Fig 9.24
Waldorf (Strand) Theatre,
Aldwych, showing the 70ft
girder supporting the second
or 'grand-circle' tier during
construction.
[Builders' Journal and
Architectural Review, 5
April 1905]

and cantilevers, creating an unbroken soffit and a clear head-height of not less than 8ft. All the steelwork was executed, and probably designed, by Dennett & Ingle.

The basic technique of improved cantilever construction employed at the Gaiety Theatre was also employed in the modestly sized Waldorf (Strand) Theatre, Aldwych (1904–5), designed as a twin to the nearby Aldwych Theatre, completed in the same year (see Fig 10.41). Executed in near-identical classical dress to designs by W G R Sprague, these theatres formed monumental bookends to Marshall Mackenzie's Waldorf Hotel (1905–7), the ensemble all part of the LCC's Aldwych/Kingsway redevelopment, London's first grand avenue to be conceived in the age of steel-framed construction. Smaller than its sibling, the Waldorf Theatre was nonetheless more ornate internally. It was also the last theatre in London to be built under the existing (1878)

regulations with three tiers,[89] the LCC having introduced legislation in July 1901 prohibiting more than this on safety grounds (Fig 9.23).[90] For the first tier the assembly was stripped down to a giant steel plate girder spanning the side walls, with cantilevers of differing reach forming the curved front of the tiers. By mid-decade this was commonplace, one publication noting 'There is nothing very special about this: it is the usual straight girder with the curved front formed by cantilevers projecting beyond.'[91] The second, 'grand-circle', tier was similar, but more elaborate, requiring a specially fabricated girder that was straight in the horizontal plane, but gently curved in the vertical to create a markedly dipped profile. So that the depth of this member was minimised, it was formed with shallow web-plates, strength coming from a massive layering of flange plates to accommodate the immense stresses in the centre of the c 70ft span. At this point, 16 plates were used,

yet the girder was only 16in. deep, tapering to a depth of just 12in. at its embedded ends (Fig 9.24). Constructional details of the third tier are lacking, but columns were entirely dispensed with throughout and the ceilings of the tiers preserved an unbroken sweep, with neither girders nor raking joists projecting below the ceiling line to interfere with clear views of the stage. This steelwork, and that framing the roof and flies was designed, supplied and erected ahead of deadline by Westminster contractors Messrs Smith, Walker & Co.

Most new West End theatres of the Edwardian boom period were built on small, constricted sites. Spatially resourceful technological innovation in the design and construction of auditoria almost invariably came from the consulting engineer or steelwork contractor, who came to be relied upon to deliver cost-effective solutions to the problems of safety and sighting. In some instances, in the pursuit of economy,

frames were stripped down to basic structural elements that were readily available or easily fabricated. In forming the tiers of Bertie Crewe's 1904 rebuilding of the generously sized Lyceum, off the Strand, rolled-steel joists were simply laid on top of deep, straight steel-plate girders spanning the width of the auditorium (Figs 9.25). The differing heights at which these girders were set gave the necessary rake, and the use of standardised steel joists, which cantilevered beyond the lower girder, avoided the expense and complications of the usual tapering cantilever girders. Nonetheless, in this case the *quid pro quo* for such cut-price steelwork was the extravagant waste of space between the floors and the ceilings of the tiers.

The apogee of Edwardian structural engineering as applied to column-free tier design came from the creative collaboration of the foremost theatre architect of the day, Frank Matcham, and a little-known engineer, Robert

Fig 9.25
Column-free tier construction in the Lyceum, Strand (1904; Bertie Crewe, architect), with rolled-steel joists cantilevering beyond the lower of two steel-plate girders. This view of 1991 shows the void below the lower balcony seating.
[BB91/17306]

Alexander Briggs (1868–1963). Together they produced the largest, most grandiose theatre London had seen: the London Coliseum, St Martin's Lane (1902–4) (Figs 9.26 and 9.27). One of the first of a new breed of family orientated variety theatre 'palaces', this enormous, costly edifice, occupying a three-quarter-acre site in the heart of the West End, was the embodiment of Oswald Stoll's entrepreneurial vision of up-to-the-minute technology in the service of comfort, splendour and spectacle. Touted as the 'Theatre De Luxe of London' it boasted a panoply of mechanical and electrical wizardry, aspects of which had never been seen before in a British or European theatre, including

a giant steel-framed 'triple-revolve' stage, electric lifts that conveyed the audience directly from the street to the upper parts of the building, and an information bureau housing telephones and telegram machines. It had structural sophistication to match. Nowhere was this more apparent than in the vast three-tier auditorium whose every dimension exceeded those of the Theatre Royal, Drury Lane, which had since 1901 vaunted London's largest auditorium, following rebuilding work by Philip Pilditch.[92] At the rear, three awesomely deep, sensuously curved balconies were carried across the full width of the auditorium by a new technique of steel 'cantilevering' that

Fig 9.26
Column-free tier construction on the grandest of scales at the London Coliseum (1902–4; Frank Matcham, architect).
[BB96/07997]

had been patented in 1902 by Matcham in conjunction with his resident engineer, Robert Alexander Briggs (Fig 9.28).[93]

The essential feature of this patent was a continuous bow girder spanning the side walls of the auditorium, supported by radial, inclined subsidiary girders, whose ends were embedded in the rear and side walls. Triangular-shaped rakers, riveted to the web of the bow girder cantilevered the balcony forward from this rigid, spider-like 'gridiron' to form the front of the circles. According to the specification, the chief advantage claimed was the avoidance of 'the irregularities of appearance and the inconveniences incidental to the construction of such

theatre parts with straight-across girders supported by the theatre side walls with angularly disposed girders connecting the same with more forward parts of the side walls, as customarily used, and to simplify improve and strengthen the construction of such theatre parts'.[94] The one publication to carry detailed description of the works in progress, the *Builders' Journal and Architectural Record*, stressed this 'kind of "gridiron" is not a cantilever construction, the whole stability depending upon the gusset plates (similar to those seen on the top side) riveted under the ends of the radial girders where they join the main girders'. Indeed, strictly, the only cantilevers were those

Fig 9.27
The exceptionally deep dress circle of the London Coliseum with the decorated underside of the upper-circle balcony seemingly floating above
[AA025408]

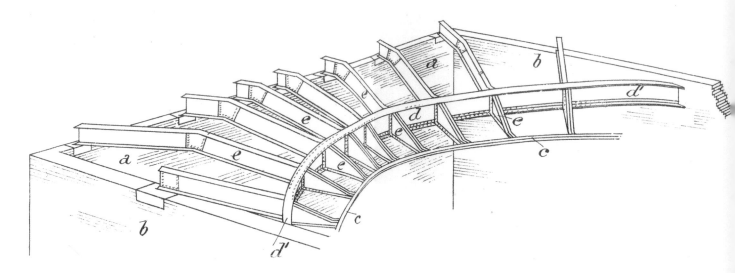

Fig 9.28
Unsupported steel-framed gallery floors in Robert Alexander Briggs and Frank Matcham's drawing for patent No. 27,146 of 1902. [© The British Library Board (04A01247P)]

members projecting outwards from the curved main girder, two of which were shown in a construction photograph (Fig 9.29). The technique, which was also intended to dispense entirely with columns, was designed and patented to debut in the Coliseum,[95] yet in its practical application there, columns were used sparingly. The main bearing for each of the three

130ft-long curved girders was taken by two lines of slender solid steel columns set 21ft in from either side wall, the cumulative loads transmitted to heavily reinforced concrete-steel grillages, 6 sq ft and over 9ft thick. The engineering contractors for this portion of the building, Richard Moreland & Son, may have introduced their patent columns as extra provision in the

Fig 9.29
The steel skeleton of the Coliseum's Grand Tier takes shape in late 1904, with employees of Richard Moreland and Son posing for the camera. The two rakers projecting from the giant curved bow girder (left) are the only cantilevered elements so far erected. [Builders' Journal and Architectural Review, 26 October 1904]

knowledge that Briggs and Matcham's principle was inadequate for a span of such unprecedented dimensions. Yet with column diameters of just 10in. on the ground floor, 9in. on the tier above, 8in. above that and 7in. above that, visual intrusion was almost nonexistent, even when the columns were encased in protective alabaster. Moreland, probably the leading firm of structural engineering contractors specialising in theatres by this date, expedited their allotted portion of the work using techniques reminiscent of American constructional efficiency. While traditional derrick poles were used to manoeuvre the heavy steelwork, the relatively novel use of pneumatic riveters saw foot-long rivets driven in at the rate of one every three seconds.

At the time of its completion, Matcham's Coliseum had almost certainly made the most extensive use of structural steel in any London theatre. Even the uppermost stage of Matcham's *pièce de résistance*, the distinctive tower rising above the main entrance and supporting the famed revolving globe, was formed around a skeleton of steel.[96] The project was of such proportions that two other steelwork contrac-

tors were brought in besides Moreland. One of these, Drew-Bear, Perks & Co. Ltd, of the Battersea Steel Works, framed the enormous, irregularly shaped space above the stage using a comprehensive grid of deep lattice girders, roof trusses and stanchions weighing some 160 tons in all (Fig 9.30). There was little new in how this was arranged and put together; it was the unprecedented size of this portion of the building that awed contemporaries:

The impressiveness of the place may be realized if we imagine entering a room 100ft. square and 70ft. high, for there is nothing between the stage and the grid, which is 71ft. – and the roof 81ft. – above the floor. The stage is free from any obstruction with the exception of two stanchions at the back which pick up the flies, grid and roof over.[97]

The stanchions, measuring a huge 18in. by 22in., 81ft tall and set 61ft apart, were mirrored by two others, built into the main proscenium wall. Besides carrying a portion of the double-skin steel and concrete auditorium roof, these

Fig 9.30
There was more steel framing the stage than any earlier London theatre. The Coliseum's roof trusses being hoisted into position, May 1904.
[Builders' Journal and Architectural Review, *26 October 1904*]

Fig 9.31
Trial erection of steel
balcony structure to
Matcham and Briggs's 1902
patent specification at
Moreland's Silvertown
works, c 1909. Subsequently
dismantled and re-erected
for Matcham's London
Palladium, 1909–10.
[Courtesy of Institution
of Civil Engineers]

embedded stanchions supported a steel bow-string girder that discreetly carried the 54ft-wide proscenium arch, the largest in the UK. Rising from the centre of the auditorium roof was a 22ft-diameter steel-framed dome, designed on the concept of the 'hit and miss' ventilator, so that one half could be opened direct to the outside air, or closed up at will. This revolving dome, similar in principle to that of an observatory, and ingeniously framed and geared to be operated manually by one man, was designed and built by William Whitford & Co., a long-established firm of structural engineers based in Millwall. In keeping with the building's Roman grandeur, the interior of the dome was draped in material to resemble a velarium.

Fig 9.32
Curved corridor at the
rear of the Palladium's
auditorium.
[AA020483]

By the time the London Coliseum was completed in 1904, Matcham had been responsible for the building or rebuilding of at least 65 British theatres.[98] Many of these, possibly the majority, will have used steel-cantilevered balconies, which he seems to have first exploited in 1892, in his Edinburgh Empire Palace.[99] Matcham, in common with his contemporaries, borrowed freely from the well-publicised, well-detailed techniques used at the Palace Theatre. Despite not being cantilever construction, his own system, co-authored with Briggs, represented the refinement of a number of strands in the structural evolution of column-free balconies. Indeed, its descent from the cantilever form

is manifest. Yet this innovation indisputably enabled construction writ large in the Coliseum, the giant three-tier auditorium providing exceptional levels of comfort, sighting and safety to a seated audience of over 3,000. Only in a handful of other Edwardian theatres, including Matcham's Palladium (1909–10), which used the same system at the end of the decade (Figs 9.31 and 9.32), were the structural limits and aesthetic possibilities of steel so eloquently or acutely observed (Fig 9.33). Here, two luxuriously deep balconies span the 100ft width of the auditorium, providing one of the most potent announcements of the influence of early steel technology over built form.[100]

Fig 9.33
The London Palladium's auditorium from the stage.
[AA020472]

10

Clubs and hotels:
opulence, proportion and planning

Clubs are no longer associations of good fellows, but associations of men united according to caste, political feeling, and governed by certain rules and regulations. The first and most important object of all, as previously observed, is to unite the economy of a chop-house with the superior accommodation of a first-rate hotel. (1844)[1]

Iron forerunners

Mid-19th-century London saw the appearance of two new types of building that set the pace in the structural application of iron within polite architecture: clubs and hotels. First to emerge were the stately clubhouses, clustered along central London's most exclusive and fashionable

Fig 10.1
Reform Club, Pall Mall (1838–41; Sir Charles Barry, architect).
[DP134000]

Figs 10.2
Reform Club, Pall Mall:
the Coffee Room.
[DP134760]

streets and providing temporary accommodation for an increasingly select clientele united by political or recreational leanings. Unlike the 18th-century coffee-houses from which this type of building emerged, the new clubs were designedly grand and varied in their liberal provision of space, typically housing one or more specialised functions, including a library, lounge, smoking room, billiard room, dining room, kitchen and, eventually, bedrooms. These basic functions almost invariably called upon structural iron to realise suitably grand spaces and complex planning. Architects of the highest rank championed the use of cast-iron floor beams in these buildings, John Nash and Charles Barry being among the first innovators in their respective United Service Club (1826–8)[2] and Travellers' Club, Pall Mall (1828–32).[3] Barry's later Reform Club, Pall Mall (1837–41), has been considered the building that set in train the whole 'club mania' (Fig 10.1).[4] It too made extensive use of long-span cast-iron beams, in this case to carry brick jack-arched floors that were non-combustible and solid, with good sound insulation. 'In this way he gave the members of the Club generously proportioned rooms of great elegance in a building,

which, to the outside observer, might well be of the most traditional timber and masonry.'[5]

In rising to the challenge of producing 'a clubhouse finer than any yet built',[6] one 'which would surpass all others in size and magnificence',[7] Barry's work provides foretastes of key future developments, both architectural and structural. One of these innovations was the provision of an exceptionally large room occupying the same relative position on two or more of the building's lower floors, subdivided by pairs of columns set close to the walls. On the ground floor of the Reform, above the basement and mezzanine floors, this is the Coffee Room. A great dining hall 117ft long and 26ft wide, extending almost the full width of the building, it is divided into three compartments by timber columns (Fig 10.2). Above this, and reproducing the same size and proportions, is the library (Fig 10.3), originally a drawing room but converted under Barry's superintendence in 1853, and above that were some of the members' and staff bedrooms that occupied the two topmost floors.[8] The fluted timber columns do more than bring spatial order to these grand rooms; they also shoulder a proportion of the floor weight, thus relieving the walls of some of

their loading. Although there is no evidence to suggest that the columns have cast-iron cores, it would be surprising if those employed in similar situations in other mid-century clubs did not, given the technique had 18th-century origins, seemingly beginning in London churches (*see* p 259). Certainly in many later clubs, and hotels, where costs were more constrained, the same facility was achieved more expeditiously using cast-iron columns sheathed in scagliola,[9] a pretence that seemingly went largely unnoticed, or at least unmentioned, by most contemporaries.[10] An early example of this may have been the extension to Sir Robert Smirke's Carlton Club (1833–6), next door to the Reform, built to rich Venetian Renaissance designs by his son Sydney in 1846–8 (demolished). Here the ground-floor Coffee Room, 93ft long, 37ft wide and 21ft high and extending the full depth of the addition, was compartmentalised into three spaces by green scagliola columns with gilded white capitals and black and gold pedestals.[11] While not disclosed in contemporary

descriptions, the presence of iron rather than timber behind the scagliola veneer can be safely deduced, given Smirke's enthusiasm for the material, and the fact that the builder was a Mr Grissell,[12] probably Henry Grissell of the Regent's Canal Ironworks.

Another architectural foretaste in Barry's Reform Club, though introduced at the behest of the building committee and not the architect, was the presence of bedrooms for the members.[13] This novel facility, together with the provision of grand principal rooms and kitchens, blurred the distinction between clubs and hotels, another new building type in which cast iron was quickly exploited. Like clubhouses, hotels had unassuming, domestic-scale origins, but in the mid-19th century their design and construction took a very different course as size, convenience and professed invulnerability to fire became the standards by which a new breed of lavish edifices vied for attention. Investment in constructional form and architectural opulence soon put some hotels on the same social standing as the best clubs. By the end of the century, hotels had outpaced most clubs in grandeur and technological sophistication. In London this trend began in the 1850s with the railway terminus hotels (mostly financed by the railways or their subsidiary companies), which provided scores of inexpensive rooms on their upper floors, but more extravagant public spaces below. Lewis Cubitt's Great Northern Hotel, Kings Cross, the third of London's main-line terminus hotels, was built in 1854 on a curved plan to follow the former line of Pancras Road. It took its functional cue from another emergent building type – the working-class Philanthropic housing block – with an accent on pragmatic non-combustibility rather than grandeur (Fig 10.4). Five storeys in height and housing some 70 bedrooms, it was constructed along industrialised fire-resisting lines, probably offering a degree of protection in excess of that demanded by the Metropolitan Building Act 1844.[14] Indeed, this was a major selling point, the original brochure confidently stating 'N.B. The hotel is fire-proof.'[15] Soon after its completion, the architect Charles Forster Hayward 'presumed that almost the only building in London constructed in fireproof fashion was the Great Northern Hotel, which was all fireproof. In that building were used ordinary iron girders, with brick arches turned one to the other'.[16] At this date the 'ordinary' iron girders were probably of cast iron. It would also be surprising if the Great Northern's grander, more distinguished forerunner, the Great Western Hotel (1851–4), had exploited the superior spanning possibilities and greater safety of wrought iron to the exclusion of cast iron. Yet the Great Western, built to designs by Philip Charles Hardwick and housing some 150 rooms, many of unrivalled size and splendour, was the metropolis's first 'Grand Hotel', prefiguring the rash of massive hotels which were built in earnest towards the end of the decade, and which were able to capitalise on the increasing availability of the new metal.

The Westminster Palace Hotel (demolished), erected in 1857–61, to designs by the brothers William and Andrew Moseley, was the first of the new 'Grand Hotels' to make extensive use of wrought-iron joists and girders (Fig 10.5). As the designers of a number of 'fireproof' hospitals and asylums built in the 1830s and 1840s, including Springfield Asylum, Tooting (1839–41), the Moseley brothers already had much experience in using cast iron, but here they were able to put the superior metal to use in a far grander building. Conveniently close to Parliament and Whitehall on the corner of Victoria Street and Tothill Street, the Westminster Palace Hotel was also the first instance of a grand London hotel that was neither owned by a railway company nor sited close to a terminus.[17] Rising six storeys in height above a basement that extended under Victoria Street, this French Renaissance pile was far larger than the Great Western, incorporating four internal

Fig 10.4
The Great Northern Hotel, Kings Cross (1854; Lewis Cubitt, architect).
[CC97/00869]

courts and housing some 450 bedrooms and sitting rooms, and a number of generously proportioned public rooms. One of these, the ground-floor Coffee Room, rivalled the equivalent room in the Carlton Club for size. As in Smirke's room this ostentatious space was subdivided by scagliola columns, in imitation of Siena marble, which supported much of the floor above on their concealed iron cores. The floors were built to a non-combustible specification that deadened noise and that had less depth than conventional brick jack-arch construction. The principal elements of this floor system comprised 7in. deep rolled I-section iron joists, spaced 2ft apart, spanning between wrought-iron box-girders and cross walls in lengths up to 17½ft. Small, closely spaced laths resting on the lower flanges supported a 5in. thickness of concrete and a ¾in. cement sand screed, the whole finished by a ¾in. layer of pure Portland cement. A suspended lath-and-plaster false ceiling, carried by binders and joists, concealed the floor structure and helped reduce sound transmission.

Fig 10.5
The Westminster Palace Hotel, Victoria/Tothill Street (1857–61; William and Andrew Moseley, architects), one of London's first 'Grand Hotels'.
[CC97/00341]

Although not of 'an entirely new principle of construction' (as proclaimed by *The Building News*),[18] this floor system was of sufficient strength and spanning capability to permit the use of many essentially non-structural partition walls throughout the building, constructed of perforated blocks rendered with lime and hair.

A technological tour de force, the Westminster Palace Hotel incorporated other innovations, including a system of forced ventilation and a hydraulic lift, its machinery supplied by Sir William Armstrong. Not all of these advances were as universally adopted as structural wrought iron, the use of which became virtually standard in all subsequent grand hotels, until ousted by steel at the close of the century. Wrought-iron members progressively supplanted timber joists and beams to provide strong, rigid floors, many of which were introduced because of the ever-present concern about fire. Indeed, a growing number of proprietary 'fireproof' floor systems reliant on rolled joists and built-up beams proliferated in hotels. But it was the potential of large built-up iron girders to span great distances that tied the use of the metal more intimately into the early stages of the building design, as architects realised the exciting spatial and planning possibilities that the use of iron presented. Lower-floor plans could be liberated from the spatial constraints of the upper floors (or vice versa), chimneys could be supported at any level, and principal staircases could be made grander.

Henry Currey's seven-storey London Bridge Railway Terminus Hotel, St Thomas Street, (1859–62, demolished), made liberal use of wrought-iron beams for reasons of spatial flexibility, not fire resistance, the architect stating 'It has not been attempted to make the building fireproof.'[19] Built-up girders were used systematically to span the generous 28ft width of the public rooms on the lower floors, bearing the weight of division walls and six-storey-high chimneybreasts alike, each girder carefully designed to sustain up to 50 tons.[20] To lessen the weight on these girders, wrought-iron girders were placed immediately underneath the division walls above: such construction, Curry believed, meant that 'in the event of any accident arising from fire … every floor should be independent and self-supporting'.[21] At the Langham Hotel, Portland Place, built in 1863–5 on a site considered to be the healthiest spot in London, the architects John Giles and James

Murray restricted 'fireproof' stone floors to the ground and basement levels, and the halls, corridors and stairways.[22] Elsewhere in the building, the same box-section built-up girders that supported the stone flags were used to carry traditional timber-framed floors (Fig 10.6). For the most part, the beams were boxed in, or concealed within the depth of the floor, but in some instances the architects chose to leave the riveted soffits exposed, as in the Ladies Reading Room or Library (Fig 10.7). Rising to a height of seven storeys above two storeys of basements (and 156ft at the top of its domed water tower), the hotel boasted some 300 bedrooms and over 30 private suites, each with a sitting room, bedroom, bathroom and dressing room. Structural iron was used most assertively in the

Fig 10.6 (top)
Langham Hotel, Portland Place, 1863–5. Renovation in 1987 exposed the wrought-iron girders used to support traditional timber floors.
[MF110/J/34]

Fig 10.7 (above)
Langham Hotel, Portland Place, 1886–5. In the Ladies' Reading Room the riveted girders were left exposed and given ornamental webs.
[BB98/08129]

composition of the impressive public spaces on the lower floors. Girders spanned the commodious Library and Drawing Room, and carried the Coffee Room, with its bow window overlooking Langham Place, over what was London's largest kitchen, a magnificent two-storey space 54ft long and 48ft wide. Most impressive of all was the salle à manger, a vast rectangular room, 120ft by 47ft, vaunting a double arcade of columns 'of composite to imitate Siena marble' that helped support the floors above.[23]

The notable exposure of the riveted ironwork in the Langham, albeit in the less important public interiors, was partly a concession to contemporary cries for the honest expression of structure, and partly a response to economic stringency. Certainly the boldly playful exterior was commended by one periodical for its refreshing absence of 'constructive sham'.[24] Alfred Waterhouse, an architect 'never afraid to use iron',[25] trod an even bolder path in his New University Club, St James's Street (1865–9, demolished c 1939), exposing much of the cutting-edge iron technology within (Fig 10.8). Throughout this 'vast-windowed, rather ecclesiastical, building',[26] comprising three floors, a mezzanine and a basement, he made extensive use of the recently patented Phillips's girders, choosing to allow many of these stronger, more efficient compound members to remain internally visible.[27] This marked one of their earliest appearances in a building of any architectural pretension. Such was their novelty that when they arrived on site, Waterhouse, who was alarmed at the wide spacing of the rivets, reverted to the old (cast-iron) practice of having each girder proof-loaded for soundness.[28] Evidently his trust was restored, as in later years he deferred entirely to the girder manufacturers or suppliers to inform him of the required depth, eschewing his former practice of doing his own calculations.[29]

Another of the era's greatest exponents of the Gothic Revival mixed traditional and modern building technology to bold and eclectic effect in a hotel then 'without rival ... for palatial beauty, comfort and convenience'.[30] George Gilbert Scott's Midland Grand Hotel, St Pancras (1868–74), like his near-contemporary Foreign Office, Whitehall (1868–73), flaunted many of its iron girders to eye-catching effect. These were of the more usual built-up I-section form, used discreetly throughout the building in support of the non-combustible floor structure spanning the brick walls, but openly celebrated in many of the principal public spaces (Figs 10.9 and 10.10). The floor structure was, for a hotel, unusual, composed of shallow corrugated-iron sheeting sprung between the widely spaced girders (up to 13ft),[31] a gentle curvature maintained by secondary wrought-iron lattice beams with curved top sections (Fig 10.11). Levelled with concrete, the corrugated arches carried standard timber-joisted floors. Below, suspended plaster ceilings hid the underside of the sheeting in all but the basement, where the entire structure was left exposed. The system as used by Scott conformed to a patent obtained by Richard Moreland & Son in 1866,[32] although corrugated-iron and concrete vaulted floor systems had already been introduced in the late 1840s.[33] Scott also gave the hotel one of the most extraordinarily grand staircases of any hotel, the stone treads and landings supported by an armature of wrought-iron tempered, but not disguised, by dun-coloured fibrous plasterwork in accordance with his remarks of 1857. Such 'honest' expression of riveted ironwork was extended to several other

rooms also but its ceremonious use in the highly public staircase was audacious, and possibly unprecedented.[34]

The 1870s and 1880s saw the continuation of the constructional trends established by this first generation of 'Grand Hotels' and clubs, marked by greater confidence and ambition in

Fig 10.8 (facing page) New University Club, St James's Street, 1865–8, demolished. Designed by Alfred Waterhouse, an advocate of iron construction, this was one of the first buildings to employ compound girders. [The Builder, 16 May 1868]

Fig 10.9 (above) Midland Grand Hotel, St Pancras (1868–74; George Gilbert Scott, architect). Terracotta dragons embellish the side of an exposed girder in the second-floor corridor. [Jonathan Clarke]

Fig 10.10 Midland Grand Hotel, St Pancras. Exposed riveted girders in the Ladies' Coffee Room on the ground floor. [OP10901]

Fig 10.11
Midland Grand Hotel,
St Pancras (1868–74):
fire-resisting floor structure
in the basement.
[Jonathan Clarke]

the application of iron framing. In some buildings 'fireproof' floors, and hence structural iron, were restricted to the principal storeys, where the threat of fire was deemed greatest, or indeed simply to the staircases and access passages in nominal compliance with the Metropolitan Building Act. At Euston Station

Hotel, Euston Road (1880–1, demolished 1963), built as a central link block connecting Philip Hardwick's Adelaide (later Euston) and Victoria Hotels (both 1839, demolished 1963), the floors were largely of timber, although the corridors used 'fireproof arching and girders'.[35] Most, however, were more comprehensively 'fireproof'. The independent annexe to the Grand Hotel, Charing Cross, between Buckingham and Villiers Street (John Fish, 1880–81), used Dawnay's filler-joist floor system, composed of rolled-iron joists embedded in concrete. Even the flat roof, intended as a 'promenade for smokers in summer time' was constructed along these lines.[36] Beyond fire-resisting floors, structural iron was often put to more ingenious and sophisticated uses. One such instance was the remodelling of the Hanover Square Club in 1875 (demolished). Giant truss girders, 11ft deep, were used to span the 70ft by 36ft ceiling of the old concert hall, creating the support for a new storey that housed accommodation for 'country and foreign members', mostly ambassadors. These trusses, fabricated and erected by Richard Moreland & Son, were filled in with brickwork, thus forming the partitions of the

Fig 10.12
The iron-framed salle à
manger of the Grand Hotel,
Northumberland Avenue
(1878–81; Francis and
J E Saunders, architects),
photographed by Bedford
Lemere in 1912.
[BL25361]

first tier of bedrooms.[37] In the seven-storey Grand Hotel, Northumberland Avenue (1878–81, Francis and J E Saunders, architects, rebuilt in near-facsimile 1986–90), the desirability of combining a range of shops with large, luxurious rooms on the ground floor and a maximum number of bedrooms on the upper floors, necessitated the substitution of some interior bearing walls with iron framing. The principal space on the ground floor was a grand salle à manger, 95ft long and 50ft wide, divided as if a basilica by 18 Corinthian columns (Fig 10.12). Iron stanchions concealed within their marble casing carried girders that supported both the back walls and the division walls of the floors above. A contemporary remarked:

By this means the first and superincumbent floors are carried over what may be called the aisles of the grand salle, and more space is got for bed-rooms. What may be called the nave or central area of the salle will be lighted for its whole length by a semicircular iron roof of rich design. Above this central portion of the salle will be a large open area, giving light not only to the salle, but to rooms on all the floors above.[38]

Iron framing carried through to the façade of the building, and to a degree affected its character (Fig 10.13). Cast-iron stanchions, possibly of H- or U-shaped sectional form, were embedded in the masonry piers, permitting a decrease

Fig 10.13
The Grand Hotel, Northumberland Avenue, rebuilt in near-facsimile (1986–90). Concealed iron stanchions helped carry the original façade, thus permitting an increase of the window area.
[Jonathan Clarke]

in their dimensions and an increase in window area.[39] This was most manifest on the ground floor, where glazed shop fronts occupied the entire width between the piers, which in effect had become shallow pilasters. Because the load-bearing role of the exterior brick walls was lessened, they could be built thinner than was customary, ranging from 2ft 8in. thick to 1ft 11in, exclusive of the stone facing, but still within the limits specified by the Metropolitan Board of Works. Plausibly, concealed spandrel girders connecting these perimeter stanchions may have been employed to give lateral stability to the façade, although there is no mention of this in available accounts. Contemporary descriptions do mention the ingenious way in which the oriel windows, with their heavy stone bases (weighing approximately 12–15 tons apiece) were carried by the ground-floor girders, which cantilevered out from the interior – a technique borrowed soon after in Matthew Wyatt's modifications to the Oxford and Cambridge Club.[40] The firm Burgoyne & Co. supplied all the ironwork, apart from that used in the fireproof floors, which was supplied by Dennett.[41]

The introduction of steel

The Grand Hotel, Charing Cross, with its interior framing brought through to the façade, marked the most extensive use of structural iron of any London hotel up to its completion in 1879. The architects F & H Francis repeated its constructional and planning formula in the still larger First Avenue Hotel, High Holborn (1882–3, demolished), which also placed spatial emphasis on a grand salle à manger, divided by two rows of load-bearing Corinthian columns. Both hotels were constructed under the technologically spirited George H Holloway, the clerk of works, who had previously worked on the new 'fireproof' East and West India Dock House, Billiter Street.[42] In each hotel, Holloway had convinced the architects to employ a form of iron and concrete staircase construction of his devising,[43] one that was also used in the architects' third and final grand hotel, the Hotel Métropole (1883–5, surviving as offices) (Fig 10.14).[44] The largest of the trio, this may have been the first British hotel to make use of structural steel, and this possibly at the behest of Holloway, who would go on to supervise the construction of a number of other buildings with early steel in them, including the Savoy Hotel and the Royal English Opera House. It was erected in two

Fig 10.14
The Hotel Métropole of
1883–5 (right) and the
National Liberal Club
(1884–7; Alfred
Waterhouse, architect), on
Northumberland Avenue
and Whitehall Place. The
former may have been the
first British hotel to employ
steel girders. The latter
concealed perhaps the most
extensive arrangement
of structural steel then
deployed in 'polite' London.
[Jonathan Clarke]

distinct, but overlapping phases: the main building, with long frontages to Northumberland Avenue and Whitehall Place was erected in 1883–4; an annexe, erected on land to the west following protracted negotiations with the Crown Estate, was built in 1884–5. When The Hotel Métropole opened in June 1885 it was reputedly unmatched for size in Europe. Looming to a height of 150ft at the top of four prominent mansard towers, this eight-storey monster packed in well over 500 bedrooms and sitting rooms on the upper storeys, and an unrivalled number of grand reception rooms on the ground floor, each with its own decorative theme. A similar constructional vocabulary as in its predecessors enabled ambitious and skilful planning, albeit on an even grander scale. Reinforced-concrete foundation walls, made with Portland cement and 'about forty miles of stout iron hooping, 2in. wide,

embedded therein' bore the great weight of the brick superstructure.[45] Some 1,100 tons of constructional iron and steelwork went into this, supplied and probably erected by Andrew Handyside & Co. of Derby. Much of this was in the form of rolled-iron joists, used in conjunction with concrete to form two fire-resisting grand staircases and some 1¼ miles of corridors, and to carry concrete jack-arch floors on the Dennett system. The trademark salle à manger, 105ft by 45ft, extended all the way from the vestibule and entrance in Northumberland Avenue to the Scotland Yard corner of the building, its two rows of Corinthian columns supporting huge girders that carried the division walls and corridors of the floors above. Iron girders were also liberally employed to span the other imposing rooms, including the library, drawing room and a secondary salle à manger. According to contemporary accounts,

Fig 10.15
The Hotel Métropole's large banqueting hall, or 'Whitehall Salon' (1886), photograph by Henry Bedford Lemere.
[BL06349]

HOTEL METROPOLE EXTENSION

SECTION ON LINE A.B

steel girders were used to span the 42ft width of the 'Whitehall Salon', a large banqueting hall distinguished by a semi-circular bay at the Whitehall Place end, and an elliptical alcove for a dais at the other, interior, end (Fig 10.15).[46] The girders derived some support from ranks of decorative cast-iron columns set close to the wall, an arrangement repeated in the basement and elsewhere in the building. It is not entirely clear why steel was used here, since other rooms were of equal or greater width. It may have been simply a matter of timing and cost: because this formed part of the extension to the original building, it seems that by 1884–5 Handysides were able to supply steel girders at a price acceptable to Gordon, the proprietor. The Francis Brothers' drawings make no reference to steel, but they do show that the girders were of

box-section form (Fig 10.16). Besides marking the first use of the new stronger material, with its superior span-to-depth ratio, this hotel was among the first in Britain to site its kitchens not in the basement, as was usual, but on the fifth floor.

By the time the Métropole, with its restricted use of steel, was nearing completion, construction of a major new London club that would win back some of the prestige lost to hotels was already under way. For the nearby National Liberal Club, Whitehall Place (1884–7), Waterhouse was commissioned to 'combine the splendour of the major apartments with the largest number of members' bedrooms of any club in London' (see Fig 10.14).[47] Members of Gladstone's Liberal Party had in 1882 secured temporary residence in part of the Grand Hotel, but at a meeting in November of that year it was

Fig 10.16
Section of the Hotel Métropole Extension by F & H Francis showing the structural methods used, but not specifying the metal of the box-section girders that span the ground-floor Whitehall Salon.
[The National Archives, UK LRRO1/2402]

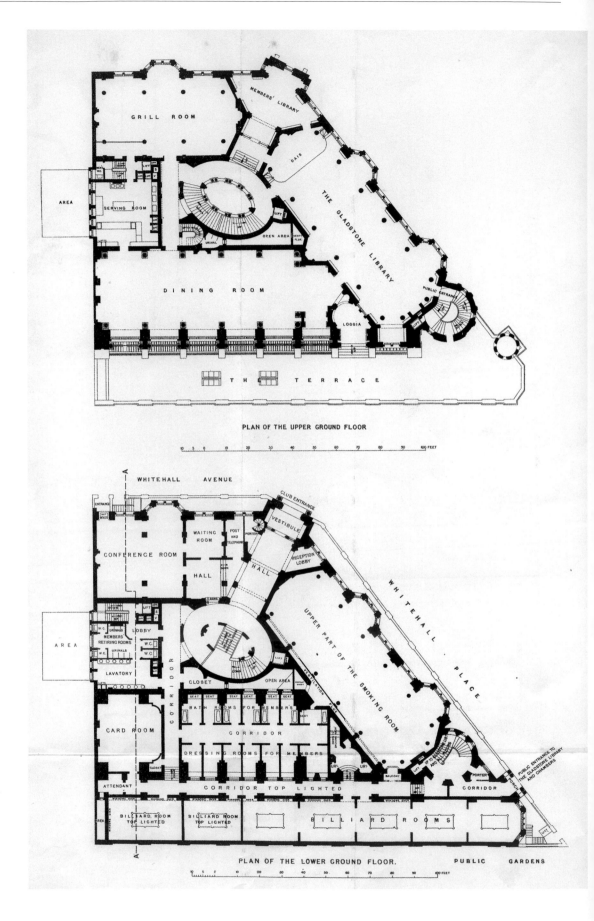

PLAN OF THE UPPER GROUND FLOOR

Fig 10.17
National Liberal Club,
Whitehall Place (1884–7),
plans showing, inter alia,
the arrangement of
detached steel stanchions
in the Smoking Room and
Gladstone Library.
[British Architect,
13 March 1885]

PLAN OF THE LOWER GROUND FLOOR.

decided that more imposing, permanent headquarters were required for the members, who numbered more than 2,500.[48] Waterhouse, by now an architect of national repute, turned in a typically virtuoso performance for the prime but awkward corner site, with a series of grand rooms stacked one above the other, all clothed in cleverly massed Renaissance elevations of Portland stone. Hiding behind this dress was the most ambitious and extensive arrangement of structural steel the metropolis had yet seen in a prestige commercial building. The 100ft-tall structure had load-bearing external walls, but a significant proportion of the floor weights were transmitted directly to the splayed concrete footings by patent steel stanchions.[49] Waterhouse stacked the majestic large spaces facing Whitehall Place above each other: an upper ground floor reading and writing room (The Gladstone Library) was set above a double-height smoking room (Fig 10.17), which in turn was superimposed on a vast wine cellar. Each of these principal rooms, measuring some 97ft by 35ft, was reliant on detached steel stanchions, standing well within the brick envelope to

'diminish the bearings of the girders which take the weight of the floor above, besides lightening the load on the walls'.[50] Spared much of their load-bearing purpose, the external walls could be punctuated by generous plate-glass windows, including huge, full-height bay windows to throw daylight deep into the sumptuous faience-clad interiors (Figs 10.18 and 10.19).

The built-up stanchions, designed and fitted by steelwork contractors William Lindsay & Co., were filled and encased in coke breeze concrete, which afforded both protection from fire and a key for rich Burmantofts faience. They were similarly employed to lavish effect in the Grill Room and in the 110ft-long Dining Room on the upper ground floor. Although the original drawings for the iron and steelwork have been lost, the general specification discloses the variety of stanchion types used, reflecting the nature of loadings they were called on to bear. For example, in support of the longitudinal wrought-iron floor girders of the Gladstone Library, there were to be Lindsay's No. 12 steel stanchions, reinforced further with No. 2 steel cores (presumably solid steel columns), fitted

Fig 10.18
The (former) Smoking Room of the National Liberal Club.
[DP134625]

with 4ft-square, 15in.-high bases; under the Dining Room and Grill Room, 'No. 6 stanchions' – a simple rolled I-section – also with 4ft-square, 15in.-high bases, were to be used (*see* Fig 2.23).[51] The use of built-up stanchions of this form in support of the library and other large-span floors was costly in terms of the additional labour required to rivet together the individual sections. Nonetheless, they were less prone to buckling, being designed to increase the moments of inertia around both axes as much as possible for the given cross-sectional area.

Given the expense of rolled-section column assembly and the cheapness of cast-iron columns (which could have done the job), the employment of these stanchions at this date is all the more remarkable. The all but universally preferred option for compressive members at this time was the tried-and-tested cast-iron column. In this respect Waterhouse, who was always willing to trial 'an invention or a process [that] seemed to him worthy of support'[52] showed typical enterprise in his approach to building structure, perhaps having been won over by the expertise of structural engineer Thomas C Cunnington of William Lindsay & Co., who may have been employed by the firm.[53] Probably for reasons of compressive strength and resistance to rusting, cast-iron columns were used to support the floors of the members' dressing and bathrooms on the lower ground floor.

Throughout this showpiece building, even on the upper storeys, an assortment of structural members was used, including rolled joists up to 19ft long, box girders, trussed girders and girders of 'T'- and 'horseshoe'-sectional form. While it is not always clear whether these were made from iron or steel, the floor structure in the principal rooms did consist of Lindsay's steel decking carried largely on steel girders (*see* Fig 2.24), 'owing to its great strength and the saving of space obtained in being able to lay spans up to 30ft. and 35ft., with a floor thickness (including concrete) of only 15in'.[54] This use of steel decking may have prompted Waterhouse to use interior steel stanchions in support in the first place. *The Builder* noted that 'while the floors may be said to be like so many diaphragms connecting and tying the building together, the weights are transmitted down the steel columns to the basement, thus relieving the walls of all *undue strain or vibration*'.[55] Some years later, (Sir) Banister Fletcher noted, in response to a paper by Cunnington, that in 'regard to the trough flooring, he had heard it

was not a success at the National Liberal Club'.[56] Whatever the flooring's structural aptitude in this building, the ceilings were hung from the underside of the decking, and the spaces created by the troughs functioned as part of the ventilation system. The ceilings were perforated, and air inlets in the troughs extracted the vitiated air, which was then cleansed by high-pressure water sprays before being re-circulated. Only the floor over the (sub-basement) wine cellar, where concrete jack-arching was employed, departed from this novel form of flooring.[57] In the 180ft tower, which served as a secondary, but 'equally exciting' staircase,[58] Dennett and Ingle's latest system of fireproof flooring was employed. The building was also extremely advanced in its heating, (electric) lighting, sewage-disposal systems and provision of hydraulic lifts.[59]

In his application of the latest practice in building and service technologies and his attention to the grandeur of the public rooms and staircase,[60] Waterhouse created the largest, most luxurious club London had seen. With over 100 bedrooms and many bathrooms and water-closets for members on the upper floors, and a kitchen occupying the whole of the Whitehall frontage of the fourth floor, it challenged the grand hotels for amenity. Not until the construction of the steel-framed Royal Automobile Club in 1908–11 was such monumental scale and splendour repeated. The use of iron and (increasingly) steel for non-combustible, spatially maximising ends in this field of building was firmly set.

Two new clubhouses closely followed the National Liberal, and like that building benefited from engineering expertise. For the Junior Constitutional Club, Nos 101–104 Piccadilly (1889–92), (Sir) Robert William Edis employed Homan & Rodgers for the structural ironwork, incorporating one of their patented 'fireproof' floors (*see* Fig 7.1).[61] Credited as one of the first buildings in London to be entirely faced in marble, this Renaissance-style work housed some 60 bedrooms in the upper three of its six storeys (including attics) (Fig 10.20).[62] It was executed at half the cost of the National Liberal,[63] but still made extensive demands on structural iron or steel in its large pubic spaces. Like Waterhouse, Edis was able to place a spacious smoking room on the ground floor – 'an accommodation [observed *The Builder*] which in a club is generally relegated to the upper floors',[64] because here, columns positioned near to the inside face of the walls could help

Fig 10.19 (facing page) The Gladstone Library of the National Liberal Club (with Whitehall Place on the left hand side of the photograph). [DP134613]

shoulder the loads borne by girders spanning 26ft (Fig 10.21). It is possible that Edis used some steel in this building, given that Homan & Rodgers are known to have begun fabricating structural members from it in the late 1880s,[65] and that, circumstantially, steel was being used in an increasing number of prominent 'public works', a broad term that, in this context at least, included clubs:

Marble as an internal wall decoration is being more largely used every year; one has only to visit, among the latest buildings, the National Liberal and the Constitutional Clubs, the National Gallery … In a few of the more recent important public works steel box girders are adopted of considerable span, carrying loads of enormous weight. Rolled joists and joist girders are also being made of steel, but not in very large quantities; iron is much cheaper and suits most purposes perhaps even better than steel.[66]

Whatever the metal, or mix of metals, Edis clearly enjoyed making a virtue out of a necessity, cladding columns throughout the building in marble to decorative (and fire-resisting) effect, one recent commentator noting:

Several of the rooms in the Constitutional Club were over 100ft long and Edis's special skill was to create impressive rooms, appropriate to an imperial age, which were not only luxurious but with appropriate furniture could be divided into intimate spaces. In particular Edis's frequent use of ceiling-high columns gave a majestic air that also broke large spaces into convenient bays.[67]

Fig 10.20 (facing page) Junior Constitutional Club, Piccadilly (1890–2), converted for the Japanese Embassy in 1989. [Jonathan Clarke]

Fig 10.21 Junior Constitutional Club: plan of first floor, showing detached perimeter columns in the smoking room. [The Builder, 3 January 1891]

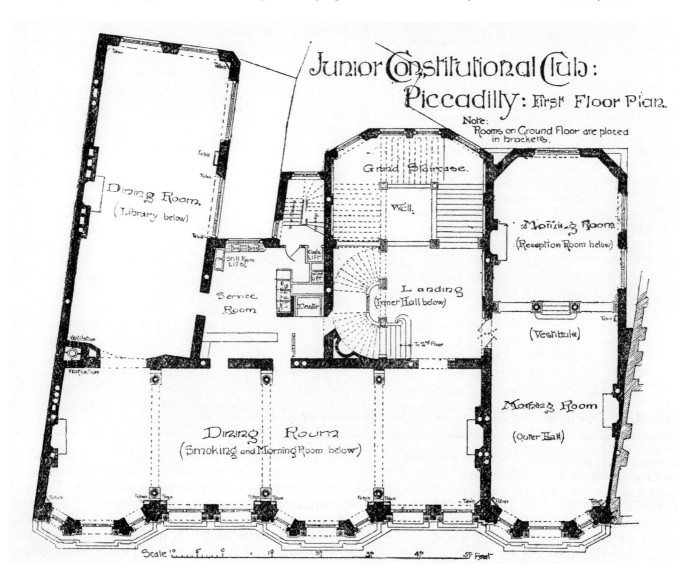

The contemporary, but smaller, New Travellers' Club, Nos 96 and 97 Piccadilly (1890–1), built to designs by Thomas Verity (1837–91) and his son Frank (1864–1937), was reported to be 'entirely of fireproof construction, on Dennett & Ingle's principle, that firm also supplying the ironwork'.[68] This was an exaggeration, but the association between the architects and the supplier, which can be traced back to the Pandora Theatre (1882–4), is significant, and it was almost certainly Dennett & Ingle who advised on the sizing and positioning of the concealed structural members in the spacious lower-floor reception rooms. Again, contemporary descriptions fail to mention steel, but it would be surprising if it was not used, for its absence at this date would imply economic stringencies.

In hotels from the late 1880s there is no such ambiguity. Indeed, it was in the grand hotels, whose size was dictated not by membership, but by building regulations, that steel was destined for more systematic use. The first instance was the Savoy on the Strand (1884–9), the creation of Richard D'Oyly Carte, impresario, chief patron of the Aesthetic Movement and the highly successful Gilbert & Sullivan operas, and, as has been seen, innovative theatre builder (Fig 10.22). Having made his fortune in the Savoy Theatre (1881), D'Oyly Carte appreciated the profits to be made from a neighbouring luxury hotel that would cater to well-heeled theatregoers, especially those from across the Atlantic. His frequent visits to the United States over Gilbert & Sullivan copyright issues had opened his eyes to the distinct superiority, in terms of comfort, services and sophistication, of the country's best hotels. Most inspiring was the Palace Hotel, San Francisco (1874–6), a vast six-storied 'caravanserai' designed by John P Gaynor, an architect who had begun his career working with Daniel Badger on iron-framed buildings in New York.[69] Soon after its completion, *The Builder* reported the advanced constructional features of the San Francisco building:

> Its general form is an immense triplicate hollow quadrangle ... All the outer and inner partition-walls, from base to top, are of solid stone and brick, built around, within, and upon a huge skeleton of broad wrought-iron bands, thickly bolted together, and of such immense size as to have required 3,000 tons of iron for this purpose alone. Thus the building is really duplex, – a huge self-supporting frame of iron, of enormous strength, within massive walls of firm-set brick and solid stone. The outer and visible walls are proof against fire; the inner and invisible frame secures against earthquakes.[70]

This grandest of hotels,[71] with its 850 bedrooms and 437 bathrooms ranged around a covered atrium, may have inspired D'Oyly Carte to reproduce a similar ratio and plan-form in the Savoy,[72] but it did not occasion self-supporting metal frame construction: that 'Londoners rubbed their eyes when the drapes were stripped off a seven-storey, steel-framed building in which concrete was used for the first time'[73] is an overstatement. Traditional bearing-wall construction, albeit with much interior columnar support, was used throughout the Savoy. It nonetheless made important steps into new constructional territory. Erected on a superb inclined site overlooking the Thames, and enclosing a quadrangle 90ft by 64ft, the building was given immensely strong fire-resisting 'filler-joist' floors on every level. These were made from 6in. by 2in. rolled-steel joists spaced 2ft apart, spanning between main beams and girders, the whole encased in mass coke-breeze and cement concrete, and finished on top with cement. The cement layer formed the floor surface in most of the rooms (marble mosaic being used in others), and on the underside plaster was applied directly to form the ceiling, thus avoiding the usual combustible lathwork. The strength of the floors was such that they comfortably spanned wide distances between the exterior envelope, with only minimal additional support from interior columns or walls. This spatial openness enhanced the usual ensemble of public rooms, such as the banqueting hall (70ft by 40ft) on the first floor of the western building, and the restaurant directly above that, where a framework of large primary girders and beams was employed to span the distances between the interior and exterior walls (Fig 10.23). This strong framework, supported at the intersections by widely spaced iron columns encased in Parian cement and faience, marked a different approach to creating large spaces. Until now the stateliest of hotel rooms had been long, tall and narrow, the floor beams partly supported by perimeter columns – the Coffee Room formula of old. Here D'Oyly Carte's design team imposed a systematic internal grid, with columns brought closer to the centre, their placement dictated by

Fig 10.22 (facing page) The Savoy Hotel, Strand (1884–9). This montage from the Illustrated London News *shows the elevation to the Thames, and the grand spaces framed by an interior grid of iron and steel. [© Illustrated London News Ltd/Mary Evans (10469560)]*

engineering considerations rather than traditional aesthetics. D'Oyly Carte demanded lower ceiling heights than was customary – just 16ft, electric lighting obviating the need for tall windows.[74] To get the same volume and sense of space, the accent had to be on breadth, not height. The employment of a regularised interior grid in the Savoy's largest spaces, probably inspired by American precedents, was but one further step towards fully framed hotel construction in London, both in terms of aesthetic acceptance and technological feasibility.

The Savoy was the first British hotel in which the trappings of comfort and sophistication designedly permeated throughout the building, and were not just restricted to the grandiose public spaces. It housed some 400 rooms, six lifts and 70 bathrooms, all made possible by adroit planning and refined construction. Such was the provision of bathrooms that George H Holloway, part of the design team, reputedly exclaimed that D'Oyly Carte must be expecting his guests to be amphibious.[75] Although far short of the Palace and other American hotels in its ratio of bathrooms to bedrooms, the Savoy was well ahead of its nearest British rival, the recently completed Hotel Victoria (now Northumberland House), Northumberland Avenue (1883–7), which had at its disposal just five

bathrooms for over 500 guests.[76] The spanning capability of the Savoy's non-combustible floors, coupled with the strength of the brickwork,[77] helped realise larger bedrooms, many of them with en suite bathrooms. In the 'River Block', for example, the floors easily spanned the 35–45ft distance between the thick exterior walls, supported at mid span by the corridor and bedroom walls, which were thinned to almost partition thickness. The floors here were also extremely shallow – approximately 9in. – which again economised on space, which was especially important given the lower ceiling heights.

There has been disagreement over responsibility for the design of the Savoy, credit going variously to Thomas Edward Collcutt (1840–1924), Arthur Heygate Mackmurdo (1851–1942) and the youthful Collings Beatson Young (1858–1922).[78] However, the evidence favours Young for this earliest phase: obituaries, and two surviving drawings of the hotel, dated 1886–7, submitted to and approved by the Metropolitan Board of Works, bear his signature, providing conclusive attribution.[79] Yet the important roles of D'Oyly Carte, who involved himself with every stage of the conception, design and construction of his properties,[80] and the energetic Holloway, who 'combined [the] capacities of clerk of works and "working master-builder'", overseeing the general contractors J W Hobbs & Co., Ltd, should not be forgotten. Without doubt Holloway had much to say in the structural design of the building, giving it his trademark fire-resisting staircases (used previously in the Grand, the First Avenue, and the Métropole, *see* p 150) and working closely with Andrew Handyside & Co., who supplied the constructional iron and steelwork, including the 'framed steel-work embedded in concrete' of the roof. *The Builder* added tartly that 'Mr. Holloway has super-added to those functions the role of architect, which is perhaps the reason why the building cannot be quoted as adding to the architectural beauty of London, except so far that it makes the Physicians' and Surgeons' Examination Hall next to it look quite a respectable piece of architecture by comparison.'[81] The riverfront was distinguished by seven storeys of verandas carried by coupled columns, the lower two storeys made of red Peterhead

granite, the others of cast iron.[82] In 1910, in the space of just 8 weeks, this whole elevation, and the mansard roof, were surgically removed, and replaced by a new steel-framed façade hung from steel lattice-girders that were firmly anchored down by a new floor and roof.[83]

As the Savoy was nearing completion in 1889, architects-cum-builders Perry & Reed,[84] with Hobbs & Co., were already pushing their scheme through for the development of the Salisbury Estate, a large, sloping tract of land next to the Savoy.[85] Covering an area almost 3 acres (1.2ha), and housing some 800 bedrooms and apartments, the Hotel Cecil, Strand (1889–95, demolished 1930) was one of the largest buildings erected in London by private enterprise and was considered the most magnificent hotel in Europe when it opened in January 1896 (Fig 10.24).[86] It consisted of four enormous blocks, linked at various points by corridors and bridges, all grouped round a central courtyard

Fig 10.24
Rising above the adjacent Savoy, the Hotel Cecil, Strand (1889–95, demolished), centre left was the largest hotel in Europe when opened. [EPW005598]

beneath which the south, east and west blocks were connected (Fig 10.25). The isolated 'pavilion' planning of the blocks afforded excellent lighting of the large showpiece spaces on the lower floors, thus avoiding light wells and glass ceilings, which would have decreased the usable space for bedrooms on the upper floors. To preserve the majesty of these spaces, interior vertical support was avoided, and so 'it became necessary to adopt a purely metal mode of construction wherever the lower floors were planned to be great rooms, and the upper parts divided into small ones'.[87] What might be described as 'cage construction' provided the structural solution to the challenge, and was employed for the greater portion of the southern block (facing Embankment Gardens) and the whole of the eastern block. Deep, built-up steel girders, weighing up to 28 tons apiece,

were used to span the great rooms, their ends 'laid loose' in cast-iron wall boxes carried on cast-iron stanchions imbedded in the brickwork of the walls (Fig 10.26). This column-and-beam arrangement was 'quite independent of the walls, which, indeed, could be removed without affecting the upper part'.[88] The stanchions, which appear to have transmitted the entire cumulative weight of all the floors, were bolted down to circular cast-iron bed plates embedded in the 6ft-deep mass-concrete raft foundation. Above the level of the girders the stanchions continued upwards as cast-iron columns, connected vertically by collars, thus 'forming one continuous column their full height'. Steel beams spanned transversely between these columns, forming the principal framing members of the 7in.-thick concrete filler-joist floors, which were expressly designed using 'no one's

Fig 10.25
Strand-level plan of the Hotel Cecil by Perry & Reed (1894), showing the south, east and west blocks.
[London Metropolitan Archives, City of London (GLC/AR/BR/19/0164)]

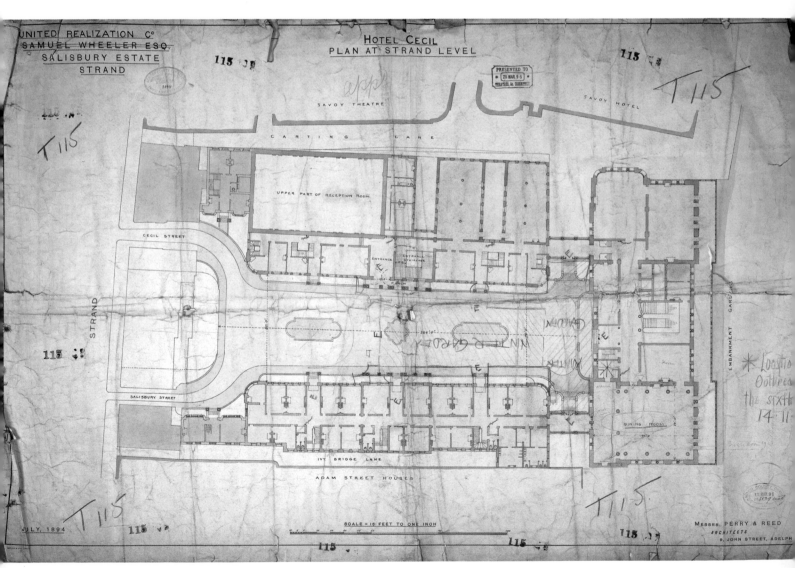

patent'.[89] This construction enabled full flexibility in the arrangement of interior brick division walls, since they were not acting in a structural capacity: indeed on any floor 'the brick division walls between its various rooms are thus independent of the walls above and below, so that all of the division-walls on a floor can be removed without affecting the general construction'.[90] In the large lower-level spaces of the south block, the traditional technique of using perimeter cast-iron columns was employed in addition to the wall stanchions, notably in the dining room (90ft by 60ft by 18ft), where the 'character of these proportions … induced the architects to depart from the Neo Classical style adopted elsewhere in the hotel'.[91] Here the columns were encased in dazzling moulded, glazed earthenware tiles and blocks, their bright yellows, blues, greens and reds contributing to the exotic style adopted throughout the principal garden floor, which was modelled on the 16th-century Emperor Mughal Akbar's palace at Fatehpur Sikri, India (Fig 10.27).[92] The roofs of the eastern and western blocks were composed of rolled-steel joists and concrete, and the giant twin domes of the

7.4.31

Fig 10.28
Demolition of the Hotel
Cecil for the erection of
Shell-Mex House (1931).
This southwards view, with
Cleopatra's Needle in the
distance, shows the ranks
of iron columns that framed
the south block.
[Lawrance Hurst]

south block – dominating the neighbouring Savoy's skyline – were framed with steel ribs.

The constructional methods used at the Hotel Cecil can thus be justly regarded as a form of cage construction, though the self-supporting exterior walls possibly helped stiffen the framework. The scale and independence of this assemblage was revealed during the building's demolition in 1931, when construction of Shell-Mex House began (Figs 10.28 and 10.29). Designed principally by Professor Alexander Kennedy, who was drafted in as consulting engineer, this framework nonetheless owed much of its detailing to Perry & Reed, *The Builder* noting 'full-sized details of construction … have been prepared in their office by them or their own staff'.[93] The employment of what must have amounted to colossal amounts of iron and steel, all of it tested and certificated by Kennedy,

demanded the combined efforts of three specialist sub-contractors, Andrew Handyside & Co., John Lysaght Ltd and Matthew T Shaw.[94]

Cage construction may also have been used for similar reasons at the Royal Palace Hotel extension, Kensington High Street (1896–7, demolished *c* 1964), erected next to Basil Champneys' main block of 1890. The principal function of the extension, designed by H S Legg & Son, was to provide an impressive ballroom and banqueting hall, with additional bedroom accommodation above. The ballroom, measuring 74ft by 48ft, was placed on the ground floor. Stacked above this on each successive floor there was a lounge, a banqueting hall and bedrooms occupying, in plan, a footprint measuring 60ft by 34ft. Huge, 36in.-deep steel girders, weighing over 100 tons each, and rounded to follow the line of the cove of the ballroom

Fig 10.29
Framing of the Dining Room of the Hotel Cecil's South Block during demolition in 1931, corroborating The Builder's *1895 remark that the column-and-beam arrangement was 'quite independent of the walls'.*
[Lawrance Hurst]

ceiling, supported the entire weight of the structure (Fig 10.30).[95] The curved ends of these girders – which must have rested on stanchions concealed in the brick 'piers' – permitted the use of curvilinear skylights which admitted natural toplight into the ballroom, a technique not dissimilar to that originated a century earlier by John Nash at the music room-cum-picture gallery at Corsham Court, Wiltshire.[96] Whether the building resting on the ground-floor girders was part framed or built along traditional bearing-wall lines is not known.

Interior framing became standard practice in hotels of all sizes and status. For the new Claridge's at No. 57 Brook Street (1894–8, Fig 10.31), the architect C W Stephens deployed the usual ranks of columns and girders to frame the principal rooms on the lower floors, continuing this upwards in support of some of the upper room suites, perhaps for the sake of spatial adaptability. Within this seven-storey, foursquare corner block, housing some 260 rooms, cast-iron stanchions were used to support some of the upper walls in addition to the

Fig 10.30
The Ballroom of the Royal
Palace Hotel Extension,
Kensington High Street
(1896–7, demolished). Steel
girders with curved ends
supported the weight of the
floors above in 'a decidedly
daring piece of construction'.
[The Builder, 24 April
1897; RIBA54876, RIBA
Library Photographs
Collection]

Fig 10.31
Claridge's, Brook Street
(1894–8).
[BL14809]

fire-resisting steel filler-joist floors used throughout the building. To spread the concentrated loads borne by these stanchions, they were seated on circular cast-iron bases, 6ft in diameter. Under these iron bases lay steel 'grillage' foundations formed of steel joists embedded in the bottom of a concrete raft which extended outwards some 12ft beyond the exterior walls. This, and all other building work, was carried out by the long-established firm of George Trollope and Sons. It was the subject of an educational inspection by Charles Mitchell accompanied by some 120 masters and members of the Polytechnic Building and Architectural Classes of Regent Street.[97] At the other end of the scale, both in dimension and social standing, the Salisbury Hotel, Green Lanes, Haringey (1898–9), by the architect G A T Middleton

and the builder John Cathles Hill (1857–1915), illustrates how internal framing techniques were becoming commonplace even in outlying, suburban buildings of this class (Fig 10.32). Far removed from the sophistication of central London hotels, this four-storey hotel-cum-gin palace exploited the strength and spanning capability of rolled-steel beams in conjunction with iron columns in a logical, orderly way to create elaborate, open-plan interiors. An internal grid of cast-iron columns, 12in. in diameter and standing on York-stone bases, carried the interior weights of the steel-framed floors, the masonry shell providing overall stability. While not inventive, this form of construction, echoing that of contemporary textile mills, shows what an ordinary builder could achieve in steel, with its straightforward and efficient means of

Fig 10.32
Framing plans of the Salisbury Hotel, Green Lanes, Haringey (1898–9; G A T Middleton, architect). [Specification v. 1. (1898); RIBA54732, RIBA Library Photographs Collection]

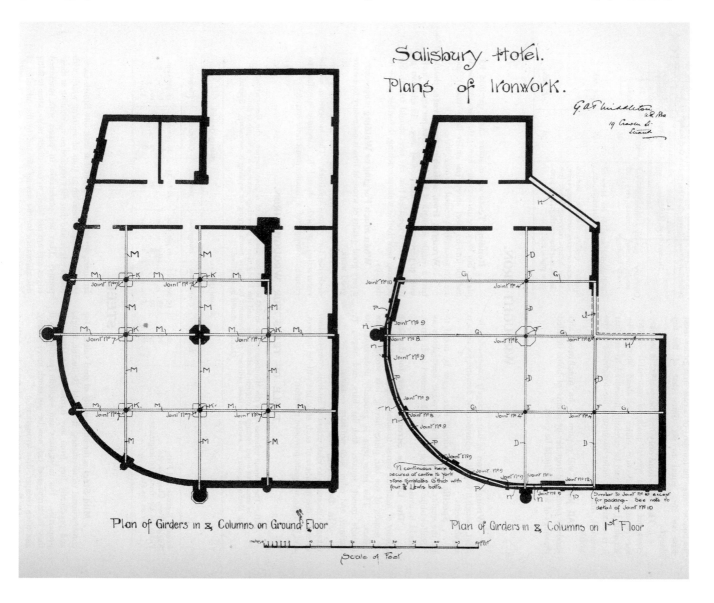

joining the simple rolled section beams to the columns through cast-iron octagonal junction boxes (Fig 10.33). Perhaps because of the simplicity of these more rigid connections, and the overall economy and repeatability of the design, this hotel featured in the first issue of the magazine *Specification*, and in a lavishly illustrated text book on building construction.[98]

The top class of hotels, however, necessarily called for more complex engineering, and it was these that led the way for further advances

in framing, with steel being used increasingly for compression members in addition to its by then widespread use for joists, beams and girders. Writing in 1898 Charles Vivian Childs noted that the standard English practice in designing iron and steelwork for large commercial buildings was to support the entire weight of the upper storeys on a framework of heavy girders ('transfer girders' in today's engineering parlance) placed at the level of the first floor.[99] He used the technique himself in one of his earlier

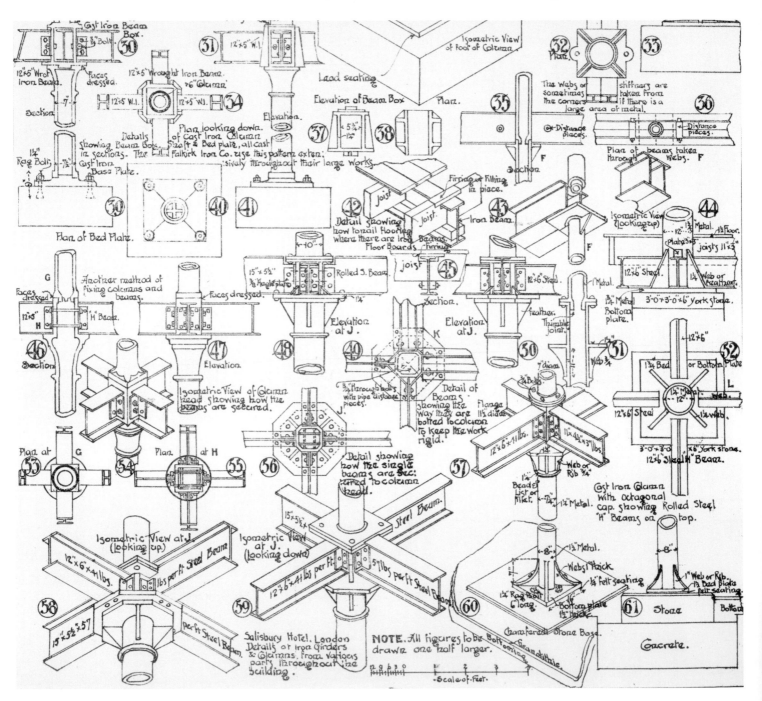

major commissions, the Hotel Russell, Russell Square (1897–1900),[100] but through dint of experience employed steel stanchions to support the girders (Fig 10.34). In contrast to the framing of the Cecil Hotel, where the deep steel girders used to span the great halls were 'laid loose' in cast-iron wall boxes carried on cast-iron stanchions, Childs effected a more rigid connection between girder and stanchion, so that the upright members were forced to take some of the overall bending. Fitted with the standard type of cap illustrated in Dorman Long and Co.'s catalogues, these compound stanchions directly supported the ends of deep steel girders to which they were either riveted or

bolted. The resultant frame was both stiffer and stronger, and more capable of safely bearing the weight of the upper seven storeys of the Hotel Russell. The double-height banqueting hall at the rear was perfectly suited to this form of construction; Fig 10.35 may show this in the process of erection. At first-floor level and above, traditional bearing-wall construction was generally used, with encased stanchions in support of wider spans. Dorman Long provided the enormous amounts of steel used in the project; Childs was at this point the firm's chief London engineer.[101] The architect behind this red brick and terracotta 'super-François-Premier chateau ... magnificently inflated to a height of

Fig 10.34
Hotel Russell, Russell Square (1897–1900). The upper seven storeys rest on rigid steel portal frames designed by the American-trained engineer Charles Vivian Childs.
[Jonathan Clarke]

eight storeys'[102] was Charles Fitzroy Doll, who planned the parallelogram-shaped block around an inner terraced garden to ensure good natural lighting throughout (Fig 10.36). Doll's surviving drawings do not include the steelwork, and we can only presume that they were completed well before Childs was brought on board – a scenario that, according to this engineer, typified English practice (*see* p 69). Doll would go on to design two other sumptuous manifestations of the same materials – including steel, and in the same vicinity: the Imperial Hotel (1905–11, demolished) and the former Dillon's bookshop (1907–8).

The same basic principle of sitting an essentially bearing-wall building on top of a largely steel-framed ground floor seems to have been exploited in another large end-of-century hotel, the Beaux-Arts Carlton Hotel (1891–9, demolished 1957–8) (Fig 10.37). Designed by C J Phipps to complement his work at Her Majesty's Theatre, which occupied an adjoining site at the corner of the Haymarket and Pall Mall, it was completed by Isaacs and Florence. The French architect Charles Mewès and his Beaux-Arts trained English colleague Arthur Joseph Davis were commissioned for the interior

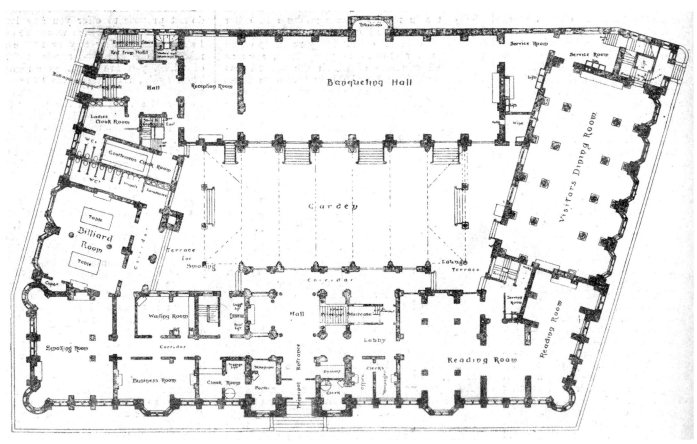

planning and design, which included a magnificent, sunken Palm Court, indirect lighting, noise-deadening double windows and other hallmarks of chic sophistication borrowed from Mewès's Paris Ritz (1897–8). A huge amount of space was allotted to the great top-lit salle à manger on the ground floor, which measured an unequalled 130ft by 50ft (Fig 10.38). Because of the great height of this saloon, a mezzanine level was introduced within it. This floor was 'inserted in a very ingenious way by slinging [it] up to the heads of the steel columns originally provided for the support of the floor above only'.[103] How this suspended floor was achieved – presumably spanning the entire distance over the restaurant without additional support in the middle – will perhaps always

remain obscure. The steelwork contract was given to Richard Moreland & Son, who also introduced their patent solid-steel columns to minimise obstruction of sightlines.[104]

The new century saw the introduction of full skeleton construction in London hotels; framing continued alongside the new, highly rationalised, technology, but a changing economic and cultural climate in the London building world meant that their days were numbered. The first metropolitan hotel to exploit the new technology was the Savoy Hotel extension (1902–4), D'Oyly Carte's long-wished-for prestige frontage to the Strand.[105] Following the great success of the original building, D'Oyly Carte had already commissioned Thomas Edward Collcutt to design a new reception hall

Fig 10.35 (facing page, top) Portal frames in course of erection at the Hotel Russell (c 1897).
[The Engineering Magazine Vol 15, May 1893 Library of Congress, General Collections]

Fig 10.36 (facing page, bottom) Ground-floor plan of the Hotel Russell.
[Builders' Journal and Architectural Review, 19 February 1902. RIBA70538, RIBA Library Photographs Collection]

Fig 10.37 (above) Carlton Hotel.
[CC97/01088]

Fig 10.38
Glazed steel-framed roof
structure of the Carlton
Hotel's great salle à manger,
resting on embedded
columns. A mezzanine floor
was subsequently introduced,
suspended from the tops of
the columns.
[Courtesy of Institution
of Civil Engineers]

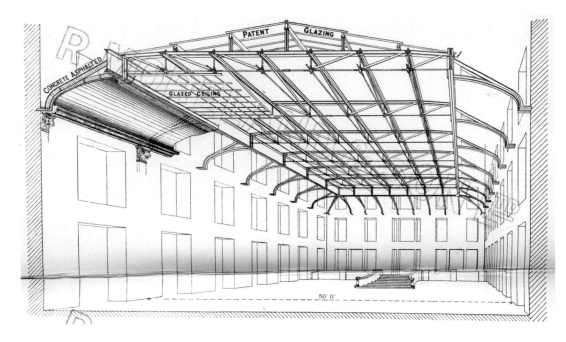

Fig 10.39
The Savoy Hotel, double-
storey steel-framed addition
erected in 1896–7 to designs
by T E Collcutt.
[© The British Library
Board (BL14120)]

with a dining room above, facing the inner courtyard (Fig 10.39). This mansard-roofed terracotta-clad structure, erected in 1896–7 by Holloway, was framed wholly in steel (Fig 10.40), *The Building News* noting that it was 'constructed with steel stanchions, girders and braces filled in with coke-breeze concrete to form the walls and floors'.[106] The efficacy and swiftness with which this block was added, probably led D'Oyly Carte to seek similar, if not more advanced, techniques to promptly realise his expansionist campaign. Indeed, the Savoy Company had a huge amount of capital in the site at the north end of the original block, and to get return on it, needed further accommodation to be put up fast. The opportunity came with the widening of the Strand in 1900, which required the demolition of a number of properties.[107] Collcutt and his former pupil and then partner Stanley Hinge Hamp were appointed to

extend the building towards the Strand, a scheme that included new east and west wings and relocating the main entrance beside the Savoy Theatre (Fig 10.41). An enormous undertaking, it comprised not so much an extension to the hotel as 'an addition to flat-land': of the 376 new rooms provided, 346 were let as offices and residential suites, with the remaining 30 intended as hotel bedrooms. Nonetheless, other components of the scheme, including an extension to the restaurant, a new courtyard and a 'Parisian' café proclaimed an extension of one of the capital's most magnificent hotels.

Executed in a little more than a year, and costing in excess of a million pounds, the Savoy Hotel extension (1903–4) went up smoothly and quickly, commencing with the east block, new entrance hall and covered courtyard, and finishing with the west block (Fig 10.42). Such rapid construction, progressing from April 1903

Fig 10.40 (below left)
Collcutt's cross-sectional drawing of the addition, steel coloured blue.
[London Metropolitan Archives, City of London (GLC/AR/BR/17/026334/01)]

Fig 10.41 (below)
Collcutt's block plan of the Savoy Hotel extension.
[London Metropolitan Archives, City of London (GLC/AR/BR/17/022092)]

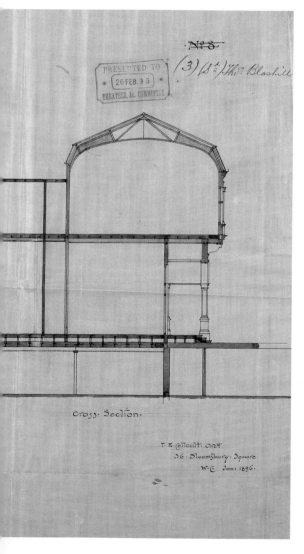

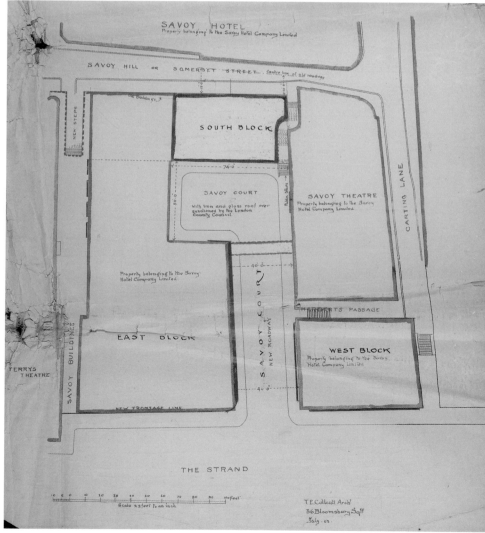

Fig 10.42
The Savoy East and West
blocks soon after completion
in 1904.
[Some Stewart Structures
(1909)]

until May 1904, resulted from the employment of an all-steel skeleton frame, erected and clothed by an American firm of contractors, James Stewart & Co. Its dynamic Canadian-born director, James Christian Stewart (1860–1942), was seemingly favoured over Holloway by the Savoy Company on the strength of an impressive record of speedy work on steel-framed buildings in the UK and before that in the United States.[108] Stewart had earned a 'great reputation'[109] for getting jobs done on time, having replaced the slow-moving original British contractor for the British Westinghouse Electric & Manufacturing Co.'s works at Trafford Park

(1900–2), and having brought the steel-framed Midland Hotel, Manchester (1899–1903),[110] to timely completion (*see* pp 75–6). The success of the Manchester hotel contract led directly to that for the Savoy,[111] where, undeterred by London's wall thickness regulations (*see* p 45), Stewart erected the 'entire framework of the building ... of steel', employing 'American methods in this direction ... up to the full limit allowed by the Municipal Building Acts in the metropolis'.[112] To this end, Stewart and his superintendents 'hustled' and 'pushed'[113] things along in their usual manner, and dispensed with the usual British builders' practice of erecting independent scaffolding, using instead a jib-boom crane located in the basement to make each storey. As the skeleton rose in height, it became the scaffold, the bricklayers standing on temporary platforms cantilevered out from the frame. The usual gantries supporting hoists were similarly disregarded; lifts were used to bring up the materials, which were stored on their appropriate floor levels, since in this form of construction the floors were typically constructed in advance of the exterior walls. The American workmen, paid in proportion to their efforts – both day and night – were organised under a highly efficient chain of command. The superintendents were in continual contact with the subcontractors, checking that delivery and job deadlines were strictly adhered to: 'the firms know that if they cannot supply the order their rivals will'.[114] As the materials arrived on site, they were marked up on blueprints showing, in minute detail, every piece of steelwork, faience and joinery. L G Peglau, superintendent for the west block, introduced the London building world to refuse chutes, a far more efficient means of disposing of building waste than manhandling sacks down ladders in the traditional manner.[115] Dorman Long & Co. Ltd, in competition with foreign firms, secured the steel contract. Some 2,500 tons was rolled at Middlesbrough, then transported by water to the firm's Vauxhall Yard, where it was fabricated to Stewart's detailed, numbered blueprints.[116] From there the steelwork was taken to the site, presumably by boat and cart.

Besides the skeleton frame, the building incorporated another advanced construction technology, also of American design. Reinforced-concrete floors on the 'Coconco' system,

manufactured in this country under licence from the Continuous Concrete Constructions Company, were swiftly laid throughout the building, complementing the rapidity with which the frames were put up (Fig 10.43).[117] Indeed, in the west block, they were fitted ahead of the exterior walls,[118] an impossibility with traditional methods of construction. These concrete draped-mesh floors 'represented the best development of the turn-of-the-century testing period', achieving a height of popularity in 1920s New York.[119] They were formed by a lattice of steel wire draped over the floor beams, thus acting as a high-strength reinforcement to the concrete, which was poured *in situ* around it. This system had the advantage over tile-arch floors of extreme rapidity of construction: in the Savoy extension an entire floor of nearly 23,000 sq ft was completed and ready for use in one week. It also had the intrinsic merit of geometric plasticity and toughness, allowing holes to be cut at any point for service pipes and skylights without compromising the floor's overall structural integrity. Tile arches on the other hand were the sum of their parts, depending on the mechanics of arch action, and would have been jeopardised by the omission of component portions. Finally, given the highly efficient structural action of the slabs – best explained by catenary action[120] – they could be extremely thin, thus increasing the building's usable space. The Savoy extension represented one of the earliest applications of the system, although the Admiralty Extension (1901–5) may have been the first British testing ground.[121]

The Savoy Hotel extension also instanced one of the first genuine collaborations between architect and engineer on a British steel-framed

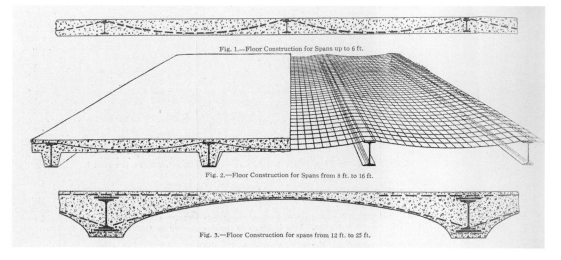

Fig 10.43
Reinforced-concrete floors on the draped mesh principle, laid throughout the Savoy at the rate of 23,000 sq ft per week. [Builders' Journal and Architectural Engineer 24 (C&S Supplement), 7 November 1906]

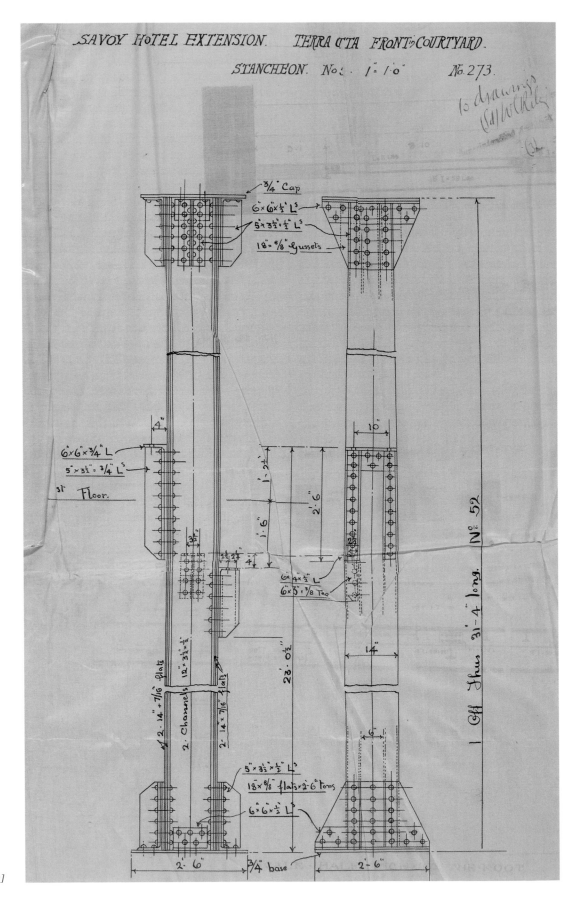

Fig 10.44
A stanchion for support
of the covered courtyard of
the Savoy Hotel extension
drawn by Collcutt, one of
the most accomplished
architects with regard to
structural steelwork.
[London Metropolitan
Archives , City of London
(GLC/AR/BR/17/022092)]

building. The *Builder's Journal and Architectural Review* noted that 'English architects have not yet accustomed themselves to rapid methods of construction … though Mr. Collcutt spends all his time with his staff in an office on the site to cope with Messrs. Stewart's marvellous appetite for details, which never seems to be satisfied.'[122] Collcutt's surviving drawings show that he, at times with Hamp, designed portions of the structural framing, including that carrying the Savoy Court carriageway and the glazed courtyard roof and archway above (Fig 10.44).[123] But the structural design of the more extensive and prominent parts of the extension – the east and west blocks – was the work of a less vocal member of the Stewart entourage, Louis Christian Mullgardt (1866–1942), the firm's in-house supervising architect, and subsequently one of early 20th-century America's most gifted designers.[124] Mullgardt's framing plan for the east block (Fig 10.45) (presaging the type of drawing that Bylander, below, was soon to exhort) was standard American practice, but was almost certainly new to the English architects, and one that palpably influenced their own drawings for the Savoy Court carriageway.[125] Born in Washington, Missouri, Mullgardt had apprenticed in the offices of, among others,

Fig 10.45
Savoy Hotel extension, framing plan of east block, drawn in 1903 by Louis Christian Mullgardt of James Stewart & Co. [London Metropolitan Archives , City of London (GLC/AR/BR/17/022092)]

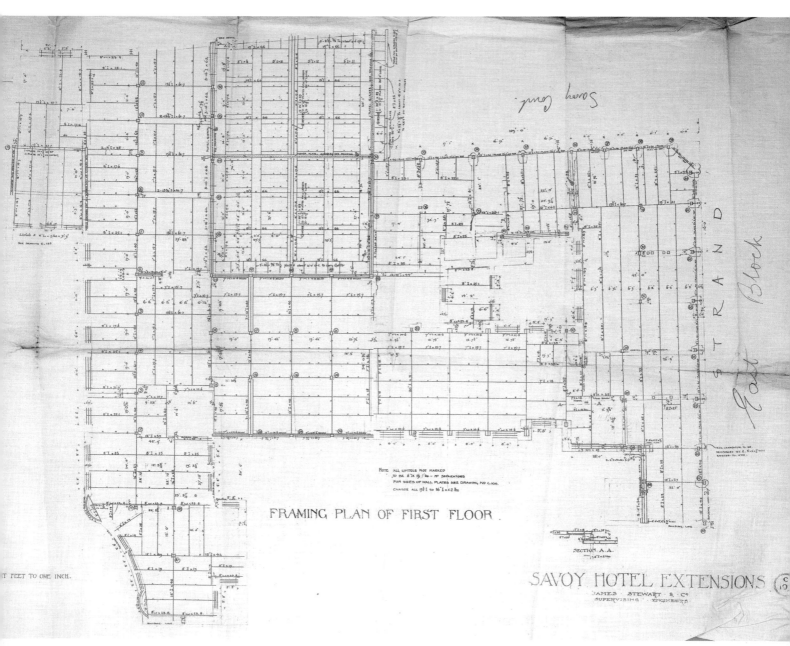

FRAMING PLAN OF FIRST FLOOR.

SAVOY HOTEL EXTENSIONS
JAMES · STEWART · & · Cᵒ
SUPERVISING · ENGINEERS

Fig 10.46
The Ritz Hotel, Piccadilly,
photographed by Bedford
Lemere, soon after its
completion in 1906. German
steel, American structural
engineering and Parisian
chic combined in Britain's
best known early steel-
framed building.
[BL19668]

H H Richardson, and James Stewart & Co., and was responsible for a number of key structures built for the 1893 Columbian Exposition in Chicago while designer-in-chief at Henry Ives Cobb. His British career as an architect-engineer began in 1902 when he moved to Manchester with Stewart, where he assisted Trubshaw on the Midland Hotel; the following year he opened offices on Somerset Street, London, and began working with Collcutt and Hamp;[126] his association with Mackmurdo (who was responsible for the interior decorations)

probably gave him his taste for the Arts and Crafts style, which he subsequently used with aplomb in his renowned series of idealised speculative framed 'California houses' for upper-middle-class clients. One of the most versatile, prolific and accomplished of all the designers featured in this book – he was also a painter, a printmaker and a sculptor – his eventual demise was also one of the most tragic: 'In 1935 Mullgardt was found wandering the streets of San Francisco with a carpetbag containing unfinished plays being dictated to him

a

b

by William Shakespeare. In 1941 he was found again, this time incapacitated, in a dreary residential hotel. He was taken to a pauper's ward at the State Hospital in Stockton, where he died the following year.'[127]

If the first direct wave of advanced American steel-frame construction methods to hit central London arrived with James Stewart & Co., then the second much larger one came with the Waring White Building Company. This consortium, founded by Samuel Waring and an American engineer, James Gilbert White, brought with them their Chief Engineer, the Swede Sven Nathanael Bylander (1877–1943), a structural engineer who had cut his teeth designing steel-framed buildings in Germany and America. Working with various architects and clients, this company was to be responsible for a catalogue of progressive, showpiece buildings in London that had both immediate and lasting impact. The building which more than any other drew the attention of British architects to the sophisticated potential of steel-frame architecture was The Ritz Hotel, Piccadilly (1904–5) (Fig 10.46), a stone-clad edifice designed by the flourishing Anglo-French partnership Mewès and Davis, and built on a gigantic skeleton framework of German steel by Sven Bylander (Fig 10.47). The steelwork was supplied by Burbacher Hütte and fabricated and erected by M A Potts and Co. of Oxford

Street under the immediate supervision of John Percival Bishop (fl 1890–1926).[128]

It was erected in an astonishingly short time: excavation began in June 1904, building work was completed by October 1905 and the hotel opened the following May. Structurally, stylistically and organisationally The Ritz was manifestly not a product of English traditions: its expertly engineered framework was consistent with contemporary practice in New York,[129] its exterior Norwegian granite and Portland stone dress consciously echoed 19th-century Parisian chic, and its interior planning – structured around a long spine corridor – was an inspired Beaux-Arts performance. This most renowned and well-documented of all early British steel-framed buildings has been extensively covered by other historians. For the purposes of this book it is necessary only to discuss certain key features of its structural design, and to re-affirm its pivotal role in construction history.[130]

Notably, the building had to conform to the London County Council Building Act 1894, so the framework had to be over-engineered to accommodate the eccentric loads of the phenomenally heavy masonry walls (39in. thick at street level, reducing to 14in. at sixth-floor level).[131] Composed of 550 individual steel columns and stanchions, joined by an assortment of some 5,200 steel beams and girders, the frame's immense strength and ability to carry

Fig 10.47 a and b
The steel skeleton of The Ritz Hotel (1904–5; designed by Sven Bylander).
[Middleton, Modern Buildings *(1906)]*

Fig 10.48 a–d
Riveted connections ensured
a framework of immense
strength.
[Middleton, Modern
Buildings *(1906)]*

all loads owed everything to the rigidity of the connections. The girders were connected to the columns by strong, well-stiffened angle-brackets riveted to their faces, the dimensions of the brackets being carefully and economically proportioned to the loads to be carried. The secondary floor beams, mostly spaced at intervals varying between 4ft and 6ft 4in., were carried by stout angle brackets, giving adequate lateral support and transferring the loads as nearly as possible to their central axis (Fig 10.48).[132] In practically every case the connections were riveted, bolts being employed in positions where field riveting was impractical.[133] Riveted, fully moment-resisting connections, with their ability to transmit stresses between the structural members, added inherent stiffness and strength to the overall frame. This crucial detail, enshrined in Bylander's own copy of Joseph Freitag's *Architectural Engineering* (1901, first published in 1895),[134] was technically in breach of the 1894 Act (*see* pp 44–5). Many years later, Bylander stated:

> According to the 1894 Building Act it was not permissible to rivet the connection between beam and pillar, because the pillars and the bressummers could not be fixed at the ends and provision had to be made for expansion by the use of oblong holes. This requirement was not insisted on by the authorities for the Ritz Hotel, and hence it was possible safely to erect this steel frame. I have always contended that the connection between the pillars and the beams must be riveted and be strong so as to assure the lateral stability of the building. Not only wind, but also loads placed eccentrically must be resisted.[135]

Bylander may have won this waiver as a concession for adhering to the wall-thickness stipulation, and in so doing helped bring British practice in line with American techniques, such was the detailed coverage the engineering of the building received. The same stipulation may also have been bypassed for the Savoy Hotel extension, but without further structural or documentary evidence this remains unknown.

The structural design and detailing of The Ritz was extremely advanced, not least in the employment of nine cantilever foundation girders fitted with rocker pins to support some of the outer columns along the rear party wall (Fig 10.49). These had to be introduced so that the columns could be supported right against the party wall, exploiting the full boundaries of the site, because Lord Wimbourne, owner of the adjoining property (Arlington House), refused to allow the foundation grillages to extend beneath his wall.[136] All the loads on the framework were thus brought down the column lines to 116 massive cast-iron bases, each with its own separate grillage foundation of steel joists set in concrete, all within the building line. It was said that this was 'probably the first use of cantilever foundations in this country';[137] individual steel grillage foundations had, however, been introduced elsewhere in London in the previous century.[138] The columns were designed with meticulous exactitude to accommodate eccentric as well as axial loading, some fifteen different sections being employed, mostly built up of channels and cover plates. Nevertheless, they were practically all of the same external dimensions, permitting the use of standard brackets for connecting the beams and girders. Manufactured by Burbacher Hütte in two-storey lengths, with perpendicular machined ends, they were joined together by field-riveted cap-and-splice plates at a height of 18in. above floor level.[139]

Fig 10.49
Cantilever foundation girders, employed to keep loads inside the site's plot. [Middleton, Modern Buildings (1906)]

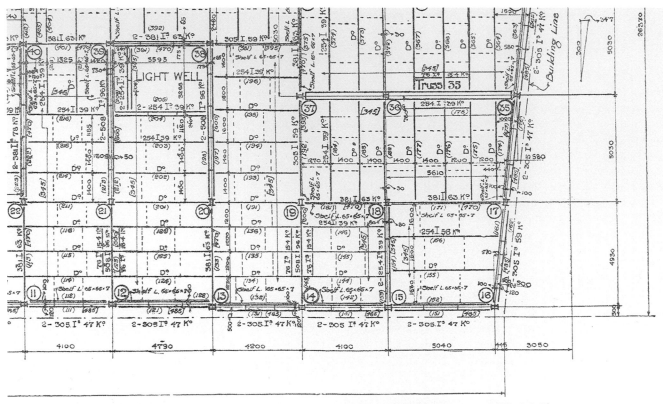

WARING WHITE BUILDING Cᵒ Lᵀᴰ
CONTRACTORS

MEWES & DAVIS & J.P. BISHOP
ARCHITECTS

RITZ HOTEL
PICCADILLY
LONDON, W.

2ᴺᴰ FLOOR FRAMING PLAN

SCALE 1 : 48

DIMENSIONS IN MILLIMETRES

DATE 2.4.04
ORDER Nᵒ 107
DRAWN BY
TRACED BY SMC
CHECKED BY VA
IN CHARGE OF J. Bylander

DRAWING Nᵒ 7

a

*Fig 10.50 a and b
Engineering and
architecture: the detailed
correspondence of Davis's
floorplans with Bylander's
framing plans.
[Middleton, Modern
Buildings (1906); Mewès
and Davis archive]*

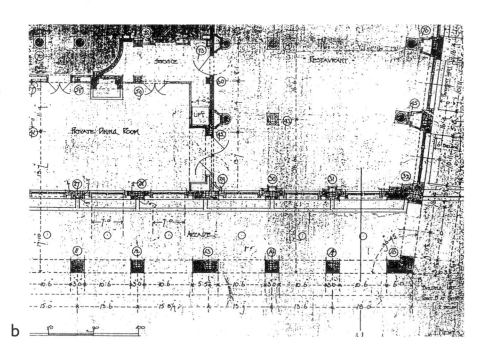

b

The Ritz's success, both structurally and architecturally, arose from a full collaboration and cooperation between architect and engineer at every stage of the design process. Nowhere is this more apparent than in a comparison of floor plans by Davis (the younger architect having taken the lead role)[140] with Bylander's exacting framing plans (Fig 10.50). Produced within weeks of each other, and possibly under the same roof, they dovetailed in their attention to precise dimensioning (Bylander using the metric system) and numbering. This was uncharacteristic of contemporary British practice, which, we have seen, typically left it to the engineer to work around the architect's completed drawings. The dimensions and coordination of the structural grid, with its columns arranged, as far as was possible, in line with the Piccadilly frontage, was a true compromise between the demands of engineering and architecture. To create the largest spaces, unencumbered by interior columns, Bylander had to design girders in spans up to 35ft. These were used most daringly above the restaurant overlooking Green Park, where giant trussed girders, a full storey in height, fulfilled the triple roles of supporting the stanchions of the floors above, carrying an enormous unbroken ceiling below, and contributing to the general stiffness of the framework. Such concessions to modular order meant that the grid stamped its repetitive, rectilinear presence on the proportioning and articulation of interior space. Externally, the main walls and arcade were of such massive proportions as to belie the building's true anatomy.

The most significant aspect of The Ritz was not its design, however, but its influence. Bylander was an extremely adept self-publicist, courting the architectural weeklies to broadcast his methods in particular, and the functional importance of the structural engineer in general. Largely through his efforts, he helped bring about a step change in public relations as regards steel-frame architecture; before The Ritz, construction technique was always a less publicised aspect of architectural endeavour. Early on *The Builders' Journal and Architectural Engineer*, which devoted more space to the construction of this building than any other, before or after, was aware of how significant The Ritz was:

> Mr. Bylander has had considerable experience in the design of steel-framed or skeleton buildings in the United States; and in view of

the novelty of this method of building in this country, and the importance of English architects and contractors becoming conversant with the principles, the article should be studied carefully. We would particularly call attention to the great exactitude of the work, every dimension being figured on the drawings and nothing left to be scaled off; the elaborate nature and number of the buildings – this being no useless expense, because the drawing-office expenses form but a small percentage of the cost of the steel, and the German steelworks are thereby enabled to execute the work at a reduced price and without preparing templates, as is usual in English practice; and finally the careful way in which the details are standardized and facility of erection studied both to secure cheapness and to aid the execution being correct as designed.[141]

The impact of this journal's detailed monthly progress reports (running from 28 September 1904 through 13 September 1905)[142] has been noted, especially the standardised drafting procedures, and Bylander's own contributions:

> in which [he] explained how to read the accompanying framing plans, described the use of standardised parts which eliminated the need for on-site templates, outlined the numbering system used to distinguish each piece of steelwork, provided factual information on loads and stresses, and, importantly, reassured the public that 'the construction practically conforms to the latest standards for steel-framed office buildings in America'.[143]

A year after its completion, The Ritz was enshrined as a textbook approach to steel-framed construction in Middleton's six-volume work, *Modern Buildings: Their Planning, Construction and Equipment*, describing it as 'probably the most thorough example of a steel-frame construction yet erected in England'.[144]

For all the added expense in calculations and materials imposed by the 1894 Act, The Ritz was a resplendent advertisement for the economic viability of steel-skeleton construction in its most systematised form. Its immediate legacy was dramatic. A letter of April 1905 to the editor of *The Builders' Journal and Architectural Engineer* asked, 'What advantage is there in adopting the American steel-frame form of building when the authority makes you put the

Fig 10.51
The Waldorf Hotel, Aldwych
(1905–7), flanked by the
Aldwych and Waldorf
Theatres.
[BL20272B]

walls of ample thickness to carry all the weights required?' The reply: 'The illogical insistence by local authorities on the same thickness of walls with a steel-skeleton construction as without such construction of course limits the advantages of a steel frame, but in many cases the saving of space and the rapidity of erection obtained by this system (and the consequent shortening of time during which capital must remain locked up without interest) warrants its employment.'[145] Middleton noted in 1906 that 'even under the existing state of affairs, a steel framework may often be used with considerable economy, as is evidenced by the number of steel-framed structures now springing up'.[146] The marked increase in commercial London buildings employing full or partial skeleton construction in the period 1905–9 possibly owes more to The Ritz than to any other single cause.

As well as demonstrating the practical and economic advantages of standardised steel-skeleton construction for large commercial architecture, The Ritz legitimised the technique in terms of status and prestige. Steel-frame architecture was fast gaining its own cachet among patrons, and was no longer simply a means to an end (large, well-lit, modular spaces, put up fast and with economy), but part of a modernising Edwardian cultural milieu that embraced progressive, stylish, cosmopolitan ideals, many of American or French derivation. Any lingering prejudice against such constructional 'falseness' within more conservative elements of the architectural profession was steadily eroded by a new set of values rooted in hard-headed commercialism. The Ritz exemplified the new trend, its constructional technique being an extension of the pursuit of technological modernity in luxury hotel building that had truly begun with the Savoy, having been foreshadowed since the mid-19th century. The success of The Ritz and a number of other pioneering buildings brought about the passing of the London County Council General Powers Act 1909 that permitted a reduction in external wall thicknesses, and gave rules for the design of framed buildings, thus fully sanctioning steel-frame architecture. In this light, Bylander's well-known but frequently misconstrued statement that 'The first steel-frame building *of importance* in London was The Ritz Hotel' cannot be dismissed as mere self-aggrandisement.[147]

Following The Ritz, Bylander's next two assignments as chief engineer with the Waring White Building Co. were the Morning Post Building (1905–7) (*see* pp 226–8) and the Waldorf Hotel (1905–7), built on Aldwych, the broad new crescent connecting the Strand with the newly formed Kingsway. Cloaked in a restrained 'Louis XVI' skin of Portland stone and Aberdeenshire granite, the Waldorf Hotel formed the centrepiece of a symmetrical composition with the Aldwych and Waldorf theatres (*see* p 132) (Fig 10.51). Commended for striking 'a dignified note in this improved heart of London',[148] both the Waldorf Hotel and the Morning Post Building were erected by the same contractors, the Waring White Building Co., but they had different architects and methods of steel framing. The architects for the steel-cage framed Waldorf Hotel were A Marshall Mackenzie and his son A G R Mackenzie. Like the awkward-shaped Morning Post (*see* p 229), the curved frontage of the Waldorf demanded an irregular framing-plan and Bylander was able to realise this more efficiently in this instance by using cage construction. In all, just 1,400 tons of structural steelwork were used, and the building was completed, furnished and equipped in the remarkably short time of 18 months. Bylander later said that the Waldorf was 'not a complete steel-frame building, the external walls being self-supporting, but the interior of the building is carried on [a] steel frame similar to the usual steel-frame structures'.[149] In essence, the exterior walls were designed to give some stability and stiffness to the frame, and *The Builders' Journal and Architectural Engineer* noted that in places both steelwork and brickwork were 'conjointly ... required to do the same work'.[150] Before the building's construction, the architects visited America and 'made a study of American requirements as well as English'.[151] In consequence, the building was exceptionally well provided with amenities and comfort: there were 176 bathrooms to the 400 bedrooms, many en suite; Otis electric passenger lifts; full noise insulation through patent 'Mack' sound-deadening room partitions; and mechanical warming and ventilation.

Given the appreciable advantages of fully framed steel construction in hotels, earlier techniques and methods incorporating iron soon vanished. So fast was the changeover that one of the last major metropolitan hotels to eschew full use of steel construction was the Piccadilly Hotel, Piccadilly, built in 1905–8 to designs by Richard Norman Shaw in partnership with William Woodward and Walter Emden (Fig 10.52).[152] Bylander was later of

Fig 10.52
The Piccadilly Hotel, Piccadilly (1905–8; Richard Norman Shaw, William Woodward and Walter Emden, architects), a skeleton-framed structure, but perhaps the last major commercial building in London to use cast-iron pillars. [OP15762]

the opinion that this was the last substantial building in London to employ cast-iron pillars.[153] However, these were employed in the lower section of the building, below first-floor level, which included three deep underground floors housing Turkish and swimming baths. Founded on cast-iron base-plates seated on mass-concrete pad foundations, these iron columns supported a huge steel skeleton superstructure, designed by Reade, Jackson & Parry, fabricated by H Young & Co., Ltd, of Nine Elms, Vauxhall, and erected by Perry & Co., of Tredegar Works, Bow.[154] Altogether, some 6,000 to 7,000 tons of steel and iron were used in the building, which was deemed 'as perfectly fire-resisting as modern resources render possible'.[155] The 'Frazzi' terracotta-slab floors[156] were perhaps less technologically advanced than some, at least compared to the draped-mesh systems employed at the Savoy Extension and the Waldorf, or the 'Columbian' reinforced-concrete slab floors of The Ritz,[157] yet it was observed

that 'one valuable advantage of this system is that the cracking of ceilings is entirely obviated, while the equally objectionable marks showing the lines of the steel joists are entirely absent'.[158]

If sophisticated steel construction had truly arrived in hotels, it was also transforming the character of clubs, though less dramatically as demand for this class of building was already on the wane. Britain's first club to be built on a full steel frame, the Royal Automobile Club, Pall Mall (1908–11), was also 'the last and grandest of the great classic club houses'.[159] With a frontage to Pall Mall of 228ft, this refined French-Renaissance-styled edifice brazenly announced its intended status as 'the club to end all West End clubs' in a street already bustling with archetypes of the previous century (Fig 10.53).[160] Its commanding Portland-stone exterior echoed the scale and grandeur of the public rooms, grand staircase and corridors within, all planned on deceptively simple Beaux-Arts lines by Mewès and Davis in conjunction

Fig 10.53
The Royal Automobile Club, Pall Mall (1908–11), in October 1910. This, the last and grandest of the great London clubs, was built on a 1600-ton framework of imported steel.
[OP03661]

with E Keynes Purchase. Complementing the clear planning was the building's repetitive steel frame, designed by Sven Bylander, who spaced each of the building's 132 stanchions at 12ft 3in. centres along the Pall Mall front, maintaining this bay module right through to the rear of the building, 'except in certain cases where straight lines could not be maintained owing to the architectural planning'.[161] This was most apparent in the framing of the large oval vestibule in the centre of the building, the hub of the design, which rose up through the first and second floors to a large skylight, before continuing upwards as an elliptical light well (Fig 10.54). Some 1,600 tons of 'foreign steel, excepting Belgian' was used in the construction of the framework, fabricated and erected by Redpath, Brown & Co. Ltd.[162] As the designs had been prepared before the passing of the London County Council General Powers Act 1909, the walls throughout had to be built irrationally thick, 'and thus a great deal of unnecessary

weight had to be carried at the first-floor level by the steelwork ... which would not have been required for purely constructional purposes'.[163] Full-skeleton construction was called upon only up to first-floor level; above that an internal cage carried all the reinforced-concrete floors and interior partitions, with partial assistance provided by the external walls.

The Royal Automobile Club forms a fitting close to this chapter, for it embodies a number of methodological and conceptual changes that had reached rapid maturation in the period 1905–10, and which were to inform practice well into the 20th century. In the pursuit of space, non-combustibility, flexibility, elegance, proportion and order, 19th-century clubs and hotels progressively pushed ferric building technologies to their creative limits. Some of the techniques that evolved were particular to the building type: for instance, the long colonnaded clubhouse coffee rooms and the vast hotel salles à manger were not simply impressive

Fig 10.54
Curved framing forming the central vestibule of the Royal Automobile Club, fabricated and erected by Redpath, Brown & Co. Ltd. The facility of curved compound beams had long been presaged in iron (see Fig 2.10).
[Modern Building Record v1 (1910)]

spaces, but crucial structural devices, framed to support the floors and walls above. Initially, the transition from iron to steel was not accompanied by any readily observable reformulation of such defining spaces, for the stronger metal was initially simply performing the inherited roles of its forebear, albeit more dramatically or efficiently. The 'Whitehall Salon' of the Hotel Métropole is one example of this. Yet with the swift evolution of more comprehensive and stronger steel framing in the 1890s, and the influx of American practices after 1900, came profound change. No longer was steel simply a useful addition to traditional 'bricks-and-mortar' construction; it became the primary structural component. In many branches of architecture, not just hotels, existing building concepts and practices had to be rethought in the face of hard engineering rationalism.

However, although the full steel frame offered spatial and planning flexibility, it was to a degree circumscribed by its own economic and structural limits. This was apparent in the systematic structural framing of the Savoy Hotel, where columns encroached deep into the interior of the grand spaces, adhering to engineering logic, not aesthetics. Later, in The Ritz and The Waldorf, and other examples, there was a discernible scaling-down in number, size and

volumetric complexity of grand public rooms, the accent being on the streamlined interlocking of sophisticated spaces rather than the traditional juxtaposition of the extravagant with the modest. The sense of uninterrupted space was instead concentrated on the grand central axis, which in the case of The Ritz ran the entire length of the building, and which benefited from the standard column spacing. Other Beaux-Arts devices, including mirrors, and lengthy diagonal views, were used to enhance what space there was. At The Ritz, the longest spanning girders were 35ft; at the Royal Automobile Club the greatest span was just 37ft – both considerably shorter than many iron forerunners, used when grand room dimensions and not regular column placement, was the main concern. The constancy and modularity of the grid demanded a particular set of decorative responses, for example, as at the Royal Automobile Club, where 'the well-laid-out interior presents a series of handsome rooms that typify the heyday of the "Period" decorator, whose task it was to clothe in faultless period dress the monotonous regularity of structural steel' (Fig 10.55).[164] It would be another two decades before the defining orderliness of structural grids was openly celebrated in London buildings with the arrival of Modernism.

Fig 10.55
Clothed in historicist garb, but blatantly intruding into the Royal Automobile Club's grand restaurant, steel stanchions demanded placements to accord with engineering logic, not aesthetics.
[Modern Building Record v1 (1910)]

11

Banks and offices

If we enter any of the large business and commercial buildings which have made London almost a rival of the commercial edifices of New York or Chicago, it will be noticed how the exigencies of trade, rapid transit, means of intercommunication between different departments have entered into the architect's design, and forced themselves upon the attention … What at one time were considered the essentials of architectural planning and design and of good taste, are entirely set aside to make room for large and uninterrupted space, the demands of passenger and goods lifts, the fitting of electric light, iron staircases, and a host of other appliances too numerous to mention … Go inside, and there is no evidence of architecture visible, except iron pillars supporting huge beams and skylights, spiral staircases, lifts caged in with ornamental grilles, mahogany doors and counters and fittings. (1895)[1]

Iron forerunners

The emergence of specialised office buildings to facilitate the administration of banking and commerce forms one of the most important chapters in London's built history. As with clubs and hotels, this class of building evolved in the early to mid-19th century from what were essentially domestic premises, and, as with those buildings, it was the emergence as a distinct building type that brought the progressive use of structural iron for pragmatic reasons. In terms of sheer quantity and variety, offices and banks ultimately had a more encompassing influence on the visual character of the metropolis than most other categories of building. Indeed, they are a defining feature of the modern city. Nowhere was this Victorian transformation more palpable than in its traditional financial centre, the City of London, which saw an explosion in the availability of office space to meet unparalleled growth in financial, commercial and legal services. Merchant, private and joint-stock banks, insurance companies, and numerous other commercial concerns competed for space on which to erect buildings that would meet their special functional requirements. From the mid-19th century, purpose-built speculative offices were built in ever-increasing numbers. Old properties were swept aside as the City steadily transformed 'from a residential area with specialist commercial and financial functions, to a financial and commercial enclave with a dwindling residue of inhabitants'.[2] Such was the pressure on available land, that it has been estimated that some 80 per cent of City buildings standing in 1855 had been rebuilt by 1905.[3] This intense redevelopment, initially concentrated around the Bank of England and the Royal Exchange, was not confined to the City, however; Westminster and other central districts also saw the multiplication of offices and banks. Rising land values and rents provoked intense development and competition, and office designers were quick to turn to structural iron to create fire-resisting buildings that maximised available light and space, and which offered flexibility in interior planning. 'Fireproof' construction was the initial – and ongoing – impetus for architects to use iron in the construction of banks,[4] but as the century wore on spatial openness, light and grandeur would assume increasing importance as justifications for iron. The story begins with banks (for which there is more evidence) but later spreads to other forms of offices, especially insurance and speculative offices, from the mid-19th century, when the functional needs and structural composition of the types became increasingly indistinct.

It seems doubtful that structural iron was used to any significant extent until the first purpose-built banks appeared in the late 1820s. With the notable exception of the Bank of England, rebuilt in 1788–1827, London banking was a private, discreet world, controlled by great dynasties of banking families who conducted their business from modified town houses. The bank itself typically occupied no more than the ground floor, while the upper floors formed living accommodation for a partner and his family, and occasionally for the manager or clerks. For various reasons, including the constant threat of bankruptcy, and unwillingness to surrender dwelling space, bankers seemed averse to investing in specialised buildings designed primarily for administrative functions.[5] Traditional timber-floored and timber-roofed masonry construction was all that was required, as in wholly domestic buildings. The typical concession of the late-Georgian banking house to its function was subtly stylistic, not structural, with the front elevation sometimes given a suitably reserved treatment, or one that signified the distinction between place of business and place of residence. Internally, there was no need, or room, for a grand banking hall, the day-to-day business of paying in bills and cashing being conducted at a long counter dividing the modestly proportioned 'shop' floor, with more private negotiations taking place in an adjoining parlour, typically modelled on a town-house drawing room.[6] Private banking was an intimate affair, dependent upon face-to-face contact between client and banker, and conducted in small and personalised office spaces.

The advent of joint-stock banking, enabled by legislative reform in 1826 and 1833, brought about a move to grander, tailor-made premises, to buildings whose identity and status was architecturally asserted in their scale and style, and whose permanency was assured by constructional solidity and resistance to fire. It was not until the passing of the second of these revolutionary acts that joint-stock banks of deposit – though not issue – were allowed within the capital, but their anticipated arrival was reflected in the construction of at least one private bank. Hoare's Bank at No. 37 Fleet Street, rebuilt to designs by Charles Parker in 1829–30, is the earliest wholly surviving purpose-built bank in London (Fig 11.1). Architecturally distinguished by a sparingly classical stone façade, it was 'a deliberate portrayal of elegance, calm and good manners ... a reassurance to customers that, whatever the meretricious appeal of the joint-stock revolutionaries, the established canons of reserve and solidity still held good with the London private bankers'.[7] In arrangement and planning it was more traditional, with entrances to each side of the main elevation, serving bank and upper apartments respectively. Inside a screen separated the discreet 'shop'-cum-banking hall from the counting-house at the rear. In its floor and roof structures Hoare's Bank foreshadowed developments in fire-resisting construction. The ground-floor banking hall was carried on shallow brick arches sprung between the flanges of heavy cast-iron 'skewback' girders spanning the walls, and the roof was framed with iron trusses.[8] Skewback, or inverted Y- or V-shaped girders were specifically cast for employment with jack-arched floors, since they provided the requisite angled abutment to resist the outward thrust of the arches. This building marks one of their earliest known appearances in the metropolis.[9] Plausibly, Hoare's Bank may have been the first London bank to make use of iron beams, but the fact that no predecessors survived even to the 1940s, coupled with the paucity of earlier detailed records, makes this claim unprovable.

The first joint-stock bank to be established in London, and hence the first to challenge the Bank of England's long-standing monopoly of joint-stock organisation, was the London & Westminster Bank. This firm's magnificent head office, designed by C R Cockerell and (Sir) William Tite, opened in Lothbury in December 1838, boldly sited opposite Sir John Soane's Bank of England. Demolished in 1908, its scale and monumental treatment marked a conspicuous departure from Georgian banking houses, bestowing it with an authority, dignity and strength of purpose rivalled only by its larger neighbour. Central to its formal expression of status was an expansive banking hall, 34ft by 50ft, flanked by galleried aisles and surmounted by a domed masonry roof that supported a large circular skylight. This skylight, together with arched windows between the supporting pendentives, provided an adroit solution to the problem of lighting a hall of such monumental proportions. The masonry framing of this dome was, according to Henry-Russell Hitchcock 'much less novel in character than Soane's shallower tile vaults and floating lanterns inside the Bank of England',[10] and there is no evidence that structural iron was used in the building's construction. Nonetheless this double-height banking hall served as a key prototype for the headquarters of many subsequent joint-stock banks, which achieved the same ends using concealed iron framing.

Glasgow, not London, was the setting for perhaps the first banking hall to exploit iron-framed top-lighting, in the form of John Gibson's celebrated and influential National Bank of Scotland, opened in 1849. There, an iron-framed cupola filled with coloured glass and supported on a framework of cast-iron girders spanning the masonry walls beguiled the customers with 'a gorgeous arrangement of sparkling colour'.[11] It was not until 1864–5 that

London was given a top-lit banking-hall of such colourful sophistication, executed by the same architect and benefiting from the superior spanning capability of wrought iron. His single-storey head office for the National Provincial Bank (now the National Westminster Bank) at No. 15 Bishopsgate took the banking hall idea to monumental proportions in a potent architectural statement of confidence and strength. Measuring some 118ft by 50ft, this hall was

Fig 11.1
Hoare's Bank, 37 Fleet Street (1829–30; Charles Parker, architect), possibly the first purpose-built bank after the Bank of England to use structural iron.
[Jonathan Clarke]

claimed to be 'the largest room in the metropolis devoted to such a purpose' (Fig 11.2).[12] It was lit by three glazed saucer-domes, each 29½ft in diameter and formed from cast-iron ribs, carried on a grid of built-up wrought-iron girders that spanned the thick masonry exterior walls. The principal, box-section girders in this framework, carried by the deep piers, must have been extremely difficult to manoeuvre, measuring 2½ft deep and some 41ft long. Nevertheless, the framing of this multi-domed ceiling in iron was quicker and cheaper than traditional masonry construction and would have been more resistant to fire than wood.

The evolution of the banking hall along increasingly commanding lines was one of the key features of the development of bank offices, reflecting the bureaucratisation of the profession: 'For the new clientele, trust in the fidelity of the joint-stock banker relied more on the scale and richness of the bank's architecture to compensate for the loss of personal contact.'[13] The arrangement epitomised by the single-storey London & Westminster Bank was extravagant, especially for less capitalised branch offices, for top lighting either precluded multiple storeys or demanded commodious light wells. A far more common approach to lighting was to open out the interior of the hall and support the floors above using cast-iron columns, permitting maximum dispersion of light from the façade. An early metropolitan example of this approach survives in the Bloomsbury branch of the London & Westminster Bank, at No. 212 High Holborn, completed in 1853 to competition-winning designs by Henry Baker (Fig 11.3). Four slender cast-iron columns placed near the centre of the 70ft-deep banking hall helped

Fig 11.2 (facing page) National Provincial Bank, Threadneedle Street and Bishopsgate (1864–5; by John Gibson). The banking hall, spanned by an iron-framed roof, was unprecedented in its scale. [London Metropolitan Archives, City of London (SC/PHL/01/003/64/398)]

Fig 11.3 The London & Westminster Bank, 212 High Holborn (1853–4; Henry Baker, architect). [Jonathan Clarke]

The use of internal columnar support within banking halls also offered excellent sightlines for managers keen to supervise the ranks of clerks and tellers. Such was the success of this simple approach that solid cast-iron columns of minimal diameter were manufactured specifically for bank buildings until the end of the century.[16] Yet banks built on the narrow plots that were typical of the City could, in some cases, dispense with them entirely. The Union Bank of London at Nos 12 and 14 Fleet Street of 1856–7 was one of the first of a new wave of impressive joint-stock banks to appear in the metropolis following their long-resisted admission to the London Bankers' Clearing House in 1854. Built to designs by George Aitchison and his son, (Professor) George Aitchison, this four-storey building echoed some of the constructional features first seen in Hoare's Bank, namely a brick-vaulted basement carried on massive skewback girders and an iron-framed roof (Figs 11.5–11.8). Its use of structural iron, however, was far more comprehensive, capitalising on the widening availability of wrought iron for building purposes. From the ground floor upwards, deep I-section wrought-iron beams were used to span the 20ft width between the side walls and carried joists of the same metal, upon which were formed shallow tile arches levelled with concrete. Known as tile creasing, this lightweight non-combustible floor structure, composed of layers of common roofing tiles bedded in cement mortar, was a popular system in mid-19th-century London, having been used in the Palace of Westminster, the House of Lords and the Waterloo Building at the Tower of London. Its invention is generally attributed to Charles Fowler, who used it for the roofs and floors of Hungerford Market (1831–3), but some written sources attribute it to George Smart, who used tile creasing in many of the private mansions he designed in the 1820s and early 1830s.[17] The five light triangulated trusses, 31ft in span, were formed from wrought-iron flats, angles and Ts, and carried the same ceramic floor structure in addition to the boarded roof-covering.[18] Externally, one of the most distinctive features of the façade was the heavy top cornice. So weighty was this that the architects anchored the back of the stonework using horizontal wrought-iron cramps and tie-rods, which passed down through the façade and were tied to the back wall by horizontal rods concealed in the floor structure. How unusual this 'tie-back' technique was at

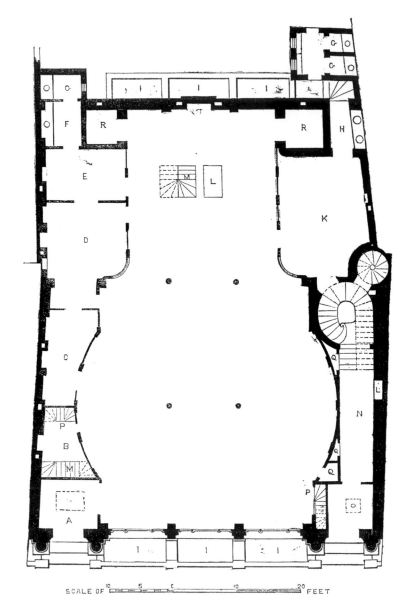

Fig 11.4

The London & Westminster Bank was one of the first banks to exploit freestanding cast-iron columns, in this case enabling curved non-load-bearing partitions. [The Builder, *18 June 1853*]

SCALE OF 10 5 0 10 20 FEET

take the weight of the manager's residence above, while huge segmental-arched windows, 'made perhaps wider than the superincumbent weight may justify',[14] threw light deep into the open interior (Fig 11.4). Curved non-load-bearing partitions created a distinctive keyhole-shaped public interior that screened off the clerical staff. The building was one of the first to embody the principles described in J W Gilbart's highly influential *Practical Treatise on Banking*, the fifth edition of 1849 being the first to encompass architectural matters. As the first general manager of the London & Westminster Bank from 1833 to 1859, Gilbart's main concerns governing bank interiors were space, light and ventilation. The *Illustrated London News* applauded Baker's bank in all these respects.[15]

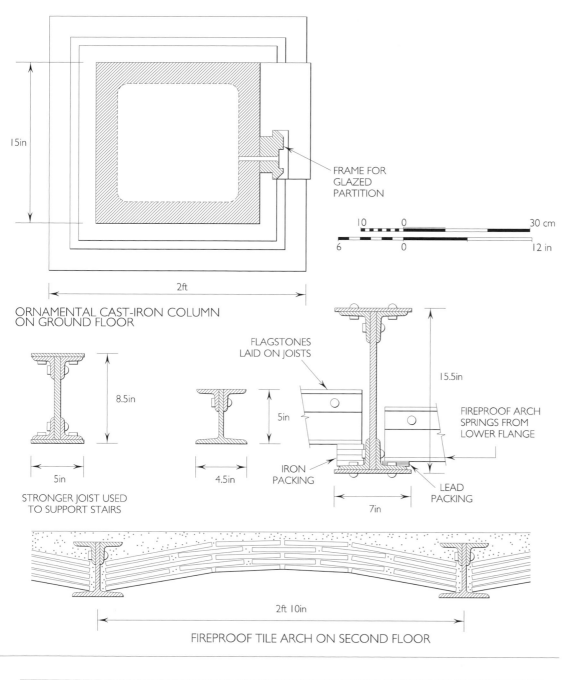

15in

2ft

ORNAMENTAL CAST-IRON COLUMN
ON GROUND FLOOR

FRAME FOR
GLAZED
PARTITION

10   0   30 cm
6   0   12 in

8.5in

5in

STRONGER JOIST USED
TO SUPPORT STAIRS

FLAGSTONES
LAID ON JOISTS

5in

4.5in

15.5in

FIREPROOF ARCH
SPRINGS FROM
LOWER FLANGE

IRON
PACKING

LEAD
PACKING

7in

2ft 10in

FIREPROOF TILE ARCH ON SECOND FLOOR

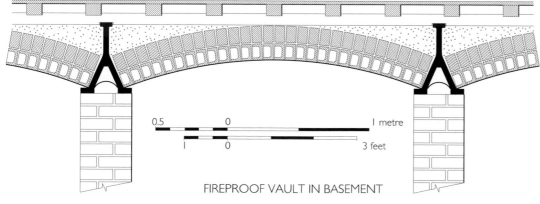

0.5   0   1 metre
1   0   3 feet

FIREPROOF VAULT IN BASEMENT

Fig 11.5
Union Bank of London,
12 & 14 Fleet Street (1856–7;
George Aitchison senior and
junior, architects), fire-
resisting floor construction
reliant on built-up wrought-
iron beams in the upper floors,
and cast-iron skewback girders
in the basement.

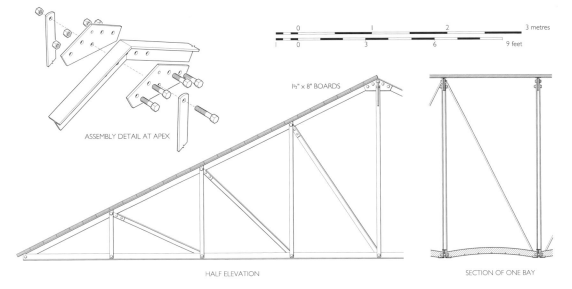

ASSEMBLY DETAIL AT APEX

1½" × 8" BOARDS

HALF ELEVATION

SECTION OF ONE BAY

Fig 11.6 (above)
Union Bank of London.
The fire-resisting floor
construction, exposed
during refurbishment in
2000. Common roofing tiles,
overlaid with concrete, span
between built-up wrought-
iron beams
[AA025293]

Figs 11.7 (above right) and
11.8 (right)
Union Bank of London. The
technologically advanced
fire-resisting roof structure,
framed entirely with rolled-
iron components.
[BB009750]

Fig 11.9 (facing page)
No. 22 Finch Lane (1845–6),
the oldest surviving purpose-
built lettable office in the
City, designed by Edward
I'Anson, junior. Extensive
glazing on the lower storeys
make the most of what light
this narrow lane offered,
and thin stone piers betoken
iron framing within.
[Jonathan Clarke]

this date is unclear, but its use later in the century was recommended.[19]

By the 1850s, purpose-built offices for insurance companies and those designed to provide space for letting to multiple occupiers were starting to leave their mark on the City. Functionally, there was little difference between the requirements of insurance offices and those of banks – above all they required a large, well-lit hall for public business, with office suites above. Notable early examples, executed in monumental masonry, include the Westminster, Life and British Fire Office in the Strand (1831–2) and the Sun Fire Office (1841–2) in Threadneedle Street, both by C R Cockerell. Speculative offices, designed to accommodate the clerical and administrative needs of various private businesses had, in contrast, no need of a public hall. Their accent was firmly on interior flexibility to meet the generalised nature of the businesses they served, and the differing suite sizes they might require. Annesley Voysey reputedly designed the first purpose-built speculative office in the City at the Lombard Street end of Clements Lane in about 1823,[20] but little is known of the appearance, plan form and construction of such early precursors of the early to mid-Victorian 'commercial chamber'. Many probably still provided dwellings above, including Edward I'Anson's (1812–88) first office building, erected in Moorgate Street in 1837.[21] Yet it is clear that by the 1850s office buildings of all types were increasingly using iron columns and beams in place of interior bearing walls to support fire-resisting floors for the purposes of greater flexibility of working spaces and greater daylight illumination. Light was an especially precious commodity in the City, an area historically hampered by narrow streets and ancient laws affecting 'light easements'. Following the repeal of taxes on windows in 1845 and glass in 1851, maximisation of window area and use of interior glazed partitions became standard practice among office builders.

Edward I'Anson, credited as 'perhaps the ablest London commercial architect' from the early 1840s to the 1870s,[22] was one of the first to use non-structural plate-glass partitions that were independent of the grids of interior cast-iron columns and exterior masonry walls. Both No. 22 Finch Lane (1845–6), the oldest surviving speculative office in the City (Fig 11.9), and Colonial Chambers, Fenchurch Street (1856–7, demolished), exploited this technique, their façades punctuated by unusually large windows. The number and size of windows in Colonial Chambers actually contravened a stricture in the Metropolitan Building Act 1855 limiting openings to no more than half the total exterior wall area. By including the wall below pavement level in his calculations, I'Anson shrewdly sidestepped the restriction.[23] Hemmed in by adjoining buildings, and with only one street frontage to the exceptionally deep plan, Colonial Chambers typified the problems of building on congested, narrow sites (Fig 11.10). I'Anson made extensive use of white-glazed, tile-lined interior 'areas' or light wells throughout the building's depth, a technique he first introduced at the rear of his Royal Exchange Buildings in 1842–4 and one still visible on the side of No. 22 Finch Lane; it quickly became standard practice.[24] His plan omits to show the positioning of the iron columns, probably because he was primarily concerned with showing the arrangement of spaces, but possibly because they were imbedded within the brick piers framing the axial passage. The same basic structural techniques were incorporated in numerous mid-18th-century buildings. John

Fig 11.10
Colonial Chambers,
Fenchurch Street (c 1857;
Edward I'Anson, architect;
demolished).
[RIBA46504, RIBA Library
Photographs Collection]

Whichcord employed interior iron columns in his block of offices on Water Lane, off Cannon Street (1858, demolished) and ornamented iron stanchions in the narrow façade (Fig 11.11). The use of cast-iron bressummers at each floor level of this elevation saved valuable space, enabling a reduction in wall thickness to no more than that required for a single storey. Because bressummers could be interpreted as constituting a foundation for a wall, the whole façade was in effect a stack of single-storey walls.[25] For fire resistance, Fox and Barrett floors with rolled-iron joists were employed;[26] these carried partition walls 'so arranged that a greater or lesser quantity [of offices] can be taken *en suite*, according to requirement'.[27] Robert Kerr employed a similar technique of interior framing and lightweight division walls in his National Provident Institution, No. 48 Gracechurch Street (1862–3, demolished c 1957) (Fig 11.12). The public offices on the ground floor were designed as 'merely one large open expanse, more than 50 feet square, the whole of the division walls above being supported on about half a dozen iron columns'.[28] These ornamental columns may have continued upwards, carrying built-up, wrought-iron box girders that framed each of the upper floors, the second of which was given over to speculative use and thus required non-structural partitions that could be moved to suit the client's taste. All of this ironwork was openly celebrated. *The Builder* delighted in the attempts made, 'without resorting to sham in any way', to decorate this ironwork, pronouncing it 'the most ornamental feature in the interior, and its effect of lightness is universally remarked upon'.[29]

The nature of interior iron framing in mid-18th-century office buildings remains poorly documented, but in one case a contemporary description does suggest the use of a full interior framework of cast-iron columns and wrought-iron beams. Nos 59–61 Mark Lane, built in 1864 for the City of London Real Property Company to designs by George Aitchison junior, was perhaps the most extensively iron-framed office building of the period in the metropolis (Fig 11.13). Because of this, and the seeming independence of the floor structure and the façade, it has been fêted by some historians as a forerunner of the American skyscraper.[30] The façade, described by *The Building News* as the 'most successful front for offices yet erected in the City', was an exquisitely detailed carved-stone Ruskinian composition of superimposed tiers of Byzantine-Gothic arches designed to maximise 'void' or window area; behind this was a 'skeleton ... wholly of iron'.[31] The great depth of the building demanded extensive use of internal areas, the suites of offices bordering these fronted with 'iron and glass only, so as to admit the greatest quantity

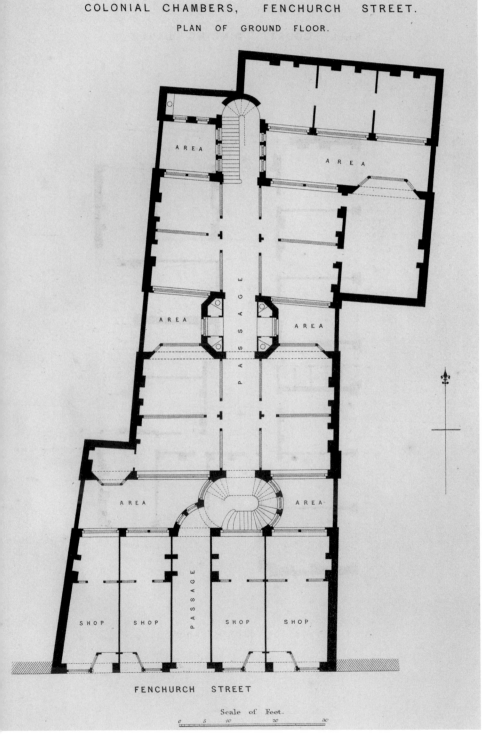

COLONIAL CHAMBERS, FENCHURCH STREET.

PLAN OF GROUND FLOOR.

AREA

AREA

AREA

AREA

PASSAGE

AREA

AREA

AREA

AREA

SHOP   SHOP   PASSAGE   SHOP   SHOP

FENCHURCH STREET

Scale of Feet.

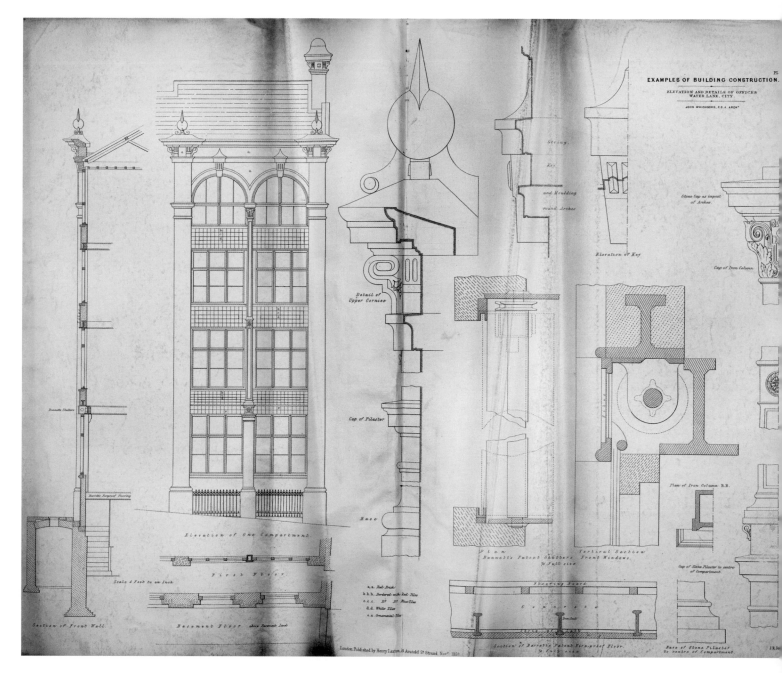

of light'.[32] Elsewhere, half-brick thick partitions were employed. The rear elevation, facing the churchyard of St Olave, Hart Street, demanded less architectural distinction, and exploited iron columns and beams to support the brickwork and large windows. Aitchison built at least one other speculative office 'on the same principle of iron construction',[33] in Mincing Lane (c 1864, demolished), and was a keen advocate of extending frameworks through to rear elevations, noting that in one of his (unidentified) buildings, 'the void is nearly 3–4ths of the whole area, or the gain is 218 feet of light'.[34]

From the 1860s banks and offices proliferated, stimulated by the Limited Liability Acts 1855 and 1862. This was accompanied by an increasing sophistication in construction, if not architectural design. Within their brick load-bearing walls, these buildings incorporated inventive, but increasingly standardised strategies for supporting multiple floors using structural iron. Built-up, wrought-iron girders with angles joining the web and flange plates, spanning between cast-iron columns or walls, enabled great latitude in the composition of internal spaces, and formed the primary support

Fig 11.11
Drawings showing iron stanchions and beams in the façade of offices in Water Lane (1858; John Whichcord, architect; demolished).
[Laxton, Examples of Building Construction ... *(1857)]*

201

Fig 11.12
National Provident
Institution, 48 Gracechurch
Street (1862–3, demolished
c 1957).

[The Builder, 3 January
1863]

for fire-resisting flooring systems. Whereas early banks and offices had mostly used techniques derived from textile mills and warehouses – segmental arches of brick spanning between iron beams – those from the 1860s tended to incorporate proprietary systems designed expressly for commercial buildings. By far the most common one up to the 1880s was the Fox and Barrett system (see p 12), employed in such prestigious buildings as F W Porter's Union Bank of London, No. 95 Chancery Lane (1865–6)[35] and P C Hardwick's Union Bank of London, Mansion House Street (1865–8).[36] Other patent systems, including Dennett's and Homan & Rodgers saw widespread application in the third quarter of the century (see Chapter 2).[37] The London & County Bank, Upper Street, Islington (c 1873, demolished), for example, used the latter system, framed into built-up beams that spanned the 26ft width of the banking hall and the floor above.[38] In most cases the floor system bore little relationship to the overall structural design of the building, often being supplied independently of the iron columns and primary girders, which were supplied and erected by the builder or engineering firm, not necessarily by the flooring company.[39] Architects frequently let the floor construction out for competitive tender once the masonry and iron superstructure was already resolved, selecting these on the basis of span and depth.[40] In this sense, the façade of the building, which in some of the mid-19th-century 'Palazzo' formulations echoed the regular spans of the segmental brick or tile arches within, was increasingly divorced from the interior structure.

If a progressively more standardised approach to floor construction was one feature of offices from the 1860s, then another was their increased size and their assimilation of other functions. This was most palpable in the City, where the huge demand for centrally located office space raised land values and rents to levels unequalled anywhere in the world. With their ready capital, new speculators such as the City of London Real Property Company, and the City Offices Company (both started in 1864) bought up old properties for redevelopment, combining neighbouring sites to realise ever-larger office blocks.[41] For the buildings to pay, developers sought to maximise the lettable space within their walls. Palmerston Buildings, No. 51 Bishopsgate (demolished) was said to have been 'the largest in extent of any recently erected'.[42] Built in 1867 to the designs of F & H Francis, architects to the City Offices Company, this building contained upwards of 260 offices. Within, spacious internal courts ensured good lighting, while thin, non-structural partition walls, enabled by internal iron framing, saved on space.[43] Two jack-arched basement levels housing strong-rooms and wine cellars yielded revenue from otherwise unprofitable space.

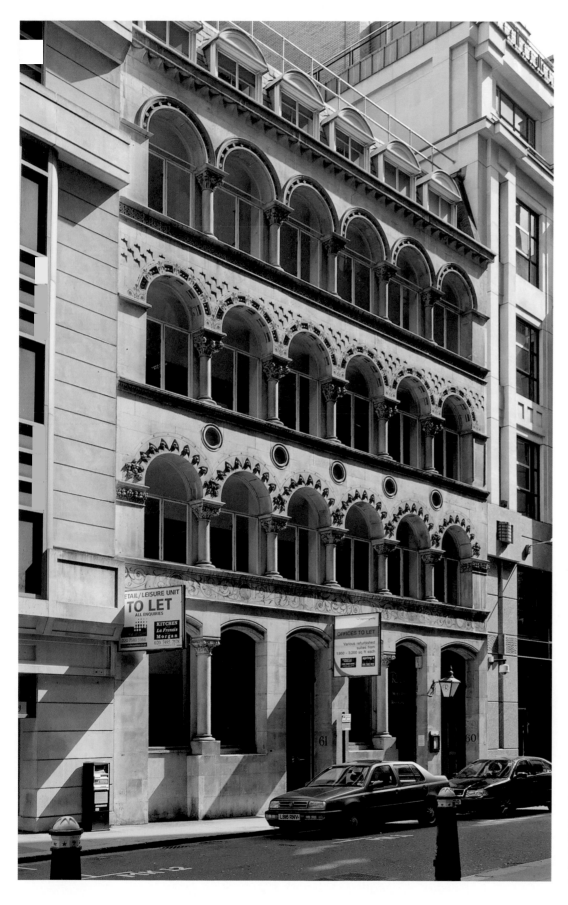

Fig 11.13
Nos 59–61 Mark Lane
(1864; George Aitchison
junior, architect),
perhaps London's most
comprehensively iron-
framed office building.
[Jonathan Clarke]

Such size and diversity came to typify speculative offices. Mansion House Buildings (Mappin and Webb Buildings), on the corner of Poultry and Queen Victoria Street (1870–1, demolished 1994), were designed by J and J Belcher with a 'fireproof' sub-basement for wine vaults, six shops on the ground floor and four upper floors for office use (Fig 11.14).[44] Contemporary comment suggests that the interior of this distinguished five-storey building was extensively framed in iron:

> The whole of the upper part is so arranged that no internal divisions are necessary, and yet that each floor can be sub-divided as may be found most convenient; the outer walls being tied in and the floors carried upon strong wrought-iron girders, with columns and stancheons [sic] to support them.[45]

This framing extended to the exterior at ground-floor level, with concealed stanchions dividing the plate-glass shop fronts and supporting bressummer girders that carried the façade above. Cast-iron stanchions built into the masonry piers of the upper levels may have been employed to take some of the interior floor loads, thus ensuring a firm bond between the façades and the interior structure. This degree of iron framing was not part of the architects' original scheme, for their working drawings described timber-framed floors on all but the first floor, which was to be carried on built-up girders. Possibly Richard Moreland & Son, suppliers of the ironwork and 'fireproof' floors, recommended that less timber be used.[46]

Fierce competition for a dwindling number of sites pushed prices out of the reach of many companies, forcing some to expand or relocate beyond the City boundary. One such was the Prudential Assurance Company, whose Ludgate Hill premises were, by the late 1870s, proving inadequate. The adjoining property was too small and too expensive to buy, so the board of directors decided that 'if an eligible site sufficiently large could be obtained in a position where land had not yet reached such high prices it would be advisable to secure it and build offices with a view to expansion as required'.[47] The site chosen was on the corner of High Holborn and Brooke Street, in Holborn, and would form the cornerstone of a prestigious office complex that was developed accretively from 1877 until 1906, mostly by Alfred Waterhouse.[48] For the first campaign, 1877–9, Waterhouse made extensive use of robust interior iron framing to create open floors on three levels, interrupted only by ranks of cast-iron columns and a central light well about 20ft by 40ft (Fig 11.15).[49] Built-up, I-section girders were employed to span transversely the 23ft space between the columns, which had

Fig 11.14 (facing page) Mansion House (Mappin and Webb) Buildings photographed shortly after completion in 1871 on a prominent site on the corner of Queen Victoria Street and Poultry. This speculative office block was demolished in 1994. [CC97/00299]

Fig 11.15 Demolition of the Brooke Street section of the Prudential Assurance Company Headquarters, Holborn Bars (1877–9) in 1932. The architect, Alfred Waterhouse, employed huge built-up I-section girders to create open floors on three levels. [Reproduced by kind permission of Prudential Group Archives, Prudential plc]

heavily reinforced capitals to spread the loads. Although the girders were encased in cement,[50] the use of traditional timber-joisted floors, framed directly onto the girders, is surprising for a building of such prestige at this date. Sheer spanning capability and not resistance to fire was clearly Waterhouse's primary concern, a deficiency he remedied in his subsequent buildings on the site.

By the 1880s huge office blocks were widespread. Many had ground-floor shops and basement storage with iron-framed, fire-resisting interiors clothed behind showy façades that vaunted often astonishing ratios of window to wall. Few if any, however, exceeded five or six storeys in height above ground level, despite the introduction into the City in 1882 of hydraulically powered lifts, a superior and cheaper alternative to the existing but rarely used steam- or gas-powered lift technology.[51] While lifts helped make five- and six-storey buildings more financially profitable (it was unrealistic to expect business clients to climb more than four or five flights of stairs), this 'vertical transport revolution'[52] did not help bring about the ever-increasing heights being reached in America. There, the office skyscraper was quickly evolved as 'the machine that made the land pay'.[53] From the passing of the 1855 Metropolitan Building Act until 1890, when the LCC introduced a General Powers Act limiting the height of buildings to 90ft, there were no legal restrictions on the heights of residential and commercial buildings in existing London streets (*see* p 250). Probably largely because of a fear of obstructing neighbouring properties' window light (a litigation minefield covered by the law of Ancient Lights),[54] City office designers of the 1880s chose not to exploit this vertical freedom. After 1894, when the LCC restricted the height of buildings to 80ft (exlusive of two mansard stories), this door was closed and remained so well into the following century.

The introduction of steel

Structural steel made its inaugural, if premature, appearance in building construction in a London bank (*see* Fig 2.13). C O Parnell's prestigious head office for the London and County Bank, Lombard Street (1860–1, demolished *c* 1965), used encased steel-plate girders to support the first floor because of their superior strength-to-depth ratio compared to wrought iron. Manufactured by Brown, Bragg and Co.,

of Sheffield, and proof-tested in the presence of the Bank's directors,[55] these 'splendid specimens of scientific construction … of unusual strength'[56] presumably spanned the external walls in a cross-configuration, gaining intermediate support from a large central iron column calculated to bear 270 tons.[57] By this means 'the whole of the space on the ground-floor [was made] available for the working of the establishment'.[58] This ground-floor space, subdivided only by 'dwarf partitions',[59] was considered 'a magnificent apartment',[60] worthy of the Italianate façade 'which we should have hardly expected to meet with out of Pall-mall'.[61] In executing this, the contractors, Jackson & Shaw, reused Portland stone from Charles Labelye's Westminster Bridge of 1738–50.[62]

Parnell, whose architectural practice was based in Pall Mall, had been appointed architect and surveyor to the bank in 1851. This seems to have been his first foray into bank design, as hitherto he had been occupied with the inspection and maintenance of the London and County's many branch premises, numbering 115 by the early 1860s.[63] Parnell died in 1865, and there is no evidence to suggest that he used steel in any of his other buildings. Other architects may have repeated this successful trial in the 1860s or 1870s, but in the absence of documentation of further instances, Parnell's Bank has to be seen as a constructional anomaly, an exceptional, daring experiment on the part of the architect and Sheffield's first manufacturer to adopt the puddling and Bessemer processes of bulk steelmaking. To call it a forerunner would be misleading, given that there was no viable economic trade in steel sections for the building trade until the 1880s. Indeed, it is not until the early 1890s that steel truly began to make its entrance in support of floors within the traditional bearing walls of banks and offices. Owing to the interchangeable use of the terms iron and steel in the architectural press of the 1880s, instances of steel construction in that decade are difficult to document, with banks and offices proving harder to pin down than other building types, possibly because they were being built in far greater numbers than the other forms of buildings considered in this book, and were thus separately less noteworthy. What little space was devoted to their description invariably concentrated on architectural form rather than construction technology. By this decade interior iron framing and fire-resisting floors were commonplace, the latter simply referred to by the proprietary name of the system, and any 'discreet revolution' involving the substitution of steel for iron went unrecognised and unreported. Such substitution was doubtless underway in the flooring systems of banks and insurance offices, especially those that were amply financed. One example might be the extension to the Bank of England, erected on the corner of Bell Yard and Fleet Street in 1886–8 to lavish Renaissance designs by (Sir) Arthur William Blomfield. *The Builder* recorded that this building 'will contain the usual arrangements of a bank on the ground-floor and basement, and a residence for the Agent on the upper floors', noting that 'the building will be fireproof throughout, Messrs Lindsay's system being used for floors and roof'.[64] By this date, W H Lindsay and Co. was producing its trademark 'trough flooring' in steel, a form of non-combustible floor that removed the need

for joists and, owing to its shallow depth, maximised available headroom. Already enjoying high-status use in Waterhouse's National Liberal Club (1884–7) it was probably used here too. Waterhouse also used it to span wide distances in his extension to the Prudential Assurance Company's headquarters in Holborn Bars, built in 1885–8 (Fig 11.16). Lindsay's sent their specification as early as 23 December 1884, and despite initial problems in achieving a sufficient depth of concrete over the decking, the floor had been completed by August 1885.[65]

One of the significant effects that structural steel had on the construction of banks and offices came in the 1890s, when steel stanchions began to replace interior cast-iron columns. This changeover was gradual compared to the speed with which flexural members in wrought iron were ousted by steel, largely because cast-iron columns were both inexpensive and strong

Fig 11.16
The Brooke Street extension of the Prudential headquarters (1885–8), showing Lindsay's steel-trough flooring. Photographed by Bedford Lemere in 1905. [BL18721–031]

(*see* Chapter 2). The type of construction seen for example in the Bank of Scotland, Bishopsgate (1896, demolished late 1970s), where cast-iron columns were used in conjunction with steel girders, was probably characteristic of a great many banks of the 1890s, if not the 1900s (Fig 11.17).[66] Designed by the bank specialist W W Gwyther[67] its spacious ground-floor banking hall, 60ft by 50ft, was made possible by steel girders carried on cast-iron columns, an assembly that took much of the weight of the four storeys above (Fig 11.18). The columns were somewhat special, 23ft high, cast vertically and turned on a lathe to give them an entasis appropriate for their classical surroundings. Four of these columns were set just inside the façade, freeing the piers of much of their load-bearing role so that windows of maximum size could be used.[68]

Despite the enduring popularity of cast-iron columns, the steel stanchion pointed to the future. A W Blomfield's little-known remodelling of 1895 to Barclay, Bevan & Co.'s premises on Lombard Street,[69] built by his former tutor, P C Hardwick, in 1864, was an early documented example in which built-up steel stanchions were employed. Designed and manufactured by Richard Moreland & Son, the tall, box-section members used there supported built-up steel girders that framed the spacious banking hall (*see* Fig 0.4). Prefabricated with generous bases to spread the loads directly onto the concrete floor and square caps onto which the lower flanges of the girders were affixed by bolts or field-rivets, they were almost certainly clothed with some fire-resisting and decorative finish. In all likelihood the framing continued upwards, for localised reinforcement of the girder above the cap in the illustration suggests it had to support at least one other tier of stanchions.

Stouter interior framing, in which steel members took the lead responsibility, was early

Fig 11.17
The banking hall of the Bank of Scotland, Bishopsgate (W W Gwyther, architect), photographed in 1896.
[BL13695]

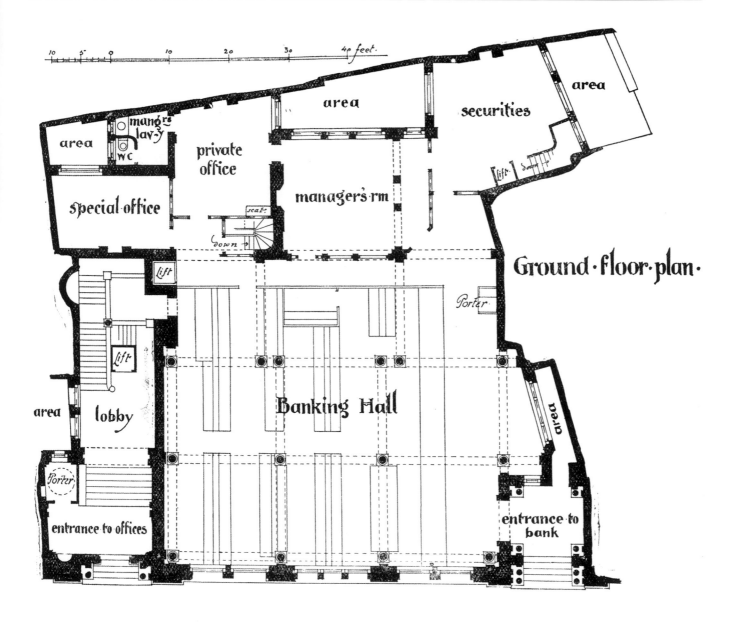

Ground · floor · plan ·

Fig 11.18
The Bank of Scotland,
Bishopsgate: the ground-
floor banking hall,
60ft-wide by 50ft-deep,
was made possible by steel
girders carried on cast-iron
columns.
[The Builder, 24 October
1896]

on employed in another branch bank, the National Provincial Bank, at Nos 207–209 Piccadilly, built in 1892–3 to designs by Alfred Waterhouse & Son. The 'superior character' of the existing building on this prominent corner site – a recently completed block of shops and residential chambers – at first induced the bank's directors to retain the upper floors and convert the ground and first floors to banking and office space, but proving unworkable, only the fire- and damp-resistant strong rooms were retained. Above those, a 'skeleton framework of iron and steel construction'[70] was built, protected by coke-breeze concrete, adorned in faience, and surmounted by a non-combustible roof of iron rafters and concrete flats and gutters (Fig 11.19).[71] The primary bearing elements

of this orderly framework were 14in. steel cruciform-section stanchions, supported for the most part by steel I-section girders, with wrought-iron girders used only sparingly. On the ground floor this structure enabled a high-ceilinged, open banking hall, well lit by huge windows in the façade, and above that four floors of residential chambers, each living space divided from the next by light coke-breeze concrete partitions. A light well above the clerk's room, with glazed-brick walls carried on the framework, ensured that the stairs and chambers at the rear of the building were fully lit. With over 30 years' experience in constructional iron and steel, it was Alfred Waterhouse himself who stipulated which metal should be used where, although a surviving note by him

209

Fig 11.19
The National Provincial
Bank, 207–209 Piccadilly
(1892–3; Alfred Waterhouse
& Son, architects), dotted
lines indicate the interior
steel framework.
[The Builder, 17 May 1893]

to the constructional engineers, Andrew Handyside & Co. informed them that they should size the beams. His note specified that 'internal girders carrying flooring, partitions, chimney etc., would need to be of mild steel, and not less than 2'–4" deep overall', but 'the girders carrying external brick wall by side of Glass roofing will do very well in iron and should be of box section not less than 1'–4" wide and need not be more than about 1'–6" deep'.[72]

More imposing than the National Provincial's banking hall was that of the Union Bank of Australia (now Banco Ambrosiano Veneto), at Nos 71–73 Cornhill, City, completed in 1896 to designs by Goymour Cuthbert. Here, the tall, aisled banking hall, with its high columns and coffered ceiling, presented the customer with a sense of monumental dignity and classical grandeur (Fig 11.20). The essential elements of this space, both visually and structurally, were

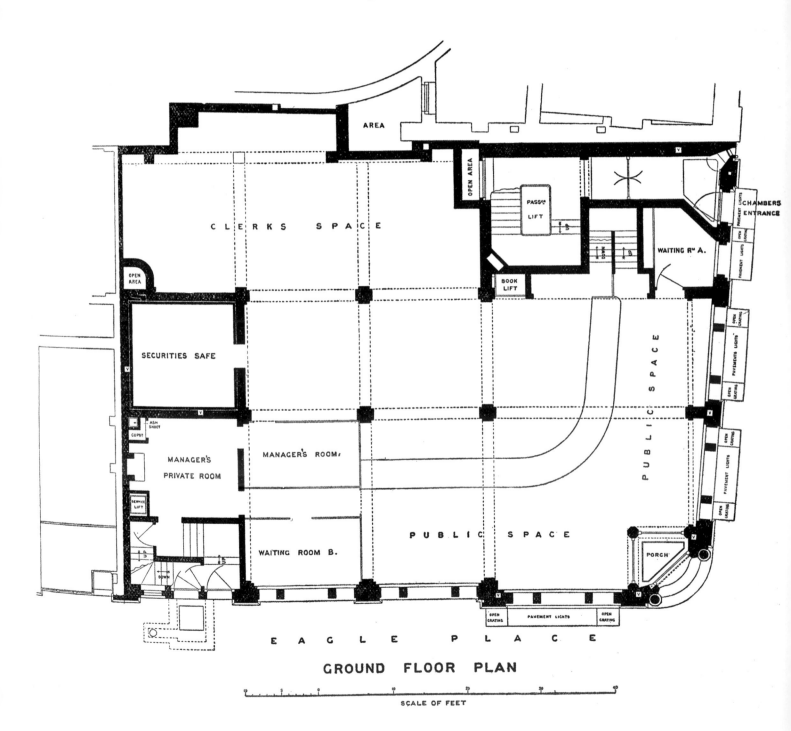

GROUND FLOOR PLAN

SCALE OF FEET

the 21½ ft-tall Ionic columns that gave support to an extensive steel-framed floor above. Concealed within these were built-up stanchions of more advanced design than those employed in the National Provincial. Measuring just 11in. in diameter, and fitted with steel caps and bases, they were formed from Lindsay's 'special angles and channels' as a square box-section, with additional reinforcement provided to the inside of the channels by 1in.-thick plate. Given great strength with the distribution of as much metal as possible away from the central neutral axis, they were tested at Kirkaldy's works where they 'admirably withstood' compressive loads of 237 tons apiece. Thomas C Cunnington, then employed by W H Lindsay, probably designed these stanchions, and the rest of the structural steelwork. An exponent of optimised, minimum-weight steelwork, Cunnington explained that:

> Where the interior of various rooms is decorated, the construction has to be confined within the limits of the proposed treatment and internal architecture. In city buildings, where space and light are of the most paramount importance, every thought and care has to be taken with a view to limiting the columns, stanchions, girders, and floors to the minimum sections consistent with their carrying power.[73]

The girders in the ground-floor ceiling were similarly designed with least depth, their flanges being unusually wide and the webs well stiffened to provide the strength commensurate with a minimal number of supporting columns.[74]

The apogee of the late Victorian banking hall was reached in the Birkbeck Bank, Chancery Lane (1895–6, demolished 1965), where extensive use of full steel framing took the form to soaring and fantastical heights. Built to designs by T E Knightley, the *pièce de résistance* of this enormous corner building was the domed banking hall set far back in the centre of the site. Typologically, this was the 'last grandiose grandchild of Soane's top-lit sequence at the Bank of England',[75] but structurally it was descended from the iron domes of the early to mid-19th century, most consciously James Bunning's London Coal Exchange of 1847–9 (demolished 1962). That structure used 32 richly decorated, slender radial ribs of cast iron to frame a dome that spanned 46ft, the interior encircled by three galleries cantilevered from the skeletal framework.[76] Exploiting the greater

Fig 11.20
The Union Bank of Australia (now Banco Ambrosiano Veneto), 71–73 Cornhill (1896; Goymour Cuthbert, architect).
[Builders' Journal and Architectural Review (Supplement) 9 December 1903, RIBA 46470, RIBA Library Photographs Collection]

tensile strength of mild steel, the Birkbeck Bank used just 16 ribs, each a simple curved I-section, 45ft in length, to frame a dome that spanned 72ft and rose to an overall height of 80ft above the circular floor (Fig 11.21).[77] Each rib was seated on a cast-iron stanchion embedded in the thick brick wall forming a 30ft-high drum, the outward thrust contained by a steel ring connecting the tops of the stanchions. The upper ends of the ribs were bolted to a smaller steel compression ring, 15ft in diameter, which formed the base of a circular iron and glass lantern. Lightweight hollow ceramic blocks and timber rafters, covered with boarding, felt and lead, spanned between each rib, forming the dome's 16 panels, each of which was pierced by an oval window. Inside, the dome was sumptuously lined with glazed tiles and Carraraware faience,[78] the ribs picked out with specially formed blocks that gripped the flanges yet allowed for thermal expansion and contraction of the metal. A 10ft-wide, steel-framed gallery, cantilevered from the stanchions and adorned

with faience corbels of giant white gryphons ridden by boys, encircled the spacious banking floor below (Fig 11.22). Below that, and supporting it, was a strong room with cast-iron, semi-circular vaulting, carefully constructed so as to 'not quite reach the sides [of the basement] so as to obviate any thrust of the arches'. To protect this vaulting 'in the extremely improbable case of [the cupola] falling', the banking floor structure was cushioned with 'noiseless' India rubber.[79]

The banking hall was set within a vast honeycomb of speculative offices, numbering about 400, in addition to those for the bank's own use. Rising to seven storeys in places, these offices and the numerous corridors serving them were of fire-resistant construction, the floors throughout carried by steel joists and girders embedded in concrete made of broken brick and cement. Two staircases served the offices, the main one rising to a height of six storeys either side of the top-lit entrance hall, between the principal entrance and the banking hall. This whole composition, with its extensive concrete landings connected by bridges, seems to have been realised as skeleton construction, probably for reasons of stability and light penetration (Fig 11.23). The end wall, screening the dome of the banking hall beyond, was a true curtain wall; steel girders carried the lantern glazing on the uppermost floor, and a wholly fenestrated first floor overlooked the cantilevered gallery, with brick infill panels of the floors in between. For the sake of appearances, glazing was positioned in front of the screen walls, giving the impression of one continuous glass wall. The designer and supplier of this steelwork is not given in contemporary sources, but it may have been Richard Moreland & Son. This firm of constructional engineers designed the steelwork, including the fully cantilevered balconies, for Knightley's other major building, Queen's Hall, Langham Place (1891–3) (see pp 124–5). The use of skeleton construction for the towering stair hall of the Birkbeck Bank, independent of the load-bearing-wall construction enclosing it, was an important technological advance, and one that in this country had rarely been presaged in iron. At this date, the preferred technique in London was still traditional bearing-wall construction throughout, with steel floor beams supported by brick walls, and more often than not, interior cast-iron columns or steel stanchions. This form of construction offered enough scope to satisfy the functional

requirements of most banks and offices, and although embodying a huge repertoire of approaches, for the most part continued to progress along increasingly rationalised lines as architects and engineering contractors gained greater confidence with steelwork. Alfred Waterhouse's final campaign of building at the Prudential Assurance Company's head offices was a tour de force of advanced interior steel framing combined with load-bearing brickwork.[80] Constructed over the period 1897–1901, and covering almost the entire block formerly occupied by the grandest of the Inns of Chancery, the Furnival's Inn development expanded the total 'footprint' of the headquarters to some 2½ acres, 'thus probably exceeding that of any insurance building in the world' (Fig 11.24).[81] The *Architectural Record* subsequently applauded its Gothic treatment as 'a successful vindication of the adaptability of that style to modern commercial purposes', and the choice of materials, noting 'The extensive use of terra cotta, faience and steel in itself suggests a due recognition of modern conditions – the acids of a great city, the possibility of diminishing the dangers from fire and the desirability of few internal walls

213

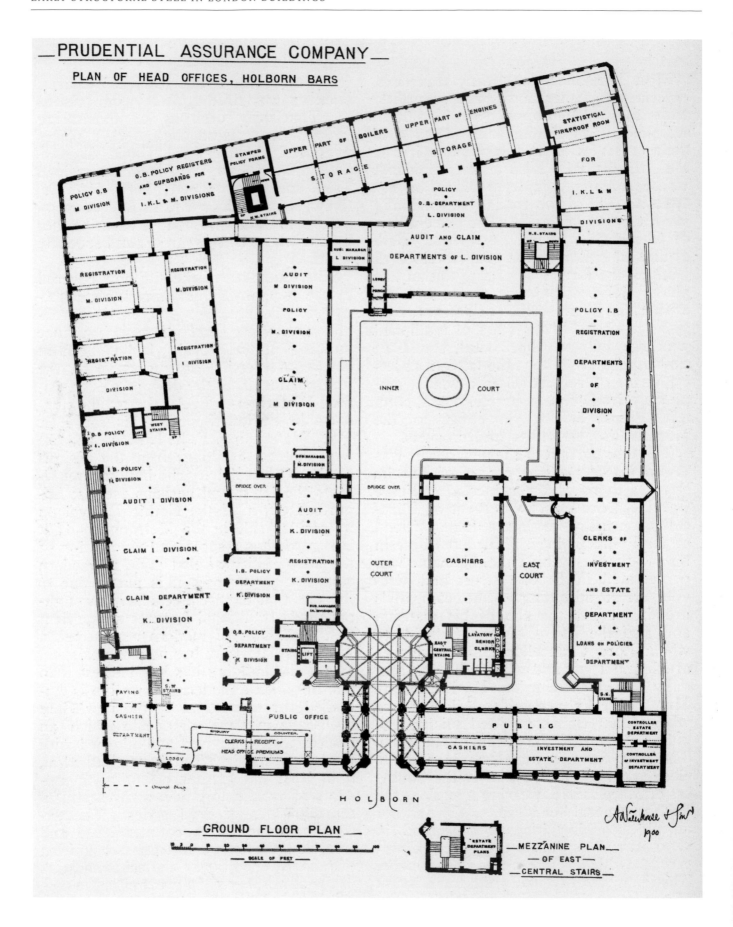

PRUDENTIAL ASSURANCE COMPANY

PLAN OF HEAD OFFICES, HOLBORN BARS

GROUND FLOOR PLAN

SCALE OF FEET

MEZZANINE PLAN
OF EAST
CENTRAL STAIRS

and piers caused by the great value of land.'[82] It was in pursuit of this last desideratum that steel proved so central to Waterhouse's conception; over 1,500 tons of it,[83] designed and erected by Handysides, was used to realise superbly open interiors that enhanced circulation, maximised natural lighting and yet looked reassuringly solid. Waterhouse reproduced the structural form of the stanchions in faceted faience sheathing, so that the built-up form is intelligible despite the loss of the steel contract drawings. Indeed, what appear to be relatives of the eight-sided 'Phoenix' column seem to have been used in some of the most opulent spaces, including the cashiers' office and public office (Figs 11.25 and 11.26). This form of column had already enjoyed much use in American offices and other buildings of the 1880s and 1890s[84] because of its great resistance against bending and buckling. Yet it was probably their decorative potential that led Waterhouse to choose them. For the 'Wood's Hotel' block to the rear of the Furnival's Inn building, erected in 1895–6, he used what appear to be steel box-section stanchions, formed from splayed and regular channels (Fig 11.27). Projecting angle-brackets

Fig 11.24 (facing page) The Prudential's headquarters (Alfred Waterhouse, architect). By 1900 the building covered some 2½ acres. [The Architectural Review 21 (February 1907)]

Fig 11.25 (left) The Cashiers' Office, Prudential Assurance Company's head offices: faience-clad built-up steel columns. [BL18721–006]

Fig 11.26 Public Office, Prudential Assurance Company's head offices. [BL18721–001]

Fig 11.27
Prudential Assurance
Company's head offices,
'Wood's Hotel' block
(1895–6), faience-clad steel
box-section stanchions
carrying steel-girder floors
with unconcealed flanges.
[BL18721–32]

Fig 11.27
Prudential Assurance
Company's head offices,
'Wood's Hotel' block
(1895–6), faience-clad steel
box-section stanchions
carrying steel-girder floors
with unconcealed flanges.
[BL18721–32]

on the column heads, covered by specially cast ceramic pieces, provide a rigid connection and spread the loads from the beams, the longer-spanning ones left unprotected from fire in an intrepid display of riveted steel structure.

Another indication of increasingly sophisticated interior framing was the use of continuous steel stanchions, spliced directly together, with beams framed into them, rather than the traditional, and weaker technique of using vertical assemblies of single-storey-high columns interrupted by beam connections. One of the first architects to exploit this technique in commercial architecture was Aston Webb, in his General Office, Nos 115–121 Tooley Street, Southwark (1899–1901), for the distillery firm Boord & Son. Here, structural steel was judiciously employed in tandem with exterior load-bearing walls to create a large, well-lit atrium complete with first-floor gallery (Fig 11.28). This visually imposing architectural space, clearly designed to impress visitors to the firm's prestige commercial headquarters, was dependent on interior steel framing of minimal sectional dimensions, fabricated by Dorman Long & Co. The atrium roof is supported by three slender, arched steel

ribs, each spliced directly to continuous stanchions (concealed by decorative wood panelling) that carry on downwards to a basement where their form is exposed. The stanchions also support a grid of beams (the outer ends of which are embedded in the exterior brickwork), which carry the gallery and interior walls. The majority of structural steel members were simple rolled sections, not complex assemblies of smaller components. The stanchions were basic I-sections, 10in. by 8in., or 'B.S.B. 19's, one of the new range of 30 standard sizes Dorman Long was producing in advance of official standardisation in 1903 (see Chapter 2). The joists and beams, at least those visible in the basement, were rolled by the Consett Iron Company, of County Durham, and are rigidly joined to the stanchion faces by riveted and bolted angles and cleats (Fig 11.29). Also illustrative of the increasing substitution of rolled beams for built-up forms by the end of the century, and of the use of these to frame curved skylights, is the Patent Office Library, Holborn, erected in 1899–1902 by the Office of Works under Sir Henry Tanner to designs by Sir John Taylor. Throughout the complex, which included

adjoining offices, the floors were formed from Homan & Rodgers' 'ordinary concrete and steel construction', consisting of rolled I-sections encased in concrete. Within the rectangular brick shell of the library itself were two tiers of steel-framed galleries carried from the walls and supported on fluted cast-iron Corinthian columns (Fig 11.30). Semi-circular arched steel ribs carried a continuous central skylight and helped frame generous clerestory windows.[85]

For prestige banks and offices the use of interior framing behind load-bearing brick or stone exteriors, using rolled-steel joists and beams, and, increasingly, steel uprights, became almost standard practice in the late 1890s and early 1900s. Another trend was the employment of concealed steel members within the façades of these buildings. Increasing scale, grandeur and compositional freedom were indubitably the result, but contemporary descriptions in the

architectural press rarely provide more than the barest outline of the techniques involved. The interior of the five-storey offices at Nos 76–78 St Paul's Churchyard, for the drapery firm James Spence & Co., is recorded as having been 'carried on steel stanchions, and the floors are of steel girders and coke-breeze concrete by Messrs. Homan & Rodgers' (Fig 11.31).[86] Built in 1898 to designs by Professor Banister Fletcher and his son Banister F Fletcher, this long-demolished building sported enormous plate-glass windows in the ground-level showrooms. It is likely that the rusticated Ionic pilasters dividing the windows concealed steel stanchions, for steel beams, cantilevered from the bressummer girders spanning these pilasters, were used to carry the four projecting bays. This framing, and that carrying the floors

within, was no doubt of sophisticated design, using a limited range of components, for by this date the Manchester-based constructional engineering firm Homan & Rodgers were advertising 'Constructional Steel Skeleton Buildings (American System)' in addition to their usual structural sections and fire-resistant floor systems.

By 1900 new and more ambitious techniques of framing that reduced or obviated the load-bearing role of the masonry walls were being tried and refined in office building. Among the earliest instances was a small, speculative office by City specialist Edward Ellis at Nos 3–4 Rood Lane (c 1899, demolished). 'Ancient Lights' easements restricted the height to just four storeys above pavement level, and demanded a stepped profile to the rear. Both this elevation and the 'Queen Anne' façade were fenestrated

Fig 11.30 (facing page)
Patent Office Library,
25 Southampton Buildings,
1899–1902.
[London Metropolitan
Archives, City of London
(SC/PHL/02/0009/75/
22809)]

Fig 11.31 (below)
Nos 76–78 St Paul's
Churchyard (1898;
Banister Fletcher and
Banister F Fletcher,
architects; demolished).
[The Builder, 20 August
1898]

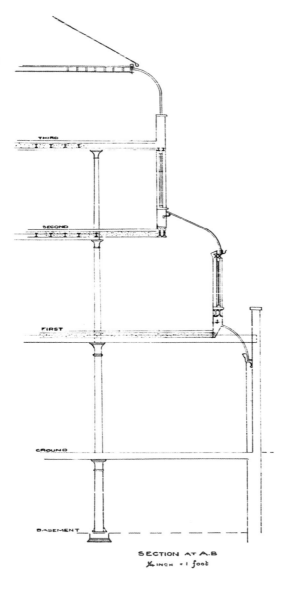

SECTION AT A.B
¼ INCH = 1 foot

continuity across the building, tying the walls together.[87] Nevertheless, given the inherent lack of stiffness in column-to-column connections, it seems likely that some rigidity derived from the brick walls and interior partitions. Trollope & Sons and Colls & Sons (later combined as Trollope & Colls), the general contractors for the project, may have designed as well as erected the framework; certainly Trollope & Colls were responsible for a number of early Edwardian fully framed buildings, and by 1912 had structural expert Oscar Faber (1886–1956) as their chief engineer.[88]

Full steel-skeleton framing may have been introduced at a similarly early date to a City office-cum-shop by Richard Moreland & Son, at Nos 50–53 St Paul's Churchyard (c 1901, demolished 1960s) for D Nicholson & Co., silk merchants. Built with a continuous curvilinear façade on a peninsula-site bounded by St Paul's Churchyard, Cheapside and Paternoster Square, this highly glazed building by architects Gordon, Lowther & Gunton, was, according to the structural engineer Richard Moreland junior, the 'nearest approach to an American building he had seen in London':

> The building was somewhat unique in a way, for there was no brickwork up to the second floor level. It was a narrow building, with glass all round, and the superstructure was carried on steel columns, which were kept back so that the goods could be displayed; the wall above was cantilevered over.[89]

This building, which was ostensibly in contravention of the 1894 Act wall-thickness regulations, must have presented a remarkable sight. A surviving drawing by the architects shows the astonishing amount of glazing on the lower two floors, made possible by solid steel columns extending from the basement to second floor at the building's perimeter, and framed rolled-steel stanchions and beams above and within (Fig 11.33). Perhaps wary of the strength and stability of Richard Moreland & Son's steelwork construction, Gordon, Lowther & Gunton commissioned the well-known engineer Arthur Walmisley to examine the strength of the solid steel columns and cantilevers at the front of the building, all of which met his approval. The 27ft-high solid steel columns, and their solid steel caps and bases, were supplied by the long-established Kirkstall Forge Works in West Yorkshire.[90]

to the maximum extent permitted under the 1894 Act regulations, and it was probably for this reason that Ellis used full framing for the upper storeys (Fig 11.32). Inside, vertically linked storey-high cast-iron columns and stanchions, inserted into the pre-existing party walls, and arranged within the interior, were used to transmit the load of each floor to cast-iron bed plates on brick spread footings. Four different column/stanchion types were used according to the location and loading: hollow circular, cruciform, H- and ⚎-section. These stanchions and columns, connected one to the other with spigot joints, carried almost all of the primary girders framing the steel filler-joist floors. The use of collar plates, which passed round the spigot and connected the webs of each pair of beams, gave the beams horizontal

Partial cage framing was used in Richard Creed's River Plate House, Nos 7–11 Finsbury Circus, City (1901, demolished and rebuilt in similar exterior style 1986–90). In this, H-section steel stanchions embedded in the external and party walls were used to carry the load of the principal beams, and less obstructive solid steel columns were employed in freestanding locations on the ground and basement floors. Secondary floor beams were, however, carried by the enclosing brickwork. The stanchions were seated on 'American style' grillages, formed from rolled-steel joists embedded in concrete.[91] The building was erected by Trollope & Sons and Colls & Sons, who, again, may have designed the steelwork, possibly subcontracting the steel columns to Richard Moreland & Son.

Despite embodying new technical developments, such buildings probably had little influence on established building practice. With next to no coverage in the architectural press, understanding of their structural design remained locked within the circle of specialist architects and contractors responsible. Even with much larger, more promoted, and more influential buildings, the metal anatomy was

Fig 11.33
Nos 50–53 St Paul's Churchyard (1899; Gordon, Lowther & Gunton, architects).
[London Metropolitan Archives, City of London (GLC/AR/BR/27/ES/09225)]

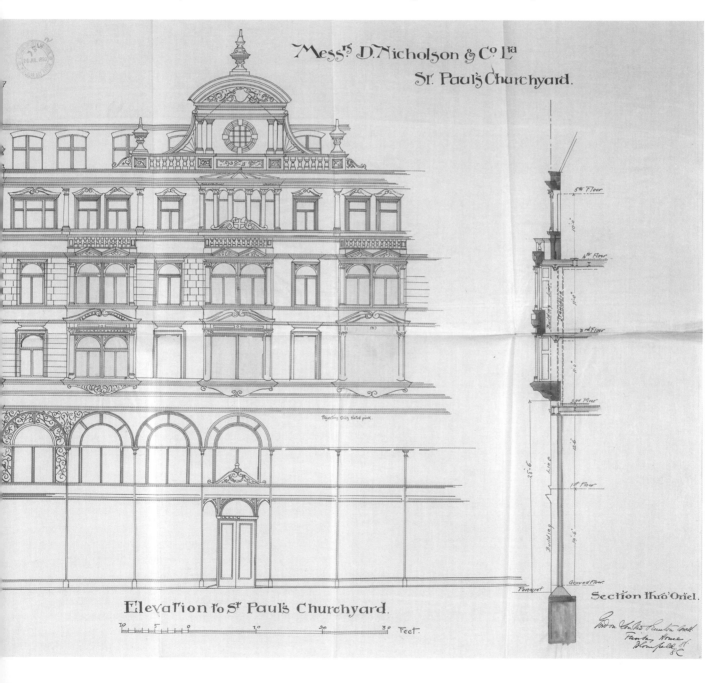

Mess.ʳˢ D. Nicholson & Co Lᵗᵈ
St. Paul's Churchyard.

Elevation to St Paul's Churchyard.

Section thro' Oriel.

invariably the least appraised aspect. When River Plate House was being completed, work had begun on what would be a considerably more renowned building by a major architect, although its concealed, partial steel-skeleton frame received virtually no attention. John Belcher's Electra House, No. 84 Moorgate, City (designed 1900, built 1901–3), 'widely illustrated, praised and influential from 1900 on',[92] effectively relied on its inner framework to see it through to completion (Fig 11.34). Early on in the construction of this mammoth block, designed to house submarine-telegraph companies, it was resolved to carry the Baroque façade on steel stanchions embedded in the granite piers, these in themselves being deemed too slender to carry safely the superimposed loads from above. Thomas C Cunnington, who had already worked with Belcher on the steel-framed roof of Colchester Town Hall (1898–1902), may have been drafted in at this point, but it seems more likely that he worked on the project from the outset. Cunnington first announced his contribution to the building in a paper before the Architectural Association:

> The front walls are carried upon stanchions within the granite piers, they (the granite piers) not being of sufficient sectional area to support the superincumbent load. These stanchions, together with the loads previously mentioned, also support the divisional walls and chimney breasts on plate girders from first floor to roof above.[93]

Henry Heathcote Statham, the editor of *The Builder* who had already deplored the use of a concealed steel skeleton for Tower Bridge, was horrified at the revelation of yet further 'sham' architecture, remarking:

> That was a terrible passage to hear read before the architectural society, and in a paper on architectural engineering. If we were going to build in that way, he did not see where the architecture came in at all. Architecture meant, among other things, not only building that was strong, but building that looked what it was. If we were going to erect a building which entirely depended upon steel piers and framing to carry the load, and then enclosed it with a skin which was intended to make the spectator think that the building was built of granite, but was nevertheless obviously insufficient as

such, he did not think there could be a greater disaster to architectural design.[94]

It was not until 6 years after the building's completion, when full steel framing was becoming commonplace and had been sanctioned in London, that a structural appraisal appeared in an architectural periodical:

> The steel framework is carried up to the roof. The 30 feet retaining wall to the basement and lower ground floor is of concrete, reinforced with indented bars. The building is carried on seven steel stanchions, each supporting from 400 to 600 tons dead weight.[95]

It seems doubtful that just seven steel stanchions were used in a building of such size and spatial complexity. Nevertheless, skeleton framing, which at the minimum carried the façade, divisional walls and much of the floor loads, seems to have been employed by Belcher to give him sufficient structural freedom for dignified elevational expression. Without concealed stanchions, the rusticated piers would have been too massive and awkward. In other buildings, the reasons for introducing the technique were often more prosaic. For the head offices of the London and Manchester Industrial Assurance Co., built on the corner of Moorgate Street and Finsbury Square in 1903–4 to designs by Gilbert & Constanduros[96] (demolished), full cage construction helped remedy the problems of the site's shaky foundations, caused by underground rail and motorised road traffic (Fig 11.35). In its brief description, *The Building News* explained that 'Owing to the excessive vibration arising from the tube railways and heavy road traffic, it was deemed advisable to erect the structure upon a concrete raft 5ft thick, reinforced with steel joists, and also to construct a steel frame of stanchions and girders taking the weight of the floors independently of the outside walls, the walls being themselves tied to the stanchions with strong ties.'[97] The self-supporting walls, generously fenestrated, were built up in Cornish granite to second-floor level, and above that in Portland stone.

At Caxton House, Tothill Street, Westminster (demolished 1970s), skeleton framing met the traditional office demands of maximum fenestration and spatial openness and flexibility (Fig 11.36).[98] Erected in 1905–6 for a major publishing company on a part of the site formerly occupied by the Royal Aquarium, the building

Fig 11.34 (facing page) Electra House, 84 Moorgate, City (1901–3; John Belcher, architect). [BB000793]

was designed to cater for other uses, including speculative offices. A concrete raft was used to provide a secure foundation for the skeleton of steel which framed the three lower floors, housing a restaurant, bank and flexible office space (Fig 11.37). This skeleton also carried the exterior walls, which had to conform to the load-bearing thickness prescribed by the 1894 Act. The four floors above this, which housed smaller office suites, although carried on the lower framework, were in themselves of conventional bearing-wall construction on the 'double corridor' plan-form, with the upper portion of all the corridors glazed as high as the ceiling.[99] Throughout, reinforced-concrete floors were used for noise- and fire-resistance, and each speculative floor was left entirely open, enabling the tenants to configure the space to their requirements. In this respect it was claimed to have been designed upon 'American lines much amplified … and meets a long-felt want in London, as large areas of floor space, exceptionally well lighted and unbroken by intermediate walls, are obtained'.[100] The structural engineer responsible for the frame is not recorded in contemporary accounts; he was possibly within the employ of the general con-

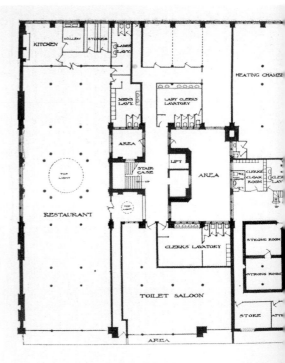

BASEMENT · PLAN

tractor, Messrs Holloway Brothers.[101] The architect, James G S Gibson (1861–1951), subsequently went on to design the steel-framed Middlesex Guildhall (1911–13).[102] As Caxton House was being erected, so too was an office-cum-hotel block now known as Carlisle House, Nos 8–10 Southampton Row, Holborn (1905–6) (Fig 11.38). Designed by Bradshaw and Gass of the Bolton firm of architects Bradshaw Gass and Hope, this eight-storey-tall stone-faced tower occupies an isolated site that was made available because of the LCC's Kingsway and Strand improvement scheme. It was planned so that offices of the Royal London Friendly Society were located on the lower floors with accommodation for the Tollard Royal Hotel sited above;[103] consequently the architects extracted as much space from the restricted site as building technology and building regulations would allow. A skeleton frame carried the turreted and mansarded block to a height of 103ft, 18ft (two storeys) of which were below the pavement level. Fabricated, erected and possibly designed by the Fulham Steel Works Co. Ltd, and resting on steel grillage foundations, this framework gave 'exceptional rigidity' to the building, which was internally divided

Fig 11.35 (facing page)
Offices of the London and Manchester Industrial Assurance Co., Moorgate Street and Finsbury Square (1903–4; Gilbert and Constanduros, architects; demolished). Full cage construction in steel was used here to resist vibrations caused by the underground railway and road traffic.
[Architect and Contract Reporter 72 (4 November 1904)]

Fig 11.36 (left)
The internal light well of Caxton House, Tothill Street, Westminster (1905–6; demolished 1970s), showing the thinly clad inner walls that maximised window area.
[BL20424]

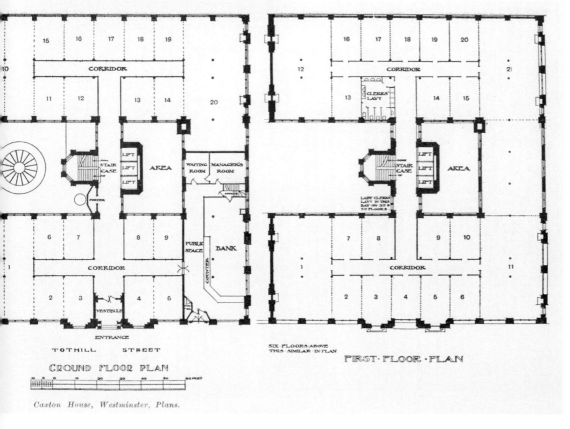

Caxton House, Westminster, Plans.

Fig 11.37
Caxton House, Tothill Street, Westminster: plans.
[The Builder, 15 September 1906]

with plaster slab partitions.[104] 'Practically, there is no load on the walls', noted *The British Architect*, 'but in accordance with the L. C. C. Building Act, the walls had to be constructed without reference to the encased steel framing, and very valuable space is wasted'.[105]

The building that really announced the arrival of sophisticated skeleton framing in English office building, and arguably Britain's most influential pre-1909 steel-framed building

after The Ritz Hotel and Selfridges, was the Morning Post Building (Inveresk House), built in 1905–7 at the corner of Aldwych and the Strand (Fig 11.39). Ironically, its construction was effectively forced by the LCC, which had demolished the newspaper's earlier offices, erected as recently as 1893,[106] for the creation of Kingsway and Aldwych. Under the terms of the LCC (Improvements) Act 1899, the LCC was required to cover the costs of rebuilding on an

Fig 11.38
Carlisle House, 8–10
Southampton Row (1905–6).
[BB000750]

adjacent site, although the owner, Lord Glenesk, had to bear any extra costs involved if the new offices were of a more elaborate character than the old. Glenesk was also required to complete the new building within 18 months from the conveyance on 12 May 1905.[107] With their proven track record of prestigious yet fast-build architecture, the dynamic Ritz team – architects Mewès and Davis, structural engineer Sven Bylander and contractors Waring White Building Co. – were perhaps an obvious choice. Within the space of 16 months they produced one of London's most elegant Beaux-Arts buildings, a huge triangular-shaped edifice facing across Waterloo Bridge and occupying 'quite the finest [site] in London'.[108] Some 950 tons of German steel, all of it rolled to order and prefabricated to exactitude, was used in building the frame, which was erected with typical rapidity, at the rate of one floor per week

Fig 11.39
The Morning Post Building (Inveresk House). One of London's most elegant Beaux-Arts buildings, it was by the same team responsible for The Ritz Hotel – that is, Mewès & Davis, Bylander and the Waring White Building Co. [Modern Building Record v2 (1910)]

Fig 11.40 (above)
The precisely engineered steel framework of the Morning Post Building. Special curved plate girders carried the rounded corners, and giant triangular gusset plates strengthened the framework against wind.
[Builders' Journal and Architectural Engineer 24 (C&S Supplement), 10 October 1906]

Fig 11.41 (right)
Similar skeletons: New York's Flat Iron Building (1901–3) was an immediate prototype for the Morning Post Building.
[© The British Library Board (F60110–01)]

(Fig 11.40).[109] In compliance with the legislation, this carried all the loads, including the thick Norwegian-granite-faced walls.

Despite the speed with which the framework went up, its overall construction time was sluggish compared to American counterparts, and could certainly be matched by conventional methods in London. The reasons for the adoption of full-skeleton framing probably owed more to the irregular, triangular-shaped site and the functional requirements of the client. Conventional bearing-wall methods would probably have failed to deal satisfactorily with the sharp angle jutting out into the Strand, forming the apex of the triangle. The use of a steel skeleton enabled full exploitation of this space at basement and sub-basement level, unbroken by the substantial brick or masonry piers that would otherwise have been required.[110] Indeed, the floor area of the building below street level was almost the same as that above.[111] The two basement levels housed the printing presses and mechanical equipment, and, to prevent vibration being transmitted to the stanchions, the walls supporting them were bedded on separate grillage foundations. Above the basement was an Advertisement Hall, occupying the ground floor; an open-plan first floor, used by editorial staff; speculative offices on the second and third floors; and composing rooms and readers' rooms on the fifth floor, which also housed the foundry, where the plates were cast.[112] The site afforded functional advantages, chiefly light and air from each of the three streets surrounding it, although it was still necessary to introduce a light well in the centre of the building. But the awkward site also created framing problems, particularly relating to the precise size and alignment of the many hundreds of beams, so that they correctly corresponded with the column grid. Practically all the beams were on a skew, with hardly any at right angles to one another, although Bylander asserted that, 'not a single measurement was scaled or taken from the site'.[113] To form the distinctive corners of the building, special curved plate girders were designed, cantilevered from the stanchions by means of huge gusset plates. These triangular solid-web knee braces also served as wind bracing, an essential feature for a building so exposed as this, standing well clear from its neighbours. This was perhaps the first British building to exploit these techniques, both of which had been conspicuously used in the framing of the Flatiron Building in New York (1901–3, D H Burnham & Company, architects), of which Bylander was undoubtedly aware (Fig 11.41). *The Builders' Journal and Architectural Engineer* noted the overall similarity of the Morning Post Building, when still at the skeleton stage, to this much larger structure, but added, reassuringly, 'but the building ... is not to be a skyscraper'.[114]

The Morning Post embodied the full standardisation and efficiency that was at the heart of this system of building. From the production of copious, fully dimensioned drawings and the use of standardised members, to the orderly contractual and erection methods, it symbolised a new mindset in the British architectural and building world. The Ritz and the Morning Post received extensive and detailed coverage in the same periodical, which monitored both buildings' progress with photographs and framing plans from start to finish,[115] and advised its readers that the constructional techniques employed would 'well repay careful study'.[116] The Morning Post building was highly influential, both structurally and architecturally. *The Builder*, in its belated, and less laudatory, notice of the building in April 1908 could observe that full-steel framing was 'now so common'.[117] Two months later *The Builders' Journal and Architectural Engineer* noted that 'the tendency to utilise steel frames for the construction of modern buildings is increasing rapidly'.[118] Despite the restrictions of the 1894 Act that continued to stipulate exterior walls of load-bearing thickness, the number of offices using full or partial steel-skeleton framing mushroomed in the period 1906–9.[119] These included some of the most outstanding works of the Edwardian era, buildings where the powerful verticality of the underlying framework was finally given architectural expression. The Royal Insurance Building, St James's Street, Piccadilly (1907–8, Belcher & Joass) was carried on a skeleton fabricated and erected by Manchester-based engineers Edward Wood and Co. Ltd. The most extreme example of the then popular neo-Mannerist style, its attenuated façades of Pentelikon marble barely covered the seven steel perimeter stanchions within, each of which supported a load of between 400 and 600 tons apiece (Fig 11.42).[120] Less extreme stylistically, but more exemplary of the glazing advantages of the steel frame, was the office building designed by the same architects for Mappin & Webb at Nos 156–162

Oxford Street (1906–8) (Fig 11.43). In this, the Pentelikon marble skin was so reduced that there could be no doubt that this was not an ordinary bearing-wall building. Within, the steel framework by Edward Wood and Co. Ltd enabled a maximum amount of clear space (Fig 11.44).[121] Other prestigious offices, including those of the British Medical Association at No. 429 Strand (1907–8, Charles Holden, architect) and Koh-I-Noor House, Kingsway (1909–10, James S Gibson and Skipwith & Gordon, architects),[122] made similar allusion to their orthogonal, full-framed underpinnings (Figs 11.45 and 11.46). However, in most cases the references were more obscure.

The functional separation of the frame and the cladding not only presented architects with a choice as to whether to articulate it or not, but also opened up a greater repertoire of elevational treatments and fenestration, which could be modified far later in the building process than was possible with load-bearing façades. Structurally, the exterior walls, carried storey by storey on the frame, were analogous to the non-structural partitions within, and as such could be removed to allow horizontal expansion with a minimum of interference. This

Fig 11.42 (facing page)
The Royal Insurance Building, St James's Street, Piccadilly (1907–8; Belcher & Joass, architects). Taut Neo-Mannerist façades express the steel skeleton within.
[Modern Building Record v1, 1910]

Fig 11.43 (left)
Mappin & Webb building, Oxford Street (1906–8; Belcher & Joass, architects).
[Modern Building Record v1, 1910]

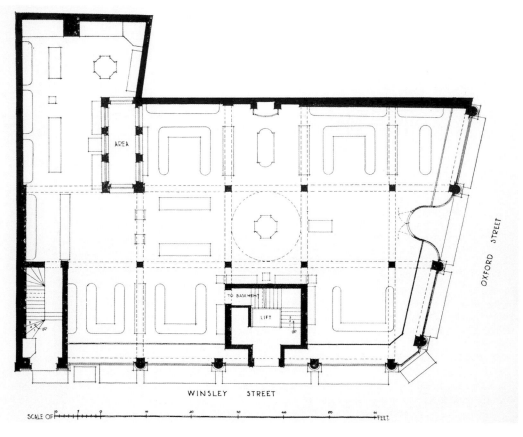

Fig 11.44
Mappin & Webb building: ground plans showing the openness enabled by full framing.
[The Architectural Review 24, November 1908]

Fig 11.45
The British Medical Association's former offices at the corner of Strand and Agar Street, now Zimbabwe House (1907–8; Charles Holden, architect).
[Jonathan Clarke]

facility had already been tried at the Guinness Storehouse (1902–4, *see* p 99), but perhaps the first London building to plan for it in the original design was the Westminster Trust Building, Broadway, Westminster (*c* 1908, demolished), rapidly erected on a giant skeleton framework by constructional engineers Redpath, Brown

and Co., Ltd (Fig 11.47). One of the last buildings obliged to comply with the wall-thickness requirements of the 1894 Act, it was nevertheless 'designed so that the walls may be supported at first-floor level on all fronts independent of any brick foundations, so that in the event of any re-arrangement of the building at ground-floor

Fig 11.46
Koh-I-Noor House,
Kingsway (1908; Gibson,
Skipwith & Gordon,
architects), another building
to exploit framed verticality
on a narrow city-centre site.
[Modern Building Record
v2, 1910]

Fig 11.47
The Westminster Trust
Building, Broadway,
Westminster (c 1908,
demolished). The giant
steel skeleton designed
and erected by Redpath,
Brown and Co., Ltd., was
intended to allow for
horizontal expansion.
[Builders' Journal and
Architectural Review,
17 June 1908]

Fig 11.47
The Westminster Trust Building, Broadway, Westminster (c 1908, demolished). The giant steel skeleton designed and erected by Redpath, Brown and Co., Ltd., was intended to allow for horizontal expansion. [Builders' Journal and Architectural Review, 17 June 1908]

level the entire masonry fascia may be taken away between ground and first floor'.[123] Speed of erection was also a key factor in the choice of skeleton construction, the laying of the noise-deadening, hollow-tile 'Kleine' floors keeping pace with the framework.[124] The involvement of the Kleine Patent Fire-Resisting Flooring Syndicate might be taken as an indication of the increasingly cosmopolitan nature of office construction in late Edwardian London, by which period the influence of Sven Bylander's precise and systematic procedures for designing, detailing and erecting frameworks had fully permeated into the practice of Redpath, Brown and Co., Ltd. *The Builders' Journal and Architectural Engineer* reproduced a drawing of one of the stanchions, which, apart from the use of imperial measurements, was nearly identical in its detailing to those produced under Bylander's direction (Fig 11.48). The publication noted:

It will be seen from the drawing that the details are prepared by this firm in accordance with the most up-to-date and most approved principles. The former method of making templates for every part of the construction is done away with, and the dimensions are set out directly upon the sections from the figures given upon the drawings. The drawing of the stanchion

detail is particularly interesting, showing connections economical in manufacture and erection. The differences in detail from the work of other designers will be apparent on close inspection.[125]

By the end of the decade, structural steel was omnipresent in the construction of metropolitan offices, as is evident in one publication's reportage of a new City block:

The office buildings recently erected by Messrs. Holland & Hannen, of London, known as 'Friars House,' Broad Street, offer an interesting example of modern methods of construction [reinforced concrete]. It is, we believe, one of the first modern buildings in the City of London to be erected *without the aid of structural steel of any kind*; in fact, there is not a single rolled steel joist in the building.[126]

This was unsurprising given its superior physical properties, falling cost and the simplification in the design of components and structures that came with standardised sections in 1903. More significant was the effect of steel framing on the conceptual basis of office construction. The progressive adoption of advanced building techniques and methodologies reliant on structural steel had, by c 1905, resulted in

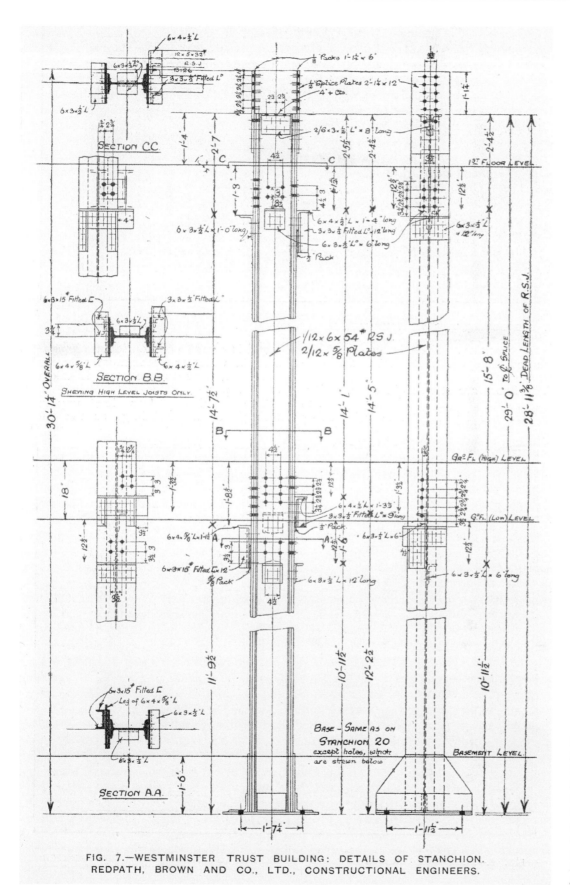

FIG. 7.—WESTMINSTER TRUST BUILDING: DETAILS OF STANCHION.
REDPATH, BROWN AND CO., LTD., CONSTRUCTIONAL ENGINEERS.

Fig 11.48
Drawing of stanchion used in the Westminster Trust Building. This level of detail was symptomatic of new design practice pioneered and publicised in this country by Sven Bylander. [Builders' Journal and Architectural Review, 17 June 1908]

the obsolescence of the masonry bearing wall for London's leading office buildings. The exacting functional requirements of this specialised form of building demanded increasingly sophisticated and flexible responses from architects and their engineering contractors, and skeleton construction in steel was the eventual and gainful conclusion. In few other building types was the development and adoption of full skeleton framing so swift, or so widespread. Even under the limitations of the London Building Act, multi-storey skeleton construction was the most comprehensive answer to the criteria demanded by commercial buildings: strength, stability and resistance to fire; economy and speed of construction; maximum admission of daylight; and the creation of the most open of interiors for the greatest freedom of arrangements. Although this technique represented the culmination of over a century's progress in iron-framed construction, and over 70 years of refinement of the purpose-built office type, its appearance in London owed a great deal to the relatively recent mass production of structural sections in steel, and the resultant move towards the design of stronger and more reliable framing elements by specialist engineering contractors. The role of skilled companies such as Richard Moreland & Son, Andrew Handyside & Co., Homan & Rodgers and W H Lindsay & Co. in designing and erecting increasingly comprehensive frameworks in the 1890s and early 1900s was crucial to preparing the ground for this revolution. Yet for all their structural sophistication and ambition, it would seem that in the great majority of these buildings, complete freedom from dependence on masonry elements was neither wanted nor realised, for steel construction was still viewed by most architects as a practical means of enhancing or augmenting the possibilities inherent in bearing-wall construction. It was not until well into the Edwardian era that steel-frame architecture came widely to be perceived as a systematic and legitimate means of accomplishing new potentialities. Inescapably, the conclusion from this must be that the crucial factor was the importation of transatlantic knowledge and techniques, for it was this period that witnessed the construction of high-profile commercial architecture that embodied virtually all the features of fully developed American skeleton construction. In particular, the well-publicised series of buildings erected by the Waring White enterprise in the face of the 1894 Act restrictions provided a

model of American constructional efficiency and organisation. Of the Morning Post Building, one publication noted that:

> There is probably no English building in which the system of steel construction, as elaborated in America, has been carried out with such logical completeness. Architecture and steelwork were designed simultaneously, architects and engineer working together and completing their scheme in every detail before a beginning was made with the building. This method is obviously more scientific and likely to lead to better results and greater economy than the practice, often adopted, of leaving steelwork till the architect's plans are complete and then getting from an engineer designs for steelwork to fit the architecture.[127]

Even before this compelling example arose, the 'American System', with its demonstrable economies and efficiency in design and time, was synonymous among contemporaries with fully framed construction. Homan & Rodgers, who as early as 1898 were advertising 'Constructional Steel Skeleton Buildings (American System)' seem to have been among the first engineering contractors to look abroad for insight, and other companies helped spread the new technologies and techniques within British practice. 'Oceanic House', built at No. 1 Cockspur Street, for the International Navigation Co. Ltd, was an example of fast-track office construction (Fig 11.49). Erected in just 6 months by the Waring White Building Co. in 1904–5 to designs by Henry Tanner junior, this neo-Mannerist speculative headquarters block was built around a framework of steel designed by Sven Bylander and fabricated and erected by Richard Moreland & Son.[128] The Building News noted, 'The building, however, has not been erected on what is known as the American system, but the outer walls carry the weight, and are not hung on the steel framing'.[129] The Builder, however, thought that it was 'being run up on the newly introduced method of American construction'.[130] Although Bylander later stated that it was 'constructed on the steel frame principle',[131] he probably meant this in terms of the floor-supporting cage within. Having already reported extensively on the construction of The Ritz Hotel, 'where the steel framework of the building was built up without the aid of outside scaffolding, and the stone and brickwork

Fig 11.49 (facing page)
Oceanic House,
photographed after the
coronation of King George V
in 1911, by when it served as
the premises of White Star.
Erected in 1904–5, it
illustrates the growing
influence of American
techniques on the design
and erection of steel-framed
offices.
[BL21285]

attached to it afterwards', *The Builders' Journal and Architectural Engineer* was perhaps most qualified in saying that in this building the 'walls are solid and take the weight'.[132] All the publications were, however, clearly mindful of the innovatory changes that were spreading from across the Atlantic. American techniques of precisely engineered frames and rapid construction were central to the widening uptake of full steel framing for Edwardian offices. However, without the pre-existing constructional know-how, and the functional refinement of the form within the City and elsewhere, it would likely not have flourished in the way that it did.

12

Stores, houses, churches, pools, fire stations and tube stations

Retail stores

*The extended application of iron as a construc-
tive material plays a very prominent part in
the modern development of street architecture.
With girders of inconsiderable section to span
large voids, and stanchions or columns doing
the duty of more massive brick piers, very
enlarged areas of light and space are obtainable, and there are examples of very admirable
buildings displaying these features of construction. From the liability of wide stone lintels to
fracture, rolled iron joists of small section are
now not unfrequently used for that purpose;
and with this advantage, that they enable the
window lights to be safely carried very near to
the ceiling.* (1877)[1]

*The blame of the paradox of the shop front
must, after the economic causes, be laid on iron
and steel ... Sham flat arches of impossible
span, and sham segmental arches with sham
abutments of ridiculous insufficiency, meet us
everywhere in streets, and only add to the
incongruity.* (1907)[2]

The introduction of iron columns, beams and
plate glass in the early Victorian period revolutionised the form of small retail merchandising
establishments, enabling the use of large windows to display goods and admit light deep into
the interior. By the late 19th century the basic
principle of using concealed metal bressummers to support the upper storeys above the
shop-front was so commonplace that barbed
comments like that of 1907 quoted above
were equally routine. Larger, more ostentatious
retail establishments invariably placed greater
demands on iron and, later, steel building technology, and in the first half of the 19th century
bazaars and arcades emerged as among the
most architecturally and structurally distinctive
forms. Providing sizeable venues for a large
number of retailers under one roof, they
exploited cast iron to create stylish top-lit, galleried interiors where customers could shop,
promenade and relax whatever the weather.
Although the origins of the arcade, which
sought to provide covered streets, lay in France,
London could lay claim to the bazaar, a form
more akin to the exchanges of the previous two
centuries, and to department stores of the later
part of the century.[3] The Western Exchange
Bazaar, Old Bond Street (1817), seems to have
been the first to use cast-iron columns, and by
mid century iron was being used for all structural elements, a development epitomised by
Owen Jones's sumptuous London Crystal Palace
Bazaar on Oxford Street (1857–8, demolished).
Other retail buildings besides bazaars and
arcades also made extensive use of iron to frame
top-lit galleried wells, corridors or showrooms:
from the 1840s, for example, men's outfitters
shops, offering ready-made garments were
increasingly arranged around such spaces,
which improved light and circulation. This configuration was not necessarily even dependent
on natural light; the flamboyant showrooms
of E Moses & Son, tailors and outfitters, on
Aldgate, were lit by an enormous 'gasolier'.[4]

From the 1860s, bazaars fell out of fashion in
the metropolis,[5] and while arcades continued
to be built, albeit sporadically, into the early
20th century,[6] the significant and influential
trend of the second half of the 19th century
was the growth of the 'big store'.[7] This phenomenon, reflecting the rise in consumer spending
among the emergent middle classes, can be
traced in a number of traditional businesses,
above all furniture, ironmongery and drapery
retailers. Expanding in piecemeal fashion at a
remarkable rate in the later Victorian era, these
establishments erected large 'emporia' and
'stores' to showcase their increasingly diversified
wares. Although many London drapery stores in
particular remained traditional in their scale

and design, others were exhibition-style edifices with interior compartments that equalled, if not exceeded, the 216,000 cu ft capacity stipulated by the 1855 Act for buildings of 'the warehouse class' – the most fitting category available. That erected by silk mercers Marshall and Snelgrove on Oxford Street in 1875–8 was the most ambitious drapery emporium thus far in London (Fig 12.1).[8] It was also perhaps the first to echo contemporary developments in Paris, where the first generation of *grands magasins*, including the famous Bon Marché, were rising on increasingly audacious frameworks of iron (*see* p 89).[9] Built to the designs of Octavius Hansard, this 160ft-long French Renaissance-styled block made extensive use of structural iron in the lower two floors for reasons of fire-resistance and spatial openness. Cast-iron columns were used to carry floors of built-up, wrought-iron girders and rolled-iron

joists, all of which was encased in concrete. Concealed bressummer girders spanning brick piers and columns carried the exterior walls, enabling huge plate-glass windows on the ground floor, which gave 'an unusual quantity of light to the counters'.[10]

The first purpose-built establishment to house an assortment of departments under one roof, acclaimed as the first department store in London, was the Bon Marché on Brixton Road (1876–7, H Parsons & W H Rawlings, architects). According to *The Builder*, it sold 'almost every imaginable article in food, furniture and dress', and was 'erected and internally arranged on the principles of some similar establishments in Paris'.[11] The Brixton store lacked the vast internal galleried courtyards and glazed roof of its Parisian namesake, which had been engineered in iron by Louis-Auguste Boileau and Moisant (of Noisiel Chocolate Factory

Fig 12.1
Marshall & Snelgrove, Oxford Street (1875–8, demolished), London's most ambitious draper's emporium and a prototype of the modern department store, with iron-framed lower floors.
[BL20468/002]

repute, *see* p 89), but interior iron framing did create open-plan showrooms that enabled sales departments on each of the four floors to flow openly and freely into one another (Fig 12.2).

The formula espoused by Brixton's Bon Marché, with its multiple interpenetrating departmental spaces offering a wide range of goods, was at once successful. Other large stores with prestigious locations in the West End, Kensington and Chelsea quickly adopted it, using structural iron to open up interiors, increase window area and create impressive spaces to beguile the customer. Many continued the traditional drapery-store practice of providing staff accommodation on the upper floors, but the majority modernised by providing at least two sales floors, typically structured around top-lit galleried courts. New stores for Peter Jones, in Sloane Square (1889–90; Perry & Reed, architects) and Harvey Nichols, Knightsbridge (1889–94, Charles Williams Stephens, architect), for example, exploited the central galleried-well form, albeit on more modest scales than French and Scottish contemporaries.[12] There is no direct evidence that either of these fashionably located buildings used steel, but it was already being exploited in at least one shop-cum-mansion block in the neighbourhood, Nos 27–30 Sloane Square, Chelsea (1892–3).

Woollands' store, Nos 1–3 Seville Street, Knightsbridge (1896–1900, demolished 1969), seems to have made at least partial use of steel-skeleton framing (Fig 12.3). *The Building News* noted 'Only sufficient brickwork is employed to comply with the requirements of the London Building Act, and consequently steel construction is largely utilised.'[13] Entirely of 'fireproof construction',[14] this was a building of five storeys, designed to provide as much natural light on all floors through the use of a circular light well and to attract passing custom through the employment of a continuous run of plate glass on the ground floor. Because living-in accommodation for staff was provided in nearby houses, more space could be devoted to retail. The lower three floors were given to shop and display space, the next two to workrooms, and the uppermost to kitchen, and dining and sitting rooms. Externally, apart from the swathe of display windows, there was little to differentiate it from conventional masonry buildings, but internally, the presence of box-section stanchions, heavily ornamented in keeping with the luxurious surroundings, hinted at advanced constructional technology (Fig 12.4). Messrs W Cubitt and Co. erected the building in three stages, to designs prepared under Henry L Florence. Throughout this period, the firm of Woollands Brothers, who had recently diversified beyond their original drapery business, continued to trade from the site. [15]

To what degree Woollands' was framed in steel may never be satisfactorily known, but it seems likely that by *c* 1900 a few provincial stores were being built around full-skeleton frames. The most famous of these was Mathias Robinson's furniture warehouse erected in Stockton-on-Tees in 1899–1901 to designs by W Basil Scott of Redpath, Brown and Co. (*see* p 99). Certainly steel construction was being extensively exploited by the larger London stores by the start of the 20th century. For the construction of the main part of the present Harrods on Brompton Road (1901–5, C W Stephens, architect) structural steel was important. A variety of techniques, traditional and modern, were used throughout the cellular complex, but for the most part they consisted of filler-joist floors carried on walls and stanchions. External walls were skeleton framed using solid steel columns to admit as much light as possible into

Fig 12.3 (above)
Woollands, Knightsbridge,
1896–1900.
[OP10358]

Fig 12.4 (left)
The interior of Woollands,
as depicted in 1922
[Drawing by F Matania,
Sphere, *2 December 1922.*
Image courtesy of The
Advertising Archives]

241

the building (Fig 12.5). Richard Moreland & Son was contracted to supply and erect the steelwork and, significantly, C V Childs was employed as consulting engineer.[16] Shortly after, Childs and Moreland's combined once more for the 3,000-ton rolled-steel cage frame of Waring & Gillow's furniture and furnishings emporium, occupying a prominent corner site in Oxford Street (1905–6) (Fig 12.6). This substantial block was designed by R Frank Atkinson, who, with assistance from Norman Shaw, gave its red brick and stone elevations a 'riotous Hampton Court baroque'.[17] *The Builder* noted matter-of-factly that 'The constructional steelwork is so arranged that the whole weight of the floors is carried on stanchions, thus relieving the brickwork of any actual weight.'[18] *The Builders' Journal and Architectural Engineer* was more impressed, exclaiming 'The great feature in the steelwork ... is that the steel frame is quite independent of the walls, stanchions having been carried up from the basement, and girders put across' (Fig 12.7).[19] In fact Childs' frame was not a true skeleton, and designedly so, enabling a huge economic saving which secured him the contract. In 1909 he explained:

For a large business premises now erected in Oxford Street two steel designs were prepared, one on the approved lines of the American steel skeleton framework [possibly by Sven Bylander], and prices were obtained from the Continent for this design. In the other case the design prepared by the writer was what might be called a semi-steel framework – that is, every main wall or floor girder in the building was carried on stanchions from basement to roof; the exterior walls were carried at the first floor level, except the main front, which being very massive was carried in three sections, at the first, third and fifth floor levels; the internal walls were sufficiently strong to support themselves, as well as the end bays of the floors. The price for this latter design, made in London and of English steel was less than half that for the skeleton design – a saving of over 25,000l. Other cases might be instanced in which the saving showed an even greater proportion.[20]

All the floors, roofs and vaults were of reinforced concrete on the 'Columbian' system.[21]

Internally, the most palpable expression of the framework was the top-lit, galleried rotunda, 54ft in diameter and 85ft tall, and easily the largest then built in a London store (Fig 12.8). Realising such huge volumes were technologically straightforward, but legislatively less so, Atkinson, and the business force behind this project, Samuel J Waring, undoubtedly had to petition the LCC to achieve this and the other large undivided spaces they required for the proposed art galleries, restaurant and reading rooms. *The Builder* noted that the 'interior ... has been specially arranged to suit the business of Messrs Warings, and is planned *on a broad basis* consistent with the London County Council's requirements',[22] suggesting, perhaps, that a waiver on the cubical extent limits stipulated by the 1894 act was granted.

To kick-start his business aspirations, Waring had formed a construction wing to Waring & Gillow Ltd; in 1904 this was taken over by the Waring White Building Co. when Waring merged his interests with American engineer James Gilbert White. This building company re-registered as Waring and White (1906) Ltd, taking over several contracts being undertaken by the Waring White Building Co. Sven Bylander was the chief engineer of the Waring White Building Co., and beyond matters of cost, it is unclear why Samuel J Waring enlisted Childs, and indeed Moreland, as engineer and steel-work contractors, respectively.

Atkinson's role in the structural aspects of this building is also of much interest to construction history. Atkinson, who had trained at the Liverpool School of Architecture, was 'once the right-hand man of J F Doyle of Liverpool'.[23] Doyle, in partnership with Norman Shaw, had designed the White Star Offices in Liverpool (1895–8), where the less restrictive building regulations allowed them to exploit fully (and make an undisguised show of) an internal frame of steel. They followed this with the Royal Insurance Offices (1897–1902) in the same city, a building dependent on a fully load-bearing, albeit highly idiosyncratic, steel frame.[24] Atkinson was an ardent disciple of Shaw, and had been trained by Doyle;[25] he was therefore almost certainly well versed in the more technical aspects of design work. He started his own practice in London in 1901,[26] and one of his first commissions was the Laurie and McConnelll store in Fitzroy Street, Cambridge (1903–4), built on a full steel frame (*see* Fig 8.26).[27] The consulting engineer for that building is not

documented, and it seems possible that the technologically minded architect had a hand in its design; at the very least it would seem that he structured his architectural designs around the exigencies of a frame from the outset, rather than simply handing a finished set of drawings to the engineer or contractor as was the usual practice.

Compared to Waring & Gillow's building, Debenham & Freebody's on Wigmore Street (1906–7) was more conservative in its interior configuration, with relatively small compartments probably designed to keep within the 1894 Act's stipulations on cubical capacity. Erected by George Trollope & Sons and Colls

& Sons Ltd to designs by William Wallace and James S Gibson, this building did not incorporate full framing, interior brick firewalls bearing some of the loads (Fig 12.9). Nevertheless, a large amount of constructional steelwork, supplied and erected by Homan & Rodgers, was used to create strong, non-combustible floors and frame the roof. Throughout the building, thin solid-steel columns, encased in concrete and marble, were employed as the principal uprights, ensuring minimal obstruction to both sightlines and daylight.[28] These qualities were especially apparent in two main galleried light wells (Fig 12.10). The new West Block of D H Evans on Oxford Street, built in 1906–9 to

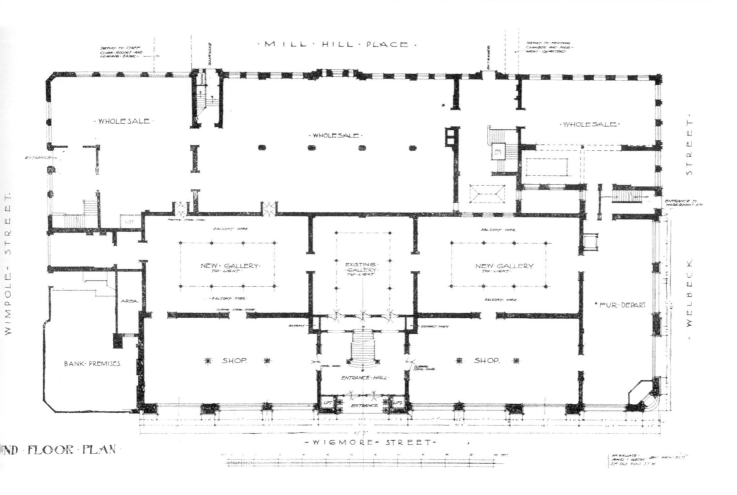

Fig 12.9 (above)
*Debenham & Freebody's,
Wigmore Street (1906–7):
small interior compartments
reflect the limits of the
Building Act of 1894*
[The Architectural Review
23, June 1908]

Fig 12.10 (left)
*The steel-framed galleried
light well at the centre of
Debenham & Freebody's.*
[The Architectural Review
23, June 1908]

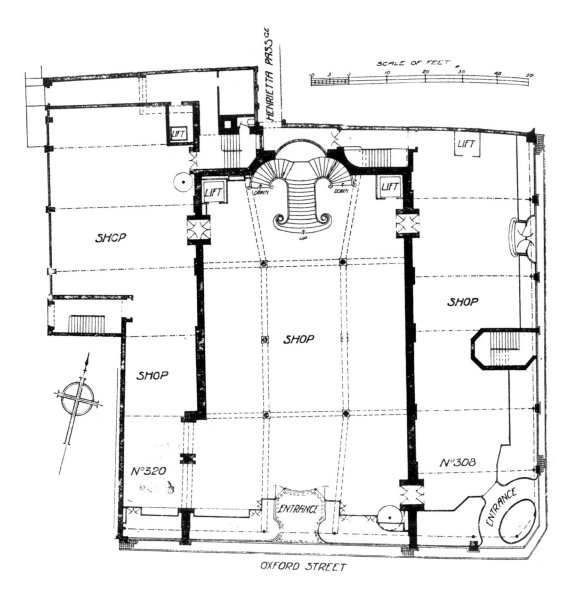

designs by John Murray 'after careful study of numerous large trade buildings in Paris, Vienna and Berlin', was more up-to-date, spatially, structurally and elevationally.[29] It fell short of the frankly exposed Art Nouveau frameworks of those cities, but boasted a continuously fenestrated ground floor and huge shop floors interrupted only by encased solid-steel columns that carried enormous floor girders, many of which were of over 40ft span (Fig 12.11). Faced with Pentelikon marble and adorned with Cippilino marble columns and pilasters, the 'general construction [was] largely carried out with steel framework'.[30]

The LCC almost certainly relaxed their rules regarding cubical capacity for the construction of the D H Evans store, in response to petitions for permission to exceed 250,000 cu ft through-

out 1907.[31] During the period 1905–6, 12 other businesses had made similar appeals, 6 of which were successful,[32] and indeed from the late 1890s such requests had became unexceptional (*see* p 294). It was in the realisation of a far grander and better publicised scheme that legislative reform, officially recognising steel-frame construction, finally came to pass. Embodying the most progressive advertising and display techniques that high-rise construction technology enabled, Selfridges, on Oxford Street (1908–9), effectively marked the direct transplantation of the 'Big Store' in its most developed, American form, direct to the heart of the West End. The proprietor, Harry Gordon Selfridge, initially enlisted fellow American Daniel Burnham to design a building that married the best features of two recently erected

Chicago stores on which he had worked: the Carson Pirie Scott Store (1906 addition) and Marshall Field's (1900–7). Both used, for the most part, steel skeleton-framing to provide multiple open-plan sales floors, and both were designed and erected under Burnham's supervision. In 1906 Burnham's office supplied Selfridge with designs for a building whose interior compartments were in excess of the LCC's absolute maximum of 450,000 cu ft. Burnham bowed out of the project when he realised the difficulty of negotiating with the LCC from across the Atlantic, and Selfridge replaced him with R F Atkinson. Throughout 1907–8 Atkinson successfully petitioned the Building Act Committee for greater cubic footage allowances (up to, but not exceeding 450,000), larger and more numerous interior wall openings (12ft by 12ft), greater window-to-wall area (exceeding one half of the

elevation), and thinner walls than those usually demanded for buildings of the warehouse class. The Committee did not, however, grant Atkinson permission to build higher than 80ft.[33]

Occupying a broadly rectangular site measuring 250ft by 175ft, the first phase of Selfridges quickly rose to a height of five storeys above ground, its rising framework looking as if it might continue upwards to reach the height of a Chicago skyscraper (Fig 12.12). Everything about it was big. Some 12,000 blueprints were prepared in the office of Waring & White (1906) Ltd, under the command of Sven Bylander, who was, unsurprisingly, brought in to engineer the colossal, 3,000-ton framework. The majority of these precisely detailed the design and erection of the steelwork, which proceeded at the rate of 125 tons per week. To create the expansive, open-plan floors that were to be among the building's most accomplished features, a

Fig 12.12
The Chicago department store arrives in London: Selfridges immediately after completion in 1909.
[Modern Building Record v1*, 1910]*

Fig 12.13
The wide-bayed steel cage
of Selfridges in course of
erection in 1908
[Modern Building Record
v1, 1910]

generous, and in this country unprecedented, 24ft by 22ft structural grid was adopted, one that would remain standard for many decades. Firmly riveted or bolted to rolled-steel girders at each floor level, this stack of continuous stanchions carried the cumulative weight of the reinforced-concrete 'Columbian' floors and all but one of the interior divisional walls to isolated grillage foundations. Only the west party wall was load-bearing; the others, made from 3in.-thick concrete blocks and placed at approximately 40ft intervals, rested on the steelwork and divided up the interior into vast, 450,000 cu ft fire-resisting compartments – one of the first, if not the first building to take advantage of the new cubical extent limits of the 1908 Act.[34] The external walls were built independently as largely self-supporting screens, but nevertheless were direct expressions of the wide-bayed steel cage within (Fig 12.13).[35] Relieved of their load-bearing role, and reduced to a row of giant Ionic pillars, piers and horizontal bands, the primary function of this masonry was to enclose what were possibly the largest plate glass windows yet sported by a British

commercial building. Some measured as large as 19ft 4in. by 12ft, and on both the Oxford Street and Duke Street frontages glazed surface area actually exceeded that of solid wall. Combined with the open-plan interiors, and two light wells, these highly fenestrated, and highly influential façades produced 'an impression of lightness and brightness'[36] through the entire building, all of which 'was devoted to either selling goods or indulging customers'.[37]

Selfridges opened amid a fanfare of publicity on 15 March 1909, just months after the London County Council General Powers Act 1908,[38] which permitted far greater cubical capacity and months before the London County Council General Powers Act 1909 ('Steel Frame Act') officially sanctioning steel-frame construction, were passed. Both beneficiary and agent of legislative change, it was London's first fully framed building to benefit from a relaxation of key clauses in the 1894 Act that had hitherto deterred, or prevented, fuller expression of this form of building. Compared to The Ritz Hotel, whose framework was masked beneath masonry of the requisite thickness, 'Selfridges, with its wide plate glass windows and near-absence of external walls, was clearly a different sort of building, achieved only after much negotiation with the LCC.'[39] Both architecturally and constructionally, it remained highly influential for years to come, one periodical noting, in 1920, 'The building gave a new scale to Oxford Street and has exercised a strong influence over the design of many big structures that have since been erected in the metropolis.'[40] Indeed, steel-framed edifices in its wake, such as Whiteley's department store in Bayswater, rebuilt on a 3,000-ton skeleton designed by Alexander Drew (1910–12; Belcher and Joass, architects),[41] Barker's on Kensington High Street (1913–14; H L Cabuche, architect) and Heal's extension on Tottenham Court Road (1912–16; Smith & Brewer, architects), were all given eye-catching façades that developed, or reproduced, the giant-order theme of Selfridges. But however influential Selfridges was, it was the inherent adaptability of its structural system that had more profound impact on the London, and British, retail environment in years to come. In terms of interior organisation (whether the open-plan, American 'horizontal' system, espoused by Selfridges, or the more traditional open-welled sales floors), exterior cladding (including veneers of high-quality stone and expanses of plate glass), massing and speed of construction, skeleton framing was uniquely felicitous, and would exert a commanding influence on the design of retail buildings in later decades, ultimately changing the face of the British high street.

Philanthropic tenements, town houses and mansion flats

Through the second half of the 19th century, iron beams and, less frequently, columns assumed increasing importance in the construction of London's larger, multi-storey, domestic buildings. The initial cue seems to have arisen from philanthropic concern for housing London's underprivileged; as early as 1857, Arthur Allom, then president of the Architectural Association, advocated 'a construction of iron girders and brick arches' in the context of 'dwellings for the labouring poor', since this 'might be considered fireproof'.[42] However, reinforced concrete seems to have answered the call for 'fireproof' flooring first. In 1862 Matthew Allen, the architect/builder for the

Improved Industrial Dwellings Company, patented a breeze concrete floor reinforced with iron rods and bars, applying it first to Langbourne Buildings, Finsbury (1863) and many subsequent blocks in the same decade, including Stanley Buildings, Kings Cross (1863–4).[43] Non-combustible floor systems, utilising wrought-iron joists or girders spanning between bearing walls may have been used in some of the multi-storeyed blocks put up by the philanthropic associations from the 1870s, although the added expense of such precautions probably mitigated against their universal adoption.[44] Wrought-iron girders saw fairly widespread (albeit concealed) application in many architect-designed town houses from the 1860s; Richard Norman Shaw, for one, employed them increasingly from the 1870s to carry interior walls, thus freeing the upper plans from the spatial constraints of the main floor, or to support chimneys so that inter-room circulation was preserved. For longer spans, which required stronger members, Shaw was quick to turn to the double-I form of compound girders, fabricated from rolled-iron joists and cover plates, rather than the older method of plates and angles.[45] In the 1880s, 'as town houses preoccupied him more, he began ... to thrust girders about without a second thought, round toplights and under outside walls, to get coherent circulation and distribution of space on each floor'.[46] Other architects or builders simply used them to create more rigid, wider-span floors on each level. For example, in the construction of the speculative high-class houses in Cornwall Gardens, North Kensington (c 1876–9) a meticulous specification demanded that all floors exceeding 17ft in span had to be stiffened with rolled-iron joists.[47] And at Stuart House, 84 Cadogan Square, Chelsea (1883–4), for example, 12in. by 6in. rolled I-sections, strengthened by 10in. plates riveted to the flanges, were used to span transversely the principal rooms on the ground, first and second floors, enabling still larger spaces uninterrupted by bearing wall partitions. Designed by Hunt & Steward and F G Knight for O L Stephen, a director of the Great Northern Railway,[48] this substantial brick house perhaps also bears witness to continuing constructional traditionalism: flitch beams and conventional timber beams are also used on the three principal floors.

Structural iron was, by this date, also entering the construction of the first generation of

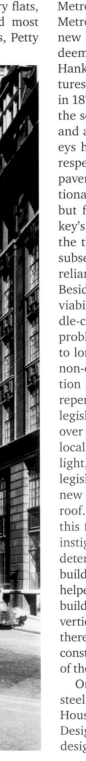

mansion flats, a form of tall, middle-class accommodation that in England remained a uniquely metropolitan phenomenon until after the Second World War.[49] Like department stores, this type of building was not recognised within any class of the London Building Acts, and like that form, it helped instigate legislative change. One of the first blocks of luxury flats, and 19th-century London's tallest and most notorious, was Queen Anne's Mansions, Petty France, Westminster, built in stages between 1873 and 1890 (demolished 1971) by a speculator named Henry Alers Hankey. Capitalising on the recently introduced hydraulic passenger lift, which enabled higher floors to command profitable rents, and exploiting a niche within the height and wall-thickness regulations of the Metropolitan Building Act 1855, whereby the Metropolitan Board of Works could sanction new buildings exceeding 100ft in height if it deemed the walls were of suitable thickness,[50] Hankey erected a series of 'Babel-like structures' of unprecedented height.[51] The first, built in 1873–5, was a 10-storey, 116ft high building; the second, in 1877, rose to 11 and 12 storeys, and a final extension in 1888–90, was 13 storeys high, its parapets and roof apex reaching respective heights of 130ft and 160ft above pavement level (Fig 12.14).[52] All used traditional bearing-wall construction throughout, but following the destruction by fire of Hankey's temporary timber residence that adjoined the timber-floored and partitioned first block, subsequent blocks used non-combustible floors reliant on wrought-iron girders and joists.[53] Besides serving as a model for the economic viability of the speculative, fire-resisting, middle-class apartment block (despite initial problems, the flats were readily let at high rents to long-term tenants), and possibly promoting non-combustible, metal-joisted floor construction in such buildings, the most far-reaching repercussion of Queen Anne's Mansions was legislative. Aghast that it had no jurisdiction over the height of the mansions, and mindful of local concerns about its adverse impact on light, air and views, the LCC rushed through legislation in 1890,[54] restricting the heights of new buildings to 90ft plus two storeys in the roof. The London Building Act 1894 reduced this figure to 80ft, plus two attic storeys.[55] By instigating one of the most significant economic deterrents to full framing, that of restricted building height, Hankey's speculation ostensibly helped delay that form of construction in those buildings that might otherwise have exploited vertical freedom (but see p 206). Nonetheless, there were some notable developments in the construction of large domestic buildings, some of them including structural steel.

One of the earliest documented examples of steel members used in this context is Riverside House in Limehouse (1887–8, demolished). Designed by Richard Harris Hill, who had designed several mission buildings, this starkly utilitarian five-storey block was built as a Scandinavian Sailors' Temperance Home. Unconcealed I-section rolled-steel girders and stringers manufactured by Dorman Long & Co. were used to support the spacious open-well staircase that served the numerous bedrooms on the upper floors. The stairs themselves were formed of incombustible concrete, possibly Stuart's 'Granolithic'. Moulded cast-iron columns in the dining and sitting rooms of the communal ground floor were used to bear the cumulative weight of the upper-floor spine corridors. The concrete floors were probably reinforced with rolled-steel joists.[56]

Already by this date, steel beams were being used in more fashionable residences for the professional classes. Here, the use of steel beams in supporting interior brick or terracotta partitions granted architects considerably more freedom in planning internal spaces than did traditional methods reliant on interior bearing. Indeed, the latter, by its nature, dictated walls continuous in a vertical plane upwards from a dedicated foundation, which therefore dictated a virtual duplication of interior layout on each successive floor. While the spatial freedom offered by this facility was anticipated in wrought iron, it was quickly taken to its logical conclusion in steel. 'Thriplands', No. 48 Kensington Court, built in 1888–9 to designs by John Slater, certainly did this (Fig 12.15). In fact there is some evidence to consider this five-storey residence as London's, if not Britain's, first substantially steel-framed house. Certainly it delighted visitors from the Architectural Association in 1889, who were awed by its constructional novelty:

> As an instance of the application of modern constructive science to domestic architecture the house is particularly interesting, iron construction of an advanced order being largely and boldly employed, both for floors and roofs, with the result of great advantages in the matter of plan. By the free use of girders and stanchions, together with Lindsay's steel decking for the ground floor, the architect has found himself unfettered by the usual necessity of supporting walls upon walls, and consequently has been at liberty to arrange his plans with considerable freedom. As an instance of the extent to which ironwork has been utilised may be mentioned the main girder, carrying a load of 120 tons over a span of 26 feet.[57]

Functional and, therefore, planning diversity were important considerations for the client, the electrical engineer R E B Crompton. The uppermost two floors were arranged as a laboratory while the ground floor housed, principally, a dining room and school room, and the first floor a boudoir and drawing room. No detailed accounts or plans of the building or its concealed structure have survived. Crompton later recollected 'My house was, I believe, one of the

Fig 12.15
'Thriplands', 48 Kensington Court (1888–9), designed by John Slater for the electrical engineer R E B Crompton.
[Jonathan Clarke]

earliest to be built in England on the modern principle of framed steel girders on which the outer and inner brickwork is supported.'[58]

That the building was at least partially skeleton framed in steel seems entirely plausible, given Crompton's background and interests, Slater's awareness of novel forms of construction, and the structural expertise of the builders/contractors, Kirk and Randall of Woolwich. Crompton, although best known as one of the pioneers of British electrical engineering, was also interested in structural engineering,[59] and it might be speculated that his involvement in establishing a new foundry at the Stanton Ironworks, Derbyshire, in 1878,[60] acquainted him with the manufacture and use of structural members. Similarly, his electric-lighting commissions of the mid-1880s, such as the new Ring Theatre, and the Burg Theatre in Vienna,[61] may have introduced him to the advanced methods of iron framing being employed in Continental theatres. John Slater, speaking to the first meeting of the Conference of Architects in 1887, believed 'the day was not far distant when steel would almost supersede iron for structural purposes'.[62] John Kirk of Kirk and Randall had previously erected a prefabricated military hospital supported by cast-iron stilts, and Kirk and Randall together had erected a cast- and wrought-iron bridge over the Regent's Canal.[63] But ultimately, it is only Crompton's reminiscences of the late 1920s (which, incidentally, were contemporary with W Basil Scott's claims about the origins of steel framing in England) that bear witness to what may be a remarkably early instance of steel-skeleton framing.

By the late 1880s mansion flats were being designed without interior load-bearing walls for ease of planning, particularly on the ground floor, which was often intended for shops. An early example was Imperial Mansions on Oxford Street (1889, demolished), built by Perry & Co. to designs by Martin & Purchase (Fig 12.16). A distinctive semi-circular building that stood at the corner of Charing Cross Road and Saint Giles High Street, it was constructed so that the ground floor and basement could be let as a whole, or be sub-divided into smaller shops (Fig 12.17). The residential chambers above were separated from the shops by a 'fireproof' floor, and were reached by private staircases and a hydraulic passenger-lift. Internal metal framing, possibly incorporating steel members, was evidently a significant aspect of the building's design:

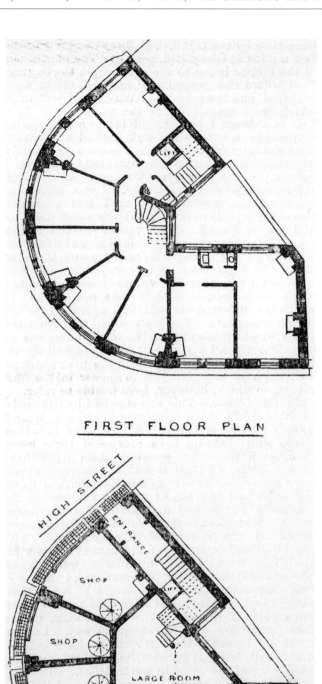

FIRST FLOOR PLAN

GROUND PLAN
Imperial Mansions, Oxford-street.

*Fig 12.16
(facing page)
Imperial Mansions,
on the junction of
Charing Cross
Road and St Giles
High Street,
photographed on
23 July 1889 soon
after completion
(demolished).
[BL09515]*

*Fig 12.17
Imperial Mansions,
built by Perry & Co.
to designs by Martin
& Purchase with
non-structural
partitions on the
lowest floors.
[The Builder, 14
December 1889]*

The internal construction is almost entirely of iron, and is carried on stanchions and columns. This was done to save the valuable space which brickwork would have taken up, and also to enable the ground-floor to be let in one large open space if desired, as all the walls on the ground and basement floors, which now form the divisions of the small shops, could at any time be cleared away without interfering with the structure.[64]

Nos 27–30 Sloane Square, Chelsea (1892–3), built by B E Knightingale under the superintendence of F G Knight certainly used steel to similar ends (Fig 12.18). This red-brick block utilised a 'considerable amount of iron and steel ... in the construction', supplied and erected by Dorman Long & Co.[65] Most of this probably went into supporting the upper storeys, with a ground-floor framework perhaps

supporting the brick-walled residential apartments. This arrangement enabled a spacious shop on the ground floor, almost completely fenestrated with large plate-glass windows. Another illustration is Wellington Court, Knightsbridge (1893–5), built by Henry Lovatt of Wolverhampton (who took an apartment there) to designs by the City-based architect M E Collins.[66] Fitted with automatic hydraulic lifts by W Waygood & Co., this high-class brick block housed five floors of self-contained suites, with each suite comprising 8 to 11 rooms, the more desirable ones complete with billiard rooms. Servants' accommodation was housed discreetly in separate mezzanines. In remarking on the 'principal novelties ... in the planning and construction', *The Builder* alluded to interior steel framing throughout the full height of the building: 'Internally, the whole construction is in steel, and so arranged that the division

Fig 12.18
Nos 27–30 Sloane Square, Chelsea (1892–3).
[DP134010]

walls can be removed in any manner to suit individual tastes.'[67] The partition walls were not described, but it seems likely that they were of terracotta, which for internal construction compared favourably to brick in terms of weight, cost and ease of erection and manoeuvre.[68]

Extensive steel framing of a more self-sufficient and ambitious form was called upon for the construction of Park Lane House (1895–7, demolished 1963), a palatial residence built for the diamond king millionaire Barney Barnato on the south corner of Great Stanhope Street (now renamed Stanhope Gate) (Fig 12.19). Richard Moreland & Son fabricated and erected the steelwork to designs of the architect Thomas Henry Smith: a photograph of its construction ahead of the enclosing brickwork shows a rigid-jointed portal-frame structure, each bent longitudinally, braced by simple I-section secondary beams (Fig 12.20). It is unclear whether the photograph was taken at ground- or first-floor level, but in any case the framing continued upwards, since I-section joists

are shown resting on the main beams, ready for laying the next floor. Skeleton framing was probably employed in parts of the five-storey building to permit internal spaciousness such as to suit the ostentatious tastes of Barnato, one of Park Lane's nouveau riche residents. Declaring 'I shall have the finest entrance hall, stairs and dining room in London',[69] he wanted the building completed and ready for occupation in time for a house-warming party on the day of Queen Victoria's Jubilee in 1897,[70] which might also explain why the expensive technology was used. When business brought him to London, he was eager to monitor its progress. In August 1896 the architect, Barnato, his wife and a party of friends were hoisted to the top of the house by the builders, Colls & Son, to see the gable apex craned into position,[71] and doubtless enjoy views of the Crystal Palace and the northern heights of Highgate.[72] Unfortunately, Barnato's subsequent suicide meant that he never saw the building's full completion.[73]

Fig 12.19
Park Lane House, built for the ill-fated millionaire Barney Barnato (1895–7, demolished 1963).
[The Builder, 10 October 1896]

Fig 12.20
Framing of Park Lane
House in course of erection
c 1895 by Richard Moreland
& Son.
[© The British Library
Board (F600077–65)]

The ground-floor plan was articulated around a 32ft-deep hall comprising a morning room, billiard room and a reception room, 24ft square, divided from the adjoining 42ft-long dining room by folding doors. On the first floor there was a gallery over the hall, with two drawing rooms (over the billiard and morning rooms), a conservatory and a large 72ft-long ballroom leading off it, complete with a minstrel's gallery at one end. The second floor contained Barnato's bedroom, en suite bathroom and rooms for his wife and children; the third and fourth floor housed, respectively, guest rooms and servant's quarters, and the basement accommodated the kitchen and offices. The cement-bonded brickwork was faced with Portland stone, all erected by Colls & Sons for a contract price of £39,940.[74] That Barnato was pleased with the design of his house is indicated by the subsequent employment of Smith to rebuild No. 10 Upper Brook Street for Woolf Joel, his nephew and co-heir. Whether the rebuilding of 1897–8 incorporated steel framing is not known, but Joel, like his uncle, never took up residence, for in 1898 he was assassinated in Johannesburg.[75]

The high fabrication and assembly costs of steel-framed construction as used in Park Lane House would have been beyond the financial reach or justification of philanthropic housing programmes of the 1890s and even 1900s. Rowton House, Churchyard Row, Newington Butts (1896–7, now the London Park Hotel), was perhaps fairly typical, employing fireproof filler-joist floors made up of coke-breeze concrete and rolled-steel I-sections throughout its seven floors, supported for the most part by cement-bonded bearing walls. The third of a series of six 'poor men's hotels' established by Disraeli's secretary W L Corry (first Baron Rowton) under the architectural command of Harry B Measures, this looming brick- and terracotta-clad quadrangle block effectively used the upper cross-walls to subdivide the sleeping quarters in the interests of fire and disease isolation. Cast-iron columns rather than steel stanchions were used on the communal ground and first floors to preserve openness of plan; these support substantial 12in. by 6in. rolled-steel beams upon which the filler-joist floors rest. Steel was also used in the construction of the roof, composed of coke-breeze slabs reinforced with steel angles. In the pitched areas of the roof, slating was nailed direct onto the 4in.-thick slabs, and in the flat areas the slabs were covered with asphalt.[76] Other Rowton Houses (the first, in Bondway, Vauxhall, was opened in 1893) employed similar techniques.

Even in the more costly, speculative mansion flats there was little incentive to use full steel-skeleton framing given that the regulation thick walls would have added to the expense. Sydney Perks, in his textbook *Residential Flats* (1905), the first British publication to be devoted exclusively to the subject, noted the obstructive effect of the 1894 London Building Act in this respect:

> The steel construction of external walls in America has attracted much attention here. Little advantage is gained by adopting the system in London on account of the present Building Act; the walls would have to be as thick for filling in the framing as if it did not exist. A new Building Act is contemplated, and no doubt this special form of construction will be provided for.[77]

One Edwardian exception, possibly for reasons of construction time, if not technological virtuosity, seems to have been Gloucester House Flats, which stood at the corner of Old Park Lane and Piccadilly (1906–7, demolished late 1960s). Replacing the late 18th-century Gloucester House, this seven-storey block of luxury flats (the largest in London) was built on a 1,400-ton steel-skeleton frame fabricated and erected by Drew-Bear, Perks & Co., Ltd, of the Battersea Steelworks (*see* Fig 0.1).[78] Notable for housing just one palatial flat on each floor, each of which enjoyed uninterrupted views of Green Park,[79] it also marked a significant step forward in British skeleton design through the incorporation of diagonal bracing to resist wind loads, a feature that was to become a standard requirement of tall buildings built under the 1909 Act. The bracing took the form of pairs of stanchions connected by horizontal and diagonal H-section bracing, forming steel towers which rested on steel grillages (Fig 12.21).[80] Riveted gusset-plate connections on the flanges provided the necessary rigidity, but the 'unnecessarily elaborate' nature of this was nevertheless criticised by *The Builders' Journal and Architectural Engineer* for being too costly.[81] This was the first skeleton-framed residential block on which

that publication had reported. Having closely followed other landmark steel-framed buildings, including the Savoy extension, The Ritz Hotel and the Morning Post Building, it was aware of the standardised efficiency and economy of American techniques. Drew-Bear, Perks & Co., Ltd did not, however, design the steelwork. Whether this was the responsibility of the architects, T E Collcutt & Stanley Hamp (who had already helped design the skeleton-framed Savoy Hotel extension, *see* pp 179–80) or a consulting engineer is unclear. Completed with an 'eggshell finish terracotta of old ivory color', and with 'green Spanish tiles' covering the mansard roof that housed the fifth and sixth storeys,[82] it flaunted an art nouveau gable (*see* Fig 0.2), and a large motor-car showroom at street level. Edward VII, however, considered it a 'monstrosity', complaining that it spoiled the view from Buckingham Palace garden.[83]

Fig 12.21
Diagonal-braced steel towers, founded on steel grillages, gave the slab-like Gloucester House Flats resistance to strong winds.
[Builders' Journal and Architectural Engineer, *28 March 1906, Contractors' Supplement*]

Ecclesiastical buildings

One shudders at the idea of naked iron in the construction of a church, and the many years during which iron girders have been in use have failed to reconcile us to their appearance, except in bridges and works of a kindred nature. (1883)[84]

The use of steel as a primary structural material in ecclesiastical buildings was relatively uncommon in the late 19th and early 20th centuries, despite iron's long yet broken history of employment for gallery supports, nave arcades and, occasionally, roof structures. Initial experimentation, in which slender cast-iron posts were called on to support galleries, began in the second half of the 18th century. At St John's, Leeds (renovated 1764), St Anne's, Liverpool (1770–2) and St James', Toxteth (1774–5),[85] such 'Cast Pillers' were boldly exposed, but perhaps more representative were the London churches of Holy Trinity, Clapham Common (1774–6) and St James', Clerkenwell (1788–92),[86] where iron pillars are concealed within classical timber columns.[87] Outside the metropolis, there was an experimental flourish in Thomas Rickman's three Liverpool churches (1813–16),[88] all of which used cast iron for their entire internal structures, designed and fabricated by the ironfounder John Cragg.[89] However, shifting, confused and sometimes contradictory dictates within architectural philosophy hampered the material's use in churches until the 1840s. In this decade, and the next, a huge colonial and home market in prefabricated 'portable' or 'temporary' churches opened up.[90] Of more direct concern in this discussion is the revived use of iron within the higher circles of ecclesiastical architecture. Structural iron, whether 'honestly' exposed, or 'dishonestly' concealed, may have been deplored by Pugin and his followers, and indeed by the great mass of the architectural profession, but at least some leading practitioners were prepared to break the mould, albeit integrating the metal within existing forms. Lewis Vulliamy, a versatile architect of French descent, deviated from conventional Gothic in his designs for the Anglican church of All Saints, Ennismore Gardens, Knightsbridge (now serving as a Russian Orthodox Cathedral). Built in 1848–9 in the short-lived Lombardic style, the interior was distinguished not only by the extensive sgraffito wall decoration, but also by the employment of cast-iron columns, each

raised on brick plinths and originally 'polished to imitate marble'.[91] Besides supporting the nave roof, these columns also carried iron girders to frame the gallery fronts on three sides (Fig 12.22). Sir Arthur William Blomfield's (1829–99) first church commission, St Paul's, Haggerston (1859–60, demolished), made similar use of iron columns, but in this instance they were fluted, with two ranks supporting a timber hammerbeam roof, and two ranks of secondary, shorter columns propping up the gallery fronts (Fig 12.23).[92] Other architects of lesser stature, posthumously classed as 'rogues', also subverted established tenets of Gothic Revivalism at this time. The noted 'low church'

Fig 12.22 (facing page)
Former All Saints' Church, Ennismore Gardens, Knightsbridge, built in 1848–9 to designs by Lewis Vulliamy and internally framed in iron.

Fig 12.23 (above)
Sir Arthur William Blomfield's first church commission, St Paul's, Broke Road, Haggerston (1859–60, demolished), used slender fluted iron columns.
[The Builder, 31 March 1860]

(1865–8, demolished). Within the circular shell of the late 18th-century building, Boileau set 12 marble-encased iron columns in the form of a cross; from their capitals sprang cast-iron ribs perforated with quatrefoils, which in turn supported a light framework of wrought iron that carried a multi-vaulted groined ceiling (Fig 12.24). As in Francois-Henri Labrouste's multi-domed reading room of the Bibliothèque Nationale, Paris (completed 1868), the weight of this iron roof seems to have been borne entirely by the columns, although the peripheral columns located just inside the walls probably derived stability from the masonry. This remarkable structure survived until 1940, when it was severely damaged by enemy action.[95]

By the 1880s British architects were using wrought iron as a principal element in the roofs of London churches. St Paul's, Hammersmith (1882–7), designed by Hugh Roumieu Gough and John Pollard Seddon, was perhaps the first British church to employ iron lattice principals.[96] Gough (who had acquired engineering knowledge from time spent in 'Government service'), in conjunction with the engineer Bartholomew Parker Bidder, employed light, arched wrought-iron trusses, carefully designed to exert no outward thrust on the clerestory walls despite the deliberate omission of iron ties (Fig 12.25). This ironwork, completed by 1882, was hidden from below by a decorated ceiling. Seddon excused their innovatory use of the unfashionable material:

> Iron for structural purposes has been hitherto rather the friend of the engineer than of the architect; indeed, to the latter, and I must own to myself, it has been almost held as an enemy. We should not have used it from choice in this instance, but the necessity for economy, and the desirability of avoiding all lateral thrust upon the lofty clerestory walls, and any visible ties which would have to cross under the vaulted ceiling, led us to its adoption ... aesthetic considerations are, in this instance, outside the question. No portion of the ironwork of the roof will be left visible, as there will be a ceiling with groining ribs below it, which it is intended to treat with coloured decorations.[97]

Wrought iron was used on a far more extensive scale in the construction of the Surrey Gardens Mission Hall (c 1885, demolished), attached to St Paul's, Lorrimore Square, Walworth. Designed

Fig 12.24
Notre Dame De France,
Leicester Place (1865–8;
L-A Boileau, architect),
demolished after bomb
damage.
[London Metropolitan
Archives, City of London
(SC/PHL/02/1155)]

architect, Enoch Bassett Keeling, employed clustered iron columns for the nave arcades at two churches begun in 1864: St Mark's in St Mark's Road, Notting Dale, and St George's on Campden Hill (both demolished).[93] Another similar example is St Mary's, Ealing, recast in 1866–74 to designs by Samuel Sanders Teulon.[94]

It took a French architect to give London a church that exploited cast iron in an altogether more ambitious and stylish manner. Louis-Auguste Boileau, one of France's leading proponents of iron architecture, had already designed a number of churches with interior iron frames when in 1865 he was given the task of converting Burford's West End panorama building in Leicester Place into a place of worship for French Roman Catholics in London

by architects Romaine-Walker & Tanner, this capacious brick and stone hall, accommodating 800 persons, employed four iron stanchions to support two axial girders, which carried the dwarf upper walls and the gabled windows. The stanchions projected above the girders, and carried wrought-iron trusses 'formed to the slopes of the roof, and constructed to prevent any outward thrust'.[98] The lower, flanking, side roofs were also partially supported by the stanchions, their iron principals bolted to the stanchions and axial girders. This 'somewhat novel' constructional arrangement 'rendered all outside buttresses unnecessary, and every inch of area was available for the floor-space of the hall'.[99] All the ironwork, fabricated and erected by Dennett and Ingle, possibly to

designs by R M Ordish,[100] was encased in wood cladding to give the effect of a lofty arcade whose wooden piers with carved caps supported a timber-framed roof.

If by the 1880s wrought iron was beginning to find an unobtrusive ecclesiastical perch in London, it was soon knocked off it. As in secular buildings, experimentation with steel was already afoot by the end of the decade, and by the beginning of the new century steel was being used in a more routine, at times structurally impudent, manner. Like wrought iron, steel was discreetly introduced for interior structural elements, as a supplementary material incorporated within or on the body of what were essentially brick or masonry buildings. One early use was in the support of domes, continuing a

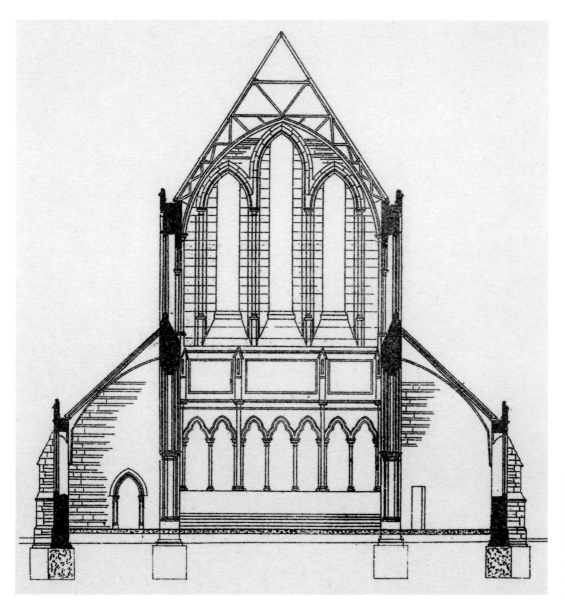

Fig 12.25
St Paul's, Hammersmith
(1882–7; Gough and
Seddon, architects), perhaps
the first British church to
employ iron lattice
principals.
[The Builder, 2 December
1882]

tradition already established by iron in the mid-19th century. For the massive, imposing Italian Renaissance St Joseph's Retreat, Highgate Hill (1887–9), the architect Albert Vicars engaged Richard Moreland & Son to frame a sizeable dome over a tall octagonal drum (Fig 12.26). They did this using just eight slight steel ribs, laterally braced by two rows of rolled-steel joists, forming a dome with a clear span of almost 40ft (Fig 12.27). At its base, a compression ring formed from steel channels resisted the horizontal thrust of the tapering ribs, while at the crown, an 8ft-diameter I-section ring took the compression (Fig 12.28).[101] Containing all the outward thrust, the lower ring beam directed the uniformly distributed weight of the whole structure onto the supporting brick piers, enabling a considerable reduction in their relative mass. This copper-clad and magnificently Italianate dome, commanding the view up Highgate Hill, supports an octagonal lantern, which sheds dramatic light on the sanctuary beneath.[102]

At St Joseph's, steel was restricted to framing the larger of the two domes. Other architects were already putting it to more extensive use in framing roof structures that covered entire naves. Richard Norman Shaw was among the

Fig 12.26 (facing page) St Joseph's Retreat, Highgate Hill (1887–9), The larger dome was framed in steel. [26615/014]

Fig 12.27 (above) Richard Moreland and Son's catalogue drawing of the dome. [Courtesy of Institution of Civil Engineers]

Fig 12.28 (left) View of the compression ring at the crown of St Joseph's dome. [AA033452]

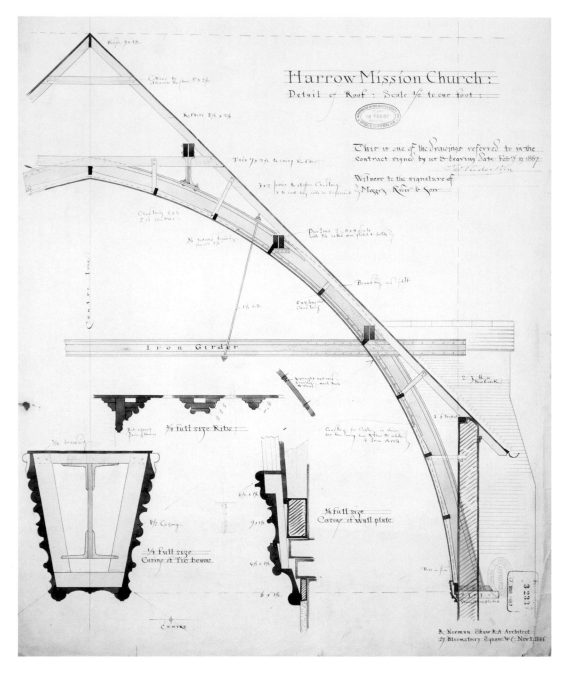

first architects to do this in London. His revised design for Holy Trinity, Latimer Road (Harrow Mission Church) (1887–9) employed steel to create a spacious tunnel lit solely at either end, a less ambitious, but necessarily less costly alternative to an earlier double-naved, groin-vaulted design conceived entirely in brick and stone.[103] To enclose the nave he used four arched steel ribs, each composed of two halves rigidly joined at the crown. Springing from either wall, their lateral thrust was held firmly in check by 70ft-long steel-girder tie-beams that spanned between brick buttresses (Fig

12.29). Because the girders rested on top of brick buttresses, they took much of the weight of the roof away from the thin brick walls.[104] The 'buttresses' were able to be pared down to a width of just 2½ft, since they functioned more as piers, without the need to resist outward thrust. The steelwork was fabricated, and probably erected, by Homan & Rodgers.[105] Shaw chose to conceal all of this structure behind wood cladding, the girders masquerading as timber tie-beams and the ribs hidden above a boarded ship's keel ceiling (Fig 12.30). Unlike the broader original design, this single-naved church did not abut the sides of neighbouring buildings. Shaw could have introduced additional light from windows in the side-walls, and, given the diminished width of the piers, very large ones at that. He purposely eschewed this: 'Evidently he knew that this would have detracted from his airy tunnel, braced by the extravagant, identical thirteen-light windows at either end, and so himself sacrificed a measure of logic to gain the perfectly articulated horizontal space he had long sought.'[106]

What may be the decade's most determined use of steel in an ecclesiastical context remains weakly substantiated. The mighty Italianate Church of Our Most Holy Redeemer, Exmouth Market, Clerkenwell (1887–91) 'could fairly claim to be the greatest non-Gothic Anglican church built in London during the reign of Queen Victoria'.[107] It might also lay claim to the first large-scale use of structural steel in a London church. According to some sources, the giant concrete Corinthian columns and their unbroken entablature enclose steel stanchions and girders that carry the groined vault (Fig 12.31).[108] This structural system is, however, dependent on the outer bearing walls for overall stability; cross girders spanning outwards from the entablature seemingly rest in the walls, although these connections are not shown on surviving published drawings. Holy Redeemer was said to be 'a fling in a modern direction'[109] for the architect John Dando Sedding, but this probably referred to style rather than structure. Certainly Sedding was not known as an innovator in the technical sense, and it might be argued that his Ruskinian background might have predisposed him against concealed structural metalwork, but it does seem plausible that steel was used in this instance to enable the strikingly open, soaring and well-lit interior. It may be that Sedding was able to achieve monumentality, Wren writ large, more expeditiously or cheaply in steel and concrete rather than in masonry. In this respect, one obituary made

Fig 12.30
Holy Trinity, Latimer Road (Harrow Mission Church). The desired effect: a tunnel-like space lit only from either end.
[AA57/00283]

reference to Holy Redeemer, noting 'In the interior all the canons of truth of construction have been abandoned, and plastered columns, plaster mouldings, and timber and plaster groinings prevail … All idea of Gothic construction had to go, and he was left only with lean forms and degenerate materials wherewith to accomplish his longed-for effects.'[110]

The 1890s saw ever more ambitious use of steel in churches, but typically none of it was conspicuous to the eye. Steel stanchions encased in concrete were used in the internal construction of the King's Weigh House Church, Mayfair (1889–91), designed by Alfred Waterhouse. Occupying a confined site, this church and its associated buildings vaunted 'Waterhouse's characteristic architectural virtues: stringency,

clarity, and mastery of plan'.[111] By placing stanchions in the external walls of the rectangular basement, from which arches sprang, Waterhouse was able to support the elliptical upper walls of the first floor without the need for heavy piers or buttresses. At the open, west end, the upper walls rested directly on four freestanding columns which originated in the basement and continued up to the gallery. Cylindrical columns, possibly of cast iron, carried the horseshoe-shaped gallery (Fig 12.32). The elliptical auditorium roof was framed by semi-elliptical steel lattice principals that sprung from the exterior walls. From this structure a false ceiling of plaster panels and wood ribs was hung. Throughout, the concrete encasing the piers and stanchions was faced with

Fig 12.31
Our Most Holy Redeemer,
Exmouth Market, said to be
framed in steel (1887–91;
J D Sedding, architect).
[DP021294]

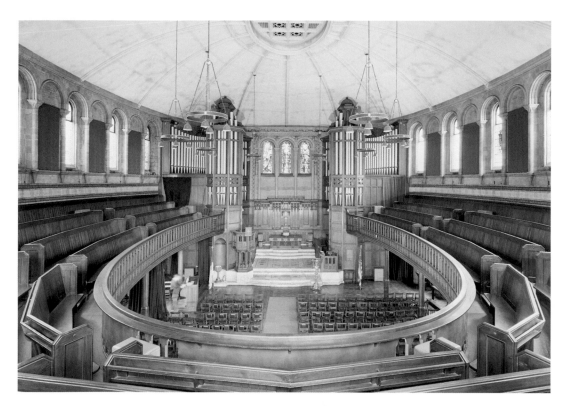

Fig 12.32 (left)
The horseshoe-shaped
interior of the King's Weigh
House Church, Mayfair
(1889–91), photographed
in 1965.
[BB65/03408]

Fig 12.33 (below)
Unconcealed steel principals
at Trinity College Mission
Hall, New Church Road,
Camberwell (1894–5; by
Richard Norman Shaw).
[Andrew Saint]

faience tiles.[112] The use of steel girders and stanchions in the adjoining schoolroom enabled a spacious attic hall on the second floor without any interior obstructions.

In the same decade Shaw designed a mission hall at Church Street (now New Church Road), Camberwell (1894–5), for Trinity College, Cambridge. For this wider-span building he repeated the basic structural vocabulary used in the Harrow Mission Room, but, significantly, chose to expose the riveted steel principals and the steel butt purlins that braced them (Fig 12.33). The form of the flattened, angular principals, unlike the arched versions in the Harrow Mission Room, left room for clerestory windows, which rested on the steel purlins, and removed the need for cumbersome, intrusive tie beams. Indeed, their action perhaps begins to approach that of a portal frame (which is itself a kind of arch), thus diminishing the true arch action and hence the net outward thrust. Consequently, elegant steel tie-rods fitted with turn-buckles to maintain the tension are all that are required to accommodate these forces. Shaw embellished the inclined spaces between the principals with wood boarding, the disjuncture accentuating their presence: he had annotated the design 'My notion is to shew the rib.'[113] A year or so later he made his, and perhaps, Victorian England's, most daring play

Fig 12.34
The 35-ton prefabricated steel dome of Brompton Oratory about to be lifted into position sometime in 1895–6.
[© The British Library Board (F60077–60)]

Fig 12.35 (facing page)
Drawing for a Presbyterian church roof in Muswell Hill (1901–3), erected by Archibald Dawnay & Sons.
[Middleton, Modern Buildings vol. 4 (London)]

ever of structural steelwork in a 'polite' context in the White Star Offices, Liverpool (1895–8).[114]

The construction of the massive dome of the Brompton Oratory, built between 1800 and 1896, called for more intricate construction than that used at St Joseph's. A 'magnificent freak among English churches',[115] this Roman Baroque Catholic church was largely the work of the architect Herbert Gribble, although the building remained incomplete at the time of his death. Most conspicuously, it lacked the shallow outer dome that Gribble intended. George Sherrin brought the work to swift completion, although the dome he gave the building was both higher and steeper than that Gribble had wished for.[116] Just 35 tons of steelwork, fabricated and erected by Joseph Westwood & Co., were used to frame the outer dome (Fig 12.34).[117] Measuring 59ft across its base it consists of eight principal trussed ribs, bearing presumably on a steel compression member at the crown, that are tied transversely by three lattice rings and a slender ring at the base. Light meridional steel members between these main ribs perform only a subsidiary structural role,

their main function being the forming of an anchor for the outer surface. The employment of three-dimensional trusses or space frames for the principal ribs resulted in a structure of greater rigidity since, being triangulated in more than one plane, they are stiff in all directions and thus less prone to buckling under eccentric wind loads. Similarly, the employment throughout of small-section members resulted in a relatively light structure, which was seemingly preconstructed at one of the manufacturer's yards before being craned into position.

The facility offered by light, prefabricated-steel trusses for enclosing large volumes while offering the maximum strength against wind forces ensured that the steel-framed church roof was set to stay. By 1900 the London steelwork contractor Archibald Dawnay & Sons had established itself as a leader in this field, and one of the firm's designs, for a Presbyterian church in the Broadway, Muswell Hill, was illustrated in one of the era's standard textbooks (Fig 12.35). Erected in 1901–3 to quirky art nouveau designs by George and Reginald Palmer Baines, this building employed four

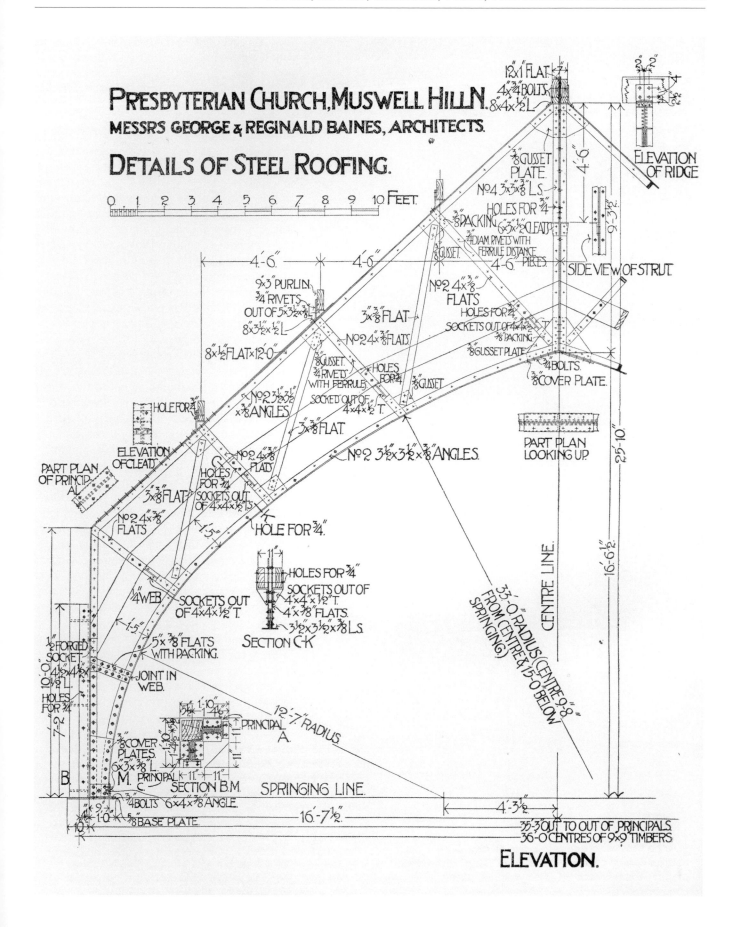

PRESBYTERIAN CHURCH, MUSWELL HILL N.
MESSRS GEORGE & REGINALD BAINES, ARCHITECTS.
DETAILS OF STEEL ROOFING.

0 1 2 3 4 5 6 7 8 9 10 FEET.

ELEVATION OF RIDGE

12"x1" FLAT
4"x3/4" BOLTS.
8"x4"x1/2" L

3/8" GUSSET PLATE.
No 4 3"x3"x3/8" L.S.
HOLES FOR 3/4"
6"x3"x1/2" CLEATS

4'-6"

9'-3 1/2"

SIDE VIEW OF STRUT.

3/8" PACKING

3/8" GUSSET.

3/4" DIAM. RIVETS WITH FERRULE DISTANCE PIECES.

4'-6"

No 2 4"x3/8" FLATS

HOLES FOR 3/4"
SOCKETS OUT OF 4"x4"x1/2" T.
3/8" PACKING

3/8" GUSSET PLATE.

3/4" BOLTS.
3/8" COVER PLATE.

4'-6"   4'-6"

9"x3" PURLIN
3/4" RIVETS
OUT OF 5"x3 1/2"x3/8" L.
8"x3 1/2"x1/2" L.

3"x3/8" FLAT
No 2 4"x3/8" FLATS

8"x1 1/2" FLAT x12'-0"

3/8" GUSSET.
3/4" RIVETS WITH FERRULE.

HOLES FOR 3/4"

3/8" GUSSET.

No 2 3 1/2"x3 1/2"x3/8" ANGLES.

3"x3/8" FLAT.

No 2 3 1/2"x3 1/2"x3/8" ANGLES.

PART PLAN LOOKING UP.

HOLE FOR 3/4"

ELEVATION OF CLEAT.

PART PLAN OF PRINCIPAL.

No 2 4"x3/8" FLATS

HOLES FOR 3/4"
SOCKETS OUT OF 4"x4"x1/2" L.S.

3"x3/8" FLAT.

No 2 4"x3/8" FLATS

1'-5"

HOLE FOR 3/4"

CENTRE LINE.

25'-10"

16'-6 1/2"

1"

HOLES FOR 3/4"
SOCKETS OUT OF 4"x4"x1/2" T.
4"x3/8" FLATS.
3 1/2"x3 1/2"x3/8" L.S.
SECTION C-K

1/4" WEB

SOCKETS OUT OF 4"x4"x1/2" T.

1'-5"

5"x3/8" FLATS WITH PACKING.

JOINT IN WEB.

1/2" FORGED SOCKET.
5'-0" 4 1/2"x4 1/2"x 1/2" L.

HOLES FOR 3/4"

7'-2"

3/8" COVER PLATES.
6"x3"x3/8" L

M. PRINCIPAL C.

B.

SECTION B.M.

3/4" BOLTS 6"x4"x3/8" ANGLE.
3/8" BASE PLATE.

4 1/2" 9"
1'-0"
10 1/2"

1'-10 1/2"

PRINCIPAL A.

12'-7" RADIUS

SPRINGING LINE.

16'-7 1/2"

33'-0" RADIUS (CENTRE 9'-8" FROM CENTRE & 15'-0" BELOW SPRINGING)

4'-3 1/2"

35'-3" OUT TO OUT OF PRINCIPALS
36'-0" CENTRES OF 9"x9" TIMBERS

ELEVATION.

intersecting steel arched trusses to roof a square, galleried nave that accommodated a 900-strong congregation. The trusses forming this large central domed square were carefully designed to place little outward thrust on the four stone piers from which they sprang.[118] Although rehabilitated as a public house in 1996, the sense of lofty spaciousness imparted by this concealed steelwork remains. Dawnay's fabricated similar structures (Fig 12.36) but the zenith of the firm's steel-roof construction, and perhaps of British steel-dome construction in this era, was the huge dome surmounting the Wesleyan Methodist Hall at Storey's Gate, Westminster (1905–11). Built to mark the centenary of the death of John Wesley (1703–91), the founder of Methodism, this vast meeting hall was erected in concrete and steel to competition-winning designs by Lanchester & Rickards. Kahn system reinforced concrete was used in the building's construction, including the 114ft-diameter inner dome over the main hall, but the larger chamfered-square outer dome was rigidly framed in rolled-steel members (Fig 12.37).[119] Clad in lead, with no intervening drum below, this huge dome forms one of the most distinctive elements of an impressive building.

Despite the ambitious use of steel to frame domes, roofs and parts of the interior structure, evident early on in such buildings as the Church of Our Most Holy Redeemer, full steel framing, in the sense of true cage or skeleton construction was never called for, nor economically viable in churches in the period under consideration. By way of contrast, in New York advanced constructional techniques normally associated with commercial buildings had, by the mid-1890s, already made the transition to church building. Pierre LeBrun's Free Church of St Mary the Virgin (1894–95), heralded as the first steel-framed church,[120] was built around a skeleton frame of rolled steel throughout. *The Building News*, in reporting this development to a British audience announced 'the skeleton system is said to have brought the cost to a figure somewhat less than would have been that of a building of equal size constructed entirely of brick or stone, while occupying less time in the execution of the work'.[121]

London never produced anything like that, at least not until the 1920s,[122] but it did turn out one exceptional building that used steel for similar reasons, and which also spoke of a more relaxed, even celebratory, attitude towards the use of frankly exposed structural steelwork in framing a place of worship. Not surprisingly this came from a Nonconformist quarter. All Saints' Mission Church, White Lion Street, Pentonville (1901–2, demolished *c* 1961) was a robust yet austere structure by R A 'Bungalow' Briggs (1858–1916), and a positively boisterous exposition of the aesthetic possibilities of unconcealed steelwork, adorned further by sub-art nouveau, wrought-iron ribbon-work

(Fig 12.38). Six box-section stanchions in a basement gymnasium carried both the floor of the church above and intricate arched steel roof principals. Deep, open-web arched girders that formed a nave arcade braced the stanchions in either row, and similar members spanning to the external brick piers gave lateral stiffness to the whole structure. All the steelwork was painted dead black, with the ribbon-work

Fig 12.36 (facing page, top) Steel principals in trial erection at the yard of Archibald Dawnay & Sons prior to framing the roof of an unidentified church, possibly also by the Baine partnership.
[© The British Library Board (F60077–51)]

Fig 12.37 (facing page, bottom) The giant steel-framed outer dome of the Wesleyan Methodist Hall, Westminster (1905–11), erected by Archibald Dawnay & Sons. [RIBA81730, RIBA Library Photographs Collection]

Fig 12.38 (left) All Saints' Mission, White Lion Street, Pentonville (1901–2; R A Briggs, architect; demolished), showing the overt use of structural steel for practical and aesthetic ends. [Builders' Journal and Architectural Review, 14 May 1902, RIBA46802, RIBA Library Photographs Collection]

picked out in gilt and transparent colours. In its report, *The Builders' Journal and Architectural Engineer* explained that 'The form of construction has been dictated by considerations of cheapness, strength and light.'[123] Another reason may have been the speed with which such a structure could be erected. The foundation stone was laid on 3 December 1901,[124] and the building was complete by 14 May 1902.[125]

Swimming baths

The proper material for the roofs of swimming-baths has been a matter of some discussion. Wood has been used with good effect; iron has been employed with economy. Wood attracts less condensation; iron is cheap. It is a matter, apparently, of individual choice.[126]

The impetus to build public baths and wash-houses stemmed from institutional and governmental aspirations to improve social and sanitary conditions among the labouring poor. In London, as in other cities, impressive numbers of baths were constructed following the Public Baths and Washhouses Act 1846, which allowed borough councils to raise funds for this purpose; other model establishments were founded as a direct result of committee activity. London's pioneering example, Goulston Square Baths and Wash-house, Whitechapel, by engineer Price Pritchard Baly, used cast-iron roof trusses on brown brick walls for utilitarian 'fireproof' reasons (1846–51, demolished). *The Builder* called it 'not simply plain and unpretentious, but downright ugly'.[127] For the next half century or so, iron ran a close second to timber in roofing the capital's increasing number of public swimming pools, and many architects chose to make open show of its decorative possibilities. For St Marylebone Vestry's new swimming baths in Seymour Place (1873–4, demolished early 1950s), Henry Saxon Snell used semi-elliptical cast-iron ribs of 41ft span to enclose a space 85ft long. These iron principals were adorned with gilded scrollwork panels, and were carried by brick piers ornamented with blue hand-painted tiles bearing representations of birds, fishes and water-fowl (Fig 12.39).[128] By the 1890s swimming pools were being built larger to accommodate increasing numbers of bathers and typically had galleries for the public to watch swimming competitions and other events. In 1894 one specialist could note that these galleries were 'generally carried on rolled iron joists as cantilevers over dressing-boxes, and continue round three sides of the hall'.[129] As in theatres, steel joists began supplanting iron for this purpose, and the metal's lightness, strength and spanning capabilities soon saw it put to the support of roofs. Hoxton Public Baths (1898–9, demolished), part of a grand civic complex erected between Pitfield Street and Hoxton Market, combined both facilities and placed greater demands on the metal (Fig 12.40). In

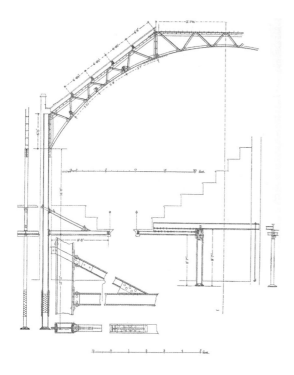

this building, 36ft-high rolled-steel stanchions embedded in the pool-hall walls supported both the cantilevered side galleries and the latticed steel roof trusses. Nine Elms Baths, Battersea, built in 1899–1901 to designs by Francis J Smith, also used prefabricated steel roof trusses, in this case erected ahead of most of the enclosing brickwork (Fig 12.41). Supplied by H Young & Co., these trusses spanned what was reputedly the largest covered pool in Britain at the time. The building, like others of the era, was designed so that by casing the pool with a temporary floor it could be converted into a public hall for nearly 1,500 people.[130] Fortunately, one of the best examples of the genre survives (Fig 12.42). Haggerston Public Baths, Whiston Road, were built in 1903–4 to designs by Alfred W S Cross, a leading architect specialising in baths and wash-houses. Erected at a cost of £60,000, the building made conscious use of showy baroque detailing in a red-brick façade that screened the 'large shed-like erections of

Fig 12.40 (left)
Hoxton Public Baths: details of the steelwork framing the roof and gallery (1898–9, demolished).
[(RIBA70539) Cross, Public Baths and Wash-Houses, 1986]

Fig 12.41 (below)
Erecting the prefabricated steel roof trusses for Nine Elms Baths, Battersea (1899–1901, demolished 1971). When opened, this was reputedly the largest covered swimming pool in Britain.
[Wandsworth Heritage Service]

Fig 12.42 (facing page)
Haggerston Baths (1903–4;
A W S Cross, architect),
photographed in 1992.
[BB92/20504]

Fig 12.43 (left)
Prefabricated steel trusses
framing Haggerston Baths
in course of erection by
constructional engineers
Homan & Rodgers.
[Builders' Journal and
Architectural Review
(C&S Supplement),
10 October 1906]

the baths proper'.[131] Steel trusses with a span of over 66ft were used to enclose the single 'first-class' pool, 100ft by 35ft, as well as three tiers of raised seating and a row of dressing boxes alongside (since removed) (Fig 12.43).[132] Prefabricated and erected by constructional engineers Homan & Rodgers, these members carried a continuous lantern that provided plentiful light and ventilation. Cross considered that steel roof principals were 'admirably adapted for swimming baths' when clad with Portland cement laid on metal lathing and finished with 'Keene's cement', as at Haggerston.[133] However, in this particular building type, reinforced concrete was soon to show its advantages in terms of reduced maintenance, one of the first metropolitan examples of this being Hammersmith Public Baths, Lime Grove (1905–7, J Ernest Frank, architect).[134]

Fire stations

Fire stations emerged as an increasingly distinctive building type in the second half of the 19th century. Following the establishment of the Metropolitan Fire Brigade in 1865, some 26 new fire stations were built between 1867 and 1871 to the designs of Edward Cresy, the body's architect. More were built in the 1870s and 1880s by his successor, Alfred Mott, who changed the house style to 'Secular Gothick'.[135] The basic requirements of an unobstructed appliance or engine room on the ground floor, with offices and accommodation for the firemen above, were probably met by girders spanning the external walls, but the constructional techniques used in these early buildings are unclear.

In 1889 control of the Metropolitan Fire Brigade transferred from the Metropolitan Board

Fig 12.44
The engine house of the
Manchester Square Fire
Station (1889; LCC
Architects' Department),
looking towards Chiltern
Street.
[AA034251]

Fig 12.45
In Manchester Square Fire
Station primary girders
pass over the capitals of
distinctive octagonal-section
columns.
[AA034254]

of Works to the newly formed LCC, marking the onset of a period of extension and upgrading of the old stations, and the construction of new ones to serve better the expanding metropolis. The first of these new buildings was Manchester Square Fire Station, Chiltern Street, Marylebone, built in 1889 to the designs of the LCC Architects' Department. A building of much architectural merit, it is also remarkable for the ambitious constructional methods used

in framing the engine house (Fig 12.44). The main, five-storey block (including attics and basement), executed in a style derived from Lewis Vulliamy that is characteristic of LCC fire stations of the period, employs a grid of box-section girders, supported on cast-iron columns, to carry the weight of the upper three floors. The stocky octagonal-section columns on the ground floor (engine house) are spaced on 13ft centres and for the most part are built into the exterior walls and piers, only three remaining freestanding. Impressive 24in. by 18in. box girders pass over the column capitals, with no apparent brackets or castings to form a rigid connection between the two (Fig 12.45). This structure, in effect a semi-cage construction, undoubtedly derives lateral stability from the infill brick walls and piers. The point loads transmitted by the three interior columns were probably originally borne by cast-iron uprights in the basement, but at some time in the early 20th century they were replaced by a steel-framed floor consisting of rolled-steel I-sections spanning between the basement bearing walls with rolled-steel stanchions giving additional support. Evidence of the original 'fireproof floor' is still faintly visible, showing that it consisted of brick jack arches sprung between cast-iron skewback girders (see p 192). Two

Fig 12.46
The structural elements used in Manchester Square Station continued in use in most of London's Edwardian fire stations, including Cannon Street (1907). [London Metropolitan Archives, City of London (SC/PHL/02/0474/ L17024)]

Fig 12.47
Rolled-steel joist sections used as beams and stanchions in Northcote Road Fire Station, Battersea (1906). [London Metropolitan Archives, City of London (SC/PHL/02/0473/ L17004)]

cast-iron columns embedded in the ground-floor entrance piers were replaced in 1969 by 10in. by 10in. universal columns, part of a widening scheme that swept aside the original stone-faced entrance arcade.[136] Whether wrought iron or steel was used for the plate girders in the engine room is not documented.

The LCC's construction programme gained pace at the end of the century, some 22 additional new stations being erected in the period 1898–1908.[137] While the styling of London's fire stations changed to Queen Anne or baroque, the constructional techniques used at Manchester Square Fire Station continued to serve as a basic prototype, steel having replaced wrought-iron for joists, beams and girders. In virtually all cases, the appliance room was internally framed with iron columns and deep, built-up I- or box-section girders, a mode of construction that, save for the use of steel, was oddly conservative compared to contemporary buildings (Fig 12.46). Only in a minority, such as Northcote Road Fire Station, Battersea (1906, demolished), was recognition given to the carrying power and simplicity of rolled-steel I-sections (Fig 12.47). Encasement of these members in fire-resisting material was rarely, if ever, undertaken, perhaps because facilities for extinguishing fires were on hand.

Underground stations

Iron girders and stanchions formed vital, visible components in the engineering of the Metropolitan Railway, which opened in 1863 as the first underground passenger railway in the world. As in fire stations, less is known about their use in the station buildings above, most of which have been replaced. The pioneers of the building type, including Paddington, Baker Street and King's Cross, designed for the Metropolitan Railway by its architect John Hargrave Stevens, or engineer-in-chief (Sir) John Fowler, were applauded for their constructional ingenuity rather than for their architectural features. *The Building News* noted that there was 'very little architectural character in any of the stations, but a wonderful amount of engineering skill and good workmanship'.[138] Neither

this report, nor that of *The Builder* (which remarked on 'the clever construction of some of them')[139] enlarged on these aspects, an omission that would characterise the great majority of the occasional, brief notices that underground stations were to receive throughout the century and beyond. Typically, the surface buildings of underground stations were single-storey brick-walled Italianate structures pierced by tall windows and doorways. Most iron went into the framing of the roofs and the domes that crowned them, although girders facilitated the construction of basement-level stores, parcel offices and lavatories.[140]

From the 1890s the application of new tunnelling technologies enabled the construction of deep-level cast-iron-lined 'tubes', which unlike the earlier 'cut-and-cover' tunnels did not impinge on streets and buildings at ground

Fig 12.48 (below) Steel-trough flooring in the Bank Station on the Central London Railway, now the Central Line (1896–8; (Sir) John Fowler, (Sir) Benjamin Baker and Basil Mott, engineers). [© The British Library Board (F60077–57)]

Fig 12.49 (facing page) A four-storey office was built above Delissa Joseph's Oxford Circus Station soon after its opening in c 1900. [© TfL from the London Transport Museum Collection (370-3-1)]

level. The City & South London Railway (now part of the Northern Line) opened in 1890 as the first of the capital's railways to employ electrical traction and a standard gauge. The surface buildings designed by Thomas Phillips Figgis may, given their date, have used steel to frame and carry the great lead-clad domes that housed the passenger lift equipment. By the end of the decade, steel-trough flooring, steel built-up girders and rolled joists were seeing extensive use in the construction of the subsurface stations and subways of the Central London Railway (now the Central Line), notably at Bank Station (1896–9, (Sir) John Fowler, (Sir) Benjamin Baker and Basil Mott, engineers) (Fig 12.48).[141] The low-slung surface stations, designed by Harry Bell Measures, prefigured subsequent developments by employing sturdy, steel-framed flat roofs that allowed for commercial development above. The first realisation of this facility was at Oxford Circus Station, where a four-storey office designed by Delissa Joseph was put up soon after the station's opening in 1900 (Fig 12.49). Joseph went on to design many of the offices over Measures' stations, including Tottenham Court Road, Lancaster Gate and Queensway.[142]

It was in the stations designed for the Underground Electric Railways Company of London Ltd (UERL, comprising what are now the Bakerloo, Piccadilly and Northern Lines) that steel was put to more ambitious and systematic use. Leslie William Green (1875–1908) was appointed architect to the UERL in 1903, and given responsibility for the 'design of stations above ground and decorative work to Stations, Tunnels, Platforms and Passages'.[143] In some 3 years Green and his assistants, Stanley A Heaps (1880–1962) and Israel Walker (1870–?), designed more than 40 stations, creating the most consistent 'house style' and distinctive corporate identity of any Edwardian transport enterprise. Each station was adapted to its site, so no two were exactly alike, but the employment of standardised techniques of steel framing permitted virtual mass production. Clad with the memorable ox-blood coloured faience tiles, these blocky frameworks enabled the most skeletal of arcaded façades with the widest possible entrances and exits and the most uninterrupted of interior spaces, to speed the flow of passengers. They also provided strong support for mezzanine floors – lit by glazed arches and flanking circular or rectangular windows – to house the lift gear and in many

Fig 12.50 (above)
The newly constructed
Hampstead Tube Station,
which opened in 1907.
[BL20042/002]

Fig 12.51 (right)
Non-combustible steel-frame
construction at Holborn
Station, erected by Waring
and White Ltd. (1906)
Glazed tiles would soon
encase the stanchions.
[Builders' Journal and
Architectural Engineer,
30 December 1908]

cases office accommodation (Fig 12.50). Fire resistance was imperative, and reinforced concrete was used to form the floors and roofs and encase the stanchions, as is shown in a rare construction photograph of Holborn Station,

erected by Waring & White Ltd (1906) (Fig 12.51). Flat roofs enabled immediate or future vertical development, facilities that were soon exploited at Leicester Square (Fig. 12.52), Knightsbridge, Hyde Park Corner and

Fig 12.52 (facing page)
By 1916, when this
photograph was taken,
a heavy, stone-faced office
block had been built on the
superstructure of Leicester
Square Station.
[BL23404]

Piccadilly Circus, among other stations. The offices above the last three stations, which according to one source are themselves steel framed,[144] were designed by Delissa Joseph. Contemporary accounts described these stations simply as 'steel framework[s] clothed with terra-cotta glazed blocks of a dark ruby colour',[145] but refurbishment work at Gloucester Road Station revealed the techniques of steel framing employed. The façade framework was formed entirely from compounded I-sections, with stanchions supporting paired 24in. by 7½in. beams, connected by flange plates to form the roof cornice. Smaller paired

beams were used to form the mezzanine level. Those forming the main entrance bressummer derive partial support from the load-bearing brick pier.[146] In fact, judicious use of traditional load-bearing techniques was a feature of most stations. Enclosed on one, two or even three sides by the party walls of pre-existing buildings, it made sense to let thick, windowless walls carry the ends of the girders of the interior skeleton. The long defunct York Road Station, Kings Cross, for example, with its complete interior skeleton and partial external skeleton, shows the rationality of this approach (Fig 12.53).

Fig 12.53
Plan of York Road Station (1906).
[© TfL from the London Transport Museum Collection. The Tramway and Railway World, 6 December 1906]

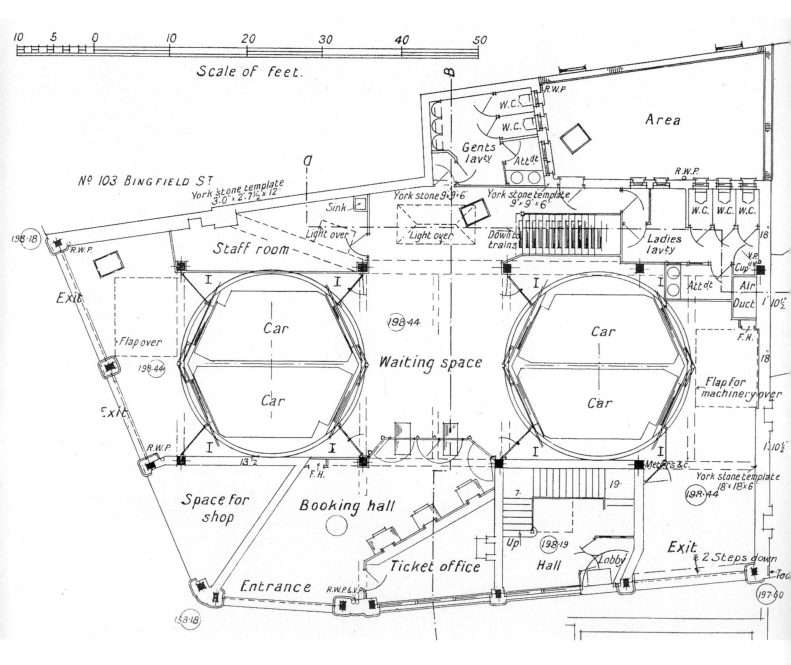

13

Industrial buildings

Multi-storey factories and warehouses

The substitution of steel for iron is greatly on the increase, mild steel having practically displaced iron in boilermaking, ship and bridge building, and it is even being used at the present time in the construction of a block of warehouses in London, requiring a total of over 8,000 tons ... (1885)[1]

Iron forerunners

The development of the internal cast-iron frame in multi-storey textile mills during the 1790s and 1800s was stimulated by the exceptional risk of fire and the need for open working floors. In London, where industry remained predominantly workshop based, and despite the notorious burning of the timber-floored Albion corn mills in 1791, the exigencies were different. London's Building Act 1774 and subsequent Acts decreed that large buildings such as warehouses had to be internally subdivided by brick walls to thwart the spread of fire. This contained the fire risk, while in the enclosed docks with their huge warehouses, the strict regulation of naked lights further reduced the risk. The type of construction favoured in northern mills, of non-combustible but heavy and costly brick jack-arches on cast-iron beams and columns was thus little used in London's industrial buildings except in particularly hazardous situations.[2] Among the latter were the tall sugar refineries of the East End, now long gone, some of which in the mid-19th century seem to have employed floors of flagstones, slate slabs or iron plates laid upon iron joists. Some warehouses designed by William Cubitt at Hay's Wharf, Tooley Street (1856–7), and New Hibernia Wharf, Montague Close (1858–60), used incombustible floors that alternated with floors of timber as a technically

questionable means of meeting the volumetric rules for compartmentation.[3]

The primary structural incentive in multi-storey industrial buildings used for storage rather than production was the attainment of greater load-bearing capability. Throughout much of the 19th century softwood timber remained the most economical, readily obtainable material for the construction of floors and roofs, yet proved deficient in withstanding the concentrated axial loads to which the columns of multi-storey warehouses were typically subjected. Cast iron was eminently suited to this role, and widely spaced cast-iron stanchions supporting timber beams and joists became the norm for London warehouses and factories throughout most of the 19th century. Where timber stanchions remained popular, as in the granaries of Bermondsey, they had to be closely spaced – an arrangement that also suited the poor foundation conditions of the Surrey marshlands, as well as reflecting such buildings' proximity to the timber ponds in the Surrey docks. Iron columns were first used to noteworthy effect in some of the big warehouses within the enclosed docks in the early 19th century, a period of concentrated investment. The overall picture during the 19th century was one of conservatism rather than innovation, even allowing for the restricted sample of buildings this conclusion is based on. London's industrial buildings have through the 20th and early 21st centuries probably seen the most intense and protracted redevelopment of any metropolitan building type, innumerable interesting examples no doubt lost with no record of their passing. Survivors form an enormously diminished reflection of the original stock. Compared to industrial buildings outside the metropolis, and to other types of buildings within it, published information concerning the use of structural iron in London's warehouses and factories in the period *c* 1800–60 is scant.

With that said, much of importance is known. One of the earliest documented large-scale uses of iron was in the enlargement of the two-storey London Dock South Stacks, which were doubled in height in 1810–11. Heavy, cruciform-section, cast-iron columns were used in preference to timber posts for carrying the loads, but still supporting the wooden crossheads that reduced the effective spans of the huge floor-beams. Another example was in the refitting in 1813–18 of six tall sugar warehouses that had been built on the north quay of the West India Import Dock in 1800–3, to plans by George Gwilt senior and junior. For reasons of

strength rather than fire-resistance, the original timber posts supporting the floors were replaced with cruciform-section columns, each fitted with detachable iron crossheads.[4] By this date, the use of cruciform cast-iron columns had already fallen out of favour in textile mills, where stiffer, hollow cylindrical sections, which could carry more load without buckling, were in widening use.[5] Hollow cylindrical columns were used early on in London, for example, at the East India Company's Pepper Warehouses, Blackwall (1807–10),[6] and at Nos 10 and 11 Warehouses, West India Docks (1808–10 and 1807, respectively),[7] but it was

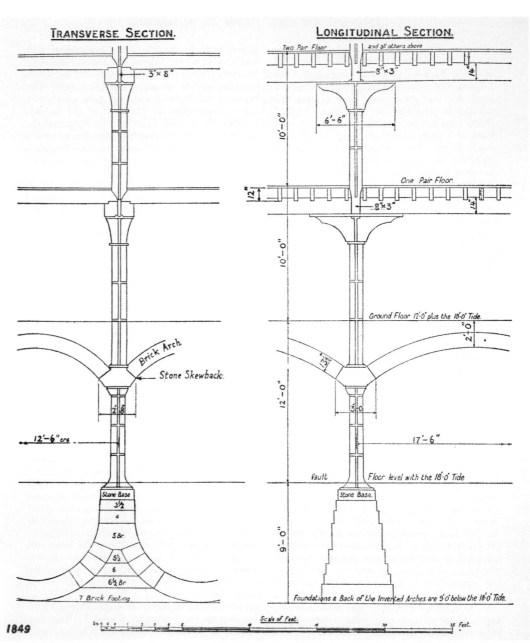

Fig 13.1
Cruciform columns in a multi-storey warehouse, London Docks (1849). [Hamilton, 'The use of cast iron in building' (1940–1) © The Newcomen Society, London]

the cruciform-section column that saw commonest use in London warehouses. Indeed the form persisted through the 19th century and beyond,[8] largely because it was cheap and easy to produce, and perhaps also casting defects were easy to see. One example of the form, with horizontal stiffeners, epitomised mid-19th-century best practice (Fig 13.1). Designed for a multi-storey warehouse at the London Docks in 1849, the bases of those columns in the upper floors were attenuated to socket into the heads of the pillars below, thus allowing the whole vertical assembly to adjust to minor changes in level or alignment without risk of fracture.[9] Dock warehouse construction, with few exceptions, continued to rely on timber beams and joists for reasons of economy and proven load-bearing capability: structural failures in fires, as at Tooley Street in 1861, and in northern mill buildings, doubtless heightened mistrust of cast iron in tensile situations among many metropolitan designers.

In fact, it was only in the generously capitalised buildings erected for the Navy that cast iron saw use in London that was in any way analogous to that of the most advanced contemporary 'fireproof' textile mills. One example

was a former warehouse for the Royal Victoria Yard, Deptford,[10] built in the 1820s when substantial funds were made available for the modernisation of the principal Royal Victualling Yards (Fig 13.2).[11] Using a system previously developed for Naval storehouses elsewhere, transverse cast-iron beams supported longitudinal joists of the same metal, on which a flat incombustible floor of York stone slabs was laid. Sockets cast on the beam webs enabled the joists to be slotted in place, enhancing the overall robustness of the floor structure, a technique extensively used elsewhere in the royal dockyards, as well as in some contemporary textile mills such as Beehive Mill, Manchester (1824).[12] As in Rennie's contemporary mills and bakery building at the Royal William Victualling Yard, Stonehouse, Devon, the shafts of the columns at Deptford were cast with grooved flanges to take removable partitions that divided the space for the purposes of grain storage.[13]

If London's industrial builders were slow or hesitant to use cast iron in beams, then the reverse seems to have been true of wrought iron. Wrought iron was quickly assimilated into the construction of London's better-capitalised, multi-storey industrial buildings following the

Fig 13.2
Warehouse interior of the 1820s at the Royal Victoria Yard, Deptford, photographed in 1963. Such extensive cast-iron framing was uncharacteristic of London warehouse construction in the first half of the 19th century.
[© London Metropolitan Archives, City of London (SC/PHL/02/0913/63/ L18482)]

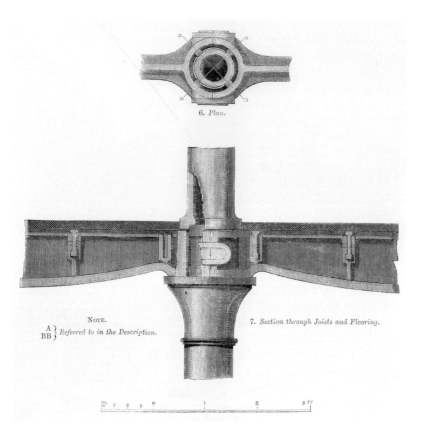

6. *Plan.*

NOTE.
A ?
BB } *Referred to in the Description.*

7. *Section through Joists and Flooring.*

Fig 13.3
Column-beam connection in the City Flour Mills, Upper Thames Street (1850–1), one of the first documented London buildings to use rolled-iron joists.
[The Builder, *17 December 1864*]

correct alignment of the beam-ends before the shrink rings were tightened. Such precision was rare in the structural engineering of most mill buildings; in this case it arose from the talents of a specialist consulting engineer, (Sir) Frederick Joseph Bramwell (1818–1903).[15]

From the late 1850s London could boast some of the most constructionally advanced multi-storey industrial buildings in the country, a phenomenon that owed much to the widening availability of built-up wrought-iron components. One of the first buildings to signal this shift was a four-storey 'fireproof' warehouse at the St Katharine Docks, erected in 1858–60 to designs by George Aitchison senior. Surviving as 'Ivory House', 'I' Warehouse, as it was originally known, used wrought-iron plate girders on cylindrical, cast-iron columns to support brick jack-arch floors.[16] Five years later, similarly constructed two- and three-storey 'fireproof' warehouses were completed at the East India Docks.[17] In their use of the more trustworthy metal, these buildings marked a distinct advance over earlier attempts at using jack-arch construction in conjunction with cast-iron beams.[18] Nevertheless, with the financial crisis of 1866, and the gradually changing requirements of transit and storage in favour of low-slung transit sheds rather than tall warehouses, multi-storey 'fireproof' construction became less suitable for London's docks, both on grounds of expense and functional suitability.

By the 1860s more rationalised design was becoming evident in buildings in the City and surrounding districts. One such was an imposing four-storey warehouse and showrooms at Nos 34–36 St John Street, Clerkenwell, erected in 1868 for the lead-and-glass merchants George Farmiloe & Sons Ltd. Replacing earlier buildings destroyed by a catastrophic fire of March 1868 (which had transformed the stock into 'hundreds of tons of lead and glass fused into one common mass'),[19] the new warehouse, designed by Lewis Henry Isaacs, was engineered to withstand the severe loads of the materials stored on site. Compound girders, formed from rolled I-sections reinforced in the top flange with plates in accordance with Julius Homan's patent of 1865 (*see* p 16), were used to carry closely spaced rolled-iron joists on the ground floor, and timber joists on the upper floors (Fig 13.4). The ground floor was designed to take a load of 1,120lb (½ ton) per square foot,[20] and massive brick-and-stone piers were built around the principal columns which

collapse of a number of textile mills with cast-iron beams (*see* p 8). One of the earliest documented uses of rolled wrought-iron joists was in the City Flour Mills, Upper Thames Street (demolished). Constructed in 1850–1 for James Ponsford, this eight-storey brick-walled edifice exploited rail-like sections as joists, suggesting that rolled I-sections were still not yet readily obtainable in London (Fig 13.3). These joists were approximately 6in. deep, with a bulbed top and a narrow rectangular bottom flange. They slotted into sockets cast on fish-bellied cast-iron beams that spanned between the columns and the exterior walls. The design of the column–beam connections was sufficiently noteworthy to receive detailed description in *The Builder*.[14] The cylindrical cast-iron columns slotted together one above the other, and the semi-circular ends of the beams were clamped around the columns using hoop-iron 'shrink rings'. These were standard techniques of jointing in cast iron, but it was the detailing of this ironwork that set it apart. To prevent the ends of the beams from touching, and hence stressing the columns under load, a small 'clearance' gap was provided under the bases of the columns above. In addition, iron keys (marked 'B' on the illustration) were used to maintain the carried the cumulative weight of the floors above; between these, auxiliary columns provided mid-span support. Huge wrought-iron bressummer girders, made up of paired I-sections riveted together between plates and supported on H-section stanchions of cast iron, took the weight of the walls above the open-sided ground floor, which faced a loading yard. All this ironwork, of Belgian origin, was supplied by William & Thomas Phillips, the first firm to market compound girders. None of the iron members were encased in incombustible material, and would never have been in this period. However, the use of timber flooring throughout is perhaps surprising, given the fate of the earlier buildings.

By the late 19th century, with the proliferation of rolled-iron joists and other sections, and proprietary fire-resisting floor systems using ceramic or concrete components, the internally iron-framed 'fireproof' factory and warehouse in London was no longer exceptional. For example, Homan & Rodgers' compound-girder floor system was used to frame the floors of a beer-bottling store of 1885 in Kentish Town, and in the lower two floors of a warehouse of 1886 in Farringdon Street, built for the printing manufacturers Messrs Harrild and Brothers. Both buildings were designed by Theodore K Green and Son.[21] Dennett and Ingle's 'fireproof' floor

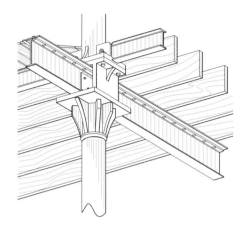

Fig 13.4
Robust floor construction using compound beams and girders at 34–36 St John Street, Clerkenwell (1868–9), an early application of Julius Homan's revolutionary patent of 1865.

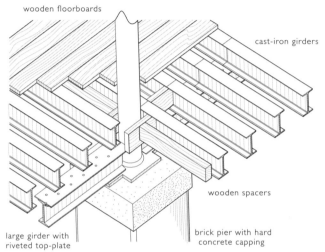

wooden floorboards

cast-iron girders

wooden spacers

large girder with riveted top-plate

brick pier with hard concrete capping

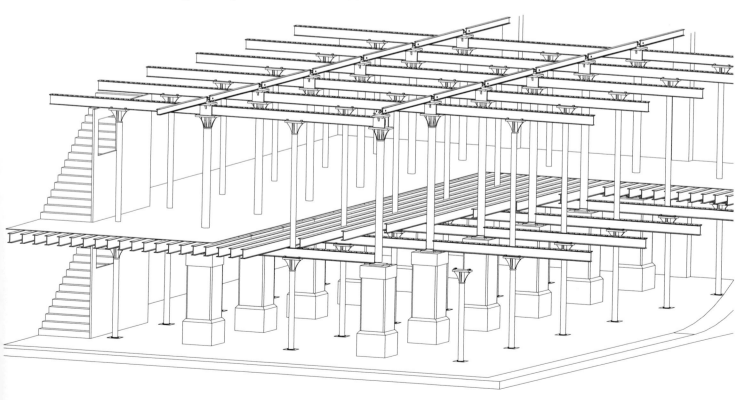

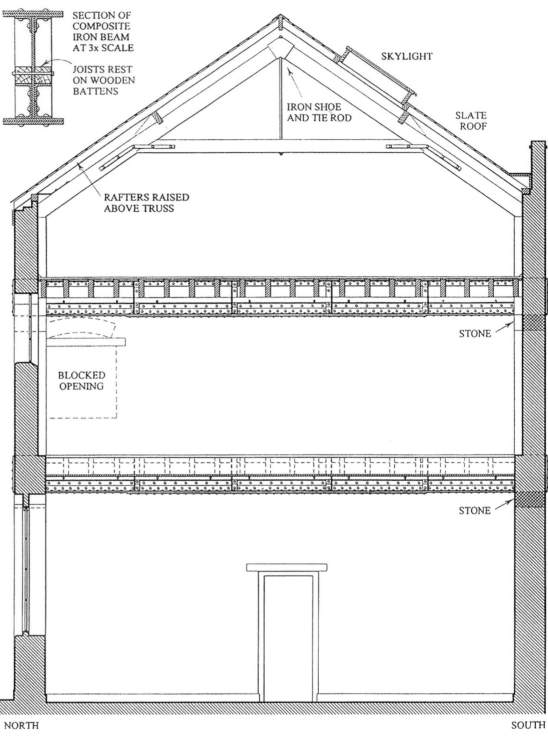

SECTION OF
COMPOSITE
IRON BEAM
AT 3x SCALE

JOISTS REST
ON WOODEN
BATTENS

SKYLIGHT

IRON SHOE
AND TIE ROD

SLATE
ROOF

RAFTERS RAISED
ABOVE TRUSS

STONE

BLOCKED
OPENING

STONE

NORTH

SOUTH

SECTION LOOKING EAST

0.5   0   1   2   3 metres

1   0   3   6   9 feet

Fig 13.5
Warehouse at 7 Caledonian Road, Kings Cross, built for Wilkinson, Heywood and Clark Ltd, varnish, colour and paint manufacturers (1885). Long-span built-up wrought-iron girders enabled strong, column-free floors.

system, employing built-up, wrought-iron beams to support shallow concrete arches, was used in the most 'risky' portions of warehouses built at King's Cross in 1885, to designs by W H Romaine-Walker and A Tanner for Wilkinson, Heywood and Clark Ltd, varnish, colour and paint manufacturers.[22] Elsewhere on the same site, strength, not incombustibility, was the primary reason for the employment of wrought-iron beams. In one warehouse block, built-up wrought-iron beams were used to span the 24ft distance between the exterior walls, unassisted by supporting columns (Fig 13.5). Shelf brackets, riveted to the beam webs, held the timber-joisted floors well within the (23in.) depth of the beams, thus maximising available space. Few British manufacturers could supply structural members of this size at competitive prices, and the structural ironwork used throughout these buildings was made by Forges de la Providence, of Charleroi, Belgium.[23]

Further conflagrations, notably the 'Great Fire' in Wood Street, City of London, which destroyed a vast swathe of warehouses in 1882,[24] may have convinced some manufacturers and warehouse-owners of the false economies of traditional timber-floored construction, although insurers, and the Metropolitan Fire Brigade, continued to approve timber. Among the greatest losses at Wood Street was that suffered by Messrs Rylands & Son. For their new building, erected in 1883–4 to imposing designs by J and J Belcher (Fig 13.6), cast and wrought iron were used extensively for interior support. This construction was perhaps then of sufficient ubiquity as to preclude comment by *The Builder*, but the younger Belcher's one-time partner J W James later recollected that it 'had to be framed with iron columns and stanchions and rolled iron joists ... thus anticipating the later steel framed buildings'.[25]

The introduction of steel

In factories and warehouses, steel, initially in the form of rolled-steel joists and beams, probably began to make its mark in London in the mid-1880s, as its availability at competitive prices increased. Some of the more adventurous textile mill designers began experimenting with steel filler-joisted floors at this time,[26] and steel joists, encased with lightweight concrete, were also slowly coming into use in London warehouses in this decade.[27] Instances of the specific use of steel are hard to document,[28] but

it is clear that some designers were sufficiently convinced of the metal's worth to employ it on a remarkable scale, if sporadically. In 1885 *The Builder* could note that more than 8,000 tons of steel were being used in the construction of a block of London warehouses, quoted at the start of this chapter. This building, which may well have exploited the material for vertical as well as horizontal framing elements, has not been identified.[29] Taken together with the goods depot of the London Tilbury and Southend Railway (1886–7), which used an astounding 20,000 tons of steel (*see* p 299), such examples show that, in London, industrial buildings were the first to exploit steel on a large scale. Most of the steel in the railway goods warehouse was used for long built-up girders, and other surviving warehouses of the period indicate that this form was chosen in

Fig 13.6
Rylands & Son's warehouse,
Wood Street, City (1883;
Belcher & Belcher, architects;
demolished).
[The Builder, *10 May 1884]*

289

architect),[30] where box-section beams on cast-iron columns carry timber floors, the whole constructional emphasis being load-bearing rather than fire-resisting. Bearing the rolling mark of Dorman Long & Co. on the corner angles, these beams were also conspicuously stencilled '… DLESBRO STEEL' at the company's Middlesbrough stockyard to differentiate them from iron (Fig 13.7). From the first year this company turned to steel, these beams endure as among the earliest of their kind in the metropolis.

It was not until the 1890s that warehouses and factories incorporating steel began to appear in increasing number. One early survivor of that decade is Gilbey House, Nos 36–48 Jamestown Road, Camden, built in 1895–6 as a warehouse and bottling plant for the famous gin distillers of that name (Fig 13.8). Designed by the company's engineer, William Hucks, this five-storey block was given an internal frame of steel stanchions and beams to support the steel filler-joist floors that bore the brunt of the heavy machinery and stock. Besides marking an early large-scale use of steel for columns as well as beams, the thick exterior walls (with piers up to

preference to wrought-iron alternatives for its diminished sectional size and greater carrying power. Deep, rolled-steel I-sections were still not widely available. A surviving example of this initial line of development is a five-storey warehouse built for the lead, glass and colour manufacturers Nicholls and Clarke, in Blossom Street, Shoreditch (1886–7, William Eve,

3ft 6in. in places)[31] are themselves of much interest for their somewhat idiosyncratic approach to reinforced concrete, being for the most part mass concrete, yet strengthened in places with hoop iron and scrap metal, allegedly including old iron bedsteads.[32]

Through the 1890s, and the first years of the following decade, single-storey columns of cast iron remained the principal interior support for most multi-storey factories and warehouses, but steel-framed floors provided superior load-capacity and resistance to vibration, and often facilitated the use of fewer columns.[33] Fully automated flour-mills illustrate this. In London, the changeover from traditional stone-milling to roller-milling in the 1880s brought about the strengthening, enlargement or replacement of traditional structures to meet the demands of a radically new technology. By the late 1890s this change was essentially complete, having left behind it a series of architecturally and constructionally ambitious buildings that had attempted to keep pace with the escalating sophistication of the machinery within. One of the most commanding survivors of this brief epoch is the huge grain silo fronting Deptford Creek, built in 1897 as Mumford's Mill, at Nos 23–25 Greenwich High Road, Greenwich (Fig 13.9). Erected to the Italianate designs of Aston Webb (1849–1930), this building exploited advanced techniques of internal metal framing, comprising a grid of substantial built-up steel girders, supported at its intersections by stout, cylindrical cast-iron columns, and at its perimeter by thick, cement-mortared exterior walls. Deep, 12in. by 5in. rolled-steel secondary beams span between these principal members, providing the support for concrete floors. This phenomenally robust structure represents, in some senses, the ultimate 19th-century extension of the principle of making the internal grid as intrinsically stable in both axes as possible, seen in embryonic form at the MSL Railway's 'London Warehouse', Manchester (*see* p 289). Such constructional strength was principally required to carry the weight of the deep grain bins, and, unconventionally, to provide non-combustible storage floors under these (Fig 13.10). It is unclear how much of this engineering can be attributed to Webb but it seems likely that Dorman Long & Co., the main supplier of the steelwork, were enlisted in matters of detail. The construction of the silo was accompanied by the insertion of steel-framed floors into an earlier mill on the site for the support of heavier

Fig 13.9
Mumford's Mill, 23–25 Greenwich High Road, Greenwich, viewed from Deptford Creek (1897; Aston Webb, architect).
[The Builder, *1 May 1897*]

roller-milling machinery. The 21ft width of that multi-storey block was spanned by compound steel beams, formed from two I-sections joined side by side by flange plates. In some situations, where loadings were reduced, plain 16in.-deep,

Fig 13.10
First-floor interior, showing hopper bases projecting through the robust steel-framed floor.
[BB99/09898]

Fig 13.11
The Bovril Company's
factory and offices in Old
Street, Islington (1898–9;
Lanchester & Rickards,
architects, with structural
engineering by the American
C V Childs; destroyed 1943)
[BL18450]

rolled I-sections are used in support of the steel joisted floors. This steelwork variously exhibits the rolling marks of both Dorman Long & Co., and the Leeds Steelwork Co., the same combination of companies responsible for supplying the steelwork for W T Foxlee's GNR warehouse (1895–8) in Manchester.[34]

A more sophisticated solution to the provision of robust manufacturing and storage space on a large scale was evident in the Bovril Company's Factory and Offices in Old Street, Islington (1898–9, demolished 1943). The first major commission of Henry Vaughan Lanchester (with the assistance of E A Rickards), this was a building of much architectural merit, both in the detailing to its deep-coloured, glazed-brick exterior (Fig 13.11), and in the quality of its principal office spaces. It was erected to keep pace with the massive demand generated by Bovril's first decade of

advertising.[35] Co-location of head-offices and manufacturing, and maximisation of available space were key objectives in the design. C V Childs was enlisted to design a steel frame that would provide sturdy, open-plan floors for manufacturing operations at the rear of the site (Fig 13.12). His solution, a robust internal framework of continuous built-up stanchions and rolled beams, encased in fire-resisting plaster and designed to withstand vibration and oscillation from the mechanised food-processing equipment, was erected ahead of the enclosing brickwork by Dorman Long & Co. (Fig 13.13). Childs's usual approach, at least for 'polite' buildings requiring large spaces at ground floor level, was to carry virtually the entire weight of a bearing-wall building on deep, built-up girders placed at first-floor level (as at the Hotel Russell). Here at least two large, undivided floors were required, and so he adopted a different form, one with greater affinities to the self-supporting skeletons he had designed in America. Nevertheless, unlike the standard American practice he was used to, in this project it is unlikely that Childs worked in full concert with the architect. Lanchester seems to have drawn up his plans, dated c 1893,[36] before the engineer was involved, as Childs did not arrive in this country until 1894.[37]

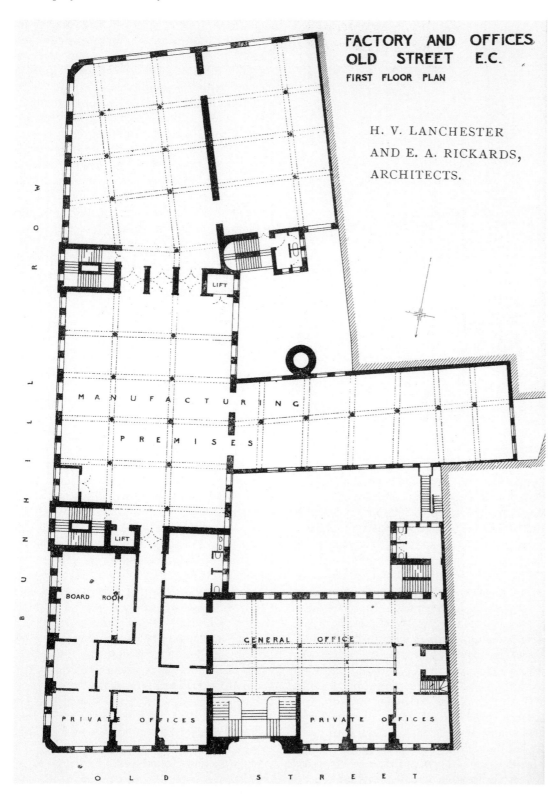

Fig 13.12
First-floor plan of the Bovril Company's factory and offices showing the internal steel grid.
[The Architectural Review 16, July 1904]

Fig 13.13 (above)
The steel framework of the
Bovril Company's factory in
course of erection early in 1898.
[The Engineering Magazine,
vol 15 (May 1898), Library of
Congress, General Collections]

Fig 13.14 (facing page, top)
The Shannon Factory, Tyssen
Street, Hackney (1902), a
'model' cabinet-making factory
designed by Edwin Sachs,
founder of the British Fire
Prevention Committee.
[Academy Architecture, 1903]

Fig 13.15 (facing page, bottom)
Section through the central
machine room block of the
Shannon Factory, showing
long-span floor girders,
drawn 1901.
[London Metropolitan Archives,
City of London (GLC/AR/
BR/06/ 019046)]

Reduction in the number of columns and the consequent maximisation of bay size to provide the most uninterrupted and hence flexible floor space for storage or manufacturing was one of the traditional, key objectives in the design of multi-storey industrial buildings. Wrought-iron beams were capable of safely spanning greater distances than cast-iron equivalents, but were difficult and costly to manufacture, whereas steel equivalents, in the form of long, deep rolled sections, were more easily procured and readily lent themselves to the making of efficient compound girders where span dictated. Larger bay sizes facilitated the expansive floor spaces demanded by up-to-date systems of production and goods handling. While building technologies reliant on timber or iron were often capable of providing economic and ergonomic constructional solutions, the substitution of steel brought about more expansive floors with fewer point supports. To control the spread of fire, the London Building Acts limited the

amount of open floor area within buildings of the 'warehouse class', but, with each revision, took account of the growing scale of manufacturing and warehousing and, implicitly, of the construction technology available to realise it: 200,000 cu ft in 1844; 216,000 cu ft in 1855 and 250,000 cu ft in 1894. The number of businesses that exploited the upper limits set by the 1894 Act multiplied, a phenomenon rooted in the increasing scale of operations, but enabled by the growing availability and use of structural steel. For example, a wool-storage warehouse of 1898 at No. 74 Back Church Lane, Whitechapel, was subdivided on each of five floors into nine compartments, seven of which were of 248,000 cu ft.[38] The use of built-up, steel beams enabled generous spacing of the cruciform-section cast-iron columns. For a growing number of enterprises, this limit was not enough, and through the same period the number of petitioners against the Act's volumetric restrictions grew steadily. The Bovril Building was a typical case. Although Childs was probably drafted in after Lanchester had worked out his basic design, the architect evidently knew that interior steel-frame construction could enable the large, unobstructed spaces required for advanced, semi-automated systems of production. The Bovril Company wanted three large undivided spaces, each well over 250,000 cu ft, in order to introduce modern methods of manufacture, and so that the workforce could be supervised easily from strategically placed supervisor's offices. Under the 1894 Act, the LCC could waive the limit up to an absolute maximum of 450,000 cu ft in special cases. At 462,000, 390,000 and 294,000 cu ft, Lanchester's intended spaces were far in excess of the Act's provisions, and Thomas Blashill, the LCC's superintending architect, turned down his application for planning permission in December 1896.[39] It is unclear how Lanchester's and Childs's designs were scaled down, yet here is an early and clear example of how the full potential of steel-frame construction to provide undivided spaces was held back by building legislation in London.

Another case where greater cubical extent was petitioned for is the Shannon Factory (now Springfield House), Tyssen Street, Dalston (1902) (Fig 13.14). Designed by the polymathic Edwin O Sachs,[40] this imposing U-shaped factory for the Shannon Company Ltd, a prestigious cabinet-making firm, represented the outcome of protracted negotiations between Sachs, the

LCC architect and the Metropolitan Fire Brigade. Sachs, at the behest of Mr Schaefer, the managing director, attempted to secure permission for large, undivided spaces for the three blocks so that the latest American and Continental methods of semi-automated production and layout could be introduced. The final compromise, Sachs's fourth revision of his original design, was a curious mixture of construction technologies that reflected the differing fire risks. The outer wings were floored with steel beams and pugged timber joists, and the central machine-room block with more advanced reinforced-concrete floors on the proprietary 'Columbian' system, which would subsequently be used in Waring & Gillow's furniture emporium, and Selfridges (*see* p 242). In this block, built-up steel girders were employed on each floor, extending transversely more than 40ft between the brick walls, supported by two central rows of steel stanchions that extended along the length of the space (Fig 13.15). On

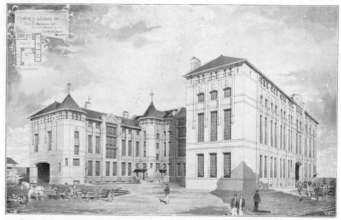

Modern Joinery & Cabinet Works

The Shannon Company's New Works in London are pronounced by leading Architects as the most modern and completely equipped wood working factories in the World.

The chief features of these factories are the Modern Machinery and Tools, the Electric Motors and Generating Station, Automatic Exhaust Fans for the removal of dust and shavings from each machine, the Heating and Ventilating Plant, Rapid Handling of Materials, Healthy Appointments and Splendid Light. All of which mean

EFFICIENCY and HIGH CLASS WORKMANSHIP.

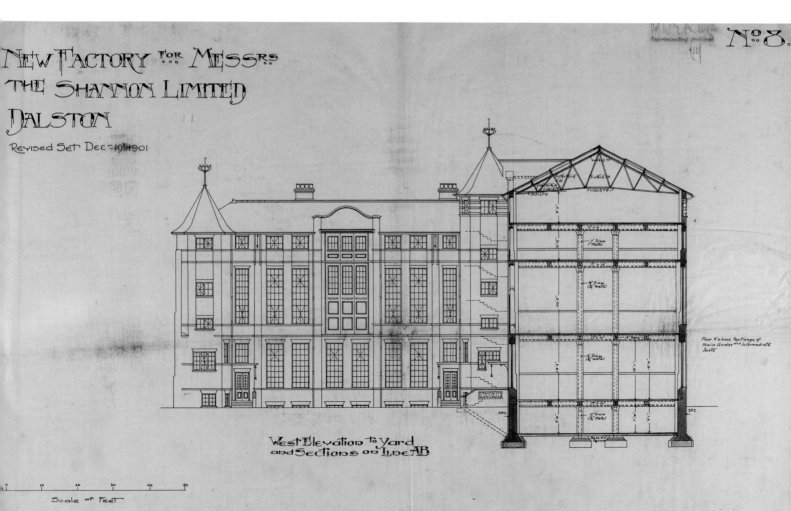

Fig 13.16
Messrs Waterlow & Sons'
stationery works,
Shoreditch, showing the
extensively glazed extension
of 1907 where traditional
brick-pilaster construction
was used in conjunction
with an 800-ton internal
steel framework.
[AA034940]

each floor of the narrower loading, packing and office blocks, a single row of cast-iron columns was used to provide intermediate support to the transverse steel girders, and in the wedge-shaped cross-cut rooms, columnar support was entirely dispensed with, the supporting brick piers extending inwards to permit the use of standardised girder lengths. All the structural metal were encased in concrete, and for extra protection the undersides of all floors were covered with 'Uralite' asbestos slabs. Perhaps the most advanced building of its type in London, if not the country, the Shannon factory sported an impressive array of fire-protection techniques and precautions including electric lighting, central heating, a full sprinkler system and a direct telephone link to the nearest fire station.[41] It was the incorporation of these measures (and perhaps Sach's founding role in the British Fire Prevention Committee) that convinced the LCC to sanction the use of the maximum cubical capacity limit of 450,000 cu ft for the central block. One periodical noted that the 'central block is particularly remarkable for the fact that it is just within the maximum

limits of cubical contents that can be accorded in the County of London, and is hence, perhaps, the largest individual factory block in the metropolis'. Of the building's architectural distinction, the same publication drew comparison with Mumford's Mill, Greenwich, noting that 'an attempt has been made to follow the example set of Mr Aston Webb, insamuch as Mr Sachs has given the factory an architectural treatment and not erected the usual factory monstrosity generally associated with our industrial work'.[42]

Notwithstanding the increased use of steel beams and stanchions for the interior framing of multi-storey industrial buildings, the transition to full-skeleton construction does not seem to have occurred to any significant extent in London before 1910. Documented examples of the large-scale use of steel, albeit sporadic, bear this out. The extension to Messrs Waterlow & Sons' printing factory in Paul Street/Scrutton Street, Shoreditch (1907), for example, is proudly recorded as incorporating some 800 tons of structural steel,[43] but the framework did not carry the thick outer walls (Fig 13.16).[44]

Erected by Trollope & Sons and Colls & Sons, this building was designed by architect Henry C Smart and used steelwork supplied by Joseph Westwood & Co., Napier Yard, Millwall.[45] Other surviving examples from former manufacturing districts ranging from Clerkenwell and Shoreditch to Old Ford, show that the exterior wall retained its structural role and pre-eminence up to and beyond 1909. The expense of entirely steel-framed construction was probably deemed unjustified by the majority of industrialists, given that walls of load-bearing thickness had to be employed anyway to satisfy the pre-1909 London Building Regulations (*see* p 45). The use of interior skeletons conferred many of the advantages of the material, namely multiple floors of great strength, stability and spatial openness. Sturdier, more self-sufficient interior frameworks in conjunction with 'pier and panel' or pilaster-wall construction, enabled the maximum window-to-wall ratios prescribed by the building regulations to be fully exploited. The occasional employment of stanchions built into the walls helped to reduce the width of the piers, but these were rarely called upon to carry the external envelope. Even in those structures exempt from Part VI of the 1894 Act, such as multi-floored warehouses built by dock companies, thick walls continued to be employed for carrying perimeter floor loads, if not for ensuring the overall stability of the structure. And even after the 1909 Act, which permitted considerably thinner walls, full framing in these contexts remained the exception rather than the rule.

Yet by *c* 1905, a profound overhaul of building methodologies in this somewhat traditionalist sector was well underway, with localised constructional traditions giving way to textbook standardisation. By this date, the most rationalised, advanced practice in London's industrial buildings was the all-steel interior skeleton used in conjunction with load-bearing enclosing walls. The use of simple rolled I-section steel members throughout, rather than built-up stanchions or girders, prefabricated from an array of smaller components, was one aspect of this design shift, as was the more sophisticated manner in which they were connected. Stanchions were spliced together above floor level, avoiding complications with the column beam connections, and beams were carried using stronger, more efficient T-section brackets, the load of the beam being brought down on to the stem and transmitted to the stanchion in

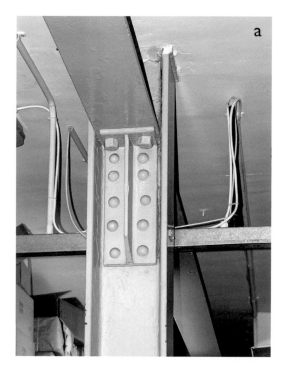

a manner greatly superior to simple angle brackets. Even unassuming buildings in the most unprepossessing of districts, such as two surviving manufactories of 1904–5 and *c* 1910 at Old Ford, built for the confectioner's Clarke, Nicholls & Combs Ltd, were the beneficiaries of this new idiom (Figs 13.17a and b). This was a

Fig 13.17 a and b
Simple yet highly developed steel detailing in two of Clarke, Nicholls & Combs Ltd's buildings in Old Ford, Tower Hamlets, one of 1904–5 (a), the other of c 1910 (b). These T-section reinforced angle cleats made for exceptionally, strong column–beam connections. Employed in The Savoy Hotel extension and The Ritz Hotel (see Figs 10.44 and 10.45), this form of jointing would later be recommended in standard textbooks on steelwork design, including Walter C Cocking's The Calculation for Steel-Frame Structures *(1917).*
[Jonathan Clarke]

Fig 13.18
John George & Son's
hardware warehouse in
Hutchings Street, Poplar,
photographed on completion
in 1908.
[BL20323]

new constructional language whose vocabulary was unequivocally and exclusively steel, and whose grammar was no longer idiosyncratic, experimental and evolutionary, but mature, rationalised and standardised (Fig 13.18). Stimulated by the standardisation of sections in 1903 (see p 33), it owed less to London's legacy of diverse constructional traditions than it did to the nationally or internationally orientated design catalogues of Britain's leading engineering firms.

Railway goods sheds and warehouses

Iron forerunners

In the second half of the 19th century goods traffic on Britain's railways grew rapidly, with revenue generated by freight overtaking passenger revenue in the 1860s, and remaining more profitable until the 1960s.[46] In many cities, especially London, where land was scarce and costly, warehousing was increasingly built above or below the goods shed, where the arrangement of railway tracks and platforms dictated broad, minimally interrupted spaces for the unloading and sorting of merchandise.

Such multi-storey goods stations demanded heavy, long-span construction. Even in those without that function, where wagons entered warehouses for more direct loading and unloading, space, and correspondingly long spans, was often needed to enable wagons to be manoeuvred by means of turntables.

The type of warehousing provided varied according to their intended purpose. Some railway warehouses were engineered to meet the storage needs of specific clients, and were tailored towards specific commodities such as grain or wool; others were designed to accommodate a wide range of goods. It was in these latter general-purpose, brick or masonry-walled buildings that structural iron saw determined application on the grandest of scales, superseding earlier techniques reliant on brick vaulting and timber framing. These buildings had to provide immensely strong, robust and usually fire-resisting floors, capable of withstanding the weight of the heaviest of commodities delivered directly by wagon, and manoeuvred by powered cranes, capstans, hoists and lifts. For example, the upper floor alone of the two-level Somers Town Goods Station of the Midland Railway Company (1883–7, demolished) used over 13,000 tons of wrought-iron girders and beams, carried by 400 iron columns which weighed some 2,000 tons.[47] Because of the loads involved in these buildings, hollow cylindrical columns were usually preferred, rather than those of cruciform section form typically used in warehouses. Many railway buildings saw increasing sophistication in the design and arrangement of structural members to provide superior floor-load capacities and wider column spacing. Nationally, one of the first to signal this development was the 'London Warehouse' of the Manchester, Sheffield & Lincolnshire Railway. In this seven-storey building, the built-up, wrought-iron girders that carried cast-iron beams supporting brick jack-arch floors were arranged longitudinally, reducing the overall number of columns by half. The exceptional width of the building was made possible by an unusually strong and stable internal framework in which specially shaped capitals cast on the cylindrical columns provided for a rigid connection between the columns and beams. Such innovatory features marked a departure from traditional warehouse or mill construction, which was more reliant on bracing from the side walls. Even so, heavy load-bearing walls were still used.[48] In London, a similar principle, creating a sturdy, three-dimensional internal frame, was exploited in the Blackfriars Goods Warehouse, Holland Street, Southwark (1871–4, demolished), designed by William Mills, Chief Engineer of the London, Chatham, and Dover Railway Company. Here a grid made up of 'probably the most remarkable collection of wrought-iron riveted girders to be met in the world'[49] supported brick jack-arch floors encased in concrete. In the basement and upper floors, this grid was supported internally by a 'forest of columns', but on the ground floor, to preserve openness for the movement of carts and wagons, only 27 principal columns were employed. These thicker columns extended through the upper floors to support the trussed wrought-iron roof. The working drawings for this structural ironwork were prepared by C M Chittick, engineer to Matthew T Shaw, the ironwork contractors. In the GNR's goods warehouses at Farringdon Road (1874–8, demolished), heavy-duty H-section wrought-iron stanchions provided a more efficient arrangement of vertical surfaces for the connection of the wrought-iron beams,[50] and the liberal use of Westwood & Baillie's corrugated wrought-iron flooring supplanted bulky jack arching, enabling more even distribution of loads. This was probably the first use of this particular patented system, presaging the steel-trough flooring of the late 1880s. The contractors for the structural ironwork were Westwood & Baillie, in conjunction with Messrs Eastwood, Swingler & Co. of Derby and Kirk & Randall.[51]

The introduction of steel

The Bishopsgate Goods Depot in Shoreditch (demolished 2003) may have been an early, yet modest, instance of the introduction of structural steel to metropolitan buildings of this class. Rebuilt as a goods shed and yard in 1877–82 by Vernon & Ewens of Cheltenham to designs by Alfred A Langley, the Great Eastern Railway Company's engineer, the depot was subsequently modified by the introduction of Lindsay's patent steel floors, probably in the mid- to late 1880s. This type of flooring, made of individual sections riveted together, was easily connected to the flanges of the supporting girders, and, given its intrinsic strength, could span great distances, frequently removing the need for supporting joists. Thomas C Cunnington, then working for W H Lindsay, seems to have carried out this alteration.[52]

The unequalled spanning and load-bearing capacity of built-up steel girders seems to lie behind their exceptionally early and extensive use in the London Tilbury and Southend Railway Company's goods depot at Commercial Road, Whitechapel (1886–7, demolished 1975). Built to complement the opening of Tilbury Dock in 1886, this enormous goods station, comprising a two-level depot surmounted by a four-storey warehouse, was a joint venture between the railway company and the East and West India Dock Company, and was designed by the chief engineers of each company, respectively, Arthur L Stride and Augustus Manning, with J Mowlem & Co. as contractors. A remarkable 20,000 tons of structural steel went into its construction, and much of it within the thick brick shell of the warehouse (Fig 13.19). This was about 500ft long and 90ft high (above street level), and varied in width from about 170ft at one end to some 210ft at the other. The internal layout was cellular, each floor subdivided into thirteen separate spaces by brick firewalls, all grouped around three enormous light wells with hydraulic lifts that rose through the full height of the building from the depot below. An interior grid of long-span girders (the longest measuring some 43ft), carried on

Fig 13.19
London Tilbury and Southend Railway Company's Goods Warehouse, Commercial Road, Whitechapel (1886–7, demolished 1975), part of a complex that used 20,000 tons of steelwork, possibly the largest amount in any building at the time of its completion.
[© The British Library Board (F60077–66)]

each level by superimposed hollow cylindrical iron columns,[53] supported the proprietary slow-burning timber-floors,[54] and framed the light wells, ensuring structural continuity across these spaces and therefore stability in both axes throughout the building's full depth and width. Bearing the maker's plate 'Arrol Brothers, Germiston Works, Glasgow 1887', these girders were formerly thought to be of wrought iron.[55] However, a trade catalogue of A & J Main & Co. Ltd, a Glasgow-based firm of engineering contractors that was incorporated with Arrols Bridge and Roof Co. Ltd suggests steel rather than iron, and a contemporary account noted that 'the steel girders … were entirely of English manufacture'.[56] Accordingly, this colossal edifice marked perhaps the largest-scale use of structural steel of any building of the 1880s or earlier.[57] By way of comparison, the steelwork of the 19th-century's biggest steel structure, the Forth Bridge (1883–1890) amounted to some 50,000 tons (2½ times as much), a good deal of it also fabricated in Arrol's Germiston works.

The Western Goods Shed in the King's Cross Goods Yard, built in 1897–9 for the GNR to provide a separate facility for outwards goods traffic, also made extensive use of structural steel. Designed by the company's chief engineer Alexander Ross (1845–1923) as a low-slung, two-level, brick-walled shed, it used internal frames of cast iron and steel in the lower floor, and steel throughout in the upper floor.[58] In the low-level shed, where uninterrupted space was of less concern, numerous cast-iron columns support the heavy non-combustible floor, formed from concrete arches turned between rolled-steel joists, which themselves rest on deep plate girders (Fig 13.20). The cast-iron columns, manufactured by W Richards and Son of Leicester, were probably cheaper than steel uprights. In the busier high-level shed, the exclusive use of steel enabled clear spans for generous track provision and non-rectilinear platform arrangements. Widely spaced stanchions seated on the girders in the floor below are used to support substantial lattice girders that carry the light wide-span steel and timber roof trusses (Fig 13.21). Generally, the stanchions are light, formed of rolled I-section with plates riveted to the flanges. They are abutted by the end posts of the lattice girders, and these connections are reinforced with substantial triangular brackets beneath the girder, riveted to the flange plates of both members. This detail, with the same function as traditional knee braces in timber-framed construction, ensured good resistance to side sway, which the slender brick piers were unable to counteract. Sturdier cruciform-section stanchions, built out of one deep and two smaller I-sections, are used to support the heavier mezzanine offices. On both levels, the ends of the principal girders are embedded firmly in the thick Gault brickwork. Ross stipulated that all of the steelwork had to be Siemens-Martin, 'of first rate quality', and 'obtained from an English Manufacturer'.[59] Dorman Long & Co., and, to a lesser extent, the Phoenix Foundry Company of Derby, supplied these materials, which were erected under the overall supervision of Ross's assistant, S Bolton.

The Western Goods Shed formed part of a nationwide programme of upgrading of key collection and distribution points, undertaken by the GNR at a time when it was facing severe competition from the GCR. Indeed, its construction proceeded to keep pace with the GCR's metropolitan goods depot, built to serve the newly opened line from the north. Designed by George Andrew Hobson and Edmund Wragge, the GCR's Goods Warehouse, Marylebone (1896–9, demolished), provided a much larger, technologically more sophisticated solution to the rapid multi-level interchanging of

Fig 13.20
Western Goods Shed, King's Cross Goods Yard, built in 1897–9 to designs by Alexander Ross, the Great Northern Railway Company's Chief Engineer.
[BB98/26504]

Fig 13.21
Upper level of the Western
Goods Shed, soon after
opening in 1899.
[The National Archives, UK
ZLIB4–36 GNR]

goods between road and rail. This austere block had 8½ million cubic feet of space on five stories, and was 'composed of an internal framework of steel enclosed within four brick walls'.[60] The cumulative loads of each steel girder floor were transmitted directly to enormous brick spread footings by 90 steel stanchions which tapered outwards through the basement storey to spread the load onto raft-like footings made of steel joists connected top and bottom by steel plates. The stanchions, each 78ft long, were of box section, carefully built up from steel I-sections and plates so that the two elements broke joint with one another, thus forming one continuous sturdy length. It was originally planned that the stanchions would be delivered by rail in one piece from the north of England, but in the event two lengths was all that was feasible. The interior of each upright, 'being inaccessible to the paint-brush', was filled with concrete to prevent rusting. Deep brackets, copiously riveted to the sides of the stanchions, carried the 2ft 9in.-deep, steel-plate girders, whose outer ends were firmly seated in the brickwork. The engineering of such details, designed to resist the tremendous shearing stresses at those points, and the use of continuous stanchions (each designed to

transmit 700 tons dead load), resulted in a phenomenally strong, rigid structure capable of bearing up to the heavy moving loads it would endure. In all, this steelwork amounted to nearly 7,000 tons.[61] The warehouse was also operationally sophisticated, 'filled up with the whole battery of mechanical apparatus then in use, and centrally heated throughout'.[62] In essence, it was this company's answer to the celebrated GNR's warehouse in Deansgate, Manchester (1895–8). That 'marvel in mild steel',[63] representing the zenith of Victorian railway technology in terms of interchanging water, road and rail traffic, was, from a purely structural point of view, in some senses less progressive than its southern rival. In W T Foxlee's Deansgate building, the box stanchions of the internal frame were of single-storey height only, riveted together via steel-plate caps that spread the load – a less elegant connection than splicing, based on outmoded practice.[64] Although much of the framework proceeded ahead of the thick exterior walls,[65] this enclosure has a significant load-bearing and bracing role. Certainly, this cannot truly be regarded as this country's 'first fully steel-framed building proper'.[66]

Despite the fact that the railways were among the earliest and heaviest users of steel, not just for track, locomotive components and other articles, but also for structural members in bridges and buildings, matters of cost and even traditionalism almost always tempered constructional zeal. Traditionally, the thick, load-bearing brick walls of multi-storey railway warehouses offered protection against theft and fire as well as stability, so there was little to be gained, economically or functionally, from extending interior frameworks to the outside so that they carried all loads. As in other multi-storey warehouses, internal steel frames provided the most cost-effective, workable solutions and it was this form of construction that held sway in London until 1910 and beyond. With single-storey railway goods sheds, functionally open to the elements along parts of the perimeter, the masonry or brick wall was more rapidly rendered obsolete, although one authority could still note in 1912 that these buildings were

usually of the fortress order, and would be capable of withstanding a moderate siege ... A large proportion of all the expense is in the building of a shell that appears to be unnecessarily strong for the things it encloses ... all that is wanted is protection for the workers and goods, etc., inside against the elements, and it seems to me that for this purpose a corrugated-iron structure would generally suffice, and easily lend itself to expansion at minimum cost.[67]

In London's multi-storey industrial buildings, the most rationalised constructional solutions were composite, and did not necessarily conclude with the full steel frame. This was also true of railway goods sheds and warehouses, but in low-rise warehouses, storage sheds, engineering factories and power stations, it was a different matter.

Dockside storage sheds

Iron forerunners

The engineering problems of designing free-standing, single-enclosure buildings were, in many respects, more readily resolved than those of multi-storey buildings designed to resist vibration, fire and severe loadings on successive floors. At their most basic, sheds consisted of a roof spanning between parallel walls or posts, enclosing unobstructed storage and handling space, with the loads borne directly by the ground. The main difficulty was initially one of providing roofs wide enough to enclose large areas, but as the functional complexity and diversity of the operations within increased, new structural solutions were required.

London's docks were the initial setting for the application of cast iron to buildings of this nature. The material began to make its mark in the 1810s, in the construction of huge top-lit or open-sided structures for the storage, sorting

Fig 13.22
Branching cast-iron columns at the New Tobacco Warehouse, London Docks (1811–14).
[Malcolm Tucker]

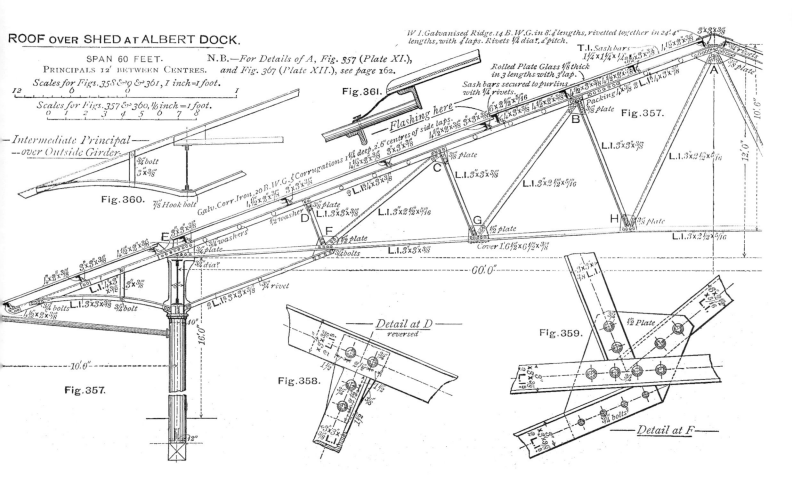

ROOF OVER SHED AT ALBERT DOCK.

and discharge of merchandise under cover. The New Tobacco Warehouse at the London Docks (1811–14) covered an area of 210,000 sq ft. Built to acclaimed designs by Daniel Asher Alexander, Surveyor to the London Dock Company, it relied on widely spaced branching stanchions of cast iron that supported the timber queen-post roofs in an idiosyncratic yet clever fashion (Fig 13.22). More sophisticated in its design and more extensive in its use of cast iron was Rennie's 1,330ft-long Rum Quay Shed, West India Docks (1813–14, demolished).[68] The beams, columns and roof trusses were all made of the metal, and the assembly provided for thermal expansion.[69] From these, and other pioneer essays in cast iron,[70] the material saw extensive and widespread use until the late 19th century, particularly in the form of supporting columns, which, unlike timber, were shown not to decay rapidly. Corrugated iron, first applied to structures by the London Docks' engineer Henry R Palmer in 1829, and produced on a commercial basis in the same year by Richard Walker in Bermondsey,

provided a lightweight, low-cost, water- and fireproof cladding and roofing material that perfectly complemented simple structural systems based on iron framing.[71] Examples of this once ubiquitous family of prefabricated structures, five of which survived until 1992, were 33 single-storey transit sheds erected in 1880–2 at the Royal Albert Dock. Engineered by Sir Alexander Meadows Rendel and erected by Lucas and Aird, they had double-span, wrought-iron roofs supported on cylindrical and I-section columns of cast iron, all enclosed by corrugated iron (Fig 13.23).[72] The resistance of cast iron to buckling in compression ensured its continued use for uprights, and wrought iron witnessed little application beyond roof-truss members. However, its tensile properties were commonly exploited from the 1840s in the trussing of timber beams with rods, or in sandwiching timber beams around rolled plates to form 'flitch beams'.[73] Timber prevailed into the 20th century as the principal constructional material for dock and other industrial sheds because of its relative cheapness. For

Fig 13.23

Contract drawings (redrawn) for the 60ft-span cast- and wrought-iron single-storey transit sheds erected at the Royal Albert Dock (1880–2; Sir A M Rendel, engineer) [Advanced Building Construction, Longmans, 1901]

example, it saw application at the West India Docks as late as 1926 in the Belfast-truss-roofed No. 7 Teak Shed.[74] Even beyond this timber continued to find favour, although the paucity of documentation and the ephemeral nature of sheds make this whole subject particularly obscure.

The introduction of steel

From the mid-19th century, as dock business shifted to transit handling, two-storey sheds came to be seen as the best type of quay accommodation. Typical examples at Victoria Dock and the Millwall Docks were brick-walled with ground floor columns supporting an open first floor with long-spanning roof trusses. Reduction, or elimination, of the obstructing interior columns at ground floor was a major design objective of the docks engineers, one that was achieved by the end of the century. With heavy enclosing walls also removed, the stability of these specialised structures became dependent on the rigidity of the connections between the perimeter columns and the wide-spanning floor beams and the roof trusses. It was in these framed structures that wrought iron probably found real service, for built-up beams of the

metal were sufficiently strong and rigid to resist the concentrated stresses as those points. An example of this was the rebuilding of No. 5 North Quay Shed at the West India Docks in 1895, to create a sturdy double-floored, open-sided structure (Fig 13.24). Built by William Whitford & Company, of the Royal Iron Works, Westferry Road, Millwall, to designs by Sir Hay Frederick Donaldson, this 338ft-long shed used wrought-iron beams spanning between large-diameter, cast-iron columns at 12ft centres to carry a timber floor and a galvanised-iron, low-pitched roof. The inherent rigidity of the framework enabled liberal provision of electric conveyors and cranes that reduced manual handling of frozen meat.[75] Steel-framed structures began to be introduced into London's commercial docks alongside those of traditional materials in the 1880s.[76] One later example followed the destruction by fire of much of what stood on the south quay of the West India Dock's South Dock in 1895. A 420ft-long two-storey fibre shed was erected in 1898–9 (demolished), framed from rolled-steel I-sections and clad in corrugated-iron. Soon after, in 1901–2, two large two-storey cut-wood sheds were put up on the west side of the Import Dock to curtail the unsatisfactory practice of

Fig 13.24
No. 5 North Quay Shed, West India Docks, 1895, demolished. A late example in iron of the double-floored, open-sided type of structure that flourished in London's docks in the late 19th century.
[Museum of London/PLA/ iv/3/13]

storing softwood in the North Quay warehouses. Designated R and S sheds, these twin, open-sided sheds were built by direct labour to plans prepared under C E Vernon (demolished) (Fig 13.25). Apart from the timber roofs, they were entirely steel framed using simple rolled I-section beams and stanchions manufactured in Germany and supplied by constructional engineers Peirson & Company. In this instance steel framing was preferred to timber, despite its cost, because it allowed fewer point supports. Both sheds were destroyed by fire in 1903, but because they had 'exactly suited' the requirements of the trade, they were rebuilt the following year to identical design, albeit with corrugated-iron rather than felt roofing to the twin-span timber Belfast trusses.[77]

Outside the enclosed docks steel-framed sheds probably began to appear in number in the 1890s. Documentation for these anonymous, architecturally mundane structures is extremely elusive and fragmentary, and it seems likely that numerous early metropolitan examples have neither survived nor been discovered or documented. One was a saw mill in Tottenham, erected by Richard Moreland to the designs of architect A Richards, possibly in the early or mid-1890s (demolished).[78] This comprised a spacious, unobstructed main compartment of 47ft clear span, flanked by two double-storey ranges, where the cut wood was presumably stored. The illustration shows rolled I-section stanchions and longitudinal beams, with heavier-section built-up transverse beams resting at their outer ends in the brick walls (Fig 13.26). The thin brick pier-and-panel walls that framed the sides of the structure were presumably more economical than stanchions

Fig 13.25
The steel-framed R and S Cut-Wood Sheds, West India Docks, erected in 1901–2 to store softwood, demolished. [Museum in Docklands (Museum of London/PLA/ iv/3/18)]

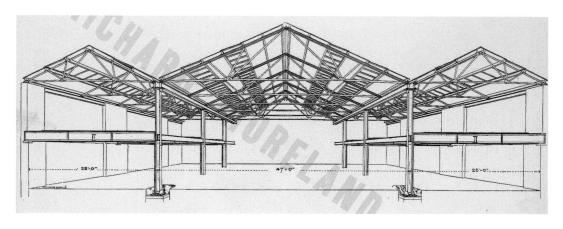

Fig 13.26
Steel-framed saw mill erected in Tottenham in the 1890s. [© British Library Board (F60077–53)]

Fig 13.27 (above)
Garage and works of the
London Motor Bus Co.
(Vanguard), Hackney
(1908–9). An early
application of a column-free
roofing system devised by
Peirson and Co., engineers.
[Builders' Journal and
Architectural Engineer, 18
March 1908]

Fig 13.28 (right)
Peirson and Co.'s system at
Hackney enclosed an area
215ft by 116ft.
[The Building News, 12
February 1909]

and essentially secure; there was little to be gained from the use of exterior stanchions, either structurally or functionally.

Outside London, the opening of the Manchester Ship Canal in 1894 encouraged experimentation in steel framing to meet an urgent demand for storage space. The design of the skeleton-framed sheds built there was of the highest sophistication (*see* pp 95–6). Local constructional engineers and contractors Edward Wood & Co. were among the first to meet this demand, fabricating and erecting many of the warehouses built for this canal and its Salford Docks in the mid to late 1890s.[79] The London firm Walter Jones & Sons, of Magnet Wharf, Bow Bridge, had erected a fully framed two-storey shed for the Manchester Ship Canal Company by 1895. In all likelihood this was entirely of steel.[80]

In London, new directions in wide-span roofing technology emerged after 1900, albeit in buildings for transport, which had little if any direct functional pedigree. The vast, low-slung sheds erected for the storage and repair of the LCC's growing fleet of tramcars were huge consumers of steel, using it in the most efficient manner for maximum enclosure at minimal expense and erection time. New Cross Depot, erected in 1905–6 to designs by the engineer E J Edwards, under the direction of W E Riley, was 'one of the largest if not the largest, car shed in the country'.[81] It used some 1,700 tons of steel stanchions, lattice girders and roof trusses to provide a top-lit, well-ventilated enclosure covering an area of some 4 acres. The stanchions, like those in railway goods sheds, were positioned on the raised gangways between the track rails. Smaller in scale but more innovative was the garage built for the London Motor Bus Co. (Vanguard) at Hackney, for here an entirely column-free, unobstructed floor area, 215ft by 116ft was engineered at low cost using 'a new departure in roof construction' (Figs 13.27 and 13.28).[82] By integrating the long-span, Pratt truss girders into the framework of the roof trusses, placing them along the ridge lines, rather than using them as separate, supporting members, a light, efficient self-supporting structure of considerable strength, rigidity and minimal depth was created. Devised by Peirson and Co., and in this case applied by architect H Yolland Boreham, this system lent itself to a whole range of single-enclosure structures, including manufacturing, signalling yet further refinements and possibilities for the coming decades.

Engineering sheds

In almost no class of building has the type changed more than in engineering works … About 20 years ago, a common system was to build very heavy walls for erecting shops, and then roof them over with bound timber couples … In the erection of modern engineering works, the latest practice is to use iron and steel largely as materials of construction; in fact, some of the best shops are almost wholly composed of iron, steel and glass. (1897)[83]

Iron forerunners

Engineering sheds emerged in number in the mid-19th century as tall, well-lit and ventilated rectangular structures to house the interlinked manufacturing processes associated with mechanical engineering and heavy manufacturing. The processes of founding, forging, fabricating and assembling were each conducted in their own particular form of shed, but constructionally there was considerable overlap between the foundry, the forge, the machine shop and the erecting shop. In some works these activities would occasionally be drawn together under one roof. Until the 1860s perhaps the majority of buildings of this kind were roofed or internally framed in timber, although their poor survival hampers meaningful generalisation.[84] One of the earliest documented large-scale uses of cast iron was in the Machine Hall of Henry Maudslay at his celebrated machinery works in Westminster Bridge Road, Lambeth (1830, demolished) (Fig 13.29). Wide-span trusses of minimal depth were used to maximise headroom and top-light, and complemented the incombustibility of the thick brick walls, but mechanically the building was less sophisticated, the components being manoeuvred across the paved floor using a swivelling crane or simple iron-wheeled trolleys. The introduction of the overhead travelling crane into this type of building brought huge improvement, and played a transformational role in its structural design.

Before 1850 cranes capable of running along tracks that extended the length of these buildings were introduced and refined; by the 1860s steam-driven overhead travelling cranes operating on both axes of the building were an established, integral feature. Offering comprehensive coverage of the entire floor for lifting and shifting the heaviest machine parts, cranes

Fig 13.29 (right)
Machine Hall of Henry
Maudslay's celebrated
machine-tool works,
Westminster Road, Lambeth,
1830, demolished. Thick
walls resisted the thrust of
the cast-iron roof trusses,
which maximised both light
and headroom.
[Charles Eck, Traité de
Construction en Poteries
et Fer, *Paris, 1836. By*
courtesy of Department of
Civil and Environmental
Engineering Library,
Imperial College, London]

Fig 13.30 (below)
Cannon Factory in the Royal
Arsenal, Woolwich (1856),
showing the position of the
former gantry beams that
originally carried an
overhead travelling crane.

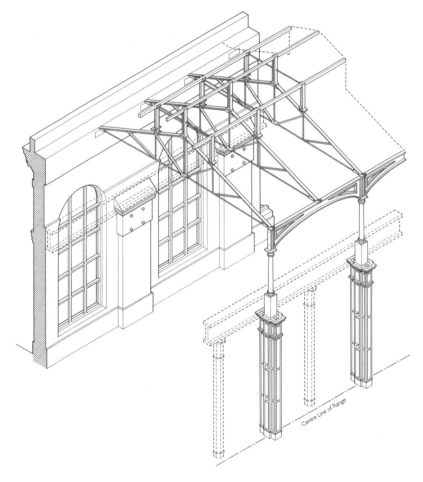

Centre Line of Range

of this sort fostered flow-line production and became the cardinal, defining component around which whole buildings were conceived and engineered.[85] London, and especially its Thames-side marine engineers, played a key role in this development. By 1854 Joseph Glynn could note that 'In designing new buildings for steam engine manufacturers the side walls are now generally made of sufficient strength to carry a line of rails upon an offset of masonry.'[86] Typically, these rails or gantry beams rested on brick piers projecting from the side walls, but the use of sturdy iron columns for support, arranged down the length of adjoining sheds, permitted multiple, intercommunicational ranges. The Cannon Foundry (Building 25) in the Royal Arsenal at Woolwich early on exploited this technique (1856–7, David Murray, architect for elevations) (Fig 13.30). In this H-plan building, enormous two-stage, cast-iron pier-stanchions were designed to support both the roof structure and the inner longitudinal girders of two gantry cranes that travelled the full length of each of three ranges.[87] Another, later, example comes from the former forge and workshop of C J Mare and Company, engineers and shipbuilders, at Nos 397–411 Westferry Road, Millwall. Originally built in 1860 to designs by William Henry Dorman and John Hughes, and probably consisting of a central, cast-iron arcade that supported an iron-trussed roof of two spans, the building was reused by Joseph Westwood & Co. from 1889 for structural fabrication work. Among the changes the new occupiers brought to the works was the insertion of two gantries, both of which had timber crane rail beams and one of which had timber stanchions (Fig 13.31). Despite being much altered, this building survives as England's only mid-19th-century shipbuilding works, and the timber gantry as the only remaining example in London of a once commonplace feature.[88]

The standard arrangement that evolved in many works was that a lofty, central crane-served machine- or erecting-shop was flanked on two, three or all sides by galleries for accommodating ancillary functions such as moulding, pattern-making and materials storage. There are few survivals; the timber-framed Garrett's Long Shop, Leiston, Suffolk (1853) survives as one of the archetypes of the aisled and galleried

Fig 13.31
Former forge and workshop of C J Mare and Company, 397–411 Westferry Road, Millwall. Built in 1860, it is the only mid-19th-century shipbuilder's works in London to survive.
[BB94/11592]

Fig 13.32 (right)
Mounting Ground, Royal Arsenal, Woolwich, built in 1887 for mounting heavy guns onto their carriages. Cast iron, wrought iron and steel were all variously employed for this non-galleried engineering shed. [BB94/14005]

Fig 13.33 (below)
Drawings for the Mounting Ground (1887), signed by Col. H D Crozier, RE. [The National Archives, UK WORK 43/1640]

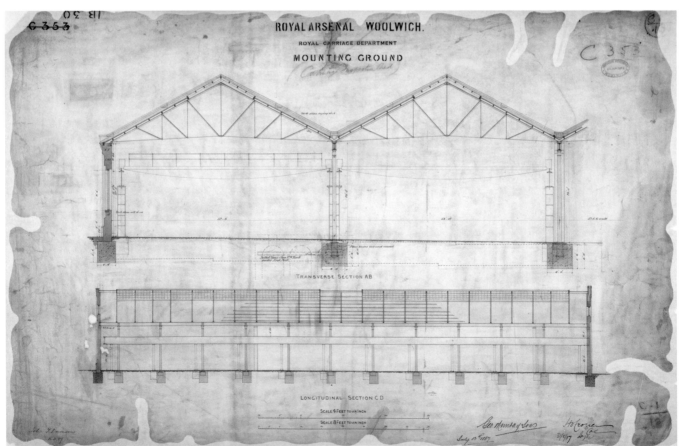

form.[89] In the late 19th century more sophisticated constructional techniques using both cast and wrought iron resulted in stronger, more rigid internal frames better able to support bigger overhead travelling cranes and withstand the dynamic forces they produced, placing less reliance on the load-bearing or bracing action of the exterior walls. The enormous, iron-framed Britannia Iron Works of William Marshall Sons & Co. at Gainsborough (1884–5) is one example of more robust construction,[90] as is the Mounting Ground at the Royal Arsenal, Woolwich, a three-range single-storey shed built in 1887 for the mounting of heavy guns onto their carriages (Figs 13.32 and 13.33). Designed by Col. H Crozier, Inspector of Works, and without galleries, this building used tall hollow cylindrical columns to support wrought-iron girders carrying the roof valleys, massive, open-webbed cast-iron stanchions to support structurally independent crane gantries, and steel principals (marked 'Stockton MCE') and struts in the 53ft-span composite roof-trusses.[91]

The introduction of steel

From its introduction in the 1890s, steel quickly came to be seen as the best material for the construction of engineering sheds, workshops and foundries. By the use of larger members in combination with new, rigid connection detailing, it gave greater robustness, providing unparalleled strength, rigidity and spanning capability to the extent that the enclosing masonry soon lost its structural role. The Royal Arsenal at Woolwich, which almost adopted Bessemer's process for making steel guns in 1858,[92] was a pioneering test-bed for the metal's application, pressing it into service to create large-volume freestanding enclosures for heavy engineering. One of the first was Building D81, a towering corrugated iron- and glass-clad structure, erected in 1891 by a local firm, Joseph Westwood & Co. (Fig 13.34).[93] With a height of 116ft, this 840-ton riveted framework was carefully engineered to withstand the wind pressures on its long sides: these were given extensive diagonal bracing, and more mightily, 12 latticed buttresses afforded wind-proof lateral stability. The framework was also called upon to resist the dynamic forces of giant overhead travelling cranes, which were crucial in the manoeuvring of some of the world's largest gun components. The cranes helped plunge red-hot gun barrels into cast-iron-lined pits filled with oil, thus

shrinking them onto their liners. This extraordinary structure was demolished in 1972, and now only the 'shrink pits' remain in the ground. Another large steel-framed enclosure latterly known as the 'Round House' (Building D73) was built in 1891–2, this time to enfold a huge radial steam crane erected in 1876 for a similar purpose (Fig 13.35). Erected by the Bristol firm

Fig 13.34 (left)
Building D81 of the Royal Arsenal at Woolwich, a colossal steel-framed structure shown in the course of erection in 1891. [© The British Library Board (F60077–59)]

Fig 13.35 (below)
The 'Round House' (Building D73), Royal Arsenal, Woolwich, shown prior to the fixture of cladding c 1892. Made by John Lysaght, Ltd, of Bristol, this giant steel structure enclosed a huge radial crane. [© The British Library Board (F60077–52)]

John Lysaght, Ltd, this 16-sided, 72-ft-high polygonal structure also no longer survives.[94]

It seems that there were no direct antecedents in structural iron for these highly imaginative skeleton structures; they were intrepid new forms born purely of steel and rapid advance in the manufacture of ordnance. Outside the Arsenal, innovative experimentation in the commercial world was more tempered by constructional orthodoxy. Many steel structures of the 1890s had structural designs that were influenced by earlier traditions of using cast-iron columns, while at the same time making determined and rationalising use of steel. A workshop built for marine engineers Caird & Rayner at Nos 779–783 Commercial Road, Limehouse, is representative of this constructional changeover and is London's only surviving 19th-century, steel-framed engineering workshop. Built on a cramped site in *c* 1896–7 for the manufacture of sea-water distilling apparatus and specialised marine-boiler heaters, it consists of a central crane-served assembly hall enclosed on three sides by galleries for the manufacture and storage of lighter components (Fig 13.36). The essential feature of this lofty, well-lit space is an internal steel frame, divided into six bays by slender (9in. by 7in.) rolled-steel stanchions and contained within a stout envelope of load-bearing brick, carrying high-level galleries, heavy gantry beams and light-steel roof trusses. The structural design was of the most logical and progressive order, the extensive use of simple rolled sections rather than compound members enabling economies of material and fabrication, savings in weight and minimal obstruction to daylight from the fully glazed roof. The modest spans involved (the stanchions run at 18ft centres either side of the 23ft-wide central space) and the use of stabilising, largely windowless outer walls permitted full exploitation of lightweight rolled sections throughout the 2½ storeys of the frame, compound stanchions being restricted to the corners. Formed from three lengths each of diminishing section and spliced together using riveted and bolted flange plates, the connection detailing of the stanchions nonetheless echoed earlier thinking in cast iron insofar as the junction was made below the level of the gallery beams.[95] Although nominally designed by architects Charles John Marshall and Charles Campion Bradley, it seems likely that this aspect of the building's design was subcontracted to a local firm of constructional engineers. The

rolling mark of Dorman Long & Co. is visible on the stanchions, but the equivalent mark of the German firm Burbach on some of the beams suggests that the structural design and fabrication was not the work of the giant Middlesbrough firm.

About 1902–3 Caird & Rayner built another crane-served galleried workshop next to this one, with steelwork supplied by the Glengarnock Steel Company of Scotland (Fig 13.37). Essentially replicating the earlier structure in scale and configuration, it did, however, incorporate a rational step forward in two aspects of its design. Continuous stanchions (British Standard Beam

Fig 13.36 (facing page)
Caird & Rayner's works at 779–783 Commercial Road, Limehouse (c 1896–7); London's only surviving 19th-century steel-framed engineering workshop. [BB000879]

Fig 13.37 (below)
The second and more refined crane-served galleried workshop erected for marine engineers Caird & Rayner (c 1902–3). [BB000887]

Fig 13.38
Yarrow & Company's works,
Cubitt Town, in 1900
(demolished). Erection of
the huge, fully steel-framed
sheds.
[Museum of London]

19, measuring 10in. by 8in.) were used to carry the gallery, crane gantry and roof trusses, this being a simpler and more economical technique than the three-stage construction of the earlier workshop. In addition, bracing struts, formed from steel angles bent into the shape of a V, were bolted between the stanchions, giving the framework far more longitudinal stability than its earlier neighbour. In the tradition of some cast- and wrought-iron-framed precursors, these struts also served to carry the crane gantry, which at the time of its construction was almost certainly traversed by an electrically powered crane. Architects Marshall and Bradley have also been credited with this aspect of the building's design, but as with the earlier workshop, it seem likely that their input here was chiefly elevational.[96]

Caird & Rayner's 1890s workshop was built on a cleared narrow site between existing buildings, while for the 1900s workshop the designers were working within the brick shell of an existing building. Although the know-how existed to erect an entirely self-supporting frame, in neither case were there any compelling practical

reasons for doing so. The curtain wall was the next major advance that steel framing conferred to production sheds, one that was quickly exploited by leading engineering firms on large sites.[97] The well-known shipbuilding firm Yarrow & Company was probably one of many such Thames-side beneficiaries. In 1898 it took over Westwood, Baillie & Company's London Yard works at Cubitt Town, clearing most of the site for redevelopment. With extensive land at its disposal, it was able to erect state-of-the-art sheds for its engineers', boiler makers' and shipbuilders' departments.[98] Structural engineers Sir William Arrol & Company were called in to erect a large group of four main conjoined workshops that together covered an 11-acre site, some 8 acres in excess of Westwood and Baillie's premises.[99] Erected in 1900, each of these 360ft-long sheds was framed entirely in steel, ahead of the interior brick division walls and exterior brick cladding (Fig 13.38). Only the lean-to to the platers' shop, which arose soon after the main ranges had been put up, used cast-iron columns. The self-supporting exterior walls, built between and around the perimeter stanchions, thus served principally as a protective screen from the elements or trespassers. The light roof trusses were entirely glazed, affording the best form of natural lighting, so the curtain walls were fenestrated with openings of traditional size and proportion; in this case brick partition and panel walls seem to have been used because of the economies inherent in building thinner brickwork. Corrugated sheet metal was restricted to the upper portion of the gable ends, since this form of cladding was best suited to works that generated their own heat, such as foundries and steel mills.[100] Internally, the sheds were fully served by electrically driven overhead travelling cranes, electric arc lamps and a battery of labour-saving machinery, driven either by steam, pneumatic or hydraulic power.[101] On a far larger scale than the Caird & Rayner workshops, here there was little need to circumscribe the lofty production spaces with mezzanine galleries, as each stage in the production process was housed in its own gigantic shed. As in America, by 1900 large production sheds had become 'a series of parallel crane-served bays that housed machines operated by electric drive'.[102] Leading firms such as Dorman Long & Co. Ltd and Redpath, Brown and Co. Ltd exploited this large-scale approach to flow-line production at their works (*see* Figs 2.16–2.18). On this scale, full steel framing, using compound and built-up members specially engineered for optimum structural efficiency, was economically worthwhile. Standardisation and repetition was at the heart of the structural design of buildings of this magnitude, each 'portal frame' or 'bent' of the frame simply repeated to provide the overall length of space required. At Yarrow's, the near-identical configuration of the frames resulted in further economies (Fig 13.39).

Arrol's, which between 1894 and 1907 designed and erected over 140 steel-framed engineering sheds, was perhaps Britain's leading specialist firm, and by 1909, with a wealth of experience behind it, could note that:

The first object in design is, of course, to build a shop suited to the business. Nowadays this practically amounts to steel structure, to all intents and purposes independent of its surrounding walls ... Columns which have to support the roof and crane, and also the structure as a whole, must be stiff enough to resist all strains due to the use of cranes and shafting, in addition to those due to external agencies, such as wind and snow. These points have been subjects of careful investigation by Sir William Arrol and Company, Limited.[103]

Sophisticated construction was of course not restricted to the big engineering firms such as Arrol's and Dorman Long. As designers and builders of their own new engineering workshops in Silvertown (1908, demolished), London constructional engineers Richard Moreland and Son were well placed to tailor structural design to their precise needs. Erected upon a reinforced-concrete raft specially designed for bearing loads and stresses exerted by overhead travelling cranes, their new workshops were fully framed in steel and enclosed an area 214ft long and 45ft wide (Fig 13.40). In a neatly pared-down design, two-stage box-section stanchions carried both longitudinal gantry girders and transverse roof girders so that the structure was fully rigid in both axes without recourse to stabilising walls. The saw-tooth roof, glazed on the north face only, ensured the even, indirect lighting conditions required for precision work. Externally the framework was enclosed in

Fig 13.39
Cross-section of Yarrow & Company's engineering sheds, showing William Arrol & Company's standardised designs. [Engineering, 5 April 1901. Courtesy of Institution of Civil Engineers]

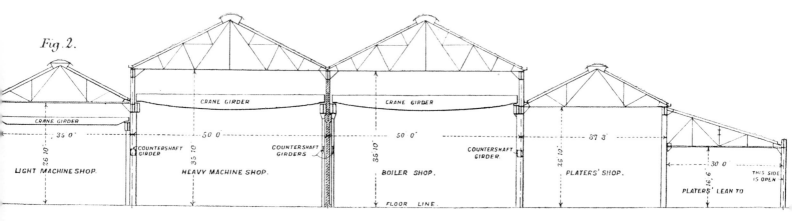

Fig. 2.

CRANE GIRDER   CRANE GIRDER

CRANE GIRDER
. 35 0'   50 0'   50 0'   57 3'
26 10"   COUNTERSHAFT   35 10   COUNTERSHAFT   35 10   COUNTERSHAFT   26 10"   30 0
            GIRDER   GIRDERS   GIRDER.   THIS SIDE
LIGHT MACHINE SHOP.   HEAVY MACHINE SHOP.   BOILER SHOP.   PLATERS' SHOP.   IS OPEN

FLOOR   LINE.   PLATERS' LEAN TO

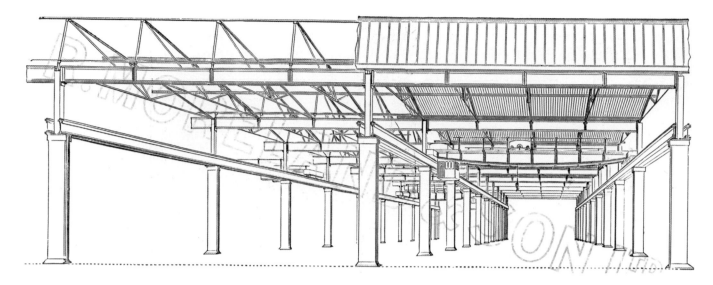

3in.-thick fire-resisting 'Frazzi' slabs jointed in cement, purposely 'left bare both inside and out, showing the pink terra cotta colour of the material'.[104] Like the company's former works in Old Street, Finsbury, which had fabricated a sizeable amount of the constructional iron and steelwork that entered into London's 19th-century buildings, this new works would fashion much of the capital's steelwork for a new century of construction.

Power stations

So far as the ideal from a fire-resisting point of view is concerned, brick, stone and concrete are the best materials of which to construct the building, but from many other aspects the use of a steel framework, filled in with brickwork, has much to recommend it. The chief advantages of such a system are that it can be constructed rapidly, and this is usually of importance; the frame and roof can be constructed and the walls filled in afterwards while the plant is being erected; the attachments for supporting the crane gantries and other plant can be made with great ease; and lastly, the construction is, on the whole, cheaper, and less excavation is required for foundations.

(1901)[105]

Taking their typological cue from the power houses of steam-powered factories, the chief features of early electricity-generating stations were a parallel arrangement of boiler and engine rooms, separated by brick firewalls, with further accommodation for ancillary functions such as coal storage. Until more efficient methods of generating electricity using fewer, larger steam engines were developed after 1890, motive power sources linked to generators could easily be housed in modestly sized brick buildings spanned by trussed roofs, or even in adapted structures. From the 1890s, increased demand for electricity, and the introduction of larger engines – principally of the Corliss and Willans type – belted to large, slow-speed generators, required more extensive, specially designed premises.[106] Incorporating overhead travelling cranes to facilitate the installation and maintenance of larger pieces of machinery, these grander edifices bore some constructional similarity to the largest engineering sheds. As in those structures, the strength and spanning capacity of iron and, from the outset, steel was exploited to provide interior frameworks that supported expansive roofs, overhead cranes, materials and equipment in the most efficient manner. The real forerunner of the type was Sebastian de Ferranti's Deptford Power Station for the London Electric Supply Corporation Ltd (1887–9, demolished).[107] Stretching electricity-generating technology to the limit to provide the West End with electricity on a scale unequalled anywhere in the world, it required correspondingly advanced methods of construction to provide a towering protective envelope to the 40,000 sq ft of generating plant. Within the load-bearing brick shell of the two-bay engine and machine hall, a central row of 100ft-high, cast-iron stanchions, each bolted together from four immense sections, carried the inner gantry girders of two overhead travelling cranes and the inner ends of the two-span arched truss roof (Fig 13.41).

Fig 13.41 (left)
Deptford Power Station
(1887–9, demolished),
the first of the great power
stations. Overhead
travelling cranes were
central to this building
type's engineering.
[The Engineer, 5 April
1889, Courtesy of Institution
of Civil Engineers]

In the adjoining two-bay engine house, separated by a massive wall, a galleried framework of cast-iron stanchions and wrought-iron girders carried the boilers and fuel. The elegant, latticed, wrought-iron roof principals were designed by Max am Ende, one of the most gifted structural engineers of the period (*see* p 116).[108]

Deptford Power Station was built at a time when structural steel sections were appearing in other industrial buildings, and it seems reasonable to suppose that at least some of the metropolitan power stations of the early 1890s made use of the new metal. However, few if any came even close to the scale of Deptford, and with less structurally ambitious objectives, steel was probably restricted to more modest uses, such as floor joists and gantry girders, corbelled from walls or carried on cast-iron columns. Despite the obvious fire risks, some power stations even persisted with timber forms of construction. The Regent's Park station of the St Pancras Vestry (1891–2, Henry Robinson, engineer), a municipally funded venture to supply electric lighting, employed trussed timber roofs to span the brick walls of the narrow engine house and boiler house.[109] In contrast, the City Road Station, Finsbury, and the Wandsworth Generating Station, were 'practically incombustible' (both *c* 1896–7 and demolished).[110] Charles Stanley Peach, an authority on power stations, designed both works in consultation with A J Lawson, chief engineer of the County of London and Brush Provincial Electric Lighting Company, Ltd. The

engine room of the City Road Station, 60ft wide and 180ft long, was divided into two bays by iron columns, the inner row supporting the roof trusses, and those flanking it carrying the inner gantry girders (Fig 13.42). The outer steel gantry girders were carried on columns seemingly independently of the side walls, although for stability they must have been tied into the brickwork.[111] This was characteristic of the

Fig 13.42 (below)
Steel girders and iron
columns in the engine room
of the City Road Station,
Walthamstow (c 1896–7,
demolished).
[The Engineer, 23 February
1900, Courtesy of Institution
of Civil Engineers]

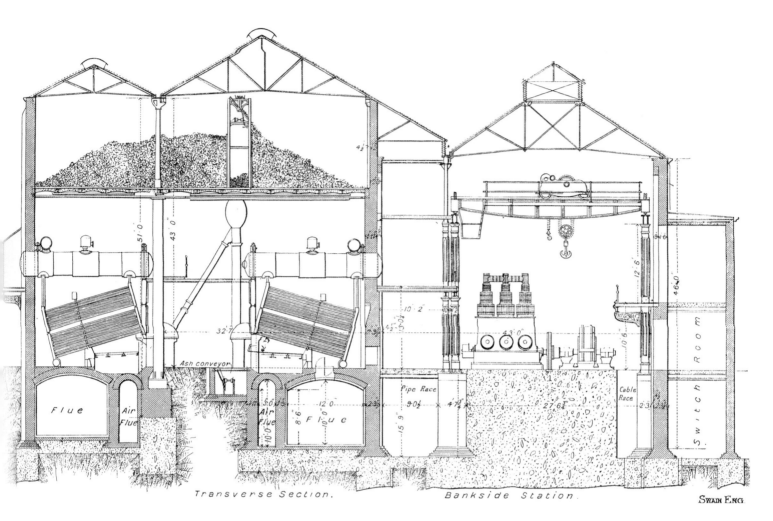

Transverse Section.   *Bankside Station.*   Swain Eng.

Fig 13.43
The first power station at Bankside, Southwark, built for the City of London Electric Lighting Company (1891–9). Iron columns supporting the crane gantry were given their own foundations, preventing movement transferring to the brick enclosure.
[The Engineer, 10 March 1899, Courtesy of Institution of Civil Engineers]

latter part of the decade in heralding a move to the increasing structural independence of the interior metal framing from the exterior walls. Certainly the first power station at Bankside, Southwark, built for the City of London Electric Lighting Company between 1891 and 1899, exemplified this development (demolished) (Fig 13.43).[112] Within the engine hall the tiers of iron columns supporting the crane gantries and side aisles were sited on brick footings independent of those on which the exterior walls stood, to isolate lateral forces generated by the overheard travelling crane. Cast with classical flutes and capitals, these columns also testify to the rising status of the engine hall as a space to be admired.

Around the turn of the century, all-steel construction in combination with load-bearing brick became customary. For example, steel stanchions embedded in shallow brick piers were used to support the overhead crane gantry in the engine house of the Central London Electric Railway's Generating Station in the City

(c 1899, demolished). In the boiler house, a riveted, galleried framework of steel supported coal bunkers high above the boilers. In both spaces, the brick walls simply took the weight of the roof. Steel was used extensively in the boiler house of the London United Electric Tramways' Generating Station at Chiswick (c 1900–1); on its opening *The Engineer* noted that it 'consists mainly of a steel frame structure'.[113]

The shift to fully steel-framed power stations, in which walls were non-load-bearing (but self-supporting), or even carried by the framework, was, however, quickly made. In London, as elsewhere, this initially owed much to American input.[114] One of Britain's earliest 'central' (meaning 'for public use')[115] stations to employ advanced methods of steel framing was the power house for the Bristol Tramways & Carriage Company, erected in 1899–1900 to designs by Horace F Parshall (1865–1932), an American electrical engineer who emigrated to London in 1893.[116] Behind dignified brick elevations designed by W Curtis Green (1875–1960),

Fig 13.44 (left)
Charles Stanley Peach and C H Reilly's design for the Central Electric Supply Company's power station in Grove Road, St John's Wood. Only one of the twelve ranges illustrated here was ever built (1902–4, demolished).
[RIBA46503, RIBA Library Photographs Collection]

a series of largely independent rigid-jointed portal frames, each longitudinally braced by steel purlins and girders, provided support for travelling cranes, coal conveyors and upper floor levels. The self-supporting brick walls were used to carry only the ends of the lowest-level floor-girders. This entire, unusually tall structure, with its coal and ash-tower extensions and 200ft steel chimney, was prefabricated by the Riter-Conley Manufacturing Company of Pittsburgh, shipped to England, and erected on a cramped Bristol site by Riter-Conley workmen.[117] Following refurbishment in the 1990s, only Green's elevations survive.

Full steel-frame construction, with its advantages of speed of erection and diminished foundation loads (and hence expense), easily and quickly lent itself to power station construction in London, a city largely built on compressible subsoil with an ever-growing demand for electrical power. Probably the first to exploit these facilities was the grand neo-Baroque design of 1902 for the Central Electric Supply Company's Power Station in Grove Road, St John's Wood (1902–4, demolished) (Fig 13.44). The intention of the architects, Stanley Peach and C H Reilly, was for it to be built on a giant steel skeleton, designed by Alexander Kennedy, Bernard M Jenkin and S T Dobson.[118] However, of the intended vast range of 12 engine rooms and boiler houses, each pair marked by semicircular gabled pavilions and 250ft-tall square, tapered chimneys, only one unit – with a mighty brick chimney – was ever realised (Fig 13.45).[119] There does not seem to have been any American input on the St John's

Fig 13.45 (below)
Erection of the skeleton-framed machinery hall and engine room of the Central Electric Supply Company's Power Station (c 1903).
[© The British Library Board (F60077–56)]

Fig 13.46
Lots Road Power Station,
Chelsea, upon completion
in 1905. Likened to 'an
elephant lying on its back
with its four legs in the air';
only two brick chimneys
survive.
[BL18925]

Wood power station, but that cannot be said for the far larger Lots Road Power Station, Chelsea, erected in 1902–5 for the Underground Electric Railways Co. of London on the banks of the Thames at the point where the Chelsea Creek runs into the river (Fig 13.46).[120] Built with transatlantic money, enterprise and expertise, all marshalled by the notorious American financier Charles Tyson Yerkes, this monumental structure served to electrify the Metropolitan District Railway and three other underground lines. The entire weight, including internal fittings and plant, of this imposing edifice was carried by a skeleton framework which transmitted loads directly to 220 concrete piers sunk to a depth of 35ft where they penetrated the London clay (Fig 13.47). Erection of the 5,800-ton framework of steel, supplied by Hein,

Lehmann and Co. of Düsseldorf,[121] commenced in May 1903, and it was completed by the end of that year, allowing installation of the turbine-alternators, boilers and other plant to proceed ahead or in concert with the brick envelope. The entire superstructure was built within a record-breaking 12 months by Mayoh & Haley of the Fulham Steel Works Co., Ltd, a firm subcontracted by the British Westinghouse Electric and Manufacturing Company, for which James Stewart & Company – already fully committed with other works, including the Savoy Hotel extension (*see* p 176) – acted as building managers.[122] To meet this timescale, 'form[ing] a record in building undertakings in this country',[123] advanced techniques of organisation and practice reminiscent of American efficiency were called forth. A large staff of

Fig 13.47
Lots Road Power Station,
Chelsea, in 1903, showing
erection of the 6,000-ton
framework of German steel.
[Builders' Journal and
Architectural Review, *15*
February 1905]

highly trained engineers and draughtsmen produced, checked and revised copious amounts of drawings, co-ordinated delivery and construction schedules, and introduced mechanised methods (including pneumatic riveting and a specially built portable electric crane) in place of hand methods wherever possible.[124]

The dimensions of the main building – 453ft by 175ft and 140ft high at the apex of the boiler-house roof – made it easily the biggest power house in Europe at the time of its completion. It was also the largest electric traction station in the world, and the first exclusively to house steam turbines – the largest ever built. Clothed with brick and terracotta, and roofed with glass and concrete, it was consciously American in its scale, massing and detailing. Its structural and architectural design seems to have been primarily the work of James Russell Chapman (1850–1934), whom Yerkes brought over from New York in 1902 to be engineer-in-chief of the UER Co.[125] In both skeleton and skin, its immediate prototype seems to have been the traction powerhouse of the Interborough Rapid Transit

Co. in New York, erected in 1902–4 to designs by McKim, Mead and White. Before emigrating from that city, Chapman may have seen the plans then in preparation. Certainly there were some notable similarities in the structural framework of the boiler house, where braced diagonal members resisted the thrust of the coal bunkers,[126] but in other respects Chapman's design was more advanced. At Lots Road, the inner stanchions framing the turbine hall gallery did not have to extend all the way upwards in support of the roof trusses because light Warren trusses, spanning between the outer stanchions and riveted to them over their full depth, resulted in portal frames of sufficient lateral stability to carry the clerestory roof without recourse to the stabilising action of side galleries. Extending the full length of the building, this gallery carried the switch-gear and busbars, and provided support for two 20-ton electric cranes (Fig 13.48).

Two other skeleton-framed power stations built around the same time as the Lots Road building also deserve mention. In each, British

Fig 13.48
The turbine hall of Lots Road Power Station in 1905.
[BL18939]

and not American structural expertise was responsible. The main block of the GWR's power station at Park Royal, near Acton (*c* 1905) employed the new technology, filled in with brick screen walls. This modest structure was designed by Messrs Kennedy & Jenkin, consulting engineers in collaboration with the architect's department of the railway company.[127] Far grander in architectural and spatial terms is the Greenwich Generating Station for the LCC, built to designs by W E Riley, the superintending architect of the LCC.[128] This, 'the largest single building erected by the early LCC and the most manifest symbol of the Progressives' policy of municipalization,'[129] was also reliant on extensive steel framing, although here the thick, buttressed walls were almost completely independent of the frame. It was

designed to supply the whole of the capital's tramway system, and was one of the largest power stations in the country when it was completed in 1910. It was put up in two phases, to meet the increasing demand for power created by the municipal programme of tramway electrification: the northern half in 1902–6, the southern in 1906–10. In fact it took almost as many years as Lots Roads took months to build, but the comparatively leisurely pace with which it was built resulted in a monument of great architectural distinction, its massive brick walls providing far greater scope for sculptural expression than the Chelsea building. In this case at least, the London builder Howard Colls was right in thinking that hurried building resulted in inferior architecture (*see* p 58). Even so, the framework of the Greenwich building was put up very quickly; the 3,000 tons of steel-work in the 220ft-long northern section was fabricated and erected in 12 months by Joseph Westwood & Co., of Napier Yard, Millwall (Fig 13.49).[130] Typically, the engine room (where a vast, lofty nave was required) used giant, longitudinally braced portal frames, distinguished by the extensive use of latticed members, whereas the boiler house (where a double-level was required to carry the coal bunkers and boilers) used a grid of compound steel stanchions and beams. In both phases, the brickwork of the boiler house was tied to the frame, while that containing the engine-house was freestanding and self-supporting, stabilised by huge external buttresses. The contract for the steelwork of the southern section went to E C & J Keay, Ltd, of Birmingham. In both blocks the brick casing was contracted to H Lovatt Ltd, of London and Wolverhampton. LCC engineers under (Sir) Maurice Fitzmaurice, chief engineer to the LCC, probably had overall structural engineering responsibility.[131]

The steel frame would become the *sine qua non* of the great majority of power stations built in the twentieth century, the anatomical skeleton of what became an increasingly specialised building type. As these pioneers of the form had shown, light, immensely strong steel frames enabled power stations to be erected rapidly, to have their plant installed ahead of the building's completion, and to bear the brunt of heavy, moving cranes and the vibrations and oscillations of machinery. They also offered spatial openness and flexibility for the re-organisation and upgrading of plant, facilities that would assume ever more importance as electrical engineering progressed and demand increased.

Fig 13.49
The skeleton of the north half of the London County Council's Greenwich Generating Station in course of erection. Erected in two phases, 1902–1906 and 1906–1910, this, the largest of the LCC's early buildings, was designed to power the capital's tramcars. [© The British Library Board (F60077–58)]

14

Conclusion: a revolution realised

This is a steel age in more senses than one – and nowhere have the days of steel been more pregnant with change than in architecture and building. Every architect and contractor knows the possibilities of steel, and looking round on present construction we see it triumphant. With few exceptions, the great buildings of the present have been accelerated by the use of steel frames, which have at least the remarkable effect of increasing the speed of construction to a degree which is, to say the least of it, dramatic.
(1910)[1]

In 1879 *The Builder* announced 'The age of steel seems drawing near very fast now.'[2] By 1909, the year in which London's authorities endorsed the steel frame as a legitimate constructional technique, the city had long passed the point of no return in its structural reliance on the metal. Whether they liked it or not, this much was palpable to contemporaries, and is certainly obvious to us today with the benefit of hindsight. Like electricity, the internal combustion engine and other innovations of the so-called second industrial revolution, steel construction was set to stay, an indispensable component of the modern city. With the biggest and fastest-growing population, and the broadest range and greatest number of buildings in Britain, London's consumption of structural steel was prodigious. Even at the dawn of the 'steel age' proper, in 1886–7, one metropolitan building alone consumed 20,000 tons of the material,[3] a huge figure by any standards, but especially impressive when it is considered that the annual production of rolled structural shapes of Germany's leading producer in the early 1870s amounted to 21,000 tons.[4] Through the 1890s and 1900s steelwork tonnages became almost routine details of architectural reportage, with figures for warehouses, hotels and offices typically in the order of thousands rather than hundreds of tons. Such figures suggest that

constructional steelwork for buildings, as opposed to that for bridges and other purely engineering structures, had occupied a more dominant share of the market at this time than is generally realised, even among economic historians. The great importance of this sector was attested by the huge stockyards and fabricating sheds of constructional giants such as Dorman Long and Redpath Brown along the banks of the Thames at Vauxhall and East Greenwich, situated cheek by jowl with those of purely London-based concerns such as Mark Fawcett (Fig 14.1). From these and many other monster steel-framed sheds, designed for maximum productive efficiency, steel poured out to meet the capital's voracious demand. In industry, commerce, entertainment and transport, buildings with steel floors, roofs and skeletons multiplied.

That steel was being used in great quantity in the quarter century before 1909, and in a growing number of metropolitan buildings, there can be no doubt. This increasing scale of use is one indicator of change, but is not in itself necessarily symptomatic of a transformation of building technology, nor of change that was visually apparent. Alterations to technique and form were not inevitable corollaries of this new structural material, at least not to begin with; these fundamental changes took time. This was true with many of London's earliest buildings to use steel, and was certainly the case with Chicago's Home Insurance Building, where steel beams were substituted for wrought iron at the last instance to span between the cast-iron columns of the upper storeys. Steel was seen as a replacement for structural elements of wrought, and later cast iron, and, like those materials, its use spread for the most part invisibly, within the floors or roofs of conventional brick buildings, or encased within sleeves of terracotta or marble. Even where it was employed in visible situations, such as train station roofs or factory floors, it was outwardly usually indistinguishable from the familiar riveted forms of wrought iron. Firms specialising in prefabricated wide-span roof structures or framed buildings for export, for example, would offer the same design in either wrought iron or steel, leaving the choice with the customer, depending on how much they were willing to pay. The Galerie des Machines (*see* p 5) provides a more dramatic insight into the design similarities between the two metals, and, further afield, in America, industrial engineers produced designs for industrial buildings that indicated maximum stresses without specifying the metal. This choice was left to the builder, depending on the market price of either wrought iron or steel.[5]

London buildings: summing up achievements in the steel age

For all its imperceptibility, structural steel did bring about change that was ultimately transformational. Inheriting the mantle of iron, steel accentuated or even concluded many of the trends begun by 'The Iron Revolution'. In terms of its impact on polite British architecture, London repeatedly led the way. The West End 'Theatreland', the largest concentration of theatres outside of New York's Broadway district, was a main focus of structural innovation in iron and steel. In the auditoria of these lavishly decorated buildings, the most palpable development was the progressive elimination

Fig 14.1
Huge engineering sheds on the banks of the Thames testified to the scale and importance of London's demand for constructional steelwork at the start of the 20th century. All these steel-framed structures, including an unusual arched shed for Mark Fawcett & Co., have long vanished.
[a) Redpath, Brown & Co. Ltd, Constructional Steel Work, *1905 edn;*
b) The Building News, *13 October 1905]*

a

b

of view-obstructing columns and pillars, achieved by cantilevering the balconies directly from columns set further back, or from the auditorium wall itself. Modestly anticipated in 1858 in the Royal Opera House, Covent Garden, and the Theatre Royal, Adelphi, and first tried in steel at the Alhambra, Leicester Square, as early as 1883, the technique matured and flourished in the stronger metal. Within a decade steel was being used to carry upper tiers to unprecedented depths. Where suitably large sites permitted, long-span steel trusses and towering stanchions were used to frame bigger roofs, wider proscenium arches, and stronger backstage structures for hoisting scenery. Theatres such as Matcham's Coliseum were triumphs of engineering, pushing steel construction to its limit for public delight and private profit.

In hotels and clubs, of which London boasted particularly grand examples incorporating iron columns and beams, steel members found ready application in the formation of even larger, more lavish or complex principal spaces. Steel girders were first used in preference to iron in spanning a commodious banqueting hall in the Hotel Métropole (1883–5), the widest of that building's ceiled spaces, while faience-clad built-up steel stanchions and girders helped carry a stack of enormous superimposed spaces in the National Liberal Club (1884–7), lessening the load on the walls so that grand bay windows were viable. Through the 1890s, ambitious iron and steel framing became standard practice in these buildings, taking the well-established British practice of long-span, wall-supported girders a stage further, framing these into stanchions to create burly self-supporting cages that carried the weight of the upper floors. Skyscraper engineer C V Childs favoured this technique rather than his trademark skeleton for the splendid Hotel Russell (1897–1900), its all-steel Dorman Long frame surely a spectacle to passersby as it rose in the heart of Bloomsbury. The Savoy (1884–9), another early recipient for constructional steel, exploited steel-skeleton framing for a diminutive extension of 1896–7. A further, much larger extension of 1902–4 marked perhaps the first large-scale use of the technique in a commercial London building. It was soon followed by the most famous metropolitan hotel of them all, the Beaux-Arts Ritz Hotel (1903–5), which sealed the future of the technology. Thereafter the steel-skeleton frame was almost an automatic choice for large hotels, and saw extensive

application in Britain's last major clubhouse, the Royal Automobile Club (1908–11).

The financial and administrative capital of both country and empire, London saw the emergence of headquarters banks and offices in their hundreds, and gave birth to the all-purpose speculative office. Interior frameworks of iron, varyingly comprehensive, but typically protected from fire, quickly established themselves as the most effective way of maximising space and window area and enabling movable, non-structural partitions for spatial flexibility. This was an area of intense structural experimentation and development, of which we have only a limited understanding given the tempo of rebuilding in the 19th century, and the rate of destruction in the 20th. More than two decades ahead of its time, steel made an appearance in a prestigious headquarters bank, saving floor depth and boosting headroom, although there is no evidence to suggest that this successful trial of 1860–1 was repeated. Yet by the early 1890s, construction technologies exploiting steel were evidently bringing improvement and refinement to those of iron, resulting in more robust, more comprehensive and wider-bayed internal frames suited to an increasing scale of building. Buildings such as the National Provincial Bank, Piccadilly (1892–3), and the Bank of Scotland, Bishopsgate (1896), exploited steel to give the public banking hall the mark of lofty dignity that we still associate with most banks. In others, compared to the previous generation, greater grandeur was sometimes achieved more expeditiously from an enriched palette of structural forms and techniques. Simple curving I-sections framed the distinctive rounded atrium of Boord's distillery office in Southwark (1899–1901), and, more spectacularly, the vast, polychromatic dome of the Birkbeck Bank, Holborn (1895–6). The lofty stair hall of the earlier building was fully framed in steel, anticipating developments at the start of the 20th century, which saw the increasing redundancy of the load-bearing wall. By this time, steel beams and stanchions were being used to carry the façades of some buildings as well as their interiors, such as Bannister Fletcher's drapers' office at Nos 76–78 St Paul's Churchyard (1898) and John Belcher's Electra House, No. 84 Moorgate, City (1901–3). Less conspicuously, City offices by lesser-known figures, such as Edward Ellis's Nos 3–4 Rood Lane (c 1899), and Gordon & Gunton's Nos 50–53 St Paul's Churchyard (c 1901), documented a move to extensive skeleton framing to extort as much space and light from cramped, awkward sites as the building regulations would allow. Steel-framing technologies provided the latitude to meet virtually any challenge provided by the site: the head offices of the London and Manchester Industrial Assurance Co., for example, were erected on a towering steel cage that rose from a thick concrete raft, cushioning it from the vibrations of the underground railway below. Structurally and architecturally, the finale of this intense phase of development was perhaps the offices of the Morning Post Newspaper, its sophisticated Beaux-Arts skin and innards hung on an expertly engineered skeleton that drew on overseas precedent. London's answer to New York's famous Flatiron Building (1901–3), its frame was designed to extract every foot of floor space from the site, to shoulder heavy printing presses and to get the building into operation fast. The Morning Post was probably not London's first office built on a full-steel skeleton, but it prefigured more clearly the heavy stone-clad edifices of coming decades.

Large department stores, emerging from mid-19th-century, iron-framed bazaars and traditional drapery emporia, were similarly brought to structural and architectural maturity under the aegis of steel, producing what may have been London's first comprehensively steel-framed example before the end of the century, Woollands', in Knightsbridge (1896–1900). In the 1900s, steel framing became de rigueur in the making of the largest, most uninterrupted and best-lit spaces on each floor. Selfridges (1908–9), the largest store the country had seen, epitomised this trend, but other West End stores, such as extensions to Harrods (1901–5) and Waring and Gillow's furniture emporium (1905–6), had used similar techniques in the pursuit of huge non-combustible volumes, and customer beguilement.

By the early years of the 20th century, London could boast many other less 'polite' buildings that had progressively evolved through iron into full steel framing. Engineering sheds, part machine and part enclosure, depended on structural advances to support overhead travelling cranes, as did London's first generation of colossal power stations, including Lots Road, Chelsea (1902–5) and Greenwich (1902–6). Retrospectively, there might seem to be a degree of teleological inevitability about this key development, but for the great majority of London buildings, full framing was neither required nor economically practicable in this period. The constructional evolution of mansion blocks, a characteristically London form of accommodation, was substantially aided by iron- and steel-framing technologies, but only one skeleton-framed example, Gloucester House Flats, Park Lane (1906–7, demolished), has been documented. Speed of construction was perhaps the major reason for the technique's invocation, a facility exploited by the forty or so underground stations erected under the auspices of Leslie William Green between 1903 and 1907. Flat roofed, they were designed to take the weight of commercial offices above, either immediately or at a later date. For practical reasons, the frameworks of these buildings often derived some support or stability from the brick enclosure, and it was not until the second great wave of 'tube' building, in the 1920s and 1930s, that full independence between frame and cladding was deemed necessary and became economically advantageous. London's fire stations, warehouses, factories and scores of other buildings made considerable use of metal framing technologies in meeting their functional exigencies, but for reasons of expense or impracticability did not, with a few exceptions, exploit the full load-bearing frame.

To focus explicitly on the development of full framing is to diminish the richness and variety of steel's contribution to London's Victorian and Edwardian architecture. In the hands of an architect such as Richard Norman Shaw, steel helped create a flawlessly articulated timber-lined tunnel for the nave of Holy Trinity, Latimer Road (Harrow Mission Church) (1887–9). For other church architects, it was a new and excitingly expedient way of realising traditional forms. Perched high on masonry superstructures, the light steel-ribbed domes of edifices such as Albert Vicars' St Joseph's Retreat, Highgate Hill (1887–9), or Herbert Gribble's addition to Brompton Oratory (1895–6) provided suitably robust, wind-resisting finales that would lastingly grace the London skyline. In one ecclesiastical building, All Saints' Mission Church, Pentonville (1901–2) by R A 'Bungalow' Briggs, the metal's artistic potential ran riot in a manner reminiscent of the curvilinear Art Nouveau of Victor Horta in Brussels. Less self-consciously, soaring steel ribs offered a straightforward means to envelop large volumes for use as drill halls, assembly halls, swimming pools and a multitude of other uses, giving their interiors distinctive curved or hemispherical profiles. Public amenities such as Nine Elms Baths, Battersea (1899–1901), and Haggerston Public Baths, Shoreditch (1903–4), benefited from cheaper constructional engineering in steel in a way that a generation earlier would not have been possible: long-span prefabricated, wrought-iron roof trusses tended to be restricted to market halls, train sheds and other privately funded ventures, where higher costs were justifiable. In general, the facilities of steel construction greatly benefited London's newly created civic authorities in their provision of public libraries and halls. These began to increase around the turn of the century, and a little later one contemporary could note that 'While it would be idle to pretend that buildings of this class could not be erected with steel, it must be conceded that the economies and conveniences of steel, particularly in the diminished area of supports, have, as well as influencing their design, influenced their utility and number.'[6]

Both qualitatively and quantitatively the most far-reaching contribution of architectural steel construction was also the most visually discreet and structurally unassuming. If 'the use of patent flooring systems incorporating wrought-iron beams became a commonplace feature of major buildings'[7] in the period c 1850 to c 1890, then the same was true for floors incorporating steel in the period c 1890 to 1909, only far more so. Truly a mass-produced item of superior quality, rolled-steel joists quickly replaced those of wrought iron as the backbone of fire-resisting flooring systems composed of incombustible materials such as clay, concrete or brick. The 1890s saw an explosion in the number of systems as more manufacturers turned to this increasingly scientific field, and with prices that were more and more competitive by comparison to conventional timber-joisted floors, their use mushroomed across a vast range of building types. By the early 1900s, the choice facing the designers of most public, commercial or institutional buildings was not whether or not to use a non-combustible floor, but which of the bewildering array of proprietary types to employ. Thus the proportion of Edwardian buildings that were not only more dependable from a fire-safety point of view, but that also had floors that were stronger, longer, more rigid, more soundproof and took up less depth was far greater than that of a generation earlier.

Taken together, the building types considered in the preceding chapters illustrate a range of architectural uses to which steel was first put by architects, engineers and builders, and the techniques they employed. The spotlight has purposely been turned away from the long- or wide-span building structures traditionally associated with Britain's 19th-century engineering skill and ingenuity: railway stations, market and exhibition halls and so forth. Instead, with the exception of buildings designed to serve industry, the focus has been on the types of structures that traditionally were the domain of architects. The most notable of these possess a high engineering component – even if it is not always immediately apparent – be it state-of-the-art steel cantilever construction in pursuit of uninhibited theatrical sightlines, or steel framing of various degrees of self-sufficiency or extensiveness in the speedy creation of large or elaborate interior volumes. Such buildings set the pace in structural innovation in steel, building on earlier developments in iron. With its superior properties, and ultimately, its mass availability, steel enabled greater refinement and application of those elemental forms such as beams and columns, arches and domes, widening the palette of constructional possibilities. Some developments were particular to certain building types, but in most cases change did not occur in isolation: it was evolutionary, not revolutionary, with new techniques spreading between the different classes of buildings. To return to steel cantilevers as an example, although we see this technology most vigorously applied in theatres, it also featured in the Birkbeck Bank, and in many swimming pools. Similarly, the favoured British technique of framing the ground storey with steel so that it carried the weight of an essentially load-bearing-wall building above saw its most widespread application in offices and hotels, but was also used systematically in fire and tube stations.

The most all-pervading transformations were also the most simple and discreet. If the use of patent flooring systems incorporating rolled-steel joists was still exceptional in 1890, within a decade it had become a basic tenet of architectural practice. By the early 1900s steel was so entwined in the methodology of everyday building construction that it was being used across the gamut of building types almost to the same degree as bricks and mortar. Schools, colleges and universities, hospitals, workhouses, barracks and prisons, public libraries, town halls and law courts, museums and galleries were just some of the many buildings whose interior spaces benefited from the substitution of piers of masonry or brick, and joists, beams and roof trusses of timber for stronger, more attenuated and fire-resisting replacements in steel. In a sense, unlike iron, whose expense militated against true mass adoption, steel engendered a constructional democratisation in buildings regardless of form or status. Revealing and instructive though it might be, a detailed investigation of the use of steel in all of these building types would run to excessive length and would be overly repetitive.

Architectural consequences

Steel may have given prevalence to such advances as gains in span and load-bearing capacity, but for the great mass of London buildings, it did not change profoundly the way they looked. Style and ornament remained superficially unchanged by the rapid strides made by building technology in the late 19th century, as a look around any theatre of the era would confirm. The volume of the auditorium, spanned by concealed steel trusses, might have been extraordinary, and the depth and sweep of the column-free tiers might have been remarkable, but for all the spatial economy and orderliness of the pared-down steel construction, historicist ornament continued to triumph. In some senses this indicates how architectural technology had outrun architectural style, or perhaps of how style was unwilling to adapt at this juncture. In another sense, it exposes our modern expectations of architectural form and technology. With hindsight, it is clear that no architectural language of sufficient clarity to unambiguously or forcefully express the clean lines of the steelwork then existed. But it should be remembered that however much building technology might have been in the throes of change, this change was itself part of what re-inforced the tempering backdrop of a historic city whose inhabitants, architects, engineers and clients included, favoured tradition and orthodoxy to change for change's sake. Conditioned as we are by a knowledge of architectural developments through the 20th century, particularly the arrival of Modernism and its particular notions of progress and rationalism, it is tempting to rate architecture of this period according to how it conforms to subsequent ideals. As commentators have observed, even the Chicago school was not intent on developing a new architecture for its own sake at this time; its assignment was to erect tall buildings that

were stable, fire-resisting and satisfied the client by making money and looking good. Ornament remained a legitimate part of this brief, well into the 20th century, and steel construction another. Arguably a more heterogeneous grouping, with a less commercially focused range of concerns, London's architects were keen to appropriate what technology had to offer without compromising traditional values such as strength of purpose, decorum and propriety. Ornament, which could really run riot as non-structural appliqué,[8] was inextricably linked with these ideals, and however much it stemmed from a monumental, stone- or brick-based tradition, it was firmly rooted in the very meaning of architecture.

That said, a structural material as potent as steel could not remain wholly mute in the stylistic developments of the time. At Boord & Son's office, for example, the Mannerist language Aston Webb employed to adorn the interior was spatially compromised by the unyielding position of steelwork; Classicism was shoehorned to fit structural dictates rather than the other way around (Fig 14.2). This idiom was in some

Fig 14.2
Structure impinging on style: door and window surrounds are squeezed between concealed steel stanchions in Aston Webb's headquarters office for Boord & Son, Tooley Street (1899–1901).
[BB99/09032]

senses the most suitable, for in any other, the distortions to proportions would have been all the more apparent. In the grand interiors of commercial buildings, such as the Savoy, or the Union Bank of Australia, or even the Salisbury Hotel, ornately coffered ceilings tacitly affirmed the gridded steel structure carrying the floor above. Within the Birkbeck Bank, steel proclaimed its authority on the rotunda composition, with an umbrella-like array of steel ribs rising from a ring of stanchions, all barely concealed by a thin faience veneer. Elsewhere, the eclectic forms of some structural steel components were exploited to accentuate the ostentation of showpiece spaces, as when Alfred Waterhouse made celebratory use of stout polygonal-section, built-up columns manufactured by William Lindsay & Co., ornamenting them with panels of polychrome faience. Other architects preferred the translucence conferred by patent solid-steel columns, using them in theatres, offices and a multitude of other contexts where maximum carrying power and minimal disruption to light, sight or space were basic requirements. For these, steel remained, and still remains, unsurpassed. Such buildings, and many others besides, tacitly acknowledged steel framing. They signified a growing realisation that there was a difference between covering construction and hiding it.

From the street, the presence of steel in London's buildings was also frequently implicit. Huge plate-glass windows on ground floors, where the legislation governing window-to-wall minima did not apply, enunciated concealed steel girders, often spanning between stanchions, and often as not forming the perimeter

a

b

framing of a fully framed ground floor and basement. The disjuncture between the transparency of these floors and the weightiness of those above was often extreme, as though the upper part was 'resting in the air … unconscious of what is, or rather, is not, below'.[9] The origins of this phenomenon lay in Georgian London, but it was in Edwardian London that it became most rampant and acute (Fig 14.3). Above glass-walled shops and showrooms, and also in buildings without these features, elevational treatment and composition often reflected the interior grid of floor beams. The strength of steel beams permitted wide spacing, but also demanded substantial piers to take concentrated floor loads. Piers helped dictate either a horizontal or vertical emphasis within the elevation, depending on their width and projection relative to that of the intervening panels. They also impacted on fenestration. Windows tended to be arranged in groups within the bays, one over the other. The freedom of varying the scheme from floor to floor was curtailed. In some buildings, such as commercial warehouses with little surface ornament, the entire width of the bay was infilled with mullioned and transomed

Fig 14.3
Nos 90–92 Oxford Street.
The riveted steel stanchions revealed by recent renovation (a and b, facing page), and original drawings (c, below), 1899, showing steel framing and concrete filler-joist floors.
[a) DP133993;
b) DP133994; c) London Metropolitan Archives, City of London (GLC/AR/ BR/22/ES/092225)]

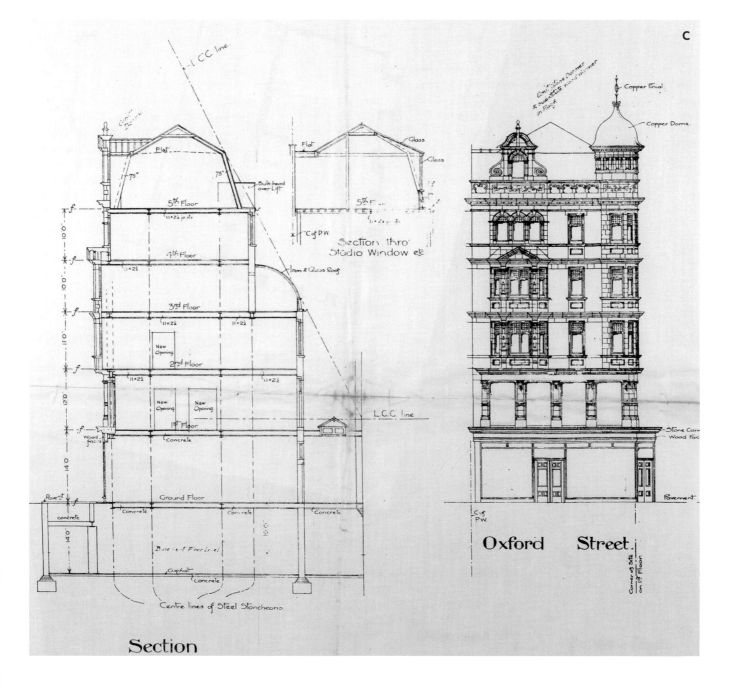

Section

Oxford Street.

fenestration – the void itself rather than its relationship to the solid becoming the most visually commanding aspect. In some districts, such as Clerkenwell and Shoreditch, it was common practice to carry these openings, and the brickwork above, on exposed rolled-steel joists that spanned between the piers.

Stanchions concealed in the exterior wall exerted a fenestral influence similar to that of solid brick piers in creating vertical bands that excluded openings. Such was the compressive strength of steel that it might be totally embedded in the thickness of the wall, or at least require only minor localised enlargement of the wall to cover it, resulting in an exceptionally narrow pier or pilaster, or even a plain wall surface devoid of window. Like pier-and-panel construction, this was certainly not a dynamic peculiar to steel – the piers of the seven-storey Grand Hotel, Northumberland Avenue (1877–9), for example, incorporated iron stanchions to lessen their width – but it was one that became more widespread. Because of their unsettling narrowness, some architects elected to give stylistic weight to concealed stanchions, at least where load was most concentrated. Bannister Fletcher for example favoured rustication for

the ground-storey portions of the stanchions of his St Paul's Churchyard building, although above that a planar surface was chosen. With the advent of full steel framing in London, the tendency to concentrate solidity in columns, piers and pilasters augmented, one contemporary noting that 'the spaces between are so treated that the emphasis is all on the voids, the solids, as far as possible, appearing to act only as their decorative framework'.[10] Neo-baroque and neo-classical, the languages seen as most suited to London's imperial identity, were also the most favoured modes for its Edwardian steel-framed buildings, not simply because they enjoyed greatest currency for monumental edifices anyway, but because they combined expression with dignity. Neo-baroque was especially rich in allegory and metaphor, and some of its stylistic devices alluded to structural underpinnings. Attached columns of solid stone or marble might have appeared 'false', giving the impression that they were doing work when they were not, but in many cases they were aesthetic symbols of support, standing directly in front of the concealed stanchion.[11] Electra House and, more strikingly, Selfridges, both employed this contrivance in a rhetorical manner. Ground floors faced in Portland stone with banded rustication ostensibly communicated the strength and solidity vested in the lower portion of traditional bearing walls. Applied to a skeleton-framed building, this architectural decoration was not inherently deceitful or in denial, for it conveyed, perhaps, the accumulated loads borne by the lower stanchions of the frame at that level. Many framed edifices employed ground-floor rustication, including the London and Manchester Industrial Assurance Co. building, Caxton House, Carlisle House, The Ritz Hotel and the Morning Post Building. The last mentioned were resplendent examples of the Beaux-Arts style, their thick skins reflecting the gridiron orderliness of the frames from which they were hung.

In this sense steel-framed construction was not completely disguised behind exuberant neo-baroque, or the trabeated expressions of Beaux-Arts classicism and other permutations of the classical revival. Aside from any coded references to the role of steel structure, the symmetry, orderliness and scale of these languages were well matched to the size and repetitiveness of the structural bays of steel. While some of the features of these buildings could not have been realised without steel (structural

Fig 14.4
The steel skeleton of Spicer Brothers' office and warehouse (Blackfriars House), New Bridge Street, being erected by Matthew T Shaw & Co. in 1913. Completed in 1917 to designs by F W Troup, it was one of the few buildings of the era to make real elevational play of the framework.
[OP01240]

failure might have resulted had the attenuated supporting arcades of buildings such as The Ritz Hotel, Caxton House and Koh-I-Noor House been built purely of stone), none of the sober attire was devised specifically for the new method of building. It fitted well, sometimes with a bit of tailoring, but it evolved outside changes in building technology. One further turn in revived style, in the mid-1900s, owed more to this dynamic. Unlike the more formal, heavy-handed and traditional articulations of mainstream neo-classicism, the language of neo-mannerism celebrated the load-bearing freedom of the exterior walls. Freed from structural duty, the elevation was open to rich and playful possibilities in form, texture and composition. Compared to orthodox neo-classicism, this was only a minor and short-lived flourish, but it did produce some of Edwardian London's most distinctive and appealing compositions. Charles Holden's 'Mannerist Trilogy', most notably the British Medical Association (latterly Rhodesia then Zimbabwe House), clearly signified a new directness and sensitivity to the underlying frame, with motifs and massing adjusted more to the attenuated character of the steelwork than to its severe repetitiveness and orthogonality. Belcher's Royal Insurance Building made clever play of the strength of the skeleton, stretching a thin-looking stone skin around it, and increasing the massing from bottom to top in a conscious inversion of conventional practice. Such visual gymnastics were pushed to an almost absurd level in Belcher's mannerist fantasia on Oxford Street (*see* Fig 11.44). To one observer, the Mappin and Webb building represented the 'ne plus ultra of light construction', an architectural endeavour whose sole purpose was to frame voids:

> It might be said, indeed, that it is not a building at all, but an erection, a decoration of architectural motives applied to a steel skeleton … architecture has departed altogether from its customary ways, to pay, as it were, an extravagant compliment to its servant, steel.[12]

The compliment continued through the Edwardian era in the designs of progressive architects, although mannerist steel-frame architecture had come to the end of this initial phase of its journey. It was sustained within neo-classicism, the most dominant language, and considered to be the most traditional English style, strands of which made noticeable

concessions to the technology before the onset of war. In many buildings, ornament was stripped away, diluted or confined only to the masonry overlying the steel grid, leaving less visual ambiguity as to whether framed construction lay behind. Commercial office buildings were the foremost recipients of this harsher clarified language, but it was soon applied to anything grand. In buildings such as Blackfriars House, New Bridge Street (1913–17, F W Troup, architect), the former Hong Kong and Shanghai Bank, No. 9 Gracechurch Street (1912–13, W Campbell Jones, architect), and Nos 34–36 Golden Square, Soho (1913–14, Leonard Stokes, architect), framed construction was instantly recognisable (Figs 14.4–14.6). The most radical,

Fig 14.5
The (former) Hong Kong and Shanghai Bank, 9 Gracechurch Street (1912–13; W Campbell Jones, architect).
[Jonathan Clarke]

Fig 14.6
Nos 34–36 Golden Square,
Soho (1913–14; Leonard
Stokes, architect).
[Jonathan Clarke]

however, was also one of the first. Kodak House, 61–63 Kingsway, 1910–11 by (Sir) John James Burnet in collaboration with job architect Thomas Tait, was a logical extension of Burnet's earlier classicism, stripping away decorative elements to articulate fully the vertical severity of the frame in a composition of extraordinary erudition (Fig 14.7). It still read

as a classical building, with a base, entablature and cornice, but it unequivocally pointed to new territory where the frame itself would form the lead motive in the overall design. Yet for socio-political and ideological reasons, this rationalist territory remained largely unexplored over the next two decades. Despite some occasional magnificent, idiosyncratic essays in

Fig 14.7
Kodak House, 61–63
Kingsway (1910–11; by
(Sir) John James Burnet
and Thomas Tait).
[Jonathan Clarke]

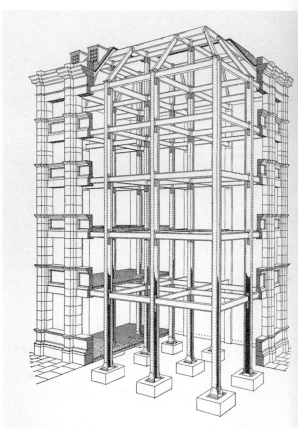

the declaratory use of steel frames (Fig 14.8), London's grand commercial and public architecture trod a more conservative path, increasingly purged of ornament, but rarely as yielding to the latent possibilities of the steel frame. Its presence remained detectable, but not immediately so (Fig 14.9).

After the 1909 Act

The passing of the London County Council General Powers Act 1909 was a long-awaited watershed in British construction history. Legislation finally caught up with technology, and thenceforward London's designers were able to operate with the kind of freedom long enjoyed by colleagues in America, and more informally, in other British cities such as Manchester. Exhaustively thought out, drafted and re-drafted, it set a new standard for other authorities to follow. Ostensibly the Act opened the floodgates to a new era of thinly clad multi-storey skeletons, but in practice this was not immediately the case. For a start, not all practitioners greeted the Act with relief. Just months before it was pushed through parliament, Childs opined that it would be a 'dead letter', increasing the cost of

steel-framed construction, and would be a source of hindrance to architects like Richard Norman Shaw, who preferred thicker walls anyway.[13] Bylander, however, believed that 'the introduction of standard stresses and a general method of design and supervision would tend to simplification and economy in design, to fairer competition, and to safer structures'.[14] For the years immediately following 1909, Childs was broadly correct. What had been intended as a facilitator to steel-frame architecture became a major obstacle to some designers. Steel construction now had to meet a set standard, and a very high one at that. The sheer number and complexity of the technical clauses, embodying highly mathematical structural design, left the majority of architects flummoxed and alienated. Four years after its passing, at a discussion of the effect and provisions of the Act, the president of the RIBA, Reginald Blomfield, divulged 'he did not know where he was ... [with such] ... relentless and implacable science', while another (unnamed) architect had 'a look of utter desolation'.[15] Much was perhaps also beyond the understanding of some of the smaller- or more empirically minded engineering contractors. Rather than cooperate with consulting structural engineers, some architects stayed well clear of skeleton construction. For those who were not deterred, there were inherent problems with the demands and bureaucracy of the new orthodoxy, not least considerable delays caused by the stipulation that district surveyors should approve the design, which entailed endless calculation checking. Some of the chief virtues of skeleton-frame construction, including speed of erection, were compromised, and as a result many designers chose to revert to building under the 1894 Act, which offered, in some senses, considerably more latitude.[16] Consequently, for the immediate years at least, the diversity in approaches to framing varied almost as much as that before the 1909 Act, with the full skeleton only one of a range of options.

Ultimately, of course, full steel-frame construction did triumph. The Kodak Building was a timely and eye-catching symbol of this success, but other, less striking, examples bore witness to this new era of architectural engineering. The first building under the new Act to be submitted and passed by the LCC was the nine-storey front block of the new Regent Street Polytechnic (1910–11) (Fig 14.10). Designed by George A Mitchell, headmaster of the polytechnic,[17] and Frank T Verity, its 1,1000-ton Dorman Long skeleton was hung with walls 'made much thinner than was requisite under the old conditions'.[18] Verity was responsible for the façade, distinguished by a giant order of Ionic columns that adumbrated the thinly clad front stanchions of the frame. Other beneficiaries of the 1909 Act, mostly commercial premises in the City, soon followed. The (former) Hong Kong and Shanghai Bank, No. 9 Gracechurch Street (1912–13) (see Fig 14.5); the Mercantile Bank of India, No. 15 Gracechurch Street (c 1913, William Wallace, architect); and London House, Crutched Friars (1912–13, architect W E Clifton)[19] all survive as representatives of the first wave of steel-framed buildings designed under the 1909 Act. Oscar Faber, chief engineer to George Trollope & Sons and Colls &

Fig 14.10
The central block of a scheme for Regent Street between Margaret Place and Cavendish Place, for the Regent Street Polytechnic, was distinguished by its strong vertical emphasis. The flanking wings were never built, but Frank Verity's façade – scaled back to five bays, went up in 1910–11, exhibiting a leanness afforded by the 1909 Act.
[RIBA70537, RIBA Library Photographs Collection]

Fig 14.11
By the 1920s the steel frame was the standard means of erecting large or important commercial buildings in London. This photograph of 1929 shows construction workers taking a lunch break on the skeleton of Chilton Court, Upper Regent Street (C W Clark, architect)
[© Hulton Archive, Getty Images (HG0106)]

Sons, Ltd, was responsible for the structural design of these three buildings. Like the architects, he represented a new generation that could apply the technology as though it was as ordinary as bricks and mortar. Bylander was right in the long run. Writing in the late 1920s, one structural engineer could preface his text book on modern steelwork: 'The number of buildings in London erected under the Steel Frame Act (as the 1909 Act is often called) has gradually increased since the date of its introduction, and now practically all the important buildings being erected in London are steel frame buildings' (Fig 14.11).[20]

London compared: necessity and innovation

This book has aimed to show how building technologies using iron evolved in London buildings over the course of the 19th century to culminate in the triumph of steel construction in the first decade of the 20th. Reinforced concrete was fast establishing itself as a rival (though often complementary) modern building material and technique, but that revolution is another story.[21] The prime factor driving the steel revolution was the British steel industry, which mass produced a commodity so light,

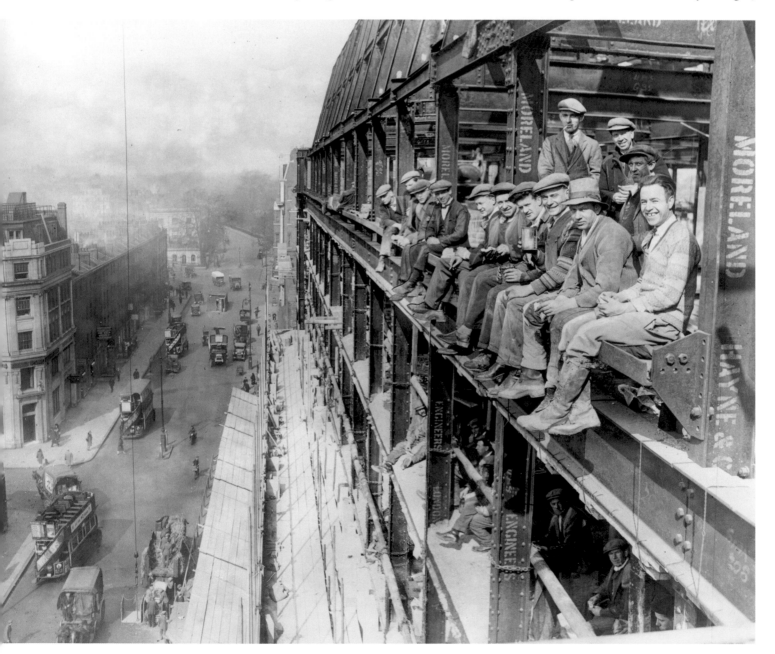

strong and useful that the spread of its constructional use became irresistible. Steel brought about many changes in the way buildings were designed and built, how they functioned, performed and looked. It accelerated change in the organisation of the building process, enhancing the role and profile of specialist contracting and engineering firms, and often hastening construction time. It also amplified the shifting relations between the design professions, notably between architects and engineers. And by giving rise to a radically different technique of building, the steel skeleton, it forced legislative change, precipitating London's first Act in almost 250 years to specifically address a building material.[22]

London was one of many cities in the industrialised or industrialising world that experienced a steel 'revolution'. American cities, notably Chicago and New York, famously underwent transformations of a speed and magnitude that put new meaning to the phenomenon of urbanisation, but it is clear that major cities in other countries, such as France, Germany and Belgium, were also quick to innovate, while others, in South Africa and Australia, for example, were eager recipients of steel-based construction technologies later, from 1900 onwards. Comparison with other countries understandably raises questions of relative technological performance and leadership; indeed these were issues that greatly exercised the minds of contemporaries. It is generally acknowledged that after the climax of the 1851 Exhibition, Britain gradually lost its industrial, economic and technological pre-eminence. In so many spheres, including iron and steel production and the application of these metals to engineering structures, the contrast between the 'heroic era' before 1850 and that after became more palpable as the century unfolded. The 1867 Paris Exhibition was in many ways prophetic of future trends, with the United States, France, Belgium and Germany dominating industrial and engineering enterprise, while Britain trailed behind. Although in 1890 Britain could proudly proffer the greatest all-steel bridge as an example of its engineering expertise, the Americans could lay claim to the high-rise, skeleton-framed office building, and the French to awe-inspiring structures such as the Eiffel Tower and the Galerie des Machines. The problem is that exceptional superlative symbols such as these invariably characterise, and typecast, national engineering endeavour,

and draw attention away from less spectacular achievements. The contributions of countries other than America to steel building technology in this period have not had the same degree of attention, with the overall result that skyscrapers epitomise both the era and America's technological prowess. In some ways, skyscraper technology became a yardstick by which other buildings of the industrialised world were measured. The Ritz Hotel, an icon of American achievement exported to London, has long been an enduring symbol in Britain's construction history, effectively diminishing what went before, and seen as presaging what came after. Yet it did not single-handedly launch steel-frame construction in London, or indeed Britain. Rather it transplanted highly developed transatlantic practice onto fertile ground long prepared by British engineers, contractors and architects, ground which had in any case seen some occasional American cultivation already, and which had given rise to sophisticated technique and structural form. In many ways The Ritz Hotel was more a technological culmination than an inauguration, for preceding it was an era of considerable creativity, inventiveness and achievement in British architectural engineering, with iron and steel exploited to highly logical and resourceful ends.

Unequivocally, London in the late 19th century was a less innovative environment for the development of steel-based building technologies for commercial architecture compared to counterparts across the Atlantic. London lagged some 10–15 years behind Chicago and New York in the introduction of the all-steel, multi-storey, skeleton-frame commercial building. The 'delay' is in fact less than that typically cited, but this detail is perhaps of less significance than what it tells us about the factors driving innovation, and the relationship of these to different types of building. As we have seen, the iron skeleton frame was seldom used, much because of the expense of the metal, which made it difficult to justify – especially to clients – outside of conditions that demanded its use over traditional forms of construction. The advent of mass-produced steel in Britain made skeleton framing feasible, initially for structures that would perform far better in steel, such as the open-sided transit sheds and warehouses of the Manchester Ship Canal. Legislative and philosophical issues aside, until the cost of steel fell enough to make skeleton framing truly viable, or even inexpensive, its use was difficult to justify in commercial buildings whose functions were already suitably met by more conventional technologies. Bylander would later note 'From 1902 to 1908 steel was cheap',[23] a period that marked the real onset of the technology for hotels, offices, power stations and other large buildings. Cost was only one factor that stimulated or retarded innovation. Across the types of buildings considered in this book, the nature, scope and sophistication of constructional steelwork was clearly tied into the strength of the stimulus to innovate.[24] In some cases this stimulus was strongest in buildings that were called on to serve new functions. Power stations and engineering sheds, whose structural evolution was closely tied to the overhead travelling crane, illustrate this

incentive. Underground stations, substantially framed and called upon to support anticipated buildings are another example. In others, perhaps the majority, the stimulus stemmed from the need to make existing buildings, or parts of buildings, perform better. Resplendent among this grouping were theatres and music halls, where the challenges of cantilevered construction engaged the energies of some of the most distinguished and brilliant engineers of the day, men such as Kennedy, Ordish and Ende. Such was the need to innovate, that some architects, such as Matcham and Briggs, faced the challenges directly, producing daring new solutions of their own. Grand social or public spaces in clubs, hotels and banks, the need for more light and openness in offices and department stores,

Fig 14.12
The Central Office of the National Bank of South Africa in course of erection in Johannesburg, November 1904
[Builders' Journal and Architectural Review, 30 November 1904]

and the need for floors of greater load-bearing or spanning capacity in factories and warehouses, demanded similar levels of innovative input, though sometimes of a less acutely focused character. Finally, in other building types, notably churches, the stimulus was never really vigorous or persistent; in churches at least, steel construction was brought into play almost as a roguish experiment, because it could be.

It is not without irony that the British constructional steel industry was kick-started by overseas competition. The failure of the indigenous wrought-iron industry to respond adequately to a burgeoning demand for joists, girders and other components for the building industry saw other nations seize the initiative and cultivate a new market, which had enormous potential. Ultimately British steel manufacturers and constructional firms appropriated this market, but in the 1860s this disjunction was symptomatic of a divorce of interests, or a lack of understanding, between producers and consumers. It will be recalled that *The Builder* bemoaned the unwillingness of British ironmasters to roll deep beams for use in buildings. It has been suggested that this lack of understanding continued through the 19th century, and contributed to Britain's dwindling engineering prowess and capacity to innovate. While British steelmakers were satisfying a market for steel structural sections, Pittsburgh steelmakers were meeting regularly with Chicago architects, tackling specific engineering problems and advancing mutual interests. One skyscraper builder could note that 'Construction and steel production are inseparably linked ... The demands of the one furnished the incentive for the colossal scale upon which the other has been developed'.[25] Yet such linkages were not totally lacking in Britain. Dorman Long, for example, responded to demand, and integrated with structural engineering to foster further demand. However, this leading company, with constructional works in London, did not 'push' for the adoption of skeleton framing to increase consumption of its products, possibly because it was aware that this would have required revision of existing building laws regarding wall thickness – and, to really make profits, height.[26] In fact it did not need to: with steelwork increasingly becoming an indispensable part of metropolitan and British buildings, its market was both huge and growing anyway without needing to press for technologies that increased consumption for each construction

project. Beyond Britain's shores, the Empire represented a market with enormous possibilities, not only in the fields of bridge-building and prefabricated structures, but also in commercial architecture, where emergent civic policies were less constrained by centuries of tradition. In South Africa for example, Johannesburg and Cape Town were already witnessing their first wave of thinly clad tall buildings in the mid-1900s, erected on frameworks supplied by Dorman Long and New York steelwork contractors Milliken Brothers (Fig 14.12).[27] In 1904 Lord Frederick Roberts

> expressed his astonishment at the transformation the town [Johannesburg] has undergone since [he] last saw it four years ago, in regard both to its commercial activity and to phenomenal growth – iron buildings having almost completely disappeared, three- and four-storey blocks having sprung up all over the town, and here and there a towering 'skyscraper' having appeared on the scene as a sort of architectural sentry, keeping guard over its Lilliputian contemporaries.[28]

Other British companies began looking to cities in Asia, Australia and South America; Edward Wood & Co., for example, set up offices in Buenos Aires in 1911, meeting a burgeoning demand for tall office buildings and other grand steel-framed edifices.[29] Cities such as these, in addition to those in America and Europe, give meaning and context to the character and chronology of London's early steel construction. It would be almost another half century before London permitted steel, and reinforced concrete, to begin changing its vertical scale, by abolishing the 100ft height limit imposed by the 1894 Act. In the same way that the restrictions to skeleton framing were not realised or felt until the 1900s, the restrictions on height did not really impinge on clients, architects or engineers until the 1950s. Indeed the majority of offices built before the Second World War did not reach the maximum allowable limit, but were 70–80ft high.[30] Unlike many other cities of the industrialised world, London cautiously appropriated what it felt was desirable from the technology without letting the technology wholly redefine it. As one witness to the late Edwardian trend for 'gigantic steel skeletons inside the gigantic wooden scaffoldings dominat[ing] the eye at many a turn' put it, 'London ... will always be London, but more so.'[31]

NOTES

Abbreviations

AR	*The Architectural Review*
A&CR	*The Architect & Contract Reporter*
BN	*The Building News*
BJ	*The Builders' Journal*
BJ&AR	*The Builders' Journal and Architectural Review*
BJ&AE	*The Builders' Journal and Architectural Engineer*
CE&AJ	*The Civil Engineer and Architect's Journal*
C&CE	*Concrete & Constructional Engineering*
GLC	Greater London Council
ILN	*Illustrated London News*
JISI	*Journal of the Iron and Steel Institute*
L&BN	*Land and Building News*
LMA	London Metropolitan Archives
MM	*The Mechanics' Magazine*
MBR	*Modern Building Record*
PICE	*Proceedings of the Institution of Civil Engineers*
TRIBA	*Transactions of the Royal Institute of British Architects*
JRIBA	*Journal of the Royal Institute of British Architects*
TSE	*The Structural Engineer*

Introduction

1   Pite 1902, 130.
2   Hanson 2004.
3   'Every day steel is supplanting iron for some one of the numerous purposes to which that metal has hitherto been applied ... It is evident that the coming century is to be an age of steel', 'The steel age'. *Manufacturer and Builder* **12**(1), January 1880, 12. *See also*, for example, Horner 1889.
4   An affinity perceived by a number of Victorian and Edwardian commentators, including Pite, who remarked 'The new "half-steel" has exactly the same claim to be interesting first and beautiful afterwards as its exact forerunner the "half-timber," ... The rivet heads, stiffeners and knees of framed girder and stanchion work are similar elements to the oak pegs, bracings and bracketings of our charming old English fronts', Pite 1902, 131.
5   'The Waring-White Building Company', *The Times*, 2 June 1906, 13.
6   'The American invasion of London', *The Times*, 20 November 1901, 8–9; *see also* a letter to the editor, 'The proposed Strand building', *The Times*, 30 November 1901, 9.
7   In internally renovating the House of Commons in 1692, Wren used two iron columns to support the short entry side gallery, and console brackets the other two. The two side galleries were subsequently deepened in 1707, carried this time by iron columns of the same appearance as the original two. Iron – probably wrought, given Tijou's trade – was selected over timber for the columns – among the earliest documented uses – because it saved space and preserved sightlines. *See* 'Wren, Sir Christopher, FRS (1632–1723) in Skempton 2002, 801–2; Saint 2007, 68–9; *The History of the King's Works* **5**, London, 1976, 60–1.
8   This term is borrowed from Thorne 1990.
9   From the 1840s German manufacturers began refining pig iron into steel using the puddling process. This met with only limited success: 'puddled steel was rarely as homogenous and hard as crucible steel, or as tough as shear steel', Landes 1969, 254.
10   Certainly, Ford and Hesketh were happy to let structural iron carry the floors and façades in their earlier City warehouses. The destruction by fire of a five-storey warehouse for wine and spirit merchants Foster and Co. in Cheapside in September 1881, 'leaving a shell of crumbled stone piers on the two frontages ... held together entirely by the framework of iron girders' revealed that the 'main façade rested on a triple-plate girder supported on cast-iron columns'. This interior iron frame, supplied and erected by Measures Brothers and Co. of Southwark Street, survived almost intact, 'while all the wooden joists and flooring were consumed', 'Fire-resisting structures', *BN*, 28 October 1881, 556.

1 Towards a structural steel: technological and official hurdles and professional diffusion

1   *The Builder* **43** (23 September 1882), 390. 'Welded iron' or 'weld-iron' is an archaic term for wrought iron.
2   'Wrought-iron and Steel in Constructional Work – Z', *BN*, **72** (12 February 1897), 229.
3   Sir William Roberts-Austen's presidential address, *JISI* (1899), quoted in Carr and Taplin 1962, 162.
4   Carr and Taplin 1962, 27.
5   Misa 1995, 15.
6   Ibid.
7   Ibid, 111.
8   Carr and Taplin 1962, 34.
9   Clarke 1990, 5–11; Clarke and Storr 1989.
10   Kent and Kirkland 1958, 104.
11   Guedes 1979, 273.
12   Tom Swailes, pers. comm.
13   Built for the Uddevalla-Vänersborg-Herrljunga Railway, it was designed by C Adelskold 1905, 527.
14   'The first steel bridge in France', *MM*, 14 December 1866, 366.
15   *MM*, 1 May 1868, 335.
16   'Steel girder bridge', *Scientific American*, 8 April 1865, 229.
17   *MM*, 27 April 1866, 265.
18   'On the construction of iron roofs', *The Builder* **24** (17 February 1866), 110. For the history of this important, extant

works, with further references, *see* Smith 2001, 232–3.

19   Steel was used in the suspension chains of this 820ft-long bridge (1867–8, demolished 1950).

20   *MM*, 1 May 1868, 335.

21   For discussion of this, *see* Hawkshaw's reply to Matheson 1882, 44–5.

22   Kent and Kirkland 1958, 104.

23   Matheson 1882, 26.

24   *The Times*, 15 September 1879, 9.

25   Rennison 1996, 20; Stratton 1999, 9.

26   This unprecedented quantity of steel had to come from a number of different companies: *c* 12,000 tons came from The Landore Siemens Steel Company near Swansea; 38,000 tons from the Steel Company of Scotland's two works near Glasgow; and 8,000 tons from David Colville and Son's Dalzell Works at Motherwell. Some 3,000–4,000 tons were returned as scrap, but, in addition, the Rivet Company Glasgow supplied over 4,000 tons of steel rivets. Chrimes 1991, 55–6.

27   Chrimes 1991, 53.

28   Shipway 1990, 102.

29   *BN* **54** (11 May 1888), 685.

30 '   A steel roof of great span', *BN* **53** (26 August 1887), 341.

31   Watson 1892, 834, as quoted in Stamper 1989, 337.

32   Siemens stated 'I had at my house in the country a terrace, and under that terrace I had a billiard-room … I put steel girders over this billiard-room, which was about 20-ft. span, and by filling in between each girder with cement and tiling and lead, I was able to gain 18 in. in height, and obtained a perfectly dry room, whereas before I had considerable difficulty in keeping the water out. This simply shows how, by the use of this stronger material, advantages in convenience and even in cost may be obtained'. Siemens, cited in Kennedy, 1880b, 175. Siemens also introduced 'an arrangement of dynamos and water-wheel, by which the power of a neighbouring stream was made to light the house, cut chaff, turn washing machines, and perform other household duties' (Munro, 1891, 131). I am grateful to John Newman for information about the build-date and architect of Sherwood, and to Kate Minnis for Siemens's dates of occupancy.

33   Allen 2000, 51.

34   'The properties of iron and its applicability to building purposes', *The Builder*, 15 May 1859, 377.

35   'Steel for Columns', *The Builder* **24** (21 April 1866), 295.

36   Kennedy 1880a; Picton 1880a. Discussions of both papers were printed in advance of, and following, the papers themselves, on

24 April 1880, 501–2, and 29 May 1880, 660–2. Picton read an earlier version of his paper, entitled 'The progress of iron and steel as constructive materials' before the Iron and Steel Institute at Liverpool, reported in full under the title 'Iron and steel as constructive materials', *BN* 3 October 1879, 396–7.

37   Kennedy 1880b; Picton J A 1880b; discussion for both papers pp 173–92.

38 '   Modern steels as structural materials', *BN* **39** (8 October 1880), 411. For the original see Beaumont 1880.

39 '   Steel for structures', *BN* **42** (3 March 1882), 205. For the original see Matheson 1882.

40   For example, 'The strength of steel', *BN* **48** (23 May 1885), 797–8; Reade 1889, also reprinted in *BN* **57** (22 November 1889), 689–91.

41   For example, Moreland 1893; Horner 1896–1898; Cunnington 1898b, 350.

42   Kent and Kirkland 1958, 104. Sections such as these were probably being rolled earlier than this: in 1869 Bernhard Samuelson acquired the North Yorkshire Ironworks at Stockton-on-Tees especially to make open-hearth steel rails, angles and plates, although the project was later abandoned because of the high phosphorous content of the local Cleveland ore. Carr and Taplin 1962, 34.

2   Constructional steelwork and its iron inheritance

1   *The Builder* **64** (14 January 1893), 322.

2   Eduard Mäurer, *Die Formen der Walzkunst und das Faconeisen, seine Geschichte, Benutzung und Fabrikation fur die Praxis der gesammten Eisenbranche dargestellt von Eduard Maurer* (Stuttgart, 1865), 33, cited in Kohlmaier and Barna von Sartory 1990, 120.

3   *The Builder* **24** (3 March 1866), 147.

4   *The Times*, 19 May 1851, 8, referring to the sudden failure of a cast-iron girder, resulting in the progressive collapse of a four-storey office block under construction in Gracechurch Street, London.

5   *See* Hamilton 1940–41, 147–8.

6   'That most of the girders were so tested not the least doubt remains, but it is now believed that the one just referred to by some accident got passed without undergoing the regular test', *The Times*, 19 May 1851, 8.

7   Sutherland 1984, 9.

8   Elliott 1992, 101.

9   Elliott 1992, 71; Sutherland 1989a 55–6.

10   Sutherland, R J M 1989a, 50; Sutherland, R J M 1990, 26–8.

11   Guillery 1995; Sutherland 1990, 26.

12   Rivington's Notes on Building Construction 1912 , 92.

13   'Rolled iron joists for building purposes', *BN* **51** (15 October 1886), 562.

14   *See* Smith 2001, 244–5, and note 15.

15   In 1844 James Kennedy of the firm Bury, Curtis & Kennedy of Liverpool and Thomas Vernon, a Liverpool shipbuilder, patented 'iron rolled in one piece having a flange on one edge projecting on one or both sides, and a rib or flange on the other edge, projecting on one or both sides'. The patent (No. 10 143, 1844) refers to bulbed T and equal-flanged joist sections, but only for use in ships' sides and decks. See J G James in discussion of Gale 1964–5, 43 and Hurst 1999, 20, and note 25.

16   For the introduction of the rolled-iron beam into French building, Steiner 1984, 72–4.

17   Gale 1967, 37. Sir William Tite (1798– 1873) seemingly used riveted T-section rolled-iron joists as early as the mid-1820s at Mill-Hill School, Middlesex (1825–27) (see Chapter 6).

18   Burnell 1854.

19   *TRIBA* **4** (1st Series; 1853–4), 41–52 (23 January 1843); 53–61 (6 February 1854); 63–74 (20 February 1854).

20   The Gracechurch building was in the process of being constructed on an embryonic, rival non-combustible floor system, one probably devised by Charles Colton Dennett, a Nottingham builder who was subcontracted for the project to act as 'a kind of superintending foreman'. According to the report, 'The walls and flooring were built so as to render them fireproof, the girders being of [cast] iron and the intermediate spaces filled with concrete', *The Times*, 19 May 1851, 8a. The Dennett 'fireproof' floor system subsequently used rolled-iron beams, and it seems likely that other systems moved over to the safer material following the Gracechurch tragedy. For further information on the Dennett system, which was patented in 1857, *see The Builder* **24** (20 June 1866), 482–3.

21   Hurst 1999, 20.

22   Guillery 1993.

23   Richardson 1991. A subsequent extension of 1879 also used the Fox & Barrett system; *see The Builder* **37** (22 November 1879), 1293.

24   Hurst 1990, 39. With the expiration of the patent in 1859, the Fox & Barrett System came into general use, for its iron, timber and concrete components were widely available. *See* Hurst 1996, 6.

25   Fairbairn 1854; also quoted in *The Builder* **12** (20 May 1854), 273.

26   *The Builder* **12** (20 May 1854), 273.

27   Elliott 1992, 97. Progress thereafter was rapid. By spring 1855 the firm had managed to produce 8in.-deep I-sections, and the following year it was rolling

beams 9in. deep. Until 1860 the Trenton Iron Works monopolised the production of rolled-iron beams. *See* Shaw 1960; Wermiel 2000, 59–62.

28 Carr and Taplin 1962, 58.

29 Gale 1964–5, 37.

30 Carr and Taplin 1962, 60.

31 J G N Alleyne, 'Manufacture of wrought-iron beams and girders', Patent No. 2107, 1859, cited in Hurst 1999, 20, and note 45.

32 C H Cottew in discussion of Scott 1929, 100.

33 Gale 1969, 63, 79–80.

34 W Basil Scott, in reply to Cottew (above) 'said he had been in correspondence with the Secretary of the Iron and Steel Institute, who . . . sent him a reprint from "The Machinery Market" of March 1, 1886, describing a visit to the works at that date, when large steel deck beams for H.M. battleships were being rolled. He was also informed that the particular rolling mill mentioned was probably installed 10 years or more before 1886', Scott 1929, 102.

35 On 16 December 1885, in response to the question 'It is the fact, is it not, that until recently the Butterley Company and Messrs. Dorman, Long, and Co. at Middlesbrough were the only people who rolled girders ?', Cleveland ironmaster, Isaac Lowthian Bell, replied 'Yes, I believe that is so', *Second Report of the Royal Commission on Depression of Trade and Industry* (1885–6), C. 4715, Q. 2152.

36 Moreland 1883–4; Moreland 1884, 291.

37 These consisted of angle iron sections riveted onto plate webs. *See* Hamilton 1940–1, 153.

38 Advertisements in *The Builder* show that by 1854 this firm was producing both built-up box girders and the more useful I-beams, made up of plates and angles. Kelly (1870) gives this location as the company's works. Shaw was formerly an employee of C J Mare Iron and Shipbuilding Works, later to become the Thames Iron Works.

39 Dibb-Fuller *et al* 1998, 24–5. I am grateful to Lawrance Hurst for bringing this article to my attention.

40 Clarke 2001a.

41 Kelly 1870, x.

42 Friedman 1995, 72.

43 Gwilt 1867, 436.

44 Matthew T Shaw and T H Head, 'Improvements in rolling iron and steel, and in wrought girders and joists, and in machinery for producing the same', Patent No. 840, 1868; Andrews 1926, 149–50.

45 'On the testing of rolled and riveted iron beams', *The Architect* **4** (29 October 1870), 247.

46 Julius Homan, 'Improvements in the construction of wrought-iron girders', Patent No. 2593, 1865.

47 *MM*, 2 March 1866, 129. *See also The Builder* **24** (3 March 1866), 147–8. This figure was subsequently shown to be erroneous; 'When compound girders were introduced in 1867 by Messrs. Homan & Phillips, a fictitious value was assigned to them owing to an error in the method of calculating their strength. We now know that they follow the same laws as any other section, and their advantage over built-up plate girders consists only in having a solid connection between the web and angles', Adams 1912, 39.

48 Ibid.

49 Gwilt 1867, 436.

50 In its report on the testing of the compound beams, at Kirkaldy's works in December 1865, *The Mechanics' Magazine* noted that both Homan & Rodgers and W & T Phillips were present, the latter described as 'the patentees of the girder', *MM*, 2 March 1866, 129. This authorship was later confirmed by Douglas P Rogers who stated that 'His firm was one of the oldest concerned in constructing fire-proof floors, Mr. Homan having acquired the patents of Mr. Phillips', Douglas P Rogers in discussion of Cunnington 1889, 365.

51 Kelly 1870; *see* note 26 in Chapter 9.

52 'Drapers' Hall, Throgmorton Street, London', *The Builder* **28** (15 January 1870), 38–40.

53 *BN* **24** (27 June 1873), 1 of advertisements.

54 In the early 1870s, Homan & Rogers devised a filler-joist floor system comprising light, 3in.-deep iron joists spaced 1ft apart and embedded in concrete; it reputedly 'held its own for many years, and the firm used it in more than 2,000 buildings', 'Reinforced concrete systems. VI.- Homan & Rodgers' System', *BJ&AE* **24** (C & S Supplement) (10 October 1906), 76–8, 77.

55 Griffiths 1883, advertisements section, 126.

56 Carr and Taplin, assisted by Wright 1962, 79.

57 'Belgian versus English iron', *Iron*, 5 March 1886, 211.

58 'Specification of ironwork', *BN* **41** (21 October 1881), 525.

59 William Farnworth, manager of the iron and tinplate works of E P & W Baldwin. *JISI* 1875, 250.

60 Unprovenanced contemporary quotation cited in Carr and Taplin, assisted by Wright 1962, 24.

61 In, respectively, the years 1858 and 1860. Tweedale 2004.

62 *The Builder* **20** (23 August 1862), 604.

63 'Steel Girders at the London and County Bank, Sheffield', *BN* **7** (17 May 1861), 422.

64 Bessemer had a vested interest in his product being seen as 'cast-steel', for he hoped to win over the Sheffield

steelmakers to his process. In 1861 he declared 'The problem we have before us is, how to produce cast-steel that will take any form in the mould or under the hammer … It is believed by the author that these desirable objects are fully accomplished by his process of converting crude molten iron into cast-steel at a single operation, which process has now been in daily operation for the last two years', Bessemer 1861, 135.

65 J R Ravenhill in discussion of Matheson 1882, 42.

66 Brown (No. 1111, 1858).

67 Tweedale 1984.

68 *See JISI* **80** (1909), 307.

69 Kennedy 1880a, 169 (emphasis added).

70 Kennedy 1880a, 163.

71 *The Times*, 27 November 1880, 6. The first 'official' blow of the new basic (Bessemer) process took place on 4 April 1879 at the Eston works, and the firm installed basic converters immediately after. Carr and Taplin, assisted by Wright 1962, 100.

72 Kennedy 1880a, 190.

73 Some Scottish ironworks may have turned to the production of rolled steel joists by this date. For example, R Swarbrick & Co., 47 Oxford Street, Manchester, and 25 Macfarlane Street, Glasgow, were advertising themselves as 'Manufacturers of Patent Girders, Rivetted Girders and Rolled Joists in Iron or STEEL'. Advertisement at the back of Beckett 1881.

74 Ewing Matheson in discussion of Kennedy 1880a, 180. The difficulties that Matheson referred to included the practical realisation that rolling steel joists required different rolls and techniques to those used for iron.

75 Matheson 1882, 3.

76 Steel ingots or blooms typically entered the rolls at a much lower temperature than piled iron, and because they were a solid mass, rather than being formed from welded layers of bar iron, considerably more power was required to force them into a given shape. Furthermore, more exact and gradual passes were required through the rolls, and for this reason alone, Skelton noted, 'large sections can seldom or never be produced alternatively in wrought iron or in steel from the same set of rolls', Skelton 1891, 222–3.

77 Burn 1961, 59–60.

78 Abé 1996, 52.

79 Bodsworth 1997, 434; Malcolm 1990, 34.

80 *JISI* (1879), 572, reported 'At the Glasgow meeting of the Institution of Mechanical Engineers in August, Mr. Edward Hutchinson, of Darlington, read a paper on "Improvements in Machinery for Rolling Iron and Steel Plates", in the course of which he described and commended the advantages of the Belgian

"Universal" Mill, adopted at the Britannia Works, Middlesbrough'. According to Burn 1961, 61, the first English Universal Mill was erected in Darlington in 1878.

81 Carr and Taplin, assisted by Wright 1962, 160.

82 Dorman Long 1901, 8–9; Bodsworth 1997, 43; according to Scott 1929, 11, both 'Dorman Long & Co., in England, and Redpath Brown & Co., in Scotland, opened constructional departments in 1886, and both firms made the first British steel joist compound girders in that year'.

83 Bates 1984, 8, 15.

84 Lawrance Hurst, 'Floors and frames in buildings from c 1850' (unpublished typescript). The first known safe-load tables were published in Humber 1868; Phillips 1870 was almost certainly the first trade literature to incorporate such data. See Thorne 1990, 59.

85 Bodsworth 1997, 434.

86 Jackson 1997, 37.

87 Hurst 1999, 25, and note 53.

88 Carr and Taplin, assisted by Wright 1962, 224.

89 ICE membership records.

90 Jackson 1997, 37.

91 Dorman Long 1901, 24.

92 Walshaw and Behrendt 1950, ch. 5 (unpaginated).

93 Payne 1979, 70, 74.

94 Advertisements in Ryland's (1895) and Ryland's (1896) shows that by the middle of the decade the 'The Leeds Steel Works Limited' was manufacturing rolled-steel joists up to 16 by 6in., in weights of 60 to 70lbs per foot run. These were marketed through their agent, H J Skelton & Co., 12 Lime Street, London.

95 Scott 1929, 11.

96 Bates 1984, 15.

97 Turner 1996, 25–30.

98 Kelly 1878, 112, 'Edinburgh advertisements'.

99 'The Works of Sir William Arrol and Co., Limited', The Engineer 89 (18 May 1900), 506.

100 BN 41 (8 July 1881), 41.

101 Kelly 1882, 30, 'London advertisements'.

102 BN 54 (13 January 1888), 95.

103 BJ&AR 2 (Contractors' Supplement, 29 October 1895), vi. (advertisement, after p 192).

104 The Builder 60 (11 April 1891), 291; Whitaker 1906, 362.

105 BN 80 (7 June 1901), 760.

106 Walter Jones 1893.

107 Alexander Penney & Co. 1892.

108 A&CR (Supplement), 25 November 1892, 17.

109 Matheson and Grant 1893, 70.

110 Quoted in PICE 69 (1982), 31 and Gale 1969.

111 Horner 1897, 449.

112 Notwithstanding their very early employment in a Cistercian monastery at Alcobaça, near Lisbon, in 1752, cast iron columns came gradually into use in the last three decades of the 18th century, particularly in supporting church galleries, and, in the 1790s, textile mill floors. Among the earliest documented uses in London was in support of a mezzanine floor at Lackington's 'Temple of Muses' bookshop, Finsbury Square, which opened in 1794. See Addis 2010, 34–7; Bannister 1950, 235, note 38.

113 Cast-iron columns, encased in stone, were used as late as 1929 for the façade of Martins Bank, Liverpool. See Yeomans 1997, 82.

114 See Hamilton 1940–1, 148–9.

115 Skempton 1959, 73.

116 Rabun 2000, 172.

117 Ibid.

118 These included Randolph and Elder Engineering Works (1858–60), Fairfield Shipbuilding Yard and Engine Works (c 1869), and Linthouse Engine Works (completed 1872). See Hay and Stell 1986, 118–27. London examples that have come to light include Farmiloe's warehouse, Clerkenwell (see pp 286–7) and the former Albion Works of Henry Pontifex & Sons, 32 York Way, Kings Cross (1867, demolished); see Richardson 1999.

119 Clarke 2001a.

120 Reade 1889, 299. Still, utility rather than structural efficiency ensured that cast-iron stanchions in many forms, symmetrical and asymmetrical, remained popular: Hasluck noted 'the exigencies of the building will probably settle the point as to whether columns or stanchions are to be employed … The stanchion is much better adapted to building in brickwork, and is usually adopted when used in that position', Hasluck 1906, 11.

121 Burnham 1959, 223.

122 Many high buildings erected in the1880s and 1890s, especially in New York, employed Phoenix columns of iron or (from 1889) open-hearth steel, including the enlarged Equitable Building, Madison Square Garden, and the World, Union Trust, Commercial Cable, and Dun buildings in the same city. See Landau and Condit 1996, 57; Wermiel 2009, 22–4.

123 Foster Milliken of the New York contracting firm Milliken Brothers; see Wermiel 2009, 23–4.

124 Norris 1971.

125 For example, Middleton wrote 'The aim of the designer should be, so far as it is consistent with rigidity, to place all the metal as far as possible from the central axis, and to use as few rows of rivets as possible', Middleton 1906, 97. The same theory was held by contemporary

American designers also; see Rabun 2000, 204.

126 In today's engineering parlance, 'they provided the highest possible moment of inertia about any central axis for their weight'. Friedman 1995, 75.

127 Wermiel 2009, 22.

128 Flanged-steel segments are illustrated in Gasley 1887 – Dorman Long's first section book.

129 'Details of structural iron and steel. 19.- Wrought-iron and steel columns and connexions', The Builder 78 (19 May 1900), 496.

130 William Lindsay reported that 'his firm had tested mild steel very fully, and that the experience thus gained had led them to make the material used in their flooring as nearly as possible approximating the best Low Moor iron, namely, having an ultimate tensile strain of from 26 to 28 tons per square inch, and giving an elongation of about 24 per cent. So tough was the material that a hole might be punched within one sixteenth of an inch of the edge of a bar without bursting or breaking the steel'. Lindsay in discussion of Slater 1887a, 742; see also Horner 1897, 449, 1898, 482.

131 See Burnham 1959, 224–5.

132 'Lindsay's steel decking', The Engineer 64 (7 October 1887), 289–90, and Horner 1898, 482–3. According to Horner, trough flooring was first patented by William Barlow in 1856, and in 1865 'Mr. Crompton patented the Lindsay section, afterwards patented by Lindsay in 1882'. The respective patents for the Lindsay section are Crompton, T R, No. 1639, 1865 and Lindsay, W H, No. 2607, 1882.

133 Moreland possibly sourced their solid steel 'rounds' from manufacturers such as Dorman Long & Co., Limited, which by 1900 was rolling rounds up to 8in in diameter in ordinary lengths up to 24ft, and a maximum length up to 60ft. See Dorman Long 1901 and 7.

134 Moreland & Son Ltd c 1905, 4–5.

135 Solid-steel columns were still referred to in BS 449–2:1969, Specification for the use of structural steel in building (BSI, 31 October 1969).

136 Moreland & Son Ltd c 1905, 4.

137 Bates 1984, 69–71.

138 'Standardised rolled section', The Builder 84 (14 March 1903), 267.

139 'Standard sections in iron and steel structure', The Builder 81 (9 November 1901), 415. According to Burn 1961, 199, this included 122 channel and angle sections, all made in the 'ordinary run of business'. The number of joist sections is not recorded, but presumably accounted for a considerable proportion of the remaining 48 sections.

140 'Standard sections in iron and steel structure', *The Builder* **81** (9 November 1901), 415. Germany, Belgium and Austria led the way in standardisation. For example, in 1883, the United Societies of German Architects and Engineers formulated a complete system of rolled-iron standard sections for shipbuilding, and by 1892 Austrian engineers had officially drawn up a list of rolled-iron and steel sections. Burn 1961, 199.

141 Bates 1984, 69–71.

142 Carr and Taplin 1962, 227.

143 *Iron and Coal Trades Review*, 24 December 1885, 889, quoted in Burn 1961, 199.

144 Skelton 1891, preface (unpaginated).

145 Woodward 1972, 8.

146 Ibid, 8. For a detailed history of the Engineering Standards Committee, *see* McWilliam 2001; Barry 1908; 'The Engineering Standards Committee', *The Builder* **85** (11 July 1903), 31.

147 BS1: 1903 *British Standard Sections* (London, February 1903).

148 BS4: 1903 Geometrical Properties of Standard Beams (London, November 1903); BS6: 104 Properties of British Standard Sections (London, July 1904).

149 *The Builder* **86** (9 January 1904), 34 (review of 'British Standard Sections Issued by the Engineering Standards Committee-Beams. London: Crosby Lockwood and Son, 1903').

150 Clarke 1999.

151 Carr and Taplin 1962, 228. These figures, based on those provided by J S Jeans for an unofficial Government advisory body called the 'Tariff Commission' in 1904, were less than those optimistically forecast the previous year. For example, the *BJ&AR* **17** (11 March 1903), 44, noted 'within a twelve-month it is believed British standard sections will be in general use throughout the country, effecting a yearly saving of millions of pounds – in structural steel alone these sections are calculated to effect a saving of £750,000 a year'. Still, by contemporary prices, this £200,000 was a substantial figure.

152 McWilliam 2001, *17–20, 29–31*.

153 Woodward 1972, 10.

3 The Continental dimension: iron and steel girder imports

1 'Belgian Competition in Iron and Steel, *A&CR* **52** (12 October 1894), 8.

2 'The Iron Industry of Great Britain', Letter to *The Times*, 30 May 1867, 7.

3 Hewitt 1868.

4 Samson Jordan, 'Revue de l'Industrie du Fer en 1867', quoted in Burn 1961, 3.

5 Guillou 1972, 16.

6 Vaizey 1974, 6.

7 Burn 1961, 5. Some two decades later, Britain was importing 'building iron (girders, beams, and pillars)' to the value of £387,005 in 1888, £520,212 in 1889 and £512,000 in 1890. *The Builder* **60** (17 January 1891), 43, citing Board of Trade Returns.

8 Burn 1961, 4. The ironwork contractor in question was John Simeon Bergheim (1844–1912), a petroleum engineer turned structural engineer, who worked with Waterhouse on many of his important projects of the 1870s and 1880s. For a list of these, *see* Appendix I of Cunningham 1992.

9 'Continental Iron Works Supplying English Markets', *Littell's Living Age* **97** (23 May 1868), 500. For details of the beams, *see* LMA HO1/SRA/A/124/012–016.

10 *See* note 28 Ch 9. W & T Phillips seemingly sourced all their structural ironwork from Belgium in this period; a lengthy list of buildings erected in the period 1867–70 that made use of their ironwork appears in Phillips, W & T 1870, 35–9.

11 Select Committee on Scientific Instruction 1868, XV, Q. 907 and Q. 1008; London Trades Council 1867, XXXIX, Q. 10,698 and Q. 14,606. Certainly, in the case of Sheffield, iron of Belgian import continued through the 1870s unabated: in 1874 the building contractor overseeing an 'immense drapery establishment' for T B and W Cockayne privileged Belgian rather than English ironwork on economic grounds. *See* 'Economy of using Belgian iron in building operations', *BN* **27** (16 October 1874), 475.

12 Letter written by 'Septimus Ledward' in *The Times*, 17 May 1867, 10.

13 *See*, for example, *The Builder* **49** (17 October 1885), 556, quoting a letter to the editor of the *Liverpool Courier*.

14 The first lines of this government-financed system, developed with the aim of giving manufacturers access to international markets, were the east–west and north–south routes crossing at Mechelen, which opened in 1835. 'Belgium', in Trinder 1992, 70.

15 The Cockerills introduced coke for smelting iron ore, puddling furnaces, hot-blast, steam-driven rolling mills and other innovations from 1823. 'Belgium', in Trinder 1992, 68.

16 Deby 1873, 396–7.

17 Ibid; Hubert 1916b, 56.

18 Fremdling 1991, 529.

19 Dawnay 1901, 387.

20 Carr and Taplin, assisted by Wright A E G 1962, 40.

21 Guillou 1972, 16.

22 *The Builder* **24** (1 December 1866), 894.

23 *JISI*, 1874, 178.

24 Deby 1876, 561.

25 'England's Iron Trade with Belgium', *JISI*, 1879, 553–4.

26 'Specification of Ironwork', *BN* **41** (21 October 1881), 525; Wolters 1886–7, 407.

27 Guillou 1972, 16.

28 'Belgian versus English Iron', *Iron,* 5 March 1886, 211.

29 'The Cleveland Iron Trade Foreman's Association', *JISI*, 1878, 279; 'Belgian versus English Iron', *Iron*, 5 March 1886, 211.

30 Misa 1995, 54.

31 Burn 1961, 360, citing Belgian metallurgist M Deby in *JISI*, 1878, 2, 149.

32 As early as 1867 *The Colliery Guardian* noted that the Belgian ironworks of 'Mesrs. Gillieaux and Co. exhibits a speciality in its Universal Mill, which is of a character quite unique, and possesses qualities that are at present little understood or appreciated in England'. 'The Belgian Ironworks', *The Colliery Guardian*, 5 January 1867, 12–13.

33 Wolters 406–7; 'Belgian Competition in Iron and Steel', *A&CR* **52** (12 October 1894), 6–8.

34 Bashforth 1962, 128.

35 Carr and Taplin, assisted by Wright A E G 1962, 58.

36 Royal Commission on Depression of Trade and Industry, 1886, C. 4715; Q. 2152.

37 *The Builder* **24** (3 March 1866), 147.

38 The Cleveland Iron Trade Foreman's Association', *JISI*, 1878, 279.

39 Wolters 1886–7, 403.

40 Ibid, 412; 'Details of Structural Iron and Steel: 4 – Manufacture of Wrought Iron', *The Builder* **78** (27 January 1900), 90.

41 Matheson 1882, 3.

42 Misa 1995, 70.

43 For example, 'Belgian Ironworks for New Buildings. Caution to Architects', *The Builder* **33** (31 July 1875), 693.

44 Walmisley 1900, 486.

45 'Rolled Iron Joists for Building Purposes', *BN* **51** (15 October 1886), 562.

46 *The Builder*, in reporting recent tests on rolled iron beams, noted that 'The rolled beams made by the Butterley Company give [values according to Fairbairn's formula] 57 to 88 as constants; the tests we have seen applied to the Belgian beams give higher results; but the difference probably depends on the fact that the flanges were smaller, and the distribution of the metal such as to enable a larger amount of work to be done in proportion to the sectional areas.' *The Builder* **24** (3 March 1866), 148. In 1894, following a visit to Marcinelle & Couillet Company's iron and steel works at Couillet, a member of the Iron and Steel Institute reported, 'The girders when finished were beautifully straight, and the material itself had all the appearance of good quality'. 'Belgian Competition in Iron and Steel', *A&CR* **52** (12 October 1894), 7.

47 Letter from 'Estimate' entitled 'Belgian and English iron in building operations', *BN* **27** (23 October 1874), 508.

48 'Rolled Iron Joists for Building Purposes', *BN* **51** (15 October 1886), 562.

49 Dawnay 1901, 387.

50 Pendred 1883, 87.

51 Skelton 1891, 289.

52 Scott 1928, 55.

53 'Belgian Ironwork for New Buildings – Caution to Architects', *The Builder* **33** (31 July 1875), 693.

54 Reade 1889, 300–1.

55 'The Belgian Competition in Rolled Girders', *Iron*, 29 January 1886, 97; also reprinted in *The Builder* **50** (6 February 1886), 256.

56 Skelton 1891, 297.

57 Bessemer 1905, 341–2.

58 Mende 2006, 2159.

59 Ibid, 2166.

60 *See* Cocroft *et al* 2005, 223.

61 For example, see advertisements in *Ryland's Iron, Steel, and Allied Trades' Directory* and *Rylands List of Merchant Exporters* for these decades. The 1884 edition of the former (Ryland 1884) and the 1886 edition of the latter (Ryland 1886) list Otto Gössell, Jun., Merchant and Agent, as the supplier of rolled-steel girders up to 16in. deep, and rolled-iron girders up to 22in. deep. By the 1901 edition, only steel sections, up to 21¾in. deep, were being offered from the 10,000 tons of stock, and by the 1908 edition, 24in. deep sections could be supplied.

62 The perception that this building pioneered the use of steel in architectural building is still widespread; for example, Curl has stated that it marked 'apparently the first major use of structural steel in a building (as opposed to bridge or other work of pure engineering)', Curl 2006, 398.

63 *See* Cunningham 1992, Appendix 1: List of Works. Although uncorroborated by contemporary descriptions, according to this source (p 226) Waterhouse's first building to incorporate structural steelwork was the New University Club (1865–9).

64 *See* Wolters 1886–7. Referring to contemporary practice in Charleroi and Liège districts, Wolters, mentions only puddling furnaces, suggesting wrought iron.

65 *BN* **48** (30 January 1885), 166.

66 'Sections of Rolled Iron and Steel', *The Builder* **41** (8 July 1881), 41: 'Messrs. Macnaught, Robertson, and Co., of Bankend, Southwark, S.E., have published a useful book of sections of rolled iron and steel joists and patent compound and other girders … Heavy rolled sections, 20in. deep by 10in. wide, are kept as well light sections …'.

67 'Belgian Competition in Iron and Steel', *A&CR* **52** (12 October 1894), 8. *See also* note 1.

68 Bussell 1997, 29; Mende 2006, 2166–7; Skelton & Co. 1908, 226.

69 'Reinforced Concrete Systems. VIII. – Broad-Flange Beams', *BJ&AE* **24** (C&S Supplement) (5 December 1906), 96–8, 97.

70 Ibid.

71 H J Skelton & Co., steel merchants and manufacturers, have been attributed as 'the pioneers in regard to the use of broad-flange beams in this country'. Ibid. *See also* 'Broad-Flange Beam', *BN* **86** (29 April 1904), 617.

4 The London building regulations

1 Middleton 1898, 780.

2 Warwick for example, was destroyed by fire in 1694, and rebuilt under an Act of Parliament modelled on that of the 1667 Act for London. Ley 2000, 2.

3 4 George III, c. 78; Fires Prevention (Metropolis) Act 1774.

4 Ibid.

5 Ley 2000, 2.

6 Castleman 1938, 12. Although Castleman interprets the mention of steel for crane-houses as a peculiar, and early constructional use of the material, the steel (and iron) referred to would be for wear-resistant components, and for areas of concentrated bearing stress, such as footstep bearings. The lawyers who drafted the legislation presumably regarded such fixtures as intrinsic to the building type, and therefore needed to be covered by the Act. I am grateful to Malcolm Tucker for arriving at this interpretation.

7 7 and 8 Victoria, c. 84; The Metropolitan Building Act 1844.

8 Where internal stairs were built 'of stone or other incombustible substance', they were to be supported by 'fire-proof constructions'; connecting landings and passages were also to be 'fire-proof' 7 and 8 Victoria, c. 84 Schedule (C) Part VI – Rule Concerning Fire-proof Accesses and Stairs to Buildings of the First and Third Classes.

9 7 and 8 Victoria, c. 84 Schedule (D) Part II – Construction of External Walls.

10 House of Commons Select Committee on the Metropolitan Buildings and Management Bill 1874: Q. 91 'Yes, there was very much iron building inside. It was constructed under the approval of the official referees under the Act of 1844, and I believe that the internal fittings consisted mostly of iron'. Walter Newall under examination by Joseph Bazalgette.

11 Magistrates had hitherto appointed District Surveyors, first introduced with the 1774 Act. *See* Guillery 2004, 283–4.

12 *L&BN* 1855a, 1.

13 By bringing outlying parishes such as Hampstead, Hornsey, Tottenham. Streatham, Tooting and Wandsworth under its jurisdiction, the 1844 Act reflected the outward spread of the metropolitan area. That some parishes still retained their rural character yet were nevertheless administered by the Metropolitan Board was a source of grievance to some: 'a gentleman having a house … cannot make the slightest alteration in it … except according to the taste and views of the district surveyor', *The Times*, 17 November 1845, 8.

14 18 and 19 Victoria, c. 122; The Metropolitan Building Act 1855.

15 *L&BN* 1855b.

16 For example 'After 1855, and still more after 1870, building regulations in London actively discouraged the use of exposed-iron construction. Fire hazards, and the dangers of oxidization, inhibited easy acceptance of the new material', Crook 1987, 124; *see also* Hitchcock 1953, 354.

17 *See* 18 and 19 Victoria, c. 122; The Metropolitan Building Act 1855, sections 56 and 112. For those buildings exempt from the Act, including 'all buildings not exceeding in extent 216,000 cubic feet, and not being public buildings, and distant at least 30 feet from the nearest street or alley, whether public or private, and at least 60 feet from the nearest buildings and from the ground of an adjoining owner', see section 6. In its commentary on this aspect of the Act, *L&BN* 1855a, 2, noted 'With the progress of iron and other new structures, a regular and ascertained code of regulations will doubtless be established in time; so that architects and builders shall not be exposed to arbitrary treatment from persons in some cases absolutely opposed to the specific novelty or improvement'.

18 Harper 1985, xviii.

19 18 and 19 Victoria, c. 122; The Metropolitan Building Act 1855, sec 15 (2); Dicksee 1906, 71.

20 Notwithstanding a further clause to this rule in 1894 to permit thermal expansion.

21 One exceptional building, a bookbindery with frontages 'formed almost wholly of cast iron and glass' to New Church Court and White Hart Street near the Strand, was permitted 'under the special sanction of the Metropolitan Board of Works'. Erected in 1879 for Diprose Brothers to designs by J & S F Clarkson, this long-demolished building made remarkable use of these materials to a degree comparable

to the cast iron fronts of Glasgow and New York. *See The Builder* **37** (24 May 1879), 557–8 and 568, 570 (illustrations).

22 Wermiel 2000, 79.

23 Metropolitan Buildings and Management Bill 1874, Q. 1451.

24 Metropolitan Buildings and Management Bill 1874, Q. 93.

25 57 and 58 Victoria, c. ccxiii, London Building Act, 1894. According to Knowles and Pitt 1972, 90, the 1894 Act was 'the most comprehensive Act on streets and buildings since the passing of the rebuilding Act of 1667'. However, this undervalues the significance of the 1774 Act, which was 'comprehensively consolidating', enforceable by district surveyors, and which 'identified seven "rates or classes" of building and laid out standards for each, including minimum wall thicknesses that varied in accordance with size', Guillery 2004, 283.

26 Knowles and Pitt 1972, 93.

27 Harper 1985, xxvi.

28 Twelvetrees 1906, 17.

29 '… if the increase in height were very considerable it is very doubtful whether the application would be granted. It must be remembered that the tendency has been towards a reduction of the height of London buildings', Cubitt 1911, 93.

30 Ibid, 23.

31 Twelvetrees noted that the collapse of the floors during construction of a large (unspecified) office building was due to the insecure support of the main girders, noting 'If these members had been incorporated into a properly-designed framework, the disaster could never have occurred', Twelvetrees 1906, 23.

32 Bylander 1913, 57.

33 Theoretically, a legal loophole to this stipulation existed, one that was pointed out by Robert Langton Cole (1858–1928) in response to an earlier paper (Gibson 1898), and one which is neatly summarised by Locker 1980, 47: 'Since the London Building Act measured the height of external walls built upon bressummers from the top of the bressummer, Cole speculated that a building with a bressummer at each floor level and a masonry enclosure could legally be considered to be enclosed with a series of one storey walls, no matter how high the building. The effect of this interpretation would have resulted in only nine inch walls for most buildings'. In effect, the toilet block of the Mount Pleasant Postal Sorting Office (1900) conformed to this interpretation (*see* note 50), although in this case constraint was immaterial, being a (Government) building exempt from the constructional provisions of the act. Thirty years before this, John Whichcord had

employed the same logic, and seemingly got away with it, for an office building on Water Lane, City of London. He brought the interior framing out to at least one wall, alternating cast-iron and stone columns supporting bressummers at each floor so that the entire wall was only that of a single-storey window. Beyond this example, there is no evidence of thinner walls being sanctioned for multi-storey commercial buildings in London before Selfridges (1908–9), and while Cole was correct in stating 'there is nothing in the London Building Act to prevent its [skeleton framing] adoption', there was much to deter it, Langton Cole 1899, 239.

34 The 1855 Act stipulated that the total area of openings and recesses in external walls should not exceed one half of the wall area. The 1894 Act re-invoked this, but discounted the ground floor and basement from the calculation – a measure designed to enable them to be left entirely open, thus providing for shop-fronts. Under the earlier Act, buildings that exploited extensive ground-floor glazing had to have the size of the openings above proportionally reduced. A small number of architects were able to sidestep this fenestral restriction, reasoning that a bressummer was a foundation for the wall above, so that the ground floor did not count. For some mid-century office buildings, Edward I'Anson seems to have taken this logic further, the extensively glazed four-storey façade of his Colonial Chambers (1857) in effect contrived as a series of superimposed shop-fronts, Locker 1984, 128–9. Perhaps because I'Anson held the post of District Surveyor (for Clapham and Southern Battersea) from 1844 until his death in 1888, he was thoroughly acquainted with the constructional possibilities of the Acts, if not with the district surveyors for the City. For I'Anson's obituary, *see BN* **54** (3 February 1888), 177.

35 Less stringent building regulations in provincial towns and cities such as Manchester enabled the maximal glazing advantages of skeleton framing to be realised, but not until the mid-1900s, when this form of construction gained currency.

36 Twelvetrees 1906, 17.

37 *See*, for example, 'London Streets and Buildings Bill', *The Builder* **65** (23 December 1893), 461–2; 'London Streets and Buildings Bill: revised form', *The Builder* **67** (21 July 1894), 35–6; 'The London Building Act, 1894', *A&CR* **52** (28 September 1894), 194–7; 'The New London Building Act II', *The Builder* **67** (6 October 1894), 233–5; 'Public buildings under the new London Act', *BN* **68** (28 June 1895), 901.

38 'Steel and iron frame construction in the United States', *The Builder* **63** (15 October 1892), 295. Two years later, in its review of William H Birkmire's *Skeleton Construction in Buildings* (New York, 1893), the same journal mused 'Whether this method of building will ever be tolerated or practised by English architects it is not easy to prophesy; in London at all events the legal limitations as to height will take away its chief or only advantage. If we do come to that, however … we counsel [the architects] … to be content with the honest steel building, and not make the matter worse by erecting a sham masonry building outside it', *The Builder* **66** (21 April 1894), 310.

39 Twelvetrees 1906, 17. As a structural engineer excited by the possibilities of skeleton construction, Twelvetrees was hardly a representative and non-partisan voice in attributing the British delay to architects. Nevertheless, there is little evidence that in London architects favourably considered skeleton framing before 1904.

40 For an account of these and other provisions, *see* 'Suggestions for the regulation of skeleton buildings', *BJ&AR* **19** (17 February 1904), 83–4.

41 *C&CE* **4** (September 1909), 232.

42 5 Edward 7, c. 209, London Building Acts (Amendment) Act 1905.

43 The London County Council (General) Powers Act, 1909, Part IV.

44 'Skeleton construction and the Amended London Building Act', *C&CE* (January 1910), 8.

45 See Landau and Condit 1996, 182. According to William J Fryer, 'Chicago did not have many restrictions and requirements relating to buildings, so an architect in that city could do much that an architect in New York would not be allowed to do', Fryer 1898, 466.

46 Thrupp 1907, 257.

47 Section 82 of the 1894 Act covered 'Special and temporary buildings and wooden structures', specifying that 'Where a builder is desirous of erecting an iron building or structure, or any other building or structure, to which the general provisions of Part VI. of this Act are inapplicable, [or, in the opinion of the Council, inappropriate, having regard to the special purpose for which the building or structure is designed and actually used], he shall make an application to the Council, accompanied by a plan of the proposed building, with such particulars as to the construction thereof as may be required by the Council'. Re-enacting the same provision in the 1855 Act (the words in parenthesis being added), it set a new, enlarged limit of 250,000 cubic feet on such buildings.

48 23 and 24 Victoria, c. 52; Metropolitan Building Act (Amendment) 1860, sec 2.

49 57 and 58 Victoria, c. ccxiii., London Building Act 1894, sec 75.

50 Declaring it 'the first example of reinforced concrete construction on an important scale undertaken in the metropolis', *The Builder* commended the 'enlightened policy displayed by H M Office of Works' in demonstrating 'the advantages which can be secured by reinforced concrete construction when unfettered by regulations such as those contained in the London Building Act', *The Builder* 1909, 394. Thereafter, the Office of Works seems to have consistently favoured reinforced concrete over steel for framed buildings, at least until the 1920s. For example, the entire framework and majority of the walls and partitions of the enormous New Money Order Offices, Holloway (1909–10), was erected on the Coignet system, *see C&CE* **5** (October 1910), 766–79.

51 'The Admiralty offices', *BJ* **1** (12 February 1895), 11.

52 'The fire in Old Bailey', *BJ&AE* **28** (26 August 1908), 147–8.

53 Leverton, in discussion of Scott 1929, 99; *see also* Scott 1944; Survey of London 2008b, 460.

54 *The Times*, 12 July 1900, 11a, as quoted in Survey of London 2008b, 460.

55 Dorman Long & Co. Ltd supplied the steel; *see* Survey of London 2012, 181–3.

56 Comprising offices and official residence block incorporating triumphal arch, Admiralty Arch was constructed by John Mowlem & Co., with steelwork supplied and fabricated by Redpath, Brown & Co. and Dorman Long & Co., *MBR* **3**, 1912, 94–6.

57 Designed by H A Collins and A J Pitcher of the Office of Works. *See* Port 1995, 136, plate 140; 271. By this date the cost-conscious Office of Works was already wedded to reinforced concrete framing (*see* note 48), a state of affairs related to the relative expense of the two techniques, and perhaps, the unreliability of steelwork deliveries: 'Non-expenditure of £7000 on the Admiralty extension in 1908/09 was ascribed to want of steel, which 'could not be delivered ... It is a difficulty that very often occurs', Port 1995, 314, n176, quoting Parliamentary Papers 1909.

58 This phrasing is adapted from that coined by Gerald Larson, who describes 'the tyranny of France's masonry tradition during the 1850s'. *See* Larson 1987, 42.

59 *The Manchester Municipal Code: Being a Digest of the Local Acts of Parliament, Charters, Commissions, Orders, Bye-Laws, Regulations, and Public Instructions and Forms within the City of Manchester* Vol. II (Manchester, 1895), 619–72; Locker 1980, 45 (cites Liverpool Improvements Act, 1882 and Glasgow Building Regulations, 1900).

60 Twelvetrees 1906, 18.

61 Ibid.

62 'Hoyle's Warehouse, Manchester', *BJ&AR* **21** (11 January 1905), 16.

63 For a brief description of this building, *see* Bowley 1966, 12. Only the elevation survives from the original building.

5 Philosophical concerns with iron, steel and framed construction

1 Driver 1875, 166.

2 Professor Robert Kerr (1823–1904) in response to Stannus 1882, 101.

3 For the 'iron problem' *see* Muthesius 1970; Muthesius 1972, 197–202; Dixon and Muthesius 1985, 106–08; Curl 1990, 213–9.

4 Crook 1987, 111.

5 Ruskin 1849, 36 and 37.

6 Ibid, 36.

7 Dr Acland, *The Ecclesiologist* XVI (1855), 249, as quoted in Crook 1987, 120.

8 As quoted by O'Dwyer 1997, 261. The comment was made to a Slade lecture in 1870, and according to O'Dwyer, appears to refer to the museum.

9 *BN* **5** (8 April 1859), 338. The writer of this article did, however, credit the beauty of Skidmore's ironwork, forecasting 'he will do more than any other man in his generation to elevate and ennoble his art, and to claim for all metals their due place in our architectural economy'.

10 Crook 1987, 125.

11 A comment by Street in 1880, as quoted in Crook 1987, 125.

12 According to Street, concealed girders were the 'enemies of all true construction'. Street in discussion of Picton 1880a, 187.

13 Crook 1987, 97.

14 Thorne 2000, 136.

15 Crook 1987, 115.

16 Thorne 2000.

17 *See* Crook 1987, 115 and 288, n101 (referencing *The Builder* 1894, 91–2).

18 'Cast iron in buildings', *The Architect* **32** (8 November 1884), 291.

19 Ibid, 291–2.

20 'I myself once tried the somewhat contradictory plan of encasing the iron columns with a thick layer of plaster ... But I am not over sanguine about it, and the system is, at least, liable to the artistic objection, that it conceals the more valuable material of the two', Lewis 1865, 116.

21 Quoted in Crook 1987, 123.

22 Statham 1898, 108.

23 Jackson 1997, 18.

24 Crook 1987, 123.

25 Locker 1980, 263.

26 The influence of the Crystal Palace on structural design is a theme developed by Robert Thorne in Thorne 1987, 22–8, 2001, 173–85.

27 Street quoted in Crook 1987, 124.

28 Quoted in Locker 1980, 30.

29 Delissa Joseph, in response to Gibson 1898, 68: 'whether they could employ in London, with any advantage, a frame of skeleton construction in warehouse work; but he could see no advantage in it, because they would be bound to cover so much of the area of framework with a solid construction, in order to meet the requirement of the Building Act. ... they would have to case in with false material so as to come within legitimate ratio of open space to solids.'

30 Howard Colls, in discussion of Dennell 1906, 46.

31 Childs 1909a, 35.

6 Professional conflicts: architect–engineer dynamics

1 'The Spirit of Architecture', *CE&AJ* **5** (August 1842), 247.

2 'Co-operation of the Architect and Engineer', *BN* **86** (10 June 1904), 821.

3 Saint 2007, 486.

4 Watson 1989, 2–3, 26.

5 Woodley 1999, 16.

6 Robert Mylne's preface to *Reports of the late John Smeaton F.R.S.* (London, 1797), as quoted by Woodley 1999, 17.

7 Preface to *Reports of the late John Smeaton F.R.S.* 1 (London, 1797), as quoted by Thorne 1991, 54.

8 'Transactions of the Institution of Civil Engineers', *MM* **28** (1838), 189.

9 Reproduced, for example, in 2009 *Charter, By-Laws, and Regulations and List of Members of the Institution of Civil Engineers (Great Britain)* (IOC, 2009), 1812.

10 Woodley 1999, 18.

11 *See*, for example, de Haan 2003, 3–19; Hodson 1992; Ionides 1998.

12 'Nash, John (1752–1835)', in Skempton 2002, 474.

13 Quoted from *Encyclopaedia Britannica*, 9 edn, in Summerson 1935 *John Nash: Architect to King George IV*. London: George Allen & Unwin Ltd. Other names, including Tom Paine, John Nash, and John Rastrick I, have been connected with the design of the Sunderland Bridge, yet the credit largely belongs to Burdon, despite the fact he sought the views of others before finalising his design. *See* Cossons and Trinder 2002, 65–8; Rennison and James 2002; 'Burdon, Rowland (1756–1838)' in Skempton 2002, 102–3.

14 The former, described by a German visitor as 'high and audaciously vaulted', closed to traffic a month after erection following

fracture of the iron, the latter collapsed into the River Tees before it opened to traffic, the cause in both being insubstantial abutments. *See* Cossons and Trinder 2002, 69.

15 Quoted in Penfold 1980, 13.
16 Cossons and Trinder 2002, 82.
17 Ibid, 82; Ruddock 1979, 156–7.
18 Letter from James Elmes, Architect, London to Thomas Hope, Esq. Hereditary Governor and Director of the British Institution on the 'state of Architecture in England', *Annals of the Fine Arts* **3** (1819), 457.
19 Gwilt 1826, 61–2.
20 Skempton 1971.
21 Clarke 2006; Survey of London 2012, 353–60.
22 Crook (1965–66).
23 Hussey 1958, 73; Parissien 2008, 153.
24 Dinkel 1983, 54–5; Diestelkamp 1990, 19–20; Saint 2007, 92–3.
25 Clarke 2013, 148–9; 152–3.
26 This lesser-known building, destroyed by fire in 1830, must have incorporated appreciable quantities of cast, and possibly wrought iron: 'Large tanks of water, with connecting pipes to every part of the building tend to give it additional security against the danger of fire; while the use of iron, where such material can possibly and properly be placed, renders the want of such assistance unnecessary', *Monthly Magazine and British Register* **40**, 2 (1815), 529.
27 Peach 2004.
28 Tucker 1981.
29 Ibid. Charles Dupin, who described this huge, open-sided structure as 'le plus remarquable de tous par sa structure', attributed the design to Rennie, although William Jessop, junior, or someone else from Butterley Company, may have been responsible. *See* 'The West India Docks: the buildings: sheds', in Survey of London 1994a.
30 Douet and Lake, 1997, 18–20.
31 Saint 2007, 76.
32 'Bage, Charles Woolley (*c* 1752–1822)', in Skempton 2002, 28–9.
33 Addis 2007, 308.
34 'Rastrick, John Urpeth, FRS' (1780–1856) in Skempton 2002, 545.
35 Tredgold 1822. *See* Sutherland 1990, 26–8.
36 In the ensuing enquiry, Whitwell declared 'In apportioning the strength of materials in architectural construction, we never approach the limits of danger; I did not consider it, therefore, necessarily following that the roof would come down because the rules had been transgressed which I had laid down against appending weights to the roof', 'Destruction of the Brunswick Theatre', *The Standard*, 13 March 1828, 3; 'Wellington, James

(fl 1827–1928)' and 'Whitwell, Thomas Stedman (d 1840)' in Skempton 2002, 768, 778.
37 'Inquiry into the Fall of the Brunswick Theatre', *MM* **9** (15 March 1828), 124.
38 Addis 2007, 350, 353.
39 House of Commons 1831, 274.
40 Thorne 1991, 54.
41 *TRIBA* **1**, 1 (1836), x.
42 http://www.oxforddnb.com/public/themes/97/97265.html.
43 Gwilt 1842, 897.
44 Saint 2007, 113.
45 'Tite, Sir William (1798–1873)', in Cross-Rudkin and Chrimes 2008, 782.
46 'The next step was made by Mr. Farrell, a smith, who was employed under Sir R. Smirke at the Custom House, and he, regarding cast-iron as a very brittle and dangerous material, proposed the use of wrought-iron, for which he took out a patent … The joist adopted by him was of this form 1, and as he could not get rolled iron, he riveted the parts together. Mr. Tite stated, that he had used the system described, many years ago in a large public school, where a cheap fireproof floor was required, with complete success, and when he last saw this construction it was as perfect as when first erected. The space between the joists was in this case about 4 feet and filled in with 4 inch and 6 inch Yorkshire landing; which gave a good upper as well as under surface, the interstices being filled with Roman cement, and the ceiling formed of the same material. This was applied to a space of 300 feet long, and on an average 15 feet wide', *CE&AJ* **17** (1854), 93.
47 'Aitchison, George (1825–1910)', Cross-Rudkin and Chrimes 2008, 568.
48 *See* Thorne 1990, 7–14; 1987, 22–8.
49 Herbert 1999, 108.
50 Thorne 2000, 137.
51 *The Engineer* **62** (17 September 1886), 233.
52 Tanner 1883, 190.
53 Sachs and Woodrow 1896–8, **3**, 102–3 [vol 3].
54 'Meeson, Alfred' in Frederic Boase, F *Modern English Biography* **2**, 1692.
55 Ibid; 'Meeson, Alfred (1808–1885)', in Cross-Rudkin and Chrimes 2008, 531–2.
56 'Letter by Archibald Dawnay to the RIBA, resigning as an ARIBA', 6 August 1872. RIBA Archive, LC/10/4/17.
57 Saint 1976, 114.
58 'Architects and Engineers', *The Builder* **27** (16 October 1869), 819.
59 Locker 1984, 139. This practice was undoubtedly not confined to office buildings, for in 1900 *The Builder* could state: 'Cases are not altogether unknown where buildings have first been designed, and plans have then been issued to various manufacturers of special types of flooring,

so that each might prepare an estimate and specification of the floors recommended. It is to be hoped that practice such as this is rare, for all parts of a building cannot be in complete harmony with one another if important features are designed by two persons, neither of whom understands the work of the other', 'Details of structural iron and steel. 25. – Floor-framing', *The Builder* **78** (30 June 1900), 649.
60 Cunnington, a prominent authority on fire-resisting flooring and constructional steelwork, was employed by William, Lindsay & Co. (established by William Henry Lindsay in 1879), from the late 1880s until the early 1900s. He worked at the firm's offices at 23 Queen Anne's gate, not its Paddington iron works, although his job title remains obscure.
61 *See* Clarke 2002, 58–61.
62 Childs 1898, 236.
63 Cunnington 1898b, 349.
64 Caws 1901, 134.
65 For example, *BJ&AR* **22** (8 November 1905), xv (advertisements at rear).
66 Caws 1901, 134.
67 *BJ&AR* **14** (9 October 1901), 139.
68 Briggs 1907, 226.
69 Horsfield-Nixon 1908, quoted in Clarke 2001b.
70 Cocking 1915, 679. Usually referred to as the 'Steel Frame Act', it also became known as 'the Engineers Act'.
71. Cocking 1915, 679.

7 American influence

1 Wood 1898b, 255.
2 E Kilburn Smith writing in the *Engineering Record*, as reported in 'An English engineer on skyscrapers', *A&CR* **72** (14 October 1904), 255.
3 For example, Gerald Larson notes that James Bogardus 'had no prior experience in construction but appears to have become familiar with iron construction during a trip to England and the continent in the period 1836 to 1840, when the English were developing prefabricated iron building systems not unlike those that Bogardus eventually patented in May 1850', Larson 1987, 40. In the following decade, architect Alexander Parris cited the iron roofs of English public buildings as exemplars in his own designs for 'fireproof' buildings. *See* Wermiel 2000, 63.
4 For example, 'Iron buildings in New York', *The Builder* **27** (3 July 1869), 529–30. This article noted the utility of iron in securing light for commercial buildings: 'An airy edifice of iron may be safely substituted for the cumbrous structures of other substances, and ample strength secured, without the exclusion of daylight. Iron, in

this respect, presents peculiar fitness ...
This material – emphatically an *American*
building material – has peculiarities of its
own, and will preserve its own
individuality.'

5   For example, Gale 1883.

6   Gass 1886b, 145.

7   Slater 1886.

8   Slater 1887a, 703.

9   *See*, for example, Cox 1891, 384–5; 'New
business buildings of Chicago', *The Builder*
63 (9 July 1892), 23–5; 'Steel and iron
frame construction in the United States',
The Builder **63** (15 October 1892), 295–6;
'Metallic construction in America', *A&CR*
72 (30 November 1894), 340–1.

10 *The Builder* **63** (17 September 1892), 220.

11 *The Builder* **63** (9 July 1892), 23–5.

12 'Skeleton construction in buildings', *The
Builder* **66** (21 April 1894), 310.

13 *See*, for example, 'Iron and steel for
building structures', *The Iron and Coal
Trades Review*, 13 December 1895, 746–7;
'The construction of an American
Building', *A&CR* **53** (Supplement, 15
February 1895), 6–8; Black 1896, 277–9;
Shankland 1898, 735–6.

14 Wood 1898b, 255.

15 Ibid, 256.

16 Ibid, 254.

17 Kelly 1898, 20 ['London Advertisements'
section]. Two years later, the same
company described itself as 'Specialists in
Steel Skeleton Construction on the
American System'. *See* advertisement in
the back of Twelvetrees 1900.

18 'Reinforced concrete systems. VI.- Homan
& Rodgers' System', *BJ&AE* **24** (C&S
Supplement, 10 October 1906), 76.

19 According to his nomination papers at the
Institution of Civil Engineers, the London-
born, American-trained Childs was at the
forefront of structural steel design. In
1889–90 Childs designed the first steel
head-frame and tipple house for coal mines
erected in America, and in 1890–91 he
worked as assistant engineer under Gustav
Lindenthal on the design and erection of
the Monongahela Bridge. From 1891–4,
before returning to London, he was Second
Engineer at the head office of the Carnegie
Steel Company, Pittsburgh, where he
designed the steel frames for 'numerous'
high-rise buildings in the Eastern States,
including the headquarter offices of the
Carnegie Steel Company in Pittsburgh
(1893–5; Longfellow, Alden & Harlow,
architects; *dem*. 1952; Fig 7.2) and the
Mutual Reserve Building, New York (1894,
William H Hume & Son, architects). In
England, Childs was elected as associate
member on 4 December 1906, aged 44; he
retired in February 1915. Institution of Civil
Engineers membership records; *see also*
Child's discussion of Purdy 1909, 209–212.

20 Childs 1898.

21 'American and English steel construction',
BN **74** (27 May 1898), 735–6.

22 'British Westinghouse works', *American
Manufacturer and Iron World*, **69**, 9 (29
August), 1141. Formed in the mid-1890s,
engineering contractors Mayoh and Haley –
who would later erect the framework of
the Lots Road Power Station, and establish
the Fulham Steel Works Co., Ltd at
Townmead Road, Fulham – had a track
record in pier construction and
reconstruction, with projects at Blackpool,
Morecambe, the Mumbles, Penarth, and
Beaumaris. I am grateful to Anne Mayoh
and Kevin Haley for this information.

23 James Stewart & Company 1909, 91.

24 Jones 1985, 210; Bowen 2010, 93.
Pneumatic riveting was not seemingly
welcomed by all of Dorman Long's
workmen: 'According to Sloan this was
greatly resisted by the on-site workers, so
they [James Stewart & Co.] telegraphed
shipyards (where the approach was
pioneered) in Scotland and Ireland for
riveters, who arrived in plentiful
numbers'. Bowen 2010, 92, citing Sloan,
W G, *Journal of Society of Western
Engineers* **7** (1902), 384–401.

25 'English and American methods of
building', *BJ&AR* **18** (9 December 1903),
251 [article runs 251–54]; also reported in
The Builder (5 December 1903), 580–2.
See also 'The Westinghouse works',
BJ&AR, **17** (29 April 1903), 156.

26 Dummelow 1949, 3–7.

27 James Stewart & Company 1909, 91.

28 Bowen 2010, 93.

29 'New Midland Hotel, Manchester', *BJ&AR*
17 (20 May 1903), 203–5; *The Builder* **84**
(24 January 1903), 95.

30 James Stewart & Company 1944, 91.

31 Ibid, 8. Mullgardt was involved with many
of these: 'Among the architectural works
of Mr. Mullgardt in Great Britain were the
designs for electric power stations for the
British Westinghouse Company, Heysham
Harbour and at Neasden, for the
Metropolitan Underground Railway of
London. He also designed a large factory
for the British Consolidated Pneumatic
Tool Company at Fraserburg, Scotland,
and two electric power stations in the
Clyde Valley, Scotland'. *The Architect &
Engineer* **35** (1913), 47. On chronological
grounds, his input into the Manchester
Westinghouse works seems doubtful, but
not implausible.

32 Heathcote 1903, 251. Heathcote conceded
'It does us no harm to be stirred up and
made to examine our methods of work.'
Some of those in attendance, including
Henry Holloway and H H Bartlett, of
Holloway Brothers, Ltd, had recently visited
America 'to study methods on the spot'.

33 'A Manchester warehouse', *BJ&AR* **19**
(30 March 1904), 153.

34 'Hoyle's Warehouse, Manchester', *BJ&AR*
21 (11 January 1905), 16.

35 'We arranged that the steelwork should be
erected by the steel engineering contractor,
so that this part should not be dependent
upon any one. We had arranged with a
mill to commence rolling for us within
twenty-four hours of receiving our
instructions, and also with a foundry to
work in three shifts of eight hours each to
get the cast work out ... From the date of
the contract to the time the whole of the
iron columns, beams and floor joists for
seven floors, inclusive of ordering, making
and erecting, was only seven weeks,
showing what can be done if one sets to
work the right way', Heathcote 1903, 265.

36 'American Skyscrapers go round the World',
The New York Times, 29 June 1902, 28.

37 'The information in regard to the forward
state of the scheme must be rather
premature, at all events, since we found
on inquiry that the architect's department
of the London County Council knew
nothing whatever of it, and such a
building could certainly not be erected in
London without a special application for
suspension of by-laws as to height, &c',
'A proposed American building in London',
The Builder **81** (23 November 1901), 453.

38 'It must have seemed obvious to Belcher
and Joass that a new style was necessary
for a stone skin hung on a steel frame, and
the verticality probably reflects their
reaction to the new construction method,
as well as to the resulting vastly higher
American buildings they had certainly
seen illustrated. We must also remember
that Belcher made a trip to New York and
California in 1899, and from what we
know of his character he would be
fascinated by the new buildings and
techniques he saw there. An examination
of the English architectural journals from
1900 on gives no other significant clue
about this verticality, nor about Joass's
sources for his detailing ... It appears likely
that the tension and vertical stretch of
these designs, especially of Mappin House,
seemed to them the best expression of a
skin of stone and glass', Service 1975, 321.

39 'Men who build. No. 65.- John James
Burnet, A.R.S.A.', *BJ&AR* **14** (9 October
1901), 138.

40 Whyte 1999, 7.

41 Ibid.

42 'First steel-frame building in Edinburgh',
BJ&AE **25** (C&S Supplement, 2 January
1907), 3. A contemporary description of
Forsyth's, including a photograph of the
partially clad framework. The contractors for
the steelwork and concrete were Fergusson,
Allen & Co., of Glasgow and London.

43 Scott 1928, 55–7. According to Scott, it was the engineer George L Allen who, 'in consultation, suggested the steel frame as a means of saving time, and his suggestion was carried out with advantage', Scott 1929, 102.

8 The evolution of the fully framed building

1 Fryer 1898, 466.
2 Condit 1964, 79.
3 The building most commonly cited as the first skyscraper to be built on a skeleton framework composed entirely of steel is the second Rand McNally Building in Chicago (1889–90, architects Burnham and Root, engineers Wade and Purdy). Of this building, Condit 1982, 126–7, wrote 'Here the frame of the high office building was wholly freed from its masonry adjuncts and built entirely of steel … Once established in Chicago, the fully developed steel frame quickly became standard'.
4 For details of these and other pioneer skeleton buildings, see Landau and Condit 1996, 157–226.
5 See Wermiel 2000, 147–50; 2009, 21–2.
6 See Friedman 2006.
7 The enormous publicity given to The Ritz Hotel – largely as a result of the contemporary and posthumous writings of its engineer, Sven Bylander – helped sustain this view through most of the 20th century. For a more recent account, see Lawrence 1990.
8 For example, Jackson has written, 'The development of the steel framed building in this country was not, as some have suggested, reliant on the importation of American construction techniques or American challenges to the building regulations, though in time these had to be revised. The Ritz Hotel and Selfridges were but part, and not the first part, of a wide group of early steel framed buildings, all of which were symbols of new techniques and ideas. The skills and experience and expertise necessary already existed in Britain', Jackson 1998, 37. Broadly correct, this view arguably discounts the (ultimately unknowable) experience gained by British designers and fabricators of the 1880s and 1890s through copying or adapting American examples, either through the mechanism of books and journals, or transatlantic visits (see Chapter 7).
9 Classic accounts of the first generation of iron-framed textile mills are provided in Bannister 1950; Skempton and Johnson 1962; Fitzgerald 1988.
10 Mainstone 2001, 294.
11 Fitzgerald 1988, 127.
12 For example, a 1906 construction photograph of Texas Mill, Ashton, Lancashire, shows a fully self-supporting iron and steel framework erected ahead of the brickwork, although in this case at least the walls appear to be carried on their own foundations, not the frame. See The Engineer 102 (28 December 1906), 652. Blakeridge Mills, Batley, Yorkshire (1912–13) used an interior skeleton of steel stanchions and encased beams, but one of the earliest fully framed examples of the genre appears to be Brookroyd Mills, Stainland c 1920, built on a skeleton of reinforced concrete. See Giles and Goodall 1992, 66.
13 See, for example, Giedion 1967, 229–30; Landau and Condit 1996, 20. This inventive, portal-framed structure had much significance besides the technological: according to Schmiechen and Carls 1999, 69, it had much influence on the design of early and mid-19th-century retail markets. It was demolished in 1962 to make way for Charing Cross Station.
14 Sutherland 1976, 104.
15 Weiler 1990, 41–2.
16 In this structure the cast- and wrought-iron plate enclosure around internal iron framing was stiffened by hollow, square-section pilasters and beams, the whole roofed with a corrugated-iron segmental barrel vault. See, for example, Fairbairn 1863, 118–123; Skempton 1961, 67; Herbert 1978, 41–2.
17 James Bogardus (1800–74) familiarised himself with iron construction during a visit to England and the Continent in 1836–40; almost certainly he witnessed, or was aware of, Fairbain's mill. In May 1850, following the construction of an all-iron factory at the corner of Duane and Centre Streets, Manhattan (1848–49), he patented his own prefabricated iron building system using cast-iron façades and bolted connections. See Gayle and Gayle 1998; Larson 1987, 40; Bannister 1956.
18 A major exception seems to have been an iron-framed military station prefabricated by H & M D Grissell in 1844–5 and exported to Mauritius. Although only one storey in height, the complex covered a huge area (the officers' quarters was 2,400 sq ft (223m²) in area, the hospital 5,200 sq ft (279m²). The Builder 9 (8 March 1851), 152. This building also marked an early use of H-section stanchions, a form that was also exploited in the Royal Dockyard slip roofs erected by George Baker & Son in the 1840s.
19 Fairbairn's fêted mill, for example, derived some stability from a transverse brick wall that supported the drive shaft of the machinery, and Laycock's celebrated Iron Palace for King Eyambo (1843) was, according to the Liverpool Times, 'a composite structure of plate and panels of iron upon a wooden skeleton merely', as quoted in Herbert 1978, 42.
20 The literature on this building is vast, but for essential structural overviews, see Addis 2007, 354–9; Mainstone 2001, 294–5; Peters 1996, 226–54.
21 Survey of London 1975, 98. William Dredge, the son and brother of the better-known engineers James Dredge (1794–1863) and James Dredge (1840–1906), also collaborated with C D Young and Co. on the Manchester Art Treasures Exhibition building of 1857, with elevational input by Salomons (1857).
22 See Skempton 1961, 70–1.
23 As Sara Wermiel has recently shown, several permanent, multi-storey skeleton-framed buildings were built in the United States in the 1850s, including the Marine Hospital in New Orleans, Louisiana (1856–60). See Wermiel 2010 'An experiment in skeleton-frame construction in the 1850s: The U.S. Marine Hospital in New Orleans', in Rinke and Schwartz eds 2010, 175–90.
24 For the classic article on this seminal building, discovered by Eric de Maré in the late 1950s, see Skempton 1961; Skempton 1959, 47; Mainstone 2001, 295–7; Addis 2007, 359–63.
25 Darley 2003, 109–11.
26 Almost certainly Ordish, and other engineers and ironfounders who had worked with Greene, would have known the Boat Store; nevertheless, it was seemingly never given any detailed reportage. In September 1860 The Times in its 'Military and Naval Intelligence' section, reported on an inspection of Royal Dockyard establishments by chief officials. At Sheerness, it was seemingly the mechanical rather than structural engineering of the fully framed building that caught their attention: 'they visited the new boat store recently built from the designs of Col. Greene, and inspected the new mode of raising and lowering boats invented by Col. Greene', The Times, 22 September 1860, 22.
27 Brick panels provided better insulation and a more weatherproof, permanent enclosure than corrugated iron, and brick-arched floors a more 'fireproof' solution than the timber floors used at the Boat Store and earlier framed buildings. Already, in mid- to late 1850s New York, Bogardus had designed and built two soaring towers for the manufacture of gun shot – the first multi-storey iron structures to support their brick enclosures – and his chief rival, Daniel Badger, was soon to apply this concept to a building, in the form of a 'fireproof' seven-storey grain

elevator for the US Warehousing Company, erected in 1860. Despite their centrality to American developments, Bogardus's retrospectively celebrated shot towers were structures, not buildings, and Badger's cast-iron-fronted grain elevators of the 1860s used cast iron exclusively for the interior framing also, ignoring or eschewing the technically superior wrought-iron beam. *See* Bannister 1957 and Larson 1987, 43.

28 The structural significance of this building was probably first brought to light by Newton 1941; *see also* Steiner 1984, 104–4; Skempton 1959, 47–8; Landau and Condit 1996, 21.

29 They were test loaded to 600lb per square foot, double the usual maximum.

30 Condit 1964, 82.

31 Steiner 1984, 104.

32 *The Builder* **23** (29 April 1865), 296–7.

33 *See* Bannister 1957, 15.

34 *See* Steiner 1984, 102–6.

35 In the event stone exterior columns were used in the belief that these would confer the necessary stability. Ibid, 59.

36 *See* Steiner 1984, 110.

37 Ibid, 110 *et seq.*

38 *See* Bannister 1957, 15.

39 Saint 1976, 14.

40 Although not mentioned in other accounts of the Menier building, according to Sharp 1993, 20: 'The repeated frame was so arranged that the uppermost floor was suspended from the roof structure, leaving the floor below completely free of columns.'

41 Larson 1987, 43.

42 *The Bombay Gazette*, 6 February 1871, 1, as quoted in Clarke 2002, 48.

43 Clarke 2002.

44 *The Architect* **2** (11 December 1869), 286.

45 Clarke 2002, 61–2.

46 Eugène-Emmanuel Viollet-le-Duc, *Entretiens sur l'architecture*, trans. Benjamin Bucknall (Boston, 1881), 128, as quoted in Larson 1987, 46.

47 Unfortunately, the truth of this may never be known, for the building is long demolished, and no records seem to survive.

48 *The Architect* **7** (18 May 1872), 256.

49 'Aberdeen Architectural Association: Structural Ironwork', *The Builder* **101** (1 December 1911), 628.

50 *The Engineer* **45** (12 April 1878), 267.

51 'Iron for building purposes', *JISI*, 1878, 274.

52 *BN* **34** (10 May 1878), 486.

53 Cooper 1991, 153, citing Manchester corporation plan register E. BK 2. 26.

54 All the older, load-bearing buildings had massive inverted arch foundations, supported by scores of piles. *See* Skempton 1961, 63.

55 Giedion 1967, 206.

56 William White, in discussion of Aitchison 1864, 105–6.

57 *See* Tucker 2000, 25, 31, 59.

58 Parkinson-Bailey 2000, 127.

59 Farnie 1980, 44.

60 Ibid, 45, 48–9.

61 Edward Wood & Co. 1914, 3. Edward Graham Wood, the company's founder, began business as a structural engineer in 1876 at Red Bank, Manchester. His business 'increasing by leaps and bounds', and with these works proving 'too small to cope with the rapidly growing trade in steel girders, etc.', he relocated to Ordsall Lane in 1890, establishing the Ocean Iron Works between the two dock systems, which were equipped with overhead travelling cranes and gas-powered machinery.

62 The aqueduct was constructed by Andrew Handyside & Co. of Derby, a firm that, along with Sir William Arrol & Co., erected the great majority of bridges and viaducts crossing the canal. For a detailed list of these structures, *see* Heywood 1894.

63 Farnie 1980, 49.

64 For example, five three-storey transit sheds had been erected on this system by 1905, to designs by W H Hunter, Chief Engineer to the Manchester Ship Canal. Twelvetrees 1907, 1–6; Mouchel, L G, and Partners, Ltd 1920, 19.

65 Jones 1985, 209.

66 Wood 1898a.

67 In this building, for example, the concrete filler-joist floors were carried on an interior framework of cast-iron columns and rolled-steel beams. Cooper 1991, 158. Constructed by Robert Neill and Sons, it was 'fireproof throughout, and lined with glazed bricks in the warehouse portions, and with faïence in the staircase and offices'. 'Warehouse for Messrs. Horrockses, Crewdson, and Co., Limited, Manchester', *BN* **75** (1 July 1898), 13.

68 For his American tour, Trubshaw was accompanied by William Towle, the Midland Hotel's Manager. *See* Dixey 'Charles Trubshaw: A Victorian Railway Architect' in Jenkinson 1993, 65–68 and Trubshaw's obituary in *The Builder* **112** (23 February 1917), 142.

69 Shaw seemingly used this technique some 20 years earlier in a bank in the City of London, probably Baring's Bank, No. 8 Bishopsgate (1880–1, demolished). *See* Saint 1976, 360 and 461, note 56.

70 In 1928 and 1929, William Basil Scott made two separate claims to have designed England's first steel-skeleton building, erected in West Hartlepool in the year 1896, Scott 1928, 1929. In 1930 he revised his recollections, saying that this building was actually erected in Stockton-on-Tees in 1898, Scott 1930. Kathryn Morrison has shown that in 1896 Mathias

Robinson did build a large branch store in Stockton-on-Tees, and that contractors from West Hartlepool erected it. Known as the Coliseum, it was destroyed by fire in December 1899. Although a newspaper article described steel girders and supports in the wreckage, recently unearthed building regulation plans show that the framework used cast-iron columns, and certainly did not carry the exterior. However, its rapidly built successor of 1900–01, designed by Barnes & Coates of Sunderland, was built on an all-steel-skeleton designed by Scott: this building, still surviving, seems to be the building which the engineer referred to. *See* Morrison 2003, 143; Scott 1944.

71 Morrison 2003, 143.

72 This building, whose skeleton structure was brought to light by Jackson 1997, predates J J Burnet's warehouse on Princes Street, Edinburgh for R W Forsyth, Ltd (1906–7), considered by some historians to be Scotland's first steel-framed building. Jackson's appraisal of the building, based on examination of detailed drawings produced in October 1898 and held in the Glasgow Dean of Guild collection, is corroborated by *The Scotsman*'s description of its new offices, which noted how the 'whole range will be constructed throughout with fireproof floors, and … steel columns running from the basement to the topmost storey with intervening girders will carry the building'. *See* Jackson 1997, 53–4, 96–101; *The Scotsman*, 13 April 1899, 8; and Sharman 1906, 203. Redpath, Brown & Co., Ltd erected the framework.

73 A full account of this building is provided by Byrne 2001.

74 Jackson 1997, 54.

75 'A Large Block of Office Premises', *BJ&AR* **21** (12 April 1905), 194.

76 *See* 'The Liverpool Dock Offices', *BJ&AR* **20** (Supplement, 28 December 1904), 25.

77 For example, Harry S. Fairhurst's first commission after setting up practice in Manchester in 1905 was India House, Whitworth Street (1905–9). Erected by Robert Neill & Sons for Lloyd's Packing Warehouses Ltd, it was built on a framework supplied and erected by Richard Moreland & Son. The frame made extensive use of solid-steel columns, one of the distinguishing hallmarks of the London company. 'Packing warehouse, Manchester', *The Builder* **91** (1 September 1906), 282.

78 For example, 'This building [the Ritz Hotel], completed in 1904, was the first fully load bearing skeletal steel-framed structure to be constructed in Britain', Peter Campbell, '1890–1910' in Collins 1983, 78.

9 Theatres and music halls: advances in safety and sighting

1 'Domestic occurrences. Particulars of the destruction of the Drury-lane Theatre', *The Gentleman's Magazine*, March 1809, 272 (article runs 271–272; no author).

2 Patented in 1773 by David Hartley (1731–1813), abolitionist MP and inventor, fireplates consisted of overlapping sheets of wrought-iron that were typically nailed to joists beneath the floorboards as a protective sheeting in multi-storey buildings, including houses.

3 Wilkinson 1825, 204, 206; Glasstone 1975, 12; Milhous *et al* 2001, 286–90.

4 Survey of London 1970, 91–3.

5 Carter 1967.

6 Foulston noted '... the whole of the framing for the Boxes, Corridors, &c, being of cast iron. The Roof (the span of which is 60 feet) is of rolled iron', Foulston 1838, quoted in Wilmore 1998, 122.

7 Guedes 1979, 270.

8 7 and 8 Victoria, c. 84; The Metropolitan Building Act 1844; 18 and 19 Victoria, *c.* 122; The Metropolitan Building Act 1855.

9 Barry 1860, 86.

10 Ibid.

11 Despite his avoidance of combustible materials, especially in the roof space, Barry made no attempt to separate the stage from the auditorium – an omission that was to bring criticism from the French architect M G Davioud in 1867 while working on a survey of London theatres. *See* Leacroft 1988, 225–6.

12 'In construction, the building displays very important points of difference with the old theatres', *The Builder* **16** (3 April 1858), 236.

13 Quoted in Saint *et al* 1982, 31.

14 Grissell, who is credited in *The Builder* as designing the eight main trellis girders that supported the roof, and who had by this date accumulated considerable technical knowledge of fabricating and erecting iron-framed buildings and bridges, almost certainly played a major role in the structural design of the tiers. For biographical information, *see* Skempton 1961, 25–6.

15 *The Builder* **16** (24 April 1858), 273.

16 *The Builder* **18** (11 February 1860), 87.

17 Ibid.

18 *The Builder* **16** (11 December 1858), 834.

19 Ibid, 833.

20 Ibid, 834.

21 Ibid, 833.

22 *The Builder* **23** (16 December 1865), 889.

23 Glasstone 1975, 38.

24 *The Builder* **25** (28 December 1867), 941–2.

25 *The Builder* **27** (26 June 1869), 509.

26 Ibid.

27 *BN* **16** (2 April 1869), 300.

28 *The Builder* **27** (22 November 1869), 872–3; *The Builder* **27** (26 June 1869), 509 and 526; *BN* **16** (2 April 1869), 299–300.

29 'The Gaiety Theatre, Strand', *The Era*, 13 December 1868, 6. Phillips's Patent compound girders were probably first used in the Gaiety Theatre, quickly followed by Her Majesty's. In the Gaiety, the 'ironwork necessary for this construction … [was] … manufactured by Messrs. W. & T. Phillips, of the Coal Exchange, at their works in Belgium, and constructed by them at the theatre,' *The Builder* **26** (26 December 1868), 941.

30 *BN* **15** (25 December 1868), 879.

31 *The Builder*, in describing the structure, noted 'The columns supporting the various tiers are carried up to a sufficient height above the gallery, and from the cap spring a series of pointed arches, supporting cornice and coved ceiling', *The Builder* **26** (26 December 1868), 941.

32 'The Gaiety Theatre, Strand', *The Era*, 13 December 1868, 6. This article also noted that Phipps 'made the science of Theatre construction a special study, and has erected the new Theatres at Bath, Bristol, Brighton, Nottingham, South Shields, and Swansea, and the "New Queen's," Long-acre … but in the "GAIETY THEATRE" he has, both in form and construction, gone still more out of the beaten track …'.

33 41 and 42 Victoria, c. 32; The Metropolis Management and Building Acts Amendment Act 1878.

34 *See* 'The construction of theatres and music-halls', *BN* **36** (16 May 1879), 528–9.

35 In theatres accommodating more than 400 people, these exit facilities had to be a minimum of 4ft 6in. wide, and 6in. wider for every additional 100 people the building could hold until a maximum of width of 9ft was reached, ibid, 528.

36 The latest surviving 'London suburban theatre with a complete pre-cantilever (hence iron-columned) auditorium' – J G Buckle's Theatre Royal, Stratford East – dates from 1884. Although unaltered Victorian theatre interiors are scarce, this does suggest a widening take-up of cantilever construction after the 1878 Act. Quotation from Earl and Sell 2000, 142. For a history of this theatre, and information on its designer, *see* Earl 1984.

37 Tanner 1883, 191.

38 *See* Clarke 2002, 58–61.

39 *See* Survey of London 1966b, 464–5.

40 Tanner 1883, 191.

41 Ibid.

42 Ibid.

43 Ibid, 192.

44 Ibid.

45 'Steel-work in the Palace Theatre, London', *Engineering* **82** (28 December 1906), 867.

46 Professor Sir Thomas Hudson Beare, reminiscing on his early professional career, noted 'My second experience as a young man, in structural engineering, was when I was an assistant to the late Sir Alexander Kennedy, M.Inst.C.E., who was then acting as Consulting Engineer in connection with the rebuilding of the Alhambra Theatre in Leicester Square, London. This was the first theatre, I think, in which structural steel was extensively used', Beare 1932, 83. Contemporary accounts in *The Builder* and *The Building News* indicate that both cast- and wrought-iron were used in the building of the new theatre in 1883, but given the interchangeability of the terms iron and steel in the architectural press at this time, and the more judicious use of the same terms in the engineering press, it seems almost certain that steel rather than iron was used. Kennedy's biography in Pike 1908, 124, notes he 'designed the entire steel structures forming the skeletons of the Alhambra Theatre and the Hotel Cecil'; his obituary in the *Engineer*, 9 November 1928, 511, noted 'While at the College too, [he was appointed Professor of Engineering at University College, London in 1874], he designed a steel arched pier for Trouville, and the steel and concrete internal structure of the Alhambra Theatre, which is said to have been the first building in which concrete slabs were used on a large scale to carry weights'.

47 Kennedy may have been brought in on the project through his connections with Thomas Hayter Lewis (1818–98) who was Professor of Architecture at University College London from 1865 to 1881. See 'Professor Hayter Lewis' [obituary], *The Builder* **75** (17 December 1898), 565.

48 The alterations made to the building in 1881 are described in detail in *The Builder* **41** (10 December 1881), 739.

49 *The Builder* **44** (16 June 1883), 810.

50 *BN* **45** (31 August 1883), 343. Again, the magazine's use of the term iron is questionable.

51 *The Builder* **44** (16 June 1883), 810.

52 Woodrow 1892, 429.

53 *The Builder* **45** (10 November 1883), 647.

54 'The architects have done their utmost to prevent a repetition of its destruction by using the now popular "fire-resisting" materials, nor would it be an exaggeration to say their work in this respect opened a new era for the London theatres. Concrete, iron and steel have been most extensively used, the iron being very wisely protected', Sachs and Woodrow 1896–98 (Vol. 1, 1896), 43.

55 Beare 1932, 83.

56 For a brief biographical notice of William Brass, *see The Builder* **54** (21 January 1888), 54.

57 *BN* **45** (31 August 1883), 343.

58 Richard Moreland & Son, whose offices and works were at No. 3 Old Street, on the corner of Goswell Road, were also one of the few contractors fabricating and erecting steelwork at this date, fabricating the lattice steel ribs of Hengler's Circus, Argyle Street in 1883 (*see* p 116).

59 *See* Moreland 1893, 320.

60 For this collaborative project, *see* Walmisley 1888, 71–6.

61 Ende 1887a, 257.

62 Ende 1887b, 284. Opened in 1884, this enormous hall, 450ft by 250ft was reputed to be the largest hall in Britain covered by a single span. Intended for agricultural shows, exhibitions and concerts, it changed its name in 1886 from the National Agricultural Hall to 'Olympia', staging its first circus performance in that year.

63 Ende 1889, 94. Lack of confidence in Ende's daring design may have been responsible for the subsequent introduction of balcony-supporting columns. According to Earl and Sell 2000, 136, these columns were removed in 1934, since when the passage of time has shown the soundness of the original structure.

64 *BN* **60** (6 February 1891), 194.

65 The Savoy was the first theatre to use electric lighting, *see,* for example, Dixon and Muthesius 1985, 91.

66 Although contemporary accounts in the architectural press and the description in Survey of London 1966a, 300, record that Collcutt designed the façades and internal decorations, and the builder G H Holloway designed and superintended all the constructional work, this was not strictly the case. In a letter to the Editor of *The Building News*, Henry Lovegrove affirmed that Collings B Young, not Holloway, was responsible for the design of the superstructure. He stated: 'Mr. Holloway is reputed to be a man of great ability as a constructor and organiser, but does not, I think, pretend to be more than the builder of the theatre . . . I am pleased to make the above statement public, as Mr. C. B. Young served his articles with me', *BN* **60** (13 February 1891), 252. This demarcation of roles was substantiated by *The Builder* **60** (14 February 1891), 127, which stated that Holloway acted 'practically as the builder'.

67 *The Builder* **58** (15 February 1890), 119, **60** (14 February 1891), 127; *Engineering* **82** (28 December 1906), 867.

68 Earl and Sell 2000, 130. James George Buckle (1852–1924), while not a significant theatre designer, was nonetheless an important force in promulgating advances in fire-resistant theatre construction. With Ernest Woodrow he co-authored an important series of articles entitled 'Theatre planning and construction', published in 1884–5 in *Building and Engineering Times*, and wrote a book on Theatre Construction and Maintenance, published in 1888. *See* Earl and Sell 2000, 269.

69 The most detailed description of the construction of this theatre is provided in Ernest A E Woodrow's 47-part series of articles on theatres, running in *The Building News* from 15 July 1892 until 28 December 1894 (*see* Woodrow, E A E 1892–94). For those dealing specifically with the Royal English Opera House, *see* 19 January 1894, 76–9, 26 January 1894, 107–10, 23 January 1894, 247–9, 23 March 1894, 391–4.

70 Woodrow 1892, 429.

71 *BN* **60** (9 January 1891), 64.

72 Sachs and Woodrow 1896–8 (Vol 1, 1896), 36.

73 Ibid, 103.

74 Woodrow 1892, 429.

75 Spain and Dromgoole 1970, 77.

76 Moreland, R & Son Ltd *c* 1905.

77 Like the Palace Theatre, the architectural authorship of this building was contested in the early 1890s. *The Builder*, which published letters from both architects concluded 'it appears that plans and designs were prepared in 1887, by Mr. C. J. Phipps, as joint architects; whilst the architecture, within and without, the acoustic arrangements, external loggias, and construction are by Mr. Knightley,' *The Builder* **65** (7 October 1893), 269. Regarding the structural design, Knightley stated in a letter that 'the skeleton sections were supplied by Messrs. Moreland, with whom Mr. Phipps had made arrangements for the iron and steel work', *The Builder* **60** (28 February 1891), 174.

78 ' Proposed new concert hall, Langham-Place', *The Builder* **60** (14 February 1891), 129.

79 '… owing to their solid nature, it would be almost an impossibility to heat them sufficiently to affect the stability of the material'. Moreland, R & Son Ltd *c* 1905, 4.

80 *The Builder* **70** (4 January 1896), 20, (27 June 1896), 563.

81 *The Builder* noted of the Camden Theatre that 'the interior is partly on the cantilever system', *The Builder* **80** (5 January 1901), 23. This information is collated from the cross-referencing of the following sources: Moreland, R & Son Ltd *c* 1905; Howard 1970; Spain and Dromgoole 1970.

82 'The circles are constructed on the suspensory [sic] principle, with no column or obstruction to sight of any kind', *The Builder* **79** (27 October 1900), 371.

83 *See* note 69.

84 This was Sprague's first independent work, with an auditorium prototypical of his preferred 'two-tier' design. Aileen Reid, draft text on forthcoming Survey of London monograph on Battersea.

85 *The Builder* **70** (4 January 1896), 20.

86 *See* Woodrow's Longitudinal Section of 1899 filed within LMA GLC/AR/BR/19/0229.

87 http://www.theatrestrust.org.uk/resources/theatres/show/1978-clapham-grand.

88 *BJ&AR* **18** (4 November 1903), 184.

89 *The Times*, 8 May 1905, 4; *BN* **89** (12 May 1905), 675.

90 Regulations made by the Council on 30 July 1901 reaffirmed the three-tier maximum specified by the Metropolis Management and Building Acts Amendment Act 1878, but on 13 November 1906 this was amended to just two tiers, applicable to all London's 'theatres, houses, and other places of public resort … except in exceptional circumstances'. *See* Cubitt 1911, 138, 636.

91 *BJ&AR* **21** (5 April 1905), 172.

92 Glasstone 1980, 15. Nonetheless, this theatre's new auditorium was hardly a match for Matcham's technologically advanced design, Survey of London 1955, 66, noting 'Among other improvements carried out under Philip Pilditch's direction, the tiers were reconstructed with steel girders and concrete floors, *using only a front row of columns to support them*'. [emphasis added]. The consulting structural engineer responsible for this was (Professor) Henry Adams (1846–1935), who in 1925 recollected 'Among other things I was responsible for [was] the steel reconstruction of the interior of Drury Lane Theatre', *TSE* **3** (December 1925), 422.

93 Patent no. 27,146, 1902. In the patent, Briggs is described as an engineer, Matcham as an architect. Articled under George Bestall Jerram, Borough Engineer and Surveyor to Walthamstow Local Board, Briggs worked briefly for that body before resigning to join Frank Matcham, 'then a promising young architect,' in the 1880s. Among the first projects he worked on were theatres at Bury and St. Helens, The Opera House, Southport and the New Empire Palace, Edinburgh, *The Bognor Post*, 2 December 1950, 9. I am grateful to Patricia Lovell for this reference.

94 Patent no. 27,146, 1902.

95 'It was during the preparation of the [Coliseum] plans that he [Briggs] encountered difficulty with the construction of the balcony and gallery, resulting in his working out an entirely new method. His partner [Matcham] was favourably impressed, and urged him to patent it, expressing a wish to be associated, to which Briggs readily agreed.

The result was a patent in their joint names, and the system, which had many advantages of the old, was used by them exclusively in the Coliseum, Sir Walter Gibbons' London Palladium and other buildings', *The Bognor Post*, 2 December 1950, 9. The patent was filed in December 1902, a month after Stoll's new Company, The London Coliseum Ltd, was floated. Glasstone 1980, 19.

96 A photograph of this framework appears in *BJ&AR* **20** (9 November 1904), 246.
97 *BJ&AR* **20** (26 October 1904), 219.
98 John Earl, 'The London theatres', in Walker 1980, 44.
99 Glasstone 1980, 18.
100 'The grand circle and gallery are constructed on the cantilever principle, the architects' patent curved girder, so successfully used at the London Coliseum and their other important buildings, being adopted', *BN* **99** (23 December 1910), 897.

10 Clubs and hotels: opulence, proportion and planning

1 *The Builder* **2** (5 October 1844), 503.
2 'In the United Service Club are two rooms of 150 feet by 50, the floors of which are constructed of cast-iron girders', 'United Service Club-House', *The Mirror of Literature, Amusement, and Instruction* **12** (4 October 1828), 210.
3 Sutherland 1990, 26.
4 Dixon and Muthesius 1985, 74–5.
5 Sutherland 1989, 50.
6 Nevill 1911, 233.
7 Timbs 1866, Vol 1, 267.
8 Woodbridge 1978, 67.
9 Certainly the technique was in use by 1831 (J U Rastrick's suggested remedial works for Nash's west front of Buckingham Palace included substantial cast-iron columns that 'may be inclosed with others of Scagliola or any other material'), and by the 1860s was commonplace: 'Columns are sometimes required to contain an iron core to support a weight above: these cores are mostly put up with the building, and the scagliola worked upon a wooden skeleton surrounding the iron core … Or, the column may be worked in a lathe in the shop by constructing the skeleton in such a way as to admit the column to be cut either in two equal parts, or one quarter out, the parts of the skeleton being pegged together in-stead of nailed, and with the centre easily re-moved, allowing space for the iron core. House of Commons 1831, 281; 'Practical remarks on Scagliola', *The Builder* **21** (28 November 1863), 839. The technique continued in use through the late 19th century; for example. Scagliola manufacturers Bellman and Ivy of Wigmore Street offered

bespoke decorative services to architects and builders, their advertisements stating 'When Columns are required to carry any weight, arrangements are made for the insertion of Iron Cores or Stancheons, or should these be already in position, the Scagliola can be fixed round without displacing them in any way'. *See*, for example, *BN* **37** (19 September 1879), advertisements section, 1.
10 In this context it is of some interest that the internal decoration of the Travellers' Club, completed in 1843 under Barry's superintendence, was criticised in *The Athenaeum*, which castigated 'the employment of affectations and unrealities, which abound everywhere – sham granite walls, sham marbled columns and dados, and sham bronze doors, sham bas-reliefs', *The Athenaeum*, 12 August 1843, 737–8, quoted in Survey of London 1960b, 403–4. However, it would seem that timber, and not iron uprights were used in at least some of the rooms of Barry's Reform Club, for in 1853, a number of internal modifications were made under the architect's super-intendence, including 'the substitution of wooden columns for scagliola in the coffee-room', Survey of London 1960b, 412. In all likelihood these scagliola replacements had cylindrical iron cores.
11 *The Builder* **6** (12 February 1848), 73.
12 *The Builder* **5** (8 May 1847), 218.
13 Woodbridge 1978, 52.
14 7 and 8 Victoria, c. 84; Metropolitan Building Act 1844.
15 Quoted in Hill 1970, 40.
16 C F Hayward, in discussion of 'Dwellings of the Labouring Poor in Large Towns', *BN* **3** (22 May 1857), 515.
17 Hill 1970, 10.
18 *BN* **4** (26 February 1858), 202. The system, as described, would appear to be that patented by Fox and Barrett.
19 'London Bridge Railways Terminus Hotel', *BN* **20** (7 March 1862), 167.
20 Hill 1970, 27; 'London Bridge Railways Terminus Hotel', *The Builder* **19** (22 June 1861), 427–9. Currey would later design the nearby St Thomas' Hospital (1868–71), which made use of 1,200 tons of Belgian structural wrought iron (*see* p 35).
21 *BN* **20** (7 March 1862), 167.
22 Despite the restricted application of non-combustible floors – a provision little in excess of the MBW strictures – some 20 years after its completion, *The Langham Hotel Guide to London* described the hotel as 'nearly fireproof as such a structure can ever be made: the halls, corridors, and stairways being all of stone', Langham Hotel 1884, 6. For a detailed account of this building, *see* Aiken 1987.

23 Langham Hotel 1904; Aiken 1987, 5.
24 'Its external façades present no constructive sham. There is, as far as we are aware, no cement work doing duty for stone. A good sound arch is turned over every opening', *BN* **12** (10 June 1865), 422.
25 Cunningham 1992, 166.
26 Chancellor 1922, 163, quoted in Thole 1992, 28.
27 *The Builder* noted that 'The building is fireproof throughout, the floors being constructed with Phillips's wrought-iron joists and laths filled in with concrete, and carried up by wrought-iron girders, the soffits of which are visible in the various rooms. The Portland stone staircase is carried on wrought-iron girders, the whole of which are visible.' Many of the rooms were both deep and wide, including a 'somewhat octagonal' morning room, 40ft by 35ft, a drawing room 45ft by 32ft, and a coffee room 48ft by 27ft, *The Builder* **26** (16 May 1868), 357.
28 Waterhouse's correspondence regarding these girders is dated 13 December 1866; *see* Cunningham 1992, 165. Although manufactured by W & T Phillips, it seems likely that it was William Brass, the general contractor, who transported and erected them, presumably under the inspection of J S Bergheim.
29 Cunningham 1992, 166–7.
30 Walford 1892, 370.
31 Sutherland 1989, 54.
32 Croad 1992, 90–1.
33 John Henderson Porter (1824–95), a pioneer of prefabricated buildings using galvanised corrugated iron, took out patents for the structural use of corrugated iron in beams and floors in conjunction with concrete (12091 and 12356 of 1848), which 'anticipated developments in fireproof flooring by 20 years', Cross-Rudkin and Chrimes, 2008, 628.
34 *See* Cruickshank 1997, 71–2.
35 *BN* **40** (27 May 1881), 606.
36 *BN* **40** (14 January 1881), 55. This report described the flat roof, covered by Claridge's Patent Asphalt as 'a most perfect piece of work of the kind': 'The rolled joists about 3in. × 1in. are supported on iron girders or binders 7ft. apart, and these carry the bed of concrete, upon which is laid the asphalte [sic] in the usual widths'.
37 *The Builder* **34** (8 January 1876), 39.
38 'The Grand Hotel, Charing-Cross: Visit of the Architectural Association', *The Builder* **37** (1 February 1879), 136.
39 *The Building News* refers to 'Iron flanged stanchions, filled up with brick in cement', suggesting perhaps an H-section or double-I-shaped cross section. 'The Grand Hotel, Charing-Cross', *BN* **36** (31 January 1879), 117.

40 'In order to avoid the necessity and expense of erecting scaffolding to support the large balconies of this club whenever they are required to view any public procession, the balconies have lately been made entirely self-supporting by a system of wrought-iron joists and built-up cantilevers, concealed from view by cement trusses and mouldings', 'Oxford and Cambridge Club', *The Builder* **45** (29 September 1883), 438.

41 *The Builder* **37** (1 February 1879), 136; *BN* **36** (31 January 1879), 117.

42 *The Builder* **37** (1 February 1879), 136.

43 'The principle which has been approved by the architect consists of supporting each step on a rolled flanged joist tailed into the wall. Through these joists are slots for hoop iron to pass through, and the concrete is filled in to encase the iron joists and bars. The steps will have Sicilian marble treads and risers, and the soffit is to be panelled in Parian cement, a trefoil moulding being run on the outer edge', *BN* **36** (31 January 1879), 117.

44 In conjunction with J E Saunders.

45 *The Builder* **48** (30 May 1885), 778.

46 Ibid, 777–8; *BN* **50** (5 June 1885), 877–8.

47 Cunningham 1992, 118.

48 'The National Liberal Club, Charing Cross', *The Builder* **44** (21 April 1883), 547.

49 *The Builder* **52** (21 May 1887), 783.

50 Ibid, 781.

51 General Specification, held by the National Liberal Club. Pages 73–80, entitled 'Constructive Ironwork &c'.

52 *A&CR* **74** (25 August 1905), 115. *See* 'Waterhouse, Alfred (1830–1905)' in Cross-Rudkin and Chrimes 2008, 628.

53 *See* p 351, note 60.

54 *The Builder* **52** (21 May 1887), 783.

55 Ibid, emphasis added.

56 Cunnington 1898b, 350.

57 'The floors are all fire-proof, of steel decking and concrete; and all the lintels and fixing-blocks are of breeze concrete', *The Builder* **50** (15 May 1886), 725.

58 Cunningham 1992, 119.

59 *See The Builder* **52** (21 May 1887), 781–4; *The Architect* **37** (24 June 1887), 373.

60 The staircase was a huge, elliptically shaped structure, supported on 'a continuous ascending colonnade of various and richly-coloured marbles', *The Builder* **48** (9 May 1885), 652. A bomb destroyed it in May 1941, but spared the rest of the building, and in 1951 a new cantilevered staircase was built to designs by Bernard Engle (d 1973). *Country Life*, 26 December 1952, 2103.

61 In 1889 Homan & Rodgers introduced a means of protecting these compound members (which by then utilised steel joists) by using hollow terracotta tiles or lintels – commonly of triangular or trapezoid form – which projected below the lower flange upon which they rested, completely enveloping the metal. Because the lintels were keyed underneath to hold plaster, they avoided the need to suspend a separate ceiling beneath. It seems likely that it was this system – designed for large spans – rather than their earlier filler-joist floor that was used at the Junior Constitutional Club. For further details of the firm's flooring systems, *see* 'Reinforced concrete systems VI – Homan & Rodgers' System, *BJ&AE* **24** (C & S Supplement, 10 October 1906), 76–8.

62 *The Builder* **60** (3 January 1891), 11.

63 'The cost of the building alone will be about 75,000*l*', *The Builder* **60** (3 January 1891), 11. The cost of the National Liberal Club, exclusive of site, was about £150,000, *The Builder* **52** (21 May 1887), 784.

64 *The Builder* **60** (3 January 1891), 11. It was only through structural necessity that the lavish, open smoking room was typically located at higher level, given that there the cumulative loads were less.

65 *See*, for example, Holy Trinity, Latimer Road (Harrow Mission Church) (1887–9).

66 Smith 1888, 293.

67 S'international Architects 1989, 32.

68 *The Builder* **60** (17 January 1891), 50. In fact *The Builder* misreported in this respect for according to the original specification the upper (bedroom) floors were framed and floored in timber. I am grateful to Lawrance Hurst for this information.

69 One of Gaynor's first works was the five-storey, cast-iron fronted Haughwout Store, erected on Broadway and Broome Street, New York City in 1857. Utilising structural members cast by Daniel Badger's Architectural Iron Works, this building also featured the first passenger elevator with an automatic brake, installed by Elisha Graves Otis. *See* Condit 1982, 84.

70 *The Builder* **36** (21 September 1878), 988.

71 'This structure claims to surpass, not only in size but in grandeur, all the hotels in Europe or America; Mr John P. Gaynor is its architect, and it is called, not without some reason, the PALACE Hotel', ibid.

72 Denby 1998, 143; *The Builder* **36** (21 September 1878), 988.

73 Jackson 1964, 20.

74 Even so, *The Builder* complained 'Here loftiness has again been sacrificed to what were deemed to be more pressing requirements, and here again electric lighting and readily obtainable cross-ventilation will mitigate the inconveniences of an apartment so proportioned, though they cannot improve its appearance', *The Builder* **57** (13 July 1889), 29.

75 Jackson 1964, 20.

76 Originally intended to be called the Northumberland Avenue Hotel, construction of this mammoth building began in the summer of 1883 by Perry & Co., of Tredegar Works, Bow, to designs by Lewis Henry Isaacs (*c* 1829–1908) and Henry Louis Florence (1843–1916). In late 1884 the possession of the building transferred from the Northumberland Avenue Hotel Company to the Building Securities Company, who replaced the existing builders with J W Hobbs & Co., of Croydon, but retained the original architects. Completed in 1887, in modified and reduced form, it still boasted nine floors, 500 rooms and two passenger lifts, but only five bathrooms. Almost all the grand spaces, including the 100ft by 42ft salle à manger, and the 56ft by 30ft coffee room were framed in the traditional manner, with perimeter columns supporting long-span cross-girders. The detail of *The Builder*'s description makes it clear that steel was not used: 'Of the ironwork used in the construction of the building, the cast-iron stanchions have been supplied by Messrs. Young & Co., and the wrought-iron girders and stanchions by Messrs. Dibley & Son', *The Builder* **50** (1 May 1886), 639.

77 'The walls are built (except the facings) with Ellistown bricks and blue lias mortar; we are informed that a sample block of the brickwork has been tested by Mr. Kirkaldy, and found to sustain 100 tons to the square foot without fracture', *The Builder* **57** (13 July 1889), 30.

78 For example, Collcutt is attributed by Boniface 1981, 47; Watkin *et al* 1984, 20; Denby 1998, 143. According to Denby 1998, 144, Collcutt and Mackmurdo were responsible for the interior furnishing and decoration of the hotel, assisted by D'Oyly Carte's wife, Helen, whose 'influence towards all modern comforts and innovations in the hotel was of considerable value'. Young is variously credited in a number of works, including Pevsner 1976, 186, and Dixon and Muthesius 1985, 82.

79 'The Late Mr. Collings Beatson Young', *JRIBA* **29**, 3 ser (23 September 1922), 610; *The Builder* **123** (18 Aug 1922), 228. The drawings, and other evidence, were uncovered through extensive research in 1978 by Margaret Walk, Archivist to the Savoy Hotel (unpublished typescript in Savoy archives).

80 For example, of the Palace Theatre, *The Building News* stated 'The theatre being built in Cambridge-circus by Mr. R. D'Oyly Carte is from plans prepared under his direction by Mr. Holloway; this gentleman carrying out the building operations', *BN* **56** (26 April 1889), 582.

81 *The Builder* **57** (13 July 1889), 30.

82 Ibid, 29.

83 The architects for these alterations, forming 'a record for British work', were T E Collcutt & Stanley Hamp; the general contractors were Leslie & Co. Ltd, of Kensington Square, and the sub-contractors for the steelwork were Andrew Handyside & Co. Ltd. of Derby. *See* 'The Savoy Hotel Extension', *C&CE* **5** (October 1910), 869–79.

84 Perry had much experience with structural ironwork. His first company, Perry and Co., of Bow, had executed The New 'Pelican' Club on the corner of Gerrard Street and Shaftesbury Avenue (1889, demolished). This modest building was constructed 'on an entirely novel principle, consisting practically of three large and lofty rooms, connected by a grand 5ft staircase, which is built in the rooms themselves, and quite open'. Designed by Martin and Purchase, it was reported that 'the building internally is principally of iron construction'. *BN* **57** (5 July 1889), 281.

85 *The Builder* **55** (20 October 1888), 285.

86 Clunn 1962, 118.

87 *The Builder* **69** (19 October 1895), 277.

88 Ibid.

89 Ibid.

90 Ibid.

91 'Restaurant Fireplace in the Hotel Cecil', *BN* **70** (24 April 1896), 599.

92 Ibid.

93 *The Builder* **69** (19 October 1895), 278.

94 Ibid, 276–7.

95 *The Builder* **72** (24 April 1897), 382; **73** (13 November 1897), 401–2.

96 *See* Clarke 2013, 150. Nash's use of coved cast-iron clerestorey lighting at Corsham (1797–1802) and subsequent buildings provided additional wall space, and a more diffuse lighting effect, but did not carry buildings above.

97 *See* 'New Claridge's Hotel', *BN* **71** (4 December 1896), 826. This source provides some information on the fire-resisting floors and roof: 'The floors throughout are fire-resisting, being constructed of steel and concrete formed of 1 of Portland cement to 6 of broken brick laid on a centring of corrugated iron, which is bent to an arched form about 4ft. 6in. in span; wood ceiling joists are notched on the steel joists, and to these fibrous plaster slabs will be fixed to form the ground for the plaster ceilings. The roof will be of steel and concrete, similar to the floors, with the exception of the iron centring.' For a history and architectural description of this building, see Survey of London 1980, 24–9.

98 *Specification* **1** (1898), 126, 128, 130, 132 (illustrations within the section entitled

99 Childs 1898, 235. It is worth noting that an early extant provincial example of this technique is the Cavendish Hotel, Eastbourne (1883–4), built to designs by T E Knightley. The ground-floor coffee room, 60ft by 33ft, was spanned by three large box girders, weighing 18 tons apiece, each calculated to bear a load of 245 tons. These girders, supplied by Rownson, Drew and Co., and presumably seated on concealed stanchions, 'form a foundation for a building four stories above', 'Cavendish Hotel, Eastbourne', *BN* **46** (13 June 1884), 912.

100 There is some disagreement among the standard secondary sources over the build dates, but a date range of 1897–1900 seems most accurate: Doll's (undated, amended) plans are stamped 17 June 1897 by the LCC Building Act Office; *The Builder* **77** (8 July 1899), 47 noted that the building was still in the process of construction, and *The Daily News*, 30 April 1900, 7, reported the hotel formally opened.

101 ICE membership records.

102 Cherry and Pevsner 1998, 326

103 *The Builder* **76** (11 February 1899), 143.

104 *The Builder* **82** (25 January 1902), 85.

105 'The Savoy Hotel Extension', *BJ&AR* **19** (22 June 1904), 292.

106 'Additions to the Savoy Hotel', *BN* **73** (5 November 1897), 651. *The Builder* exclaimed 'The construction is entirely of steel in the form of stanchions, girders to floors, and trusses to the roof', *The Builder* **71** (11 July 1896), 34.

107 Most notable among the buildings demolished to make way for the Strand widening was Simpson's Divan and Tavern, a well-known eatery erected in 1848 by John Simpson, a caterer. Weinreb *et al* 2008, 839. Other demolished buildings included 'Beafort Buildings', 'Rimmels' and 'French's'. 'The Savoy Hotel Extension', *BJ&AR* **19** (22 June 1904), 294.

108 For examples of this company's work *see* James Stewart & Co. 1909, 1944.

109 'A Palace in the Strand', *Out and Home* (Savoy Archives contemporary cuttings file, nd, *c* May 1904).

110 'New Midland Hotel, Manchester', *BJ&AR* **17** (20 May 1903), 203–5; *The Builder* **84** (24 January 1903), 95.

111 James Stewart & Co. 1909, 59.

112 *The British Clay Worker* **13** (May 1904), 68. The *BJ&AR* **19** (22 June 1904), 292, offered more comment, noting that 'although the steelwork carried all the weight the London County Council would not allow thinner brick walls. In Manchester the by-laws allow such walls to be built, and the London by-laws ought to make

this provision so as to avoid waste'. This stipulation must have been a source of irritation to Stewart, who was used to cladding American frameworks with thin skins of brick and terracotta.

113 These terms were used to describe Stewart's on-site methods by *BJ&AR* **19** (22 June 1904), 292.

114 *BJ&AR* **19** (22 June 1904), 292.

115 The wooden chute ran the entire height of the building, and was provided with openings at each floor. Just three men, constantly shovelling the dust and rubble down the chute, were needed compared to the 15 or so required by the sack method. Ibid, 292.

116 'The details of this steelwork were of the most complicated nature, and it reflects great credit on Messrs. Dorman, Long that the work was so successfully carried out'. Ibid, 292.

117 In England, this was known as 'Johnson's Wire Lattice system' of reinforced concrete construction, and was manufactured from *c* 1904 by Richard Johnson, Clapham & Morris of Manchester, *BJ&AE* **24** (C & S Supplement, 7 November 1906), 87. Introduced 'in response to suggestions of American engineers engaged in the construction of the Savoy Hotel extension', who 'had been accustomed to using continuous wire lattice in the United States' (ibid.), the key figure in the floor system's introduction in this country was probably Louis Christian Mullgardt, not Stewart. *See* note 121.

118 Information kindly supplied by Lawrance Hurst.

119 Friedman 2010, 119–20.

120 Ibid, 121.

121 BJ&AR **19** (22 June 1904), 298 . However, the British licensees, Johnson, Clapham & Morris, had previously used their own type of wire lattice for reinforcing a concrete conveyor cover alongside No. 9 Dock, Manchester Ship Canal.

122 Ibid, 292.

123 LMA GLC/AR/BR/17/022092.

124 Seemingly only one contemporary publication, *The British Clay Worker*, credited him with the work, noting 'The steel structural work has been in the hands of Mr. L.C. Mullgardt, F.A.I.A., and American methods in this direction have been employed up to the full limit allowed by the Municipal Building Acts in the metropolis', 'Notes by F.R.I.B.A – the new Savoy', *The British Clay Worker* **13** (May 1904), 68. The *BN* **87** (9 September 1904), 357, referred to him insofar as he chose to employ the lattice-wire reinforced concrete floor system. Mullgardt remained uncredited in James Stewart & Co. 1909, 59, an omission rectified in the subsequent James Stewart & Co. 1944, 86 and 89.

125 This schematic, dimensioned, drawing, dated September 1903, and annotated 'First Floor Framing Plan', is stylistically very different to the type of floor plan, showing scaled joists and girders, that Collcutt produced in 1896. LMA GLC/AR/BR/17/022092.

126 *Architect and Engineer* **35**, 1913, 47.

127 Verplanck 2001, 7.

128 *BN* **88** (3 March 1905), 312.

129 'The working stresses used were in accordance with the practice in New York at that time, as given in the Carnegie Steel Company's Handbook, issued 1897 …', Bylander 1937, 3.

130 The most detailed modern accounts of the construction are provided by Binney 2006, 90–97; Lawrence 1990, 30–1; Montgomery-Massingberd and Watkin 1980, 31–47.

131 *BJ&AR* **20**, 30 November 1904, 286; Lawrence 1990, 30.

132 '[N]o beam is milled at the ends to fit the lower flange of the beam on which it is resting, as usual in English practice, but strong connections are used to keep the beams laterally supported and to transfer the loads centrally to the beams', *BJ&AR* **21** (22 March 1905), 150.

133 'Steel and Concrete at the Ritz Hotel, London', *C&CE* **1** (January 1907), 450.

134 Binney 1999, 74. This source actually gives the date of 1902 for Freitag's book, which survives in the company archives of Bylander 2000 Ltd (formerly Bylander Waddell Partnership Ltd) in Harrow. In fact there were two editions of this American book, published in 1895 and 1901, the second revised and re-published in 1904 and subsequent years.

135 Bylander 1937, 2.

136 *BJ&AR* **21** (22 March 1905), 150.

137 *BJ&AR* **20** (28 September 1904), 165.

138 For example, an unidentified warehouse in Peckham, built before 1898 for Jones & Higgins. The structural engineer, T C Cunnington, based the design on those used in support of the 16-storey Marquette Building, 140 South Dearborn Street, Chicago (1893–4; architects Holabird and Roche). For brief details of the Peckham warehouse, *see* Cunnington 1898b, 349; for the steel skeleton-framed Marquette Building, *see* Condit 1964, 120–3.

139 'Steel and Concrete at the Ritz Hotel, London', *C&CE* **1** (January 1907), 448.

140 Binney 1999, 23: 'The roles of Mewès and Davis were often inseparable, but Davis, as the partner in London, clearly had a major role in developing building plans.'

141 *BJ&AR* **21** (22 March 1905), 148.

142 *BJ&AR* **20** (28 September 1904), 164–5; (2 November 1904), 235–7; (30 November 1904), 286–9; **21** (1 March 1905), 111–12; (22 March 1905), 148–56;

(12 April 1905), 184–6; (24 May 1905, 284–6; **22** (13 September 1905), 146–9; *BJ&AE* **23** (10 January 1906), 22–3; (18 April 1906), 210 and preceding two (unpaginated) pages.

143 Lawrence 1990, 31.

144 Middleton 1906, 134.

145 *BJ&AR* **21** (5 April 1905), 177.

146 Middleton 1906, 134.

147 Bylander 1937, 3, emphasis added.

148 'The Waldorf Hotel, London', *AR* **23** (March 1908), 176–83; *MBR* **1** (1910), 159.

149 Bylander 1937, 7.

150 *BJ&AE* **25** (C & S Supplement 27 February 1907), 22.

151 'The Waldorf Hotel, London', *AR* **23** (March 1908), 176.

152 *BJ&AE* **23** (30 May 1906), 53; Saint 1976, 378.

153 Bylander 1937, 2.

154 'The Piccadilly Hotel, London', *AR* **24** (October 1908), 199–208.

155 Ibid, 199.

156 *BJ&AE* **25** (Fire Supplement, 13 March 1907), 23.

157 *See* 'Steel and Concrete at the Ritz Hotel, London', *C&CE* **1** (January 1907), 453–4.

158 'The Piccadilly Hotel, London', *AR* **24** (October 1908), 199.

159 Fleetwood-Hesketh 1971.

160 Service 1977, 163.

161 'The New Royal Automobile Club', *C&CE* **5** (July 1910), 467.

162 'Foreign steel, excepting Belgian, was permitted, and the whole of the material was inspected during manufacture at the different stages, and also during the process of building up in the shops by Messrs. Robert Hunt & Co., without whose approval as regards quality and workmanship no material was allowed to be delivered on site', ibid, 472.

163 Ibid, 467.

164 Survey of London 1960a, 417.

11 Banks and offices

1 *BN* 68 (1 February 1895), 150–1.

2 Bradley and Pevsner 2002, 101.

3 Holden and Holford 1951, 173; Black 2000, 357.

4 The use of fire-resisting materials in banks was so commonplace by the early 1850s that the *Banking Magazine* could state that fires in banks were 'one of the rarest of accidents', *Banking Magazine* **12** (1852), 547 as quoted in Booker 1990, 80.

5 Ibid, 11.

6 Black 2000, 357.

7 Booker 1990, 38.

8 A photograph showing the jack-arch construction is included in National Monuments Record Building Index File 94565. The use of an iron roof, presumably of cast-iron members, was

noted in *The Gentleman's Magazine* **99** (December 1829), 637, which stated simply 'Its roof and rafters are iron; the front and sides of Portland stone'.

9 Another early documented use of skewback girders in London was in the Springfield Asylum, Glenburnie Road, Wandsworth (1839–41). Inverted Y-shaped and V-shaped girders were used extensively throughout Jesse Hartley's Albert Dock Warehouses, Liverpool (1843–46).

10 Hitchcock 1954, 353.

11 *ILN* **15** (1849), 11, as quoted in Booker 1990, 74.

12 *The Builder* **23** (25 November 1865), 834.

13 Black 2000, 371.

14 *The Builder* **11** (18 June 1853), 392.

15 Booker 1990, 83.

16 Locker 1980, 57, citing Campin 1896, 133.

17 For a discussion of the origins of tile creasing and its historic uses, see Hurst 1996, 289–90, 2009, 1–2.

18 Clarke 2001a. *See also* Kelsall 2001.

19 Richard Moreland noted 'Perhaps the best fireproof buildings are composed of brick or stone, with iron beams and columns, properly tied together with rods built into the walls, and brick arches for the floors', Moreland, R 1884, 295.

20 I'Anson 1864–5, 31; Bradley and Pevsner 2002, 112. Annesley Voysey, who died from influenza in Jamaica in 1839, was grandfather of the more renowned C F A Voysey (1857–1941).

21 Hitchcock 1954, 375.

22 Ibid.

23 18 and 19 Victoria, c. 122; The Metropolitan Building Act 1855; *The Builder* **14** (20 December 1856), 691.

24 For a detailed history of both of these buildings, *see* Jefferson Smith 1997.

25 Locker 1984, 129.

26 Laxton 1857, plate 80.

27 *The Builder* **16** (27 March 1858), 216, original emphasis.

28 *The Builder* **21** (3 January 1863), 13.

29 Ibid. The plain columns were fitted with separately cast delicately foliated capitals, carved brackets and enriched base mouldings; the box girders were adorned with perforated cast-iron panels, the junction boxes over the columns were ornamented with medallion plates, and the rivet heads on the underside of the girders fitted with cast-iron paterae. The general contract was undertaken by 'Mr. Myers' (probably George Myers, A W N Pugin's builder), while 'Mr. Grissell, Mr. Shaw, and Mr. Barrett [were] severally … connected with the ironwork'. Of these latter three, one may have been Henry Grissell, of the Regent's Canal Ironworks (who supplied the ironwork for Greene's Sheerness Boat Store (1858–60) and another Matthew

Turner Shaw, later of Matthew T Shaw & Co.; however, the Mr Barrett cited cannot have been James Barrett of Fox & Barrett, since he died in April 1859.

30 Hitchcock 1949, 73, remarked 'Except that the masonry screen is self-supporting, it is, therefore technically not unlike a skyscraper façade. There is no wall as such and although the exterior masonry skeleton is arcaded, it expresses rather clearly the interior skeleton of metal'. More recently, Crook 1987, 97, mused 'Ruskin's theory of the wall-veil – the façade as expressive skin, an idea derived from medieval Italy – had been turned to novel advantage. The future development of the Chicago skyscraper had been foreshadowed, in miniature, in a London City back street'.

31 'New Offices, Mark-Lane, City', *BN* **11** (19 February 1864), 134.

32 Ibid.

33 Ibid.

34 Aitchison 1864, 102.

35 *The Builder* **24** (24 February 1866), 136.

36 *BN* **15** (17 January 1868), 44.

37 For example, Homan & Rodgers flooring was used in a five-storey speculative office block erected *c* 1877 on the corner of Coleman and Gresham Streets, designed by J & W Wimble (demolished). *See The Builder* **36** (23 December 1878), 193.

38 *The Builder* **32** (10 January 1874), 29. This building was designed by a Mr Chancellor, possibly Frederick Chancellor (1825–1918).

39 Hurst 1990, 37–8.

40 Locker 1984, 139.

41 Bradley and Pevsner 2002, 113.

42 'City Offices, Old Broad Street, and Bishopsgate-Street', *BN* **14** (22 February 1867), 142.

43 Ibid.

44 Bradley and Pevsner 2002, 115, 578.

45 *The Builder* **29** (1 July 1871), 506.

46 *The Building News* could report that 'The shop-front forming part of our illustration is the only one which has been carried out in accordance with the architects' plans,' *BN* **25** (4 July 1873), 8.

47 Prudential Assurance Company. Board Minutes **5**, 4 November 1875, 186, as quoted in Cunningham *c* 1991, 7.

48 Cunningham 1992, 111–6, 166.

49 Cunningham 1992; Cunningham *c* 1991.

50 *The Commercial World*, 2 June 1879, 144–6.

51 Turvey 1993–4, 147.

52 Ibid.

53 This famous comment by Cass Gilbert, architect of the Woolworth Building, New York (1910–13), is quoted in Landau and Condit 1996, xiii.

54 A complex subject, covered by such works as Kerr 1865, Fletcher 1902 and Roscoe 1904.

55 The girders 'were tested … with a dead weight of 100 tons in the centre, causing a deflection of 1¾ inch, and leaving but a slight permanent set', 'Steel Girders at the London and County Bank, Sheffield', *BN* **7** (17 May 1861), 422. According to *The Builder*, the deflection was even less: 'Two of these, [girders] … deflected, we are informed, only 1 inch each with a distributed weight of 200 tons, *The Builder* **20** (23 August 1862), 604.

56 'New City Buildings', *BN* **7** (26 April 1861), 359.

57 'With the external walls, one large column in the centre sustains the superstructure, containing twenty-five rooms, some of large dimensions', *The Builder* **20** (23 August 1862), 604.

58 *The Builder* **20** (23 August 1862), 604.

59 'New City Buildings', *BN* **20** (26 April 1861), 359.

60 *British Almanac and Companion* (London, 1862), 273, quoted in Black 2000, 365.

61 *The Builder* **20** (23 August 1862), 604.

62 Ibid; Weinreb, B et al, eds 2008, 1009.

63 Black 2000, 364.

64 *The Builder* **52** (21 May 1887), 763.

65 Cunningham, *c* 1991, 14.

66 Steel columns were sometimes in place of those of cast iron. The London and County Bank, Lombard Street (1907, demolished *c* 1965), rebuilt to the designs of W Campbell Jones on the site of Parnell's earlier building is one example. In its construction, which had to proceed nocturnally to permit daytime banking business, the existing load-bearing walls were retained, heavy steel girders and columns being inserted within the shell by H Young & Co., of Nine Elms Ironworks. Swiss Cippilino marble cylinders, sawn in half and hollowed out, were used to encase four 15½ft-high steel columns situated in the banking hall. *AR* **21** (June 1907), 323.

67 William Warlow Gwyther (1829–1903) was a prolific but curiously unsung architect. In the 1890s, he rivalled Alfred Waterhouse as a freelance bank specialist of truly national standing, with many prestigious commissions besides the Bishopsgate building. Nevertheless, this 'safely classical architect … appears to have made little impact on colleagues in his own profession', Booker 1990, 83.

68 All constructional iron and steelwork was supplied and erected by Joseph Westwood & Co., of Millwall. The 'fireproof' floors were constructed on 'Picking's twin-arch principle'. *The Builder* **71** (24 October 1896), 334.

69 Barclays Group Archives: library C65 (a fragmentary memorandum found in the 'time capsule' recovered during the complete rebuilding of the site in 1960–61) and Barclays Group Archives: 38/639 (a book of valuation reports incorporating a ground floor plan by Blomfield dated 7 July 1896). I am grateful to Nicholas Webb, Archivist for Barclays Group Archives, for informing me of these records.

70 *The Builder* **64** (27 May 1893), 408.

71 Ibid*; BN* **66** (18 May 1894), 698.

72 Jackson 1997, 46, citing RIBA Drawings Collection, WATA[40] and RAN[41].

73 Cunnington 1898a, 327.

74 Cunnington also employed this type of girder in the ceiling of the newsroom of Shoreditch Central Library, Hackney (1895–7, architect H T Hare). Cunnington 1898a, 327.

75 Taylor 1965.

76 For an appraisal of the structural development of iron-framed domes, *see* Sutherland 2000.

77 'New Buildings for the Birkbeck Bank Chambers', *BN* **83** (4 July 1902), 9.

78 Doulton and Company introduced Carraraware, a matt-glazed stoneware with a crystalline glaze similar in appearance to Carrara marble, in 1888. The Birkbeck Bank, which made extensive use of the material both externally and internally, was its first major showcase. *See* Stratton 1993, 103–4.

79 'The New Birkbeck Bank', *The Times*, 23 June 1902, 12.

80 Following renovation of the Prudential Corporation head offices it was confirmed that 'The Waterhouse buildings are among the earliest examples of steel construction: load-bearing brick exterior walls with terracotta ornamentation, internal riveted steel columns and girders built up from angles and plates, and rolled-steel beams supporting, in the main, filler joist floors', Atling 1993, 6.

81 Mansford 1907, 135.

82 Ibid, 135.

83 Jackson 1997, 45.

84 Phoenix columns of iron were used exclusively in 24 offices, banks, churches and industrial buildings in the period from 1887 to 1890, including the Union Trust Building in New York, although only the Hoyt Building in the same city used Phoenix columns of steel. *See* Misa 1995, 60.

85 *The Engineer* **88** (18 August 1899), 175. Other details are given in the same publication for 29 July 1898, 5 and 10 January 1902, 38–40.

86 *The Builder* **75** (10 August 1898), 172.

87 Locker 1984, 288.

88 Faber 1989, 9.

89 Richard Moreland junior, in a discussion of Cunnington 1902, 85.

90 Extracts from a report by A T Walmisley to Messrs Gordon & Gunton on the steelwork supplied and erected at premises in St Paul's Churchyard in Moreland, R & Son Ltd *c* 1905, 6–7.

91 Locker 1984, 289.

92 'Belcher and Joass', in Service 1975, 317.

93 Cunnington 1902, 68–9. The steelwork was supplied and erected by W Lindsay & Co. *MBR* 1, 1910, 81.

94 Statham in reply to Cunnington 1902, 85.

95 *A&CR* **82** (8 October 1909), 232.

96 Horace Gilbert (1860–1932) and Stephanos Constanduros (b1871) set up practice at No. 43 Finsbury Square in 1902; over a 30-year career they produced a wide variety of buildings mostly in the West End of London, including cinemas, flats, garages and offices. One of their most accomplished works was Harcourt House, 19 Cavendish Square, a steel-framed office and apartment block erected in 1907–10 on the site of the famous 18th-century mansion of that name (demolished 2007).

97 *BN* **86** (10 June 1904), 833. Locker 1984, 290, notes that J Greenwood Ltd were the builders.

98 Gibson stated 'The necessity for an abundance of light has left its mark on the design', *The Builder* **91** (15 September 1906), 324; Locker 1984, 291.

99 *The Builder* **91** (15 September 1906), 324.

100 *MBR* 1, 1910, 77.

101 *The Builder* **91** (15 September 1906), 324.

102 Gibson, Skipwith and Gordon, in collaboration with H T Wakelam, engineer and county surveyor. The building was designed with self-supporting outer walls and an interior steel frame, executed by Redpath, Brown and Co. Ltd. *See* 'Middlesex Guildhall, Westminster', *The Builder* **105** (14 November 1913), 519–23.

103 Lingard 2007, 78.

104 *The British Architect*, 1 March 1907, 146.

105 Ibid.

106 Designed by H O Cresswell. *See BJ&AR* **17** (29 April 1903), 156–7.

107 Inveresk Group, *The Morning Post Building, Inveresk House, 1 Aldwych* (booklet privately published by Lund Humphries, 1978).

108 *MBR* 1, 1910, 16.

109 *BJ&AE* **24** (C&S Supplement, 7 November 1906, 86.

110 Chatley c 1909, 73.

111 Bylander 1937, 8.

112 *AR* **23** (May 1908), 127–37.

113 'Everything was completely dimensioned, the length of every beam was calculated, and every shop detail was made to those dimensions, and all the steelwork was made in the shop from the details without assistance and special templates set out in the shop.' Bylander in discussion of Brown (1915). I am grateful to Ann Robey for alerting me to this quotation. Still later Bylander recalled 'I introduced a method which I afterwards used with advantage for skew dimensions of connecting beams to pillars', Bylander 1937, 8.

114 *BJ&AE* **24** (C&S Supplement, 7 November 1906), 85.

115 *See* C&S Supplements of the *BJ&AE* for 18 July 1906, 10 October 1906, 7 November 1906 and 27 February 1907.

116 *BJ&AE* **25** (C&S Supplement, 27 February 1907), 21.

117 *The Builder* **94** (25 April 1908), 485, as quoted in Locker 1984, 292.

118 'The Westminster Trust Building', *BJ&AE* **27** (17 June 1908), 510.

119 Others include No. 42 New Broad Street (1907, Ernest Flint, architect); the Union Bank of Scotland, Cornhill (1909, architect unknown, demolished); William Deacons' Bank, 127 Cheapside (1909, T H Smith, architect); General Accident and Assurance Co., Aldwych (1909, John Burnet, architect). Among the lesser-known works that Sven Bylander and the Waring White Company produced in this period was the stripped-classical British Electrical Federation Offices, Kingsway (*c* 1907, W Brunton, architect) above Holborn Underground Station).

120 *BJ&AE* **28**, 30 December 1908, 13; *A&CR*, 8 October 1909, 232.

121 *AR*, December 1908, 254–9.

122 The constructional steelwork for Koh-I-Noor House was undertaken by Drew Bear, Perks & Co. Ltd, *MBR* **2**, 1910, 102; *The Studio* **48** (1910), 339.

123 'The Westminster Trust Building', *BJ&AE* **27** (17 June 1908), 510.

124 The Kleine floor system incorporated thin, longitudinal strips of steel 'reinforcement' in the mortar joints between the hollow clay blocks; once concrete had been poured over the assembly, each joint acted as a slender reinforced concrete beam. It was first patented in Germany in 1892 by Johann Franz Kleine, who subsequently filed a number of British patents that refined the system, namely No. 2335 (1893), No. 18,295 (1896) and No. 21,046 (1905). *See* Elliott 1992, 47; Tappin, 2002, 93.

125 'The Westminster Trust Building', *BJ&AE* **27**, 17 June 1908, 510.

126 *C&CE* **3** (May 1908), 113, emphasis added.

127 Chatley *c* 1909, 77–8.

128 *BN* **89** (16 June 1905), 855; Middleton 1906, vol 2, 32.

129 *BN* **89** (16 June 1905), 855.

130 *The Builder* **89** (4 November 1905), 468.

131 'I used the same system [the steel frame principle] for the Morning Post, Waldorf Hotel, Oceanic, and Park Side buildings, all in London, and the Calico Printers' Association building in Manchester, and others', Bylander 1913, 71.

132 'There is a considerable amount of steelwork at these shipping offices, but the walls are solid and take the weight', *BJ&AR* **22** (26 July 1905), 44.

12 Stores, houses, churches, pools, fire stations and tube stations

1 Knight, S 1877, 19. I am grateful to Geraint Franklin for allerting me to this quotation.

2 Horsburgh 1907, 697. In some ways Horsburgh was wrong to blame metal for this development; Georgian shop-fronts often worked on the concealed (timber) bressummer principle. Information kindly provided by Peter Guillery.

3 Morrison 2006, 301–2.

4 Morrison 2003, 128.

5 Among the last true bazaars erected in London were The Corinthian Bazaar, off Oxford Circus (1867–8, Owen Lewis, architect) and St Paul's Bazaar, situated between St Paul's Churchyard and Paternoster Row (opened 1874). Morrison 2006, 301–2, 2003, 99.

6 For example, Piccadilly Arcade built 1909–1910 by G Thrale Jell and its younger, plainer sibling, the Princes Arcade (1929–1933).

7 Although itself anachronistic and ahistorical when applied to the 19th century, as a categorisation this term is perhaps preferable to 'department store', which works only as a functional rather than architectural categorisation. I am grateful to Kathryn Morrison for this discrimination.

8 Morrison 2003, 134.

9 For a detailed account of the structural development of French department stores, *see* Chapter 2 of Steiner 1984; also Saint 2008, 151–61.

10 *The Builder* **36** (20 July 1878), 752.

11 *The Builder* **35** (24 March 1877), 289.

12 Morrison 2003, 139.

13 *BN* **78** (23 February 1900), 263.

14 Ibid.

15 Survey of London 2000, 33.

16 Institution of Civil Engineers membership records.

17 Pevsner 1957, 534.

18 *The Builder* **89** (4 November 1905), 474.

19 *BJ&AE* **23** (C&S Supplement, 20 June 1906), 29.

20 Childs 1909b, 92.

21 For a description of this system, *see* 'The Columbian Fire-Resisting System', *BJ&AR* **19** (18 May 1904) (Trade and Craft section, xiii, bound at rear of volume).

22 *The Builder* **89** (4 November 1905), 474, emphasis added.

23 Saint 1976, 351.

24 Saint 1976, 357–60. For the Royal Insurance Offices, Shaw made alterations and improvements to (1896) competition designs by J F Doyle.

25 Service 1977, 198.

26 Ibid.

27 Jackson 1997, 54, citing Dowsett 1986 'Laurie McConnell store, Cambridge – a case history', National Structural Steel Conference.

28 *AR* **23** (June 1908), 362–9; *BJ&AR* **28** (3 December 1908), 30 and 140; *The Builder* **96** (20 March 1909), 346; *MBR* **1**(1910), 20–1.

29 *The Builder* **97** (18 December 1909), 670. Party-wall rules seem to have been waived, *The Draper's Record*, 11 September 1909, 637.

30 *The Builder* **97** (18 December 1909), 670; *BN* **98** (21 January 1910), 98.

31 Lawrence 1990, 28.

32 Ibid.

33 Ibid, 33.

34 Bylander, 1909, 9.

35 Although Bylander himself declared 'The Selfridge Store was designed as a complete steel frame building, with the exception that the external walls were self-supporting', he also noted 'The ground-floor piers were built sufficiently large in blue brick to carry the external wall as well as the load from the floors', Bylander 1913, 71. *BN* **97** (2 July 1909), 13, had earlier noted 'The outer walls, however, are constructional and the outer columns and piers carry their due proportion of weight without steel reinforcement.'

36 Bylander 1909, 280.

37 Morrison 2003, 162.

38 The London County Council General Powers Act, 1908. Part III. Amendment of London Building Act 1894.

39 Lawrence 1990, 40.

40 *AJ* **51** (18 February 1920), 222, as quoted in Lawrence 1990, 41.

41 'Whiteley's New Premises', *AR* **31** (March 1912), 164–78.

42 *BN* **3** (22 May 1857), 516.

43 For a description of Langbourne Buildings ['the model upon which all the subsequent buildings had, with minor variations, been constructed'] including the system of reinforcing used in the floors, *see* 'Dwellings for the Industrial Classes, Finsbury' *The Builder* **21** (21 March 1863), 198. *See also* Survey of London 1980, 93–8 (quoted above); Tarn, 1968, 43–59.

44 T Roger Smith, in discussing the subject of 'Dwellings of the Labouring Poor in Large Towns', 'did not think it would be warrantable, in reference to the buildings under consideration, to incur the heavy extra expense of making them fireproof', *BN* **3** (22 May 1857), 515.

45 Saint 1976, 114.

46 Ibid, 115.

47 Survey of London 1986, 154.

48 *The Architect* **32** (16 August 1884), 103; Cherry and Pevsner 1999, 579.

49 Until then, the only other British cities in which tall buildings in multiple occupation were built were Edinburgh and Glasgow. Gray 1985, 28.

50 The Metropolitan Building Act 1844 and an amending Act of 1862 restricted the height of buildings in new London streets, but as far as existing streets were concerned, there was no limit beyond an implicit requirement in the London Building Act 1855 that special consent was needed for residential and commercial buildings above 100ft. *See* Chapter 4.

51 'Queen Anne's Mansions and Milton's Garden', *The Builder* **35** (2 June 1877), 556.

52 Dennis, 2008, 233–). *See also* Mangeot 1939.

53 Full details of construction are provided in Hamilton *et al* 1964, 143–50.

54 53 and 54 Victoria, c. 218, The London County Council General Powers Act 1890.

55 Dennis, 2008, 238.

56 Survey of London 1994a, 403–4.

57 *The Builder* **56** (1 June 1889), 411.

58 Crompton 1928, 139.

59 Survey of London 1986, 75.

60 Greater London Coucil Historians File, 24 October 1977, 1.

61 Survey of London 1986, 75.

62 Slater, J 1887b, 673. Slater's paper, 'New materials and inventions', read before the first meeting of the Conference of Architects held by the RIBA, also 'alluded to the slowness with which novelties were introduced into buildings, and contrasted our conservatism in this respect with the eagerness with which fresh notions were taken up in America.' The same source also reported that 'He believed wrought-iron columns would in the future largely be used in this country, as the material resisted tension better and was more reliable than cast iron, and even steel was coming into use for this purpose'. A more detailed report of Slater's paper appeared in *The Builder* (Slater 1887a), and another version in *The Architect* **37** (6 May 1887), 268–9.

63 From *The Builder* **26** (11 April 1868), 265–6 (Victoria Hospital, Suez); and *The Builder* **36** (23 March 1878), 294, 296 (New Bridge over the Regent's Canal, at Gloucester Gate, Regents Park).

64 *The Builder* **57** (14 December 1889), 425.

65 *The Builder* **64** (7 January 1893), 15.

66 Survey of London 2000, 61.

67 *The Builder* **68** (18 May 1895), 381.

68 *See* Friedman 1995, 42.

69 Raymond 1897, *B.I. Barnato: a memoir* (London: Isbister and Co.; Capetown: Juta & Co.), 197, quoted in Crook 1999, 184.

70 Jackson 1970, 229.

71 *The Cabinet & Art Furnisher*, July 1897, 25.

72 Jackson 1970, 162.

73 Barnett Isaacs Barnato (1852–1897) threw himself overboard from the Scot on 14 June 1897, on his way from Cape Town to Southampton, and thus did not live to complete the mansion he commenced building in 1895. *The Compact Edition of the Dictionary of National Biography* **2** (Oxford: OUP, 1975), 129.

74 *The Builder* **71** (10 October 1896), 290.

75 Survey of London 1980, 204.

76 *BN* **73** (22 October 1897), 576; *BJ&AR* **13** (20 March 1901), 111–14 and (29 April 1903), 160.

77 Perks 1905, 164.

78 Even at this late date, *The Building News* for one could display terminological imprecision in its reference to the building's 'skeleton ironwork', *BN* **90** (16 March 1906), 382. Contrastingly, *The American Architect and Building News* (see note 82) noted that 'It is a steel framed structure with brick filling.'

79 'Gloucester House Flats: Notable Piccadilly Sale', *The Times*, 9 April 1932, 9; Sparrow 1907, 29.

80 Frank N Jackson, in replying to some of the points raised during discussion of his paper 'Modern steel building construction', noted 'Years ago, in a building known as Gloucester House, at the bottom of Park Lane, a complete system of tower bracing against wind was adopted. There they had put vertical stanchions in the outer wall, and internal stanchions about fourteen feet away, and these, he thought, were joined with vertical diagonal bracing with rigid gusset plate connections on the flanges', Jackson and Dicksee 1913, 487.

81 'A detail of connections … show its elaborate nature, and the peculiar way in which the plates have been riveted on the stanchions, projecting one beyond the other, completed with a small angle iron', *BJ&AR* **23** (contractors' supplement, 28 March 1906), 18.

82 *The American Architect and Building News* **91** (5 January 1907), 17. I am grateful to Isobel Watson for this reference.

83 *Country Life*, 25 April 1963, 903.

84 Letter to *The Builder* from architect Edward John Tarver (1842–91), *The Builder* **45** (17 November 1883), 649.

85 Hitchcock 1987, 170–1. *See also* Bannister 1950, 282, and Hughes 1964, 7, the latter also providing other pioneer examples, including Tetbury Church, Gloucestershire (1777–81), where the slender cast-iron columns were encased in decorative timber.

86 Information provided by Robert Bowles, Alan Baxter Associates.

87 The practice of concealing iron cores within timber pillars had earlier, less ambitious origins; the original pulpit of Nicholas Hawksmoor's St George's, Bloomsbury (*c* 1730), for example, was

iron-cored. Saint 2008, 500, note 21. The technique remained popular through the early 19th century and later; John Nash for example employed a ring of iron cored timber columns to support the audacious timber-framed polygonal roof of the 'Rotunda' ballroom erected in 1814 in Carlton House Gardens, a building he hoped would be reused as a church. *See* Clarke 2006.

88 St George's, Everton (1813–14), St Michael's, Toxteth (1814–15) and St Philip's, Hardman Street (1815–16). These are described in Hughes (1964, 136–45). *See also* Bannister 1950, 245–6.

89 Cragg's dominant role in this tripartite project and its undervalued importance to iron architecture is examined in Saint 2007, 73–5.

90 One of the earliest pioneers of the form was a church fabricated for shipment to Jamaica by London manufacturer Peter Thompson in 1846, 'whose "pilaster supports [were] of cast iron, on which [were] fixed the frame roof, of wrought iron, of an ingenious construction, combining great strength with simplicity of arrangement"', Herbert 1978, 98, quoting the *Illustrated London News*, 28 September 1844. Numerous examples followed throughout the 1850s, but despite the attempts of some manufacturers to produce works with more ornamented or even architectural character, Victorian ecclesiastical opinion tended to find their utilitarian, industrialised aesthetic unacceptable. The same criticism – itself part of the wider condemnation of the use of iron in the fifties – was not spared for the 'temporary' churches manufactured for the burgeoning home market, despite, in some instances, the enlistment of architects' services. In the wake of Samuel Hemming's St Paul's, Kensington (1855) – London's first temporary church in iron – a spate of similar structures employing cast-iron columns, and clad in corrugated iron, followed in the Metropolis, enduring well into the late 19th century.

91 *The Ecclesiologist*, August 1849, 64, as quoted in Survey of London 2000, 187.

92 *The Builder* **18** (31 March 1860), 200–1.

93 *See* Curl and Sambrook 1973, 1999.

94 Five pairs of cast-iron columns, each with Corinthian capitals of wrought iron were installed to support a new gallery and timber roof, 'giving the church a greater sense of height and space, and vastly improving the ventilation of the building', Rosoman 1980, 165.

95 A detailed account of the Notre Dame De France Roman Catholic Church is given in Survey of London 1966b, 482–6. Its replacement, retaining the name and the

circular plan of the Notre Dame De France Roman Catholic Church, opened in 1955.

96 According to the architects' own belief, *The Builder* **43** (2 December 1882), 727. Another church to employ wrought-iron trusses in this decade was All Saints' Mission Church, Boothen, Stoke-on-Trent, erected in 1887 to designs by Lynam & Rickman, architects; *see The Architect* **37** (14 January 1887), 24.

97 *The Builder* **43** (2 December 1882), 727.

98 *The Builder* **49** (19 December 1885), 875.

99 Ibid.

100 For his last 5 years, R M Ordish worked for Dennett and Ingle of No. 5 Whitehall. Clarke 2002, 60; 'Ordish, Rowland Mason (1824–1886)' in Cross-Rudkin and Chrimes 2008, 590–92.

101 *BN* **64** (24 February 1893), 261; *A&CR* **49** (supplement, 17 February 1893), 22–4.

102 Cherry and Pevsner 1998, 661.

103 Saint 1976, 289.

104 *The Architect* **42** (18 October 1889), illustration after p 222 showing Shaw's Cross Section of the church.

105 Saint 1976, 428.

106 Ibid, 289.

107 Leonard 1997, 173.

108 'The late Mr. Sedding tried very hard to induce his clients to adopt that style of classic which prevailed in Rome during the pontificate of Alexander VII. He succeeded in one instance – the Holy Redeemer, Clerkenwell, where steel stanchions and girders are enclosed in columns and entablatures cast in concrete', Maskell 1905, 134. *See also* Day 1911, 19, and Clarke 1966, 57. I am grateful to Rosalind Woodhouse for providing me with these references.

109 *Holy Redeemer Church and Parish Chronicle*, May 1898.

110 'The late J. D. Sedding', *The Architect* **45** (8 May 1891), as quoted in Boucher 1988, 35.

111 Survey of London 1980, 88.

112 *The Builder* **60** (9 May 1891) 372.

113 Saint 1976, 338.

114 For a description of this building, *see* Saint 1976, 357–9.

115 Dixon and Muthesius 1985, 224.

116 The lantern, however, was designed by Sherrin's young assistant, Edwin Rickards (1872–1920). Survey of London 1983, 54–5; Metcalf 1972, 138–9.

117 Ibid.

118 Middleton 1906, vol 4, 132.

119 For the building's construction, see 'The Wesleyan Hall, Westminster', *C&CE* **5** (October 1910), 720–31.

120 Landau and Condit 1996, 422, note 29.

121 'Steel frame construction for churches', *BN* **70** (29 May 1896), 774–5.

122 St Catherine's Church, Hammersmith (1922–3; destroyed 1940), was reputedly the first skeleton-framed church in England. Designed by Robert Atkinson (1883–1952)

in collaboration with Oscar Faber, the steelwork was fabricated and erected by Archibald Dawnay & Sons. A photograph showing the steel skeleton appears in Archibald D Dawnay & Sons, Ltd, 1927, 121.

123 *BJ&AR* **15** (14 May 1902), 195.

124 *The Times*, 3 December 1901, 4.

125 *BJ&AR* **15** (14 May 1902), 195. Further information about the building can be found in Survey of London 2008b, 390–1.

126 Allsop 1894, 34.

127 *The Builder* **9** (8 February 1851), 90. The building is described in *The Builder* **4** (3 October 1846), 470.

128 'S. Marylebone New Swimming-Bath', *BN* **26** (5 June 1874), 687; 'St. Marylebone New Swimming Bath', *The Builder* **32** (6 June 1874), 474–5. The builders for the whole works were Perry Brothers.

129 Allsop 1894, 33.

130 Loobey 2002, 117; Nine Elms Baths, Battersea (adaptability to public hall with 1,500 capacity). *See* Survey of London 2013, 267.

131 Cross 1906, 11.

132 This 'amphitheatre system', with the dressing-boxes placed on the level of the top row of the spectators' galleries rather than under the gallery staging, was designed to bring 'each occupant of a box under the direct observation of the bath attendant'. Haggerston seemingly represented one of the earliest uses of this system: 'As far as I am aware, there are in England only two or three other swimming baths having their galleries constructed on this, the amphitheatre system, viz., in those recently erected at Battersea … and in the Corporation Baths in Leicester'. Cross 1906, 12.

133 Ibid, 50.

134 Visually, Frank's interior design for the roof of this first-class main pool was a virtual carbon-copy of Cross's Haggerston building, but constructionally it was very different: 'The architect considered that even if the roof trusses had been carried out in steel and covered with 2in. of some non-conducting external coating, such a structure would be less fire-resisting than ferro-concrete, and the ribs showing in the ceiling of this bath are true structurally. Moreover, steel roof trusses would have required periodical painting; but with reinforced concrete the necessity for maintenance hardly arises', *BJ&AE* **30** (31 December 1909), 648.

135 Weinreb and Hibbert 1993, 487.

136 London Metropolitan Archives, GLC/AR/ PL/18/0201.

137 'Fire and fire extinction', *Encyclopaedia Britannica*, 9 edn, 1912, 413.

138 *BN* **9** (10 October 1862), 268 as quoted in Lawrence 1994, 10.

139 *The Builder* **21** (10 January 1863), 23.

140 For example, the surviving Farringdon Street Station (1865–6), which replaced a temporary brick-walled terminus with timber queen post roofs, employs shallow wrought-iron bowstring roof trusses carried on cast-iron columns and brick side walls. *See* Survey of London 2008a, 366.

141 'The Central London Railway', *The Engineer* **86** (4 November 1898), 439–42.

142 Lawrence 1994, 28; Stratton 1993, 119–21.

143 Quoted in Powers 1987, 10. *See also* Leboff 2002.

144 Powers 1987, 10. This is given credence in at least one case, Knightsbridge Station, where the building above was constructed by the Waring White Building Co., Ltd – a firm particularly active in steel-framed architecture.

145 Eg *The Transport & Railway World*, 6 December 1906, 527.

146 Evans 1997.

13 Industrial buildings

1 *The Builder* **49** (12 September 1885), 349.

2 Tucker, 'Warehouses in dockland', in Carr 1986, 22.

3 Ibid, 28.

4 Guillery 1999, 79; Tucker, 'Warehouses in dockland', in Carr 1986, 26.

5 The first textile mill to employ hollow-cylindrical columns was reputedly the Salford Twist Mill, Greater Manchester, built in 1802 to designs by George Lee; by the end of the decade this form had almost fully ousted the cruciform section in buildings of this type. For details of the Salford Twist Mill, *see* Fitzgerald 1988, 128–9 and Skempton and Johnson 1962; for early developments in Greater Manchester *see* Williams with Farnie 1992, 58–65.

6 Survey of London 1994b, 655–8.

7 Survey of London 1994a, 291–3.

8 Securely dated examples show that this form was still being used in London warehouse construction in the late 1890s, in conjunction with wrought-iron and steel beams. For example, a five-storey warehouse built in 1898 in Tower Hamlets for the storage of wool utilised cruciform section columns to support built-up wrought-iron or steel beams carrying timber-joisted floors. Only in some portions of the building were rolled-steel stanchions used, and these may well be the result of later rebuilding. *See* Alan Baxter & Associates 2000; also Adams 1894, 108–20, on the working procedures for the design of such stanchions.

9 *See* Hamilton 1940–41, 151–3. Although Hamilton suggests this column assembly may well have been designed by Sir John Rennie (1794–1874), a more likely candidate is J Nesham, who was engineer to the London Docks for most of the 1840s.

10 Renamed as the Royal Victoria Victualling Yard in 1858.

11 For a historical outline of the Royal Victoria Yard *see* Leyland 1901a, 1901b.

12 Williams with Farnie 1992, 61–2. For example Boat House No. 6, Portsmouth naval dockyard (1845–6), designed by Captain Roger Steward Beatson (1812–1896).

13 Information kindly provided by Jonathan Coad, 1989.

14 'The Mode of Connecting Iron Columns in Tiers', *The Builder* **22** (17 December 1864), 916–18.

15 Although Hitchcock 1954, 526, noted that James Ponsford built the mill 'without benefit of either engineer or architect', the authorship of the structural design is firmly attested by *The Builder* **22** (17 December 1864), 916–18.

16 This scheme was proposed in 1856, shortly after Fairbairn 1854, which illustrates a similar floor structure.

17 Guillery 1999, 83.

18 The fallibility of this form of construction, and the fallacy of the (misnamed) term 'fireproof', was underscored in 1861 when the Great Fire of Tooley Street, Southwark, consumed much of Hay's Wharf. One of the direct outcomes of this disaster was the establishment in 1864 of the Wharves and Warehouses Committee by the London Fire Offices, which specified modifications to existing buildings, and best practice in new buildings, albeit with no strictures against the constructional use of cast iron.

19 *The Builder* **26** (14 March 1868), 200.

20 Survey of London 2008a, 228.

21 *The Builder* **48** (7 February 1885), 198 and *BN* **51** (24 December 1886), 982.

22 Smith 2000.

23 Many of the rolled-iron joists used in support of the concrete-arch floors and spanning openings bear the rolling mark 'Cie BELGE PROVIDENCE Δ' (the stamp of Forges de la Providence, Charleroi), and it was probably this manufacturer who supplied the built-up beams also.

24 The fire began in the premises of Foster, Porter, and Co., wholesale hosiers sand mercers in the early morning of 8 December 1882, and quickly spread to engulf 'the labyrinth of streets between Aldersgate and the Guildhall'. *The Times* also noted 'Thirty years ago the buildings were smaller, and the streets narrower, but in the past few years huge buildings have been built for use as warehouses, and in these the amount of material stocked is immense'. 'The Great Fire in the City', *The Times*, 9 December 1882, 6.

25 James's obituary, *JRIBA* **21**, December 1913, 75; as quoted in Service 1975, 311. Further details in 'The Late Great Fire in Wood Street: Extensive New Building Works', *The Builder* **44** (21 April 1883), 552.

26 Among the earliest documented examples are Palm Mill, Oldham (1884); Atlas No. 6 Mill Bolton (1888) (*see* Williams with Farnie 1992, 109) and No. 1 Mill, Ferguslie Mills (*see* Jones 1985, 162).

27 Tucker, 'Warehouses in dockland', in Carr 1986, 29. By way of comparison, a warehouse of 1885 at 8–10 Henry Street, Liverpool, has steel filler-joisted floors. *See* Giles 1998, 21.

28 One of the earliest known uses of steel joists in buildings of this type is a two-storey brick warehouse in Ealing put up for Charles Steel, a local landowner and fruit grower. No. 53 Northfield Road, Ealing has been dated to the mid-1880s, and it has been suggested that the 5in. by 12in. rolled I-section joists supporting the first floor may be steel, chiefly on the basis of their large sectional size and length. *See* Wilson 1999.

29 Of many warehouses under construction in 1885, perhaps the best candidate for the source of this quotation was a monster block of nine six-storey warehouses (including basement and attic), completed in 1887 at St Paul's Churchyard, City (demolished). Eventually unified behind a harmonising Portland stone and granite façade, they were originally independently conceived in two separate groupings of six and three by Delissa Joseph (centre block and north wing) and Frederick Hemmings (south wing), respectively. Contemporary description refers only to the 'large amount of constructional ironwork in the buildings, which was supplied by Messrs. Williams & Co.' [possibly Charles Williams & Co., Ferry Ironworks, Cubitt Town, Isle of Dogs], but this might well be another example of terminological imprecision. Certainly, the scale and prestige of this well-capitalised project suggest that 8,000 tons of steel could have been used here. *See The Builder* **51** (9 October 1886), 518; **52** (5 March 1887), 348 (quotation above); *BN* **48** (27 June 1885), 898; and *BJ&AR* **3** (3 March 1896), 57 (illustration of exterior).

30 *The Builder* **52** (16 April 1887), 591.

31 'Tonic for Gilbey's', *AJ* **190** (16 August 1989), 12.

32 *See* Whitehead 2000, 68–9; Stratton 1999, 15; Jackson 1998, 24.

33 For example, the internal construction of the Putney Wharf Mills (c 1905, demolished) comprised 'principally of steel girders and iron columns, the latter bearing also the shafting for driving the machinery'. This large, architecturally distinguished

complex, was designed by Herbert Hillier for the manufacture and storage of ropes and twines. *BN* **88**(31 March 1905), 457.

34 Clarke 2002.

35 *See* 'Bovril and its New Home in London', *Grocery* **2** (February 1900), 115–28.

36 'Design for the Bovril Warehouse, Old Street, London', *c* 1893, RIBA Library: RAN 28/B/15.

37 Institution of Civil Engineers membership records.

38 *See* Alan Baxter & Associates 2000.

39 Letter from H V Lanchester, 12 Great James Street, London, W.C. to the Superintending Architect, London County Council, dated 22 December 1896. Reply letters from Thomas Blashill, Superintending Architect, dated 8 and 29 January 1897. London Metropolitan Archives, E/BOV/V/001.

40 Among his many achievements, Sachs wrote *Modern Opera Houses and Theatres* (1896–98) and founded both the British Fire Prevention Committee and the periodical *Concrete and Constructional Engineering* (established 1897 and 1906, respectively), Hurst 1998.

41 Much of this information is derived from Robinson 1999, 20–3. For contemporary accounts, *see A&CR* **68** (8 August 1902), 90–91; *BJ&AR* **17** (15 July 1903), 320.

42 *BJ&AR* **17** (15 July 1903), 320.

43 Joseph Westwood & Co., Limited, 1910, 18.

44 Trollope and Colls microfilmed drawings archive, Roll 4, building No. 93. The surviving drawings indicate that the trabeated brick envelope was load-bearing.

45 Joseph Westwood & Co., Limited, 1910, 18.

46 Until 1965 it remained some 30–50 per cent higher. Boyes, 'Freight traffic', in Simmons and Biddle 1997, 169.

47 'The Somers Town goods station of the Midland Railway', *The Builder* **50** (29 May 1886), 777–8. The British Library now occupies the site.

48 Pacey 1968, 366–9. The only survivor of a group of four warehouses and two goods sheds built by the MS&LR and the LNWR, this 'magnificent piece of industrial architecture' was described by Owen Ashmore as 'perhaps the best surviving example of the railway warehouse in the country'. In 2000–1 it was converted for mixed retail and residential uses. Ashmore 1969, 200.

49 'Blackfriars goods warehouse: iron for structural uses', *The Builder* (8 November 1873), 878–9; 'Goods Warehouses, Blackfriars – London, Chatham, and Dover Railway', *The Engineer* (31 October 1873), 284–5. William Mills was assisted by resident engineers J A C Hewitt and R Barker; the general contractors were Messrs Hill, Keddell & Waldron.

50 *The Builder* **33** (19 June 1875), 549–50, noted 'The ironwork, of which there will be upwards of 1,500 tons in sub- and superstructures, is somewhat peculiar in character. In numerous instances, where clear space on the ground-floor is a desideratum, very strong wrought-iron stanchions, with an interior web for extra strength, take the place of cast-iron columns. At the upper angles they have brackets, giving the girders an extra bearing of about 3 ft. on each side. The stanchions are 2 ft. by 1 ft 5in., and built of four 2¼in. plates, strengthened by angle irons 3½in. by ¾in. . . A number of the stanchions have been tested up to 700 tons each.'

51 *The Builder* **32** (9 May 1874), 397; **33** (19 June 1875), 549–50; **36** (21 September 1878), 995–6.

52 Cunnington 1898b, 349.

53 'The intention had been to use cruciform cast-iron stanchions which, on the goods station platforms, were to have been encased in terracotta. But Stride, the LTSR engineer, found that these stanchions would not support the weight of the warehouse. The final arrangement was that circular cast-iron columns were used throughout', Smith 1980, 9.

54 'Evans and Swains flooring, which consisted of wooden planks fastened to close set timber joists usually plastered on the underside was used for its fire resisting qualities, but the E&WIDCo failed to plaster the underside of the timbers'. Smith 1980, 9. For a discussion of this 'slow-burn' floor system, *see* Lawford 1889, 50–1.

55 Smith 1980, 9.

56 'Belgian v. English iron', *The Times*, 26 February 1886, 4.

57 The photograph reproduced in Fig 13.19 is taken from the trade catalogue of A & J Main & Co. Ltd (nd, *c* 1904), a Glasgow-based firm of engineering contractors that specialised in iron and steel structures. By the date of its publication, it had merged with Arrols Bridge and Roof Co. Ltd, with the Bridge and Girder Department located at the Germiston Iron Works.

58 Clarke 1998.

59 Ibid, notes 8 and 11.

60 Hobson and Wragge 1900–1, 98.

61 *The Builder* **76** (18 March 1899), 272; (17 June 1899), 597; Hobson and Wragge 1900–1, 98–101; *A&CR* **61** (Supplement 1, 16 June 1899), 18.

62 Simmons 1978, 165.

63 *The Engineer* **86** (2 September 1898), 225.

64 Clarke 1993.

65 Stratton 1999, 15, Fig 9.

66 Ibid, 12.

67 West 1912.

68 'Designed to allow gauging to be done under cover, it was an open-sided structure and a *tour de force* of early cast-iron construction. Baron Charles Dupin, who found much to admire at the West India Docks, singled out the Rum Quay Shed, the design of which he attributed to John Rennie, as 'le plus remarquable de tous par sa structure'. The iron … for the shed was supplied by the Butterley Company, through William Jessop, junior. William Pillgrem supervised the erection of the shed, which cost approximately £13,000'. Survey of London 1994a, 303.

69 Tucker, 'Warehouses in Dockland', Carr 1986, 26.

70 Among the earliest recorded structural uses of cast iron was in the South Stacks at London Docks (1810, demolished), where cruciform-section stanchions were employed, albeit still with wooden crossheads. The low physical survival rate of such buildings, and the inadequacies of contemporary documentation hamper assessment of this formative episode of London's early iron construction, but those instances that are known, including the Smithery at Woolwich Dockyard (1814–15), testify to the increasingly more sophisticated exploitation of the metal in the 1810s for low-rise industrial buildings. I am grateful to Malcolm Tucker for sending me his unpublished appraisal of the skin floor at London docks; Tucker 1981.

71 Herbert 1978, 32–5; also Mornement and Holloway 2007.

72 Guillery 1995b.

73 Tucker, 'Warehouses in Dockland', in Carr 1986, 27.

74 Survey of London 1994a, 307.

75 Ibid, 300.

76 Steel-truss roofs, supported on iron columns, were used in the construction of a number of single-storey sheds of the 1880s. At the D Yard of the Millwall Docks, 'K and L Sheds were of steel and iron with double- and triple-span roofs' (both constructed in 1883, demolished). 'M2 Shed was built by William Whitford & Company in 1887 and let to timber merchants … It was a steel-and-iron building, about 500ft by 100ft, with 20ft headroom', Survey of London 1994a, 359.

77 Ibid, 306, 309.

78 Moreland, R & Son Ltd *c* 1895.

79 Edward Wood & Co. Ltd 1897.

80 Walter Jones 1893. The majority of the works illustrated in this catalogue are of steel.

81 'New Cross Depot of London County Council Tramways', *The Tramway & Railway World*, 9 August 1906, 122.

82 *BJ&AE* **27** (18 March 1908), 250. *See also BN* **96** (12 February 1909), 251.

83 Biggart 1897, 398–9.

84 A well-documented example is the fabrication of parts for Brunel's 'Great

Eastern' in the late 1850s, which took place within timber sheds at Millwall Iron Works and Napier Yard. *See* Survey of London 1994a.

85 Fitzgerald 2003.

86 Glynn 1854 *The Construction of Cranes and Machinery*, 2 edn, as quoted in Fitzgerald 1990, 186.

87 Calladine 1994; Survey of London 2012, 171–5.

88 Information provided by Tom Ridge and listed building description.

89 Illustrated in Alderton and Booker 1980, 166.

90 *See* Briden 2002.

91 The columns, stanchions and possibly all other structural metal were supplied by John Lysaght Ltd of Bristol. *See* Guillery 1994; Survey of London 2012, 180–1.

92 The Royal Gun Factory's rolling mills were set to make Bessemer steel guns, but production at the foundry was halted by a change in management that favoured wrought iron over steel. *See* Survey of London 2012, 173.

93 This date is gleaned from the list of main services and new constructions given in Hogg 1963, 898. I am grateful to Rob Kinchin-Smith for this reference.

94 Ibid; Survey of London 2012, 179.

95 Information provided by Malcolm Tucker and Tom Ridge.

96 For further information see unpublished reports by Tom Ridge on former Caird & Rayner engineering workshops, 777–785 Commercial Road (2000, 2002, 2003); copies available in English Heritage library.

97 Outside of the metropolis, probably the most expansive range of sheds the country had ever seen came with the construction of the British Westinghouse Electric & Manufacturing Co.'s works (1900–2) at Trafford Park (*see* pp 74–75), which included a 900ft by 440ft machine shop.

98 Survey of London 1994a, 43, 535; *BN* **5** (April 1901), 469.

99 *The Engineer* **91** (5 April 1901), 353.

100 Bradley 1999, 149.

101 For a description of the mechanised operations in the works, *see Engineering*, 5 April 1901, 441–6.

102 Bradley 1999, 153.

103 Arrol & Company 1909, 140–1.

104 *BJ&AE* **30** (31 December 1909), 661.

105 Wordingham 1901, 43.

106 Christopher Green, 'power station', in Trinder 1992, 590–1.

107 Cochrane 1986.

108 *The Engineer* **66** (26 October 1888), 355; **67** (5 April 1889), 286–7 (illustrations) and 293; (31 May 1889), 471 (illustration).

109 'St. Pancras Electric Lighting. – Regents Park Station', *The Engineer* **73** (19 February 1892), 45 and 148 (illustration).

110 *The Engineer* **89** (23 February 1900), 197.

111 'The Engineer (23 February 1900), 197–200.

112 'City of London Electric Lighting Company's Station, Bankside', *The Engineer* **87** (10 March 1899), 232–3.

113 'The London United Electric Tramways', *The Engineer* **92** (19 July 1901), 62.

114 In America, by the start of the new century steel skeleton-frame construction had entered the vocabulary of the power-station designer, early examples being the Kingsbridge power station of the Third Avenue Railroad Company of New York (1900–1) and the Waterside Power Station of the New York Edison Company (1901). Peach 1904, 290.

115 Green, 'power station', in Trinder 1992, 591.

116 Having moved to London as a chief designer with the General Electric Company of America, in 1895 Parshall began practice as a consulting engineer, designing and supervising many important British electrical tramway installations. During his distinguished career, he designed new types of electric motors, wrote several standard works on electric motor design, and was consultant (and chairman, in 1913) to the Central London Railway, the Yorkshire and Lancashire Power Companies and the Barcelona Traction Light and Power Company. *See* obituaries in *The Engineer* **154**, 1932, 637; *The Times*, 22 December 1932, 14.

117 'The Power House of the Bristol Railway', *The Engineering Record*, 21 September 1901, 277–8.

118 'The building is skeleton steel construction, with substantial external walls. Messrs. Kennedy and Jenkin and S.T. Dobson are the engineers …' Peach 1904, 298.

119 *AR* **16** (October 1904), 174, 180; Brockman 1974, 100–1; Peach 1904, 296–7.

120 A large body of literature exists for this building, of which the following has been consulted principally here: Fyfe 1904, 221–5; 'Chelsea Power Station of the Underground Electric Railways Company', *The Tramway and Railway World* **17** (9 February 1905), 97–128; Stamp 1979; *Lots Road Generating Station* (London Transport pamphlet, 1965); Galicki 1989; Burgess 1995.

121 Fyfe 1904, 224.

122 *See The Tramway and Railway World*, 9 February 1905, 97–128. Besides Lots Roads, James Stewart & Company acted as Building Managers for other British Westinghouse Electric and Manufacturing Company projects, including Motherwell Station and Yoker Station for the Clyde Valley Electrical Power Company, both 'models of their kind'. James Stewart & Company 1909, 127.

123 *BJ&AR* **21** (15 February 1905), 84.

124 Ibid, 84–6; 277–8.

125 Chapman is cited in most contemporary articles, but among the many assistant engineers who contributed to the design, the names J W Towle (engineer in-charge at the station), Gilbert Rosenbush, and Z E Towle are recorded in primary sources. Anne Upson/Circadian, 'Lots Road Power Station and Land at Thames Avenue Development: Environmental Statement Appendix C2' (Unpublished Standing Building Assessment), 6; Fyfe 1904, 225.

126 Ibid, 39.

127 Peach 1904, 296.

128 Edwards was the executive architectural officer; the foundations and chimneys were built by G W Humphreys, the Council's manager of works; and the engines, boilers and machinery were erected under the direction of A L C Fell and J H Rider of the Tramways Department. 'The London County Council Tramway Power Station at Greenwich', *Engineering*, 2 March 1906, 272.

129 Saint, Victorian Society tour notes, August 1981, quoted in Guillery 2000, 3.

130 Joseph Westwood & Co., Limited, 1910, 12. Barges were used to transport the steelwork to a temporary timber pier, where a derrick unloaded it onto trucks that were hauled along a metre-gauge track by steam winches. For details of the erection procedure, *see* 'The London County Council Tramway Power Station at Greenwich', *Engineering* **81** (30 March 1906), 411–12.

131 Guillery 2000, 4, 7.

14 Conclusion: a revolution realised

1 Introduction to *MBR* **1**, 1910, 10.

2 *The Builder* **37** (10 May 1879), 509.

3 *See* Fig 13.21.

4 *See* p 39, Chapter 3.

5 'In 1892 engineer John Seaver recommended that the designer of a one-storey mill building specify maximum strains before the kind and quality of iron or steel were chosen. Seaver thought it best to let the builder specify the material, depending on the market price', Bradley 1999, 144.

6 Horsburgh 1907, 692.

7 Thorne 1990, 12.

8 'The unsatisfactory appearance of much of our town architecture is due to the unacknowledged use of steel: to its concealment behind a mask of architectural drapery … When features are no longer constructional … [but] … applied as decoration, there is nothing but good taste to limit their multiplication', Charles 1904, 62, 63–4.

9 Hill 1909, 4.

10 Ibid, 7.

11 In the Victorian period, this device was typically regarded as concealment rather than allegory: 'Attempts to conceal the iron stanchion by a weak ashlar pilaster of stone, frequently having a base and capital of still weaklier design, has a disastrous effect upon the composition', Knight 1877, 19.

12 Ibid, 4.

13 Locker 1984, 57, citing *PICE* **178**, 1909–10, 211. *See also* Childs 1909c, 75.

14 *The Times* (Engineering Supplement), 19 February 1913, 24.

15 Discussion of Jackson and Dicksee 1913, *JRIBA* (10 May 1913), 477.

16 Locker 1984, 294–7.

17 George, the son of Charles Frederick Mitchell (d 1916), architect, of London, was appointed assistant master at the polytechnic in 1892 and headmaster in 1904. Together they were responsible for Mitchell's Building Construction and other well-known textbooks. *See The Dictionary of Scottish Architects, 1840–1940* (http://www.scottisharchitects.org.uk/).

18 'The Polytechnic: the history of the premises, and their re-building', *The Polytechnic Magazine*, April 1912, 38; 'Current architecture: the Regent Street polytechnic', *AR* **31** (February 1912), 104–16.

19 *See* Faber, 1913, 408–7; also Discussion of Jackson and Dicksee 1913, *JRIBA* (10 May 1913), 480.

20 Etchells 1927, 9.

21 *See*, for example, Newby 2001.

22 The Act of 1667 for rebuilding the City of London, specified, for the first time, that external and party walls were to be of brick or stone. *See* Guillery 2004, 48–9, and Hurst 2006, 1634–5.

23 Bylander 1937, 12. In fairness, Bylander's comment cannot be used as a comparison with relative costs before 1902, the date he arrived in Britain, but standardisation of British structural sections in this era had a marked impact on reducing costs.

24 For example, *see* Bowley 1966, 5–7.

25 Starrett 1928, 161–2; as quoted in Misa 1995, 84.

26 For example, Burn 1961, 290, has suggested that steelmakers, constructional firms and civil engineers may have been obstructive and unimaginative; 'Together these groups might doubtless have done much; among other things they might have persuaded the local authorities who imposed building restrictions to look with less (and uniform) suspicion upon the new material'. On this point, Bowley 1966, 14, states 'Nor did the steel industry take much active interest in building; in any case the advantages in terms of increased demand for steel as a result of revision of the building regulations might well not have been spectacular'.

27 South Africa's first steel skeleton-framed building was an eight-storey warehouse in St Georges Street, Cape Town, erected by Milliken Brothers in 1901–2 for J W Jagger & Co. According to *Men of the Times* (Anon 1906), 'there were many objectors to this style of building, and it was not without much trouble that the plans were passed by the Town Council, and that, once passed, many other buildings have been erected on similar lines both in Cape Town and in other towns of the Colony'. Through the 1900s, there was considerable rivalry between Milliken Brothers and Dorman Long & Co., which established an office in Cape Town in 1903, followed 3 years later by a local company called Dorman Long and Co. (South Africa) Limited. This information, and the quotation, is provided by Rosenthal 2001, 43.

28 Lord Roberts in *Commercial and Industrial Transvaal* (1904), as quoted in Rosenthal 2001, 49–50.

29 This company's trade catalogue of 1913, entitled *Steel Structures* (Edward Wood & Co. Ltd 1913) details many of these grand Buenos Aires buildings, including a seven-storey furnishing warehouse and showrooms for Maple & Co. (Buenos Aires), Ltd and nine-storey offices for The Argentine Steam Navigation Company.

30 Simon 1996.

31 'London's New Buildings. Goliaths of Steel', *The Observer*, 29 March 1914, 18.

BIBLIOGRAPHY

Abbreviations

AR	*The Architectural Review*
A&BN	*The Architect & Building News*
A&CR	*The Architect & Contract Reporter*
BN	*The Building News*
BJ&AR	*The Builders' Journal and Architectural Review*
CE&AJ	*The Civil Engineer and Architect's Journal*
C&CE	*Concrete & Constructional Engineering*
EH	English Heritage
IAR	*Industrial Archaeology Review*
JISI	*Journal of the Iron and Steel Institute*
JRIBA	*Journal of the Royal Institute of British Architects*
JSAH	*Journal of the Society of Architectural Historians*
MM	*The Mechanics' Magazine*
PICE	*Proceedings of the Institution of Civil Engineers*
PIME	*Proceedings of the Institution of Mechanical Engineers*
RCAHMS	Royal Commission on the Ancient and Historical Monuments of Scotland
RCHME	Royal Commission on the Historical Monuments of England
RIBA	Royal Institute of British Architects
TRIBA	*Transactions of the Royal Institute of British Architects*
TNS	*Transactions of the Newcomen Society*
TSE	*The Structural Engineer*
Trans SE	*Transactions of the Society of Engineers*

Publications, unpublished reports and trade literature

Abé, E 1996 'The technological strategy of a leading iron and steel firm, Bolckow Vaughan & Co. Ltd: late Victorian industrialists did fail'. *Business History* **38**(1), 45–76

Adams, H 1894 *The Practical Designing of Structural Ironwork*. London: E&FN Spon

Adams, H 1912 *The Mechanics of Building Construction*. London: Longs, Green & Co

Addis, B 2007 *Building: 3000 Years of Design Engineering and Construction*. London: Phaidon Press

Addis, B 2010 'The iron revolution: how iron replaced traditional structural materials between 1770 and 1870', *in* Rinke, M and Schwartz, J (eds) *Before Steel: The Introduction of Structural Iron and its Consequences*. Zürich: Verlag Niggli AG, 33–48

Adelskold, C 1905 'The first steel bridge – a letter from its designer'. *Scientific American*, 30 December, 527

Aiken, K 1987 'The Langham Hotel', unpublished report in RCHME Buildings File BF081692, December 1987

Aitchison, G 1864 'On iron as a building material'. *TRIBA*, 97–107

Alan Baxter & Associates 2000 'No. 74 Back Church Lane, London Borough of Tower Hamlets, E1', Unpublished report, August 2000

Alexander Penney & Co. 1892 'Descriptive catalogue' (August 1892, Trade Catalogue in British Library Trade Literature Collection)

Allen, J S 2000 *A History of Horseley, Tipton: 200 Years of Engineering Progress*. Ashbourne: Landmark Publishing

Allsop, R O 1894 *Public Baths and Wash-Houses*. London: E&FN Spon

Andrews, E S 1926 'Production of structural steel sections'. *TSE* **4** *(New Series)*, May, 148–53

Anon 1906 *Men of the Times: Old Colonists of Cape Colony and Orange Free State*. Joahnnesesburg: Transvaal Publishing Co.

Appleby-Frodingham Steel Company 1962 *Appleby-Frodingham Steel Company: A Branch of the United Steel Companies, Ltd*. Scunthorpe, Lincolnshire: Appleby-Frodingham Steel Company

Archibald D Dawnay & Sons, Ltd 1927 'A handbook containing a collection of tables & data for the design of constructional steelwork'

Arrol & Company, Sir W 1909 *Bridges, Structural Steel Work and Mechanical Engineering Productions by Sir William Arrol & Company, Ltd*. Dalmarnock Ironworks, Glasgow. London: Engineering

Ashmore, O 1969 *The Industrial Archaeology of Lancashire*. Newton Abbot: David & Charles

Atling, D, *et al* 1993 '"Future assurance": the redevelopment of Holborn Bars'. *The Arup Journal* **28**(3), 3–7

Bannister, T C 1950 'The first iron-framed buildings'. *AR* **107** (April), 231–46

Bannister, T C 1956 'Bogardus revisited. Part I: the iron fronts'. *JSAH* **15** (December), 12–22

Bannister, T C 1957 'Bogardus revisited. Part II: the iron towers', *JSAH* **16** (March), 11–19

Barlow, S and Foster, G 1957 'Universal beams and their application'. *The Engineer* (8 November), 673–5

Barrett, Mr 1854 'On the French and other methods of constructing iron floors', *CE&AJ*, **17**, 90–8 [incl discussion]

Barry, E M 1860 'On the construction and rebuilding of the Italian opera house, Covent Garden'. *The Builder* **18** (11 February), 85–7

Barry, J W 1908 'Standardisation in British engineering practice' *in* Pike, W T (ed) *British Engineers and Allied Professions in the Twentieth Century*. Brighton: W.T. Pike [Pike's new century series; 24]

Bashforth, G R 1962 *The Manufacture of Iron and Steel, Vol 4 The Mechanical Treatment of Steel*. London: Chapman & Hall

Bates, W 1984 *Historical Structural Steelwork Handbook: Properties of U.K. and European Cast Iron, Wrought Iron and Steel Sections Including Design, Load and Stress Data Since the Mid 19th Century*. London: British Constructional Steelwork Association [Series: Publication/British Constructional Steelwork Association; 11/84]

Beare, T H 1932 'Developments during the past 60 years in the manufacture of the materials employed by the structural engineer'. *TSE* **10** (February), 80–3

Beaumont, W W 1880 'Modern steel as a structural material'. *MM* **4** (October), 109–30

Beckett, R 1881 *Beckett's Provincial Builders' Price Book and Surveyors' Guide ... To Which are Added: The Employers' Liability Act of 1880. Sections and Schedules of the Liverpool and Manchester Building Acts, etc.* London: E&FN Spon

Bessemer, H 1861 'On the manufacture of cast steel, and its application to constructive purposes', *PIME* **12**, 133–46 [discussion 146–57]

Bessemer, H 1905 *Sir Henry Bessemer, F.R.S. An Autobiography*. London: Offices of 'Engineering'

Biggart, A S 1897 'Some recent examples of steel structural work adopted in various types of buildings'. *Iron and Steel Trades Journal* **27** (March), 398–9

Binney, M 1999 *The Ritz Hotel, London*. London: Thames & Hudson

Binney, M 2006 *The Ritz Hotel, London* (rev. and expanded edn). London: Thames & Hudson

Black, A 1896 'The sky-scraper'. *BJ&AR* **4** (9 December), 277–9

Black, I S 2000 'Spaces of capital: bank office building in the City of London, 1830–1870'. *Journal of Historical Geography* **26** (July), 351–75

Bodsworth, C 1997 'Historical metallurgy: Dorman Long and Co. Ltd', *Ironmaking and Steelmaking* **24**(6), 434–37

Boniface, P 1981 *Hotels & Restaurants: 1830 To The Present Day*. London: HMSO

Booker, J 1990 *Temples of Mammon: The Architecture of Banking*. Edinburgh: Edinburgh University Press

Boucher, R K 1988 *A Celebration of One Hundred Years of Worship at the Church of Our Most Holy Redeemer with St. Philip, Clerkenwell 1888–1988*. London: Senate House, University of London

Bowen, B 2010 'The building of the British Westinghouse electric and manufacturing plant, Trafford Park, Manchester, 1901–2: an early example of transatlantic co-operation in construction management'. *Construction History* **25**, 85–100

Bowley, M 1966 *The British Building Industry: Four Studies in Response and Resistance to Change*. Cambridge: Cambridge University Press

Bradley, B H 1999 *The Works: The Industrial Architecture of the United States*. Oxford: Oxford University Press

Bradley, S and Pevsner, N 2002 *The Buildings of England: London* **1**: *The City of London*. New Haven, CT: Yale University Press

Bradley, S and Pevsner, N 2003 *The Buildings of England: London* **6**: *Westminster*. New Haven, CT: Yale University Press

Briden, C 2002 'Britannia iron works, Gainsborough: architectural and historic appraisal'. Unpublished survey report, October

Briggs, M S 1907 'Here and there. Iron and steel and modern design. – 1'. *AR* **21** (April), 221–6

Brockman, H A N 1974 *The British Architect in Industry, 1841–1940*. London: Allen & Unwin

Brooks, C (ed) 2000 *The Albert Memorial: The Prince Consort National Memorial, Its Contexts and Its Conservation*. New Haven, CT: Yale University Press, 134–59

Brown, W E A 1915 'The architect and structural engineering'. *Concrete Institute Transactions & Notes* 5 pt3–6, 863–85 [discussion, 885–917]

Burgess, J M 1995 '90 years of power: a history of Lots Road generating station, London underground 1905-1995'. London Transport pamphlet

Burn, D L 1961 *The Economic History of Steelmaking, 1867–1939: A Study in Competition*. Cambridge: Cambridge University Press

Burnell, H H 1854 'Description of the French method of constructing iron floors'. *TRIBA* 4 Ser 1; 1853–4, 35–41 [discussion 41–52 (23 January 1843); 53–61 (6 February 1854); 63–74 (20 February 1854)]

Burnham, A 1959 'The rise and fall of the phoenix column', *Architectural Record* **125**, April, 223–5

Bussell, M 1996 'The era of the proprietary reinforcing systems'. *PICE: Structures and Buildings, Special Issue: Historic Concrete* **116** (3/4), 295–316

Bussell, M 1997 *Appraisal of Existing Iron and Steel Structures*. Ascot: The Steel Construction Institute

Bylander, S 1909 'Steel and concrete at the Selfridge stores, London'. *C&CE* **4**, 9–26

Bylander, S 1910 'Constructional steelwork'. *BN* (15 April), 506–7

Bylander, S 1913 'Steel frame buildings in London'. *Concrete Institute Transactions & Notes* **5**(1) (13 February), 55–106 [discussion 107–125; correspondence **5**(2), viii–x]

Bylander, S 1937 'Steelwork in buildings – thirty years' progress'. *TSE* **15**(1), 2–25 [discussion **15**(2), 128–32]

Byrne, M P 2001 *A History of Market Street Storehouse*. Dublin: A. & A. Farmar

Calladine, A 1994 'Building 25 (The "Armstrong" Gun Factory), The Royal Arsenal, Woolwich', RCHME Survey Report NBR No. 92394

Campin, F 1896 *Constructional iron and steel work as applied to public, private, and domestic buildings: a practical treatise for architects, students, and builders*. London: C. Lockwood and Son

Carr, J C and Taplin W, assisted by Wright A E G 1962 *History of the British Steel Industry*. Oxford: Blackwell

Carr, R J M (ed) 1986 *Dockland: An Illustrated Historical Survey of Life and Work in East London*. London: North East London Polytechnic in conjunction with the Greater London Council

Carter, R 1967 'The Drury Lane theatres of Henry Holland and Benjamin Dean Wyatt'. *JSAH* **26**(3), 200–16

Castleman, H G 1938 'The development of construction under the London building acts'. *TSE* **16**(3), 111–22 [discussion: **16**(10), 328–33]

Caws, F 1901 'Iron and steel construction' [review of W N Twelvetrees's *Structural Iron and Steel: A text book for Architects, Engineers,*

Builders, and Science Students, London, 1900]. *JRIBA* **26** January, 134

Chancellor, E B 1922 *Memorials of St James's Street.* London: Grant Richards

Charles, E M 1904 'The development of architectural art from structural requirements and nature of materials' (unpublished typescript, awarded RIBA Silver Medal, 1905; RIBA Archive; X(079)E 72.03:69)

Chatley, H *c* 1909 *Steel Construction: An Easy Introduction to the Science of Designing and Building in Steel.* The Illustrated Carpenter and Builder Technical Series No. 19 (John Dicks Press, nd, *c* 1909)

Cherry, B and Pevsner, N 1998 *The Buildings of England: London 4: North.* London: Penguin

Cherry, B and Pevsner, N 1999 *The Buildings of England: London 3: North West.* London: Penguin

Childs, C V 1898 'American and English practice in architectural steel construction'. *The Engineering Magazine* **15** (May), 233–40

Childs, C V 1909a 'Steelwork 1'. *A&CR* 8 January, 35–6

Childs, C V 1909b 'Steelwork IV', *A&CR* 5 February, 91–2

Childs, C V 1909c 'Steelwork III', *A&CR* 29 January, 75–6

Childs, C V 1909d 'Steelwork II', *A&CR* 15 January, 43

Chrimes, M 1991 *Civil Engineering 1839–1889: A Photographic History.* Stroud: Alan Sutton; /London: Thomas Telford

Clarke, B F L 1966 *Parish Churches of London.* London: Batsford

Clarke, J 1993 'The Great Northen Railway Company's goods warehouse', unpublished report, The Ironbridge Institute, April

Clarke, J 1999 'Boord & Son's Distillery Offices', EH Architectural Survey Report NMR No. 98660, April

Clarke, J 2001a 'Nos 10–15 Fleet Street: Structural Aspects', English Heritage Architectural Investigation Report NBR BI No. 106808, October

Clarke, J 2001b 'The Way to White City: The Porte Monumentale and Overhead Exhibition Halls of the Franco-British Exhibition, 1908', English Heritage Survey Report, NBR No. 106787, September

Clarke, J 2002 'Like a birdcage exhaled from the earth: Watson's Esplanade Hotel, Mumbai and its place in structural history'. *Construction History* **18**, 37–77

Clarke, J 2006 'Cones, not domes: John Nash and regency structural innovation'. *Construction History* **21**, 43–63

Clarke, J 2013 'Pioneering yet peculiar: John Nash's contributions to late Georgian building technology' *in* Tyack, G (ed) *John Nash: Architect of the Picturesque.* Swindon: English Heritage, 147–62

Clarke, J F 1990 'Mild steel and shipbuilding – the failure of Bessemer steel' *in* Lang, J (ed) 1990 *Metals and the Sea.* London: Historical Metallurgy Society, 5–11 [Papers presented at the March 1990 Conference of the Historical Metallurgy Society held at the National Maritime Museum, Greenwich, London]

Clarke, J F and Storr, F 1989 'The introduction of the use of mild steel into the shipbuilding and marine engine industries'. *Occasional Papers in the History of Science and Technology* 1, 1–97

Clunn, H P 1962 *The Face of London.* London: Spring Books

Coad, J 1989 *The Royal Dockyards 1690–1850: Architecture and Engineering Works of the Sailing Navy.* Aldershot: Scolar Press

Cochrane, R 1986 *Cradle of Power: The Story of the Deptford Power Stations.* London: Central Electricity Generating Board, South Eastern Region

Cocking, W C 1915 'Calculations and details for steel-frame buildings from the draughtsman's standpoint'. *Concrete Institute Transactions & Notes* **5**(3), 676–708 [discussion 709–25, 726–68]

Cocking, W C 1917 *The Calculations for Steel-frame Structures.* London: Scott, Greenwood and Son.

Cocroft W, Tuck C, Clarke J and Smith J 2005 'The development of the Chilworth gunpowder works, Surrey, from the mid-19th century'. *IAR* **27** (November), 217–34

Collins, A R (ed) 1983 *Structural Engineering: Two Centuries of British Achievement,* 2 edn. Chislehurst: Tarot Print Ltd [the Institution of Structural Engineers anniversary publication]

Colvin, H 2008 *A Biographical Dictionary of British Architects, 1600–1840.* New Haven, CT: Yale University Press

Condit, C W 1964 *The Chicago School of Architecture: A History of Commercial and Public Building in the Chicago Area, 1875–1925.* Chicago, IL: University of Chicago Press [revised and enlarged edn]

Condit, C W 1982 *American Building: Materials and Techniques From the First Colonial Settlements to the Present,* 2 edn. Chicago, IL: University of Chicago Press

Cooper, A V 1991 'The Manchester commercial textile warehouse, 1780–1914: a study of its typology and practical development' (unpublished PhD thesis, University of Manchester, 4 vols)

Cossons, N and Trinder, B 2002 *The Iron Bridge: Symbol of the Industrial Revolution,* 2 edn. Chichester: Phillimore [with a foreword by Eric DeLony]

Cox, A A 1891 'The Godwin bursary: a tour in the United States', *TRIBA* **7** (new series, 1890–91 session), 351–88

Croad, Z 1992 'Fireproof floors' (unpublished dissertation, Architectural Association)

Crompton, R E B 1928 *R.E. Crompton: Reminiscences.* London: Constable

Crook J M 1965 'Sir Robert Smirke: a pioneer of concrete construction'. *TNS* **38**, 5–22

Crook, J M 1987 *The Dilemma of Style: Architectural Ideas from the Picturesque to the Post-Modern.* London: John Murray

Crook, J M 1999 *The Rise of the Nouveaux Riches: Style and Status in Victorian and Edwardian Architecture.* London: John Murray

Cross, A W S 1906 *Public Baths and Wash-Houses.* London: B T Batsford

Cross-Rudkin, P S M and Chrimes, M (eds) 2008 *A Biographical Dictionary of Civil Engineers in Great Britain and Ireland, Volume 2: 1830–1890.* London: Thomas Telford

Cruickshank, D 1997 'Gothic revivalist'. *Masters of Building: The Midland Grand Hotel. AJ* **20** (November), 59–76

Cubitt, H 1911 *Building in London. A Treatise on the Law and Practice Affecting the Erection and Maintenance of Buildings in the Metropolis.* London: Constable

Cunningham, C *c* 1991 *A History and Appreciation of the Prudential Chief Office Buildings in Holborn Bars.* London: Prudential

Cunningham, C 1992 *Alfred Waterhouse, 1830–1905: Biography of a Practice.* Oxford: Clarendon Press [Series: Clarendon Studies in the History of Art; with Prudence Waterhouse]

Cunnington T C 1889 'Fireproof flooring as applied to buildings generally'. *BN* 15 March, 361–5

Cunnington T C 1898a 'Constructional steelwork'. *The Builder* 2 April, 327–9

Cunnington T C 1898b 'Constructional steelwork', *The Builder* 9 April, 347–50

Cunnington T C 1902 'Architectural and constructional engineering'. *BJ&AR* **22** (January), 396–9 [also published in *The Builder* 25 January, 80–6; *A&CR* **24** (January), 65–9]

Curl, J S 1990 *Victorian Architecture.* Newton Abbot: David & Charles

Curl, J S 2000 *A Dictionary of Architecture.* Oxford: Oxford University Press

Curl, J S 2006 *A Dictionary of Architecture and Landscape Architecture*. 2 edn. Oxford: Oxford University Press

Curl, J S and Sambrook, J 1973 'E. Bassett Keeling, architect'. *Architectural History* **16**, 60–9

Curl, J S and Sambrook, J 1999 'E. Bassett Keeling – a postscript'. *Architectural History* **42**, 307–15

Darley, G 2003 *Factory*. London: Reaktion

Dawnay, A D 1901 'Constructional steelwork as applied to building'. *The Builder* **20** (April), 385–90

Day, E H 1911 *Some London Churches*. London: A R. Mowbray

de Haan, D 2003 'The iron bridge – new research in the ironbridge gorge'. *IAR* **XXVI**, 3–19

Deby, J 1873 'On the rise and progress of the iron and steel industries in Belgium'. *JISI* **1**, 391–401

Deby, J 1876 'Report on the progress of the iron and steel industries in foreign countries'. *JISI* **1**, 525–655

Denby, E 1998 *Grand Hotels: Reality and Illusion: An Architectural and Social History*. London: Reaktion

Dennell, R A 1906 'American methods of erecting buildings'. *JRIBA* **XIII**, Ser 3 (1905–6), 29–44

Dennis R 2008 '"Babylonian Flats" in Victorian and Edwardian London'. *The London Journal* **33**(3), 233-4

Dibb-Fuller, D, *et al* 1998 'Windsor castle: fire behaviour and restoration aspects of historic ironwork'. *TSE* **76**(19), 367–72

Dicksee, B 1906 *The London Building Acts 1894 to 1905*, 2 edn. London.

Diestelkamp, E 1988 'Building technology & architecture 1790–1830', *in* White, R and Lightburn, C (eds) 1988 *Late Georgian Classicism: Papers Given at the Georgian Group Symposium 1987*. London: Georgian Group, 73–91

Diestelkamp E 1990 'Architects and the use of iron', *in* Thorne, R (ed) *The Iron Revolution: Architects, Engineers and Structural Innovation, 1780–1880*. London: RIBA, 15–23

Dinkel, J 1983 *The Royal Pavilion Brighton*. London: Philip Wilson

Dixey, S J 'Charles Trubshaw: a Victorian railway architect', *in* Jenkinson, D (ed) *Bedside Backtrack: Aspects of Britain's Railway History: An Anthology*, 65–8

Dixon, R and Muthesius, S 1985 *Victorian Architecture: With a Short Dictionary of Architects*, 2 edn. London: Thames and Hudson [Series: World of art]

Dobson, D B 1900 *Coloured Elementary Plates on Building Construction*. Glasgow: John W. Morgan & Co

Dorman Long 1901 *Description of the Works of 1900 Steel Sections, Manufactured by Dorman, Long, & Co., Limited, and Bell Brothers, Limited, Middlesbrough*. Middlesbrough: McCorquodale & Co., Ltd.

Douet, J and Lake, J 1997 'Royal Naval Dockyards. Thematic list review. Summary report and recommendations'. Unpublished report, English Heritage

Driver, C H 1875 'On iron as a constructive material'. *TRIBA* **25** (1874–5), 165–183

Dummelow, J 1949 *1899–1949*. Manchester: Metropolitan-Vickers Electrical Company [Publication to commemorate the golden jubilee of the Metropolitan-Vickers Electrical]

Dunkeld, M, *et al* (eds) 2006 *Proceedings of the Second International Congress on Construction History*, 3 vols. Ascot: Construction History Society [Queen's College, Cambridge University 29 March–2 April 2006; main author: International Congress on Construction History]

Dunn, W 1891 'Iron joists and stanchions', *The Builder* 20 June, 483–4

Earl, J 1984 'J G Buckle: a note on the Theatre Royal, Stratford East'. *Theatre Notebook* **1**(**4**), 62–4

Earl, J and Sell, M (eds) 2000 *The Theatres Trust Guide to British Theatres, 1750–1950: A Gazetteer*. London: A & C Black

Edward Wood & Co. Ltd 1897 *Illustrated Catalogue* [trade catalogue, nd, but dated by the British Library Trade Literature Collection (former Patent Office Library) as 1897]

Edward Wood & Co. Ltd 1914 *A Record of Progress* [1914 trade catalogue in British Library Trade Literature Collection]

Edward Wood & Co. Ltd 1913 *Illustrated catalogue of steel structures manufactured and erected by Edward Wood & Co. Ltd* 13 edn

Elbaum, B 1986 'The steel industry before World War 1', *in* Elbaum, B and Lazonick, W (eds) *The Decline of the British Economy*. Oxford: Clarendon, 55–81

Elliott, C D 1992 *Technics and Architecture: The Development of Materials and Systems for Buildings*. Cambridge, MA: MIT Press

Ende, M A 1887a 'On the constructive ironwork in a new theatre'. *The Engineer* 23 September, 257

Ende, M A 1887b 'On the constructive ironwork in Terry's new theatre'. *The Engineer* 7 October, 283–5

Ende, M A 1889 'Structural ironwork in the Terry, Court, and Garrick theatres'. *The Engineer*, 1 February, 94–5

Etchells, E F 1927 *Modern Steelwork: A Review of Current Practice in the Application of Structural Steelwork to Building and Bridges*. London: Nash and Alexander

Evans, B 1997 'Electric refurbishment'. *AJ* **206** (27 November), 59–61

Faber, J 1989 *Oscar Faber: His Work, His Firm and Afterwards*. London: Quiller Press

Fairbairn, W 1854 *On the Application of Cast and Wrought Iron to Building Purposes*. London: John Weale

Fairbairn, W 1863 *Treatise on Mills and Millwork Vol. 2: On Machinery of Transmission and the Construction and Arrangement of Mills*. London: Longman, & Roberts

Farnie, D A 1980 *The Manchester Ship Canal and the rise of the Port of Manchester, 1894–1975*. Manchester: Manchester University Press

Fitzgerald, R 2003 'Buildings for mechanical engineering & foundries' [lecture delivered to the Institution of Structural Engineers History Study Group, 1 February, unpublished]

Fitzgerald, R S 1988 'The development of the cast iron frame in textile mills to 1850'. *IAR* **X**(2), 127–45

Fitzgerald, R S 1990 'The anatomy of a Victorian crane: the Coburg boiler shop crane and its technological context'. *IAR* **XII**(2), 185–204

Fleetwood-Hesketh, P 1971 'The royal automobile club'. *Country Life* **14** (October), 966–9

Fletcher, B 1902 *Light and Air*. London: B T Batsford, 4 rev edn.

Fletcher, B 1915 *The London Building Acts* (5 edn). London: B T Batsford

Foulston, J 1838 *The Public Buildings, Erected in the West of England as Designed by John Foulston*. London: J. Williams

Fred West, F 1912 *The Railway Goods Station: A Guide to its Control and Operation*. London: E&FN Spon

Fremdling, R 1991 'The puddler – a craftsman's skill and the spread of a new technology in Belgium, France and Germany'. *Journal of European Economic History* **20**(3), 529–67

Friedman, D 1995 *Historical Building Construction: Design, Materials, and Technology*, London. W.W. Norton

Friedman, D 2006 'Building code enforced evolution in early skeleton buildings', *in* Dunkeld, M *et al* (eds) *Proceedings of the Second International Congress on Construction History*, Ascot: Construction History Society **2**, 1171–87

Friedman, D 2010 *Historical Building Construction: Design, Materials, and Technology*, 2 edn. London: W.W. Norton [Series: Norton professional books]

Friedman, T 1997 *Church Architecture in Leeds 1700–1799*. Leeds: The Thoresby Society [Series: The publications of the Thoresby Society. 2nd series; no. 7], 52–6

Fryer W J 1898 'A review of the development of structural iron', *in* Real Estate Record Association (ed) *A History of Real Estate, Building and Architecture in New York City During the Last Quarter of a Century*. New York: Record and Guide, 455–83

Gale, A J 1883 'American architecture from a constructional point of view'. *TRIBA* **33** (1882–3), 52–5 [includes discussion]

Gale, W K V 1967 'The rolling of iron'. *TNS* **37** (1964–5), 35–46

Gale, W K V 1969 *Iron and Steel*. Harlow: Longmans [Series: Industrial archaeology 2]

Gale, W K V 1973–4 'The Bessemer steelmaking process'. *TNS* **46** (1973–4), 17–26

Galicki, M 'Lots Road power station, Lots Road, Chelsea, London SW10'. English Heritage typescript report, 1989

Gasley, W J B 1887 *A Hand-Book of Steel & Iron Sections Manufactured by Dorman, Long & Company*. Newcastle-upon-Tyne: Andrew Reid

Gass, J B 1886a 'Some American methods'. *TRIBA* **2**, new series (1885–6 session), 129–44

Gass, J B 1886b 'The Godwin bursary: portions of a report of a visit to the United States of America and Canada'. *TRIBA* **2**, new series (1885–6 session), 145–52

Gayle, C and Gayle, M 1998 'The emergence of cast-iron architecture in the United States: defining the role of James Bogardus'. *APT Bulletin: The Journal of Preservation Technology* **29**(2), 5–12

Gibson, R W 1898 'Fireproof buildings in the United States'. *JRIBA* 10 December, 49–64

Giedion, S 1967 *Space, Time, and Architecture. The Growth of a New Tradition*, 5 edn. Cambridge, MA: Harvard University Press

Giles, C and Goodall, I H 1992 *Yorkshire Textile Mills: The Buildings of the Yorkshire Textile Industry, 1770–1930*. London: HMSO

Giles, G 1998 'The historic warehouses of Liverpool'. RCHME Survey Report, September

Glasstone, V 1975 *Victorian and Edwardian Theatres: An Architectural and Social Survey*. London: Thames and Hudson

Glasstone, V 1980 *The London Coliseum*. Cambridge: Chadwyck-Healey. [Series: Theatre in focus; Somerset House, in association with the Consortium for Drama and Media in Higher Education]

Gray, A S 1985 *Edwardian Architecture: A Biographical Dictionary*. London: Duckworth

Griffiths, H W 1883 *Iron and Steel Manufacturers of Great Britain, and Brand Book of British Iron and Steel*. London.

Guedes, P (ed) 1979 *Encyclopaedia of Architectural Technology*. New York: McGraw-Hill

Guillery, P 1993 'Armoury House and Finsbury Barracks, City Road, Islington', RCHME Report, NBR Index No. 92360

Guillery, P 1994 'The Royal Arsenal, Woolwich: Rapid Survey', RCHME Rapid Survey Report, NBR Index No. 92394

Guillery, P 1995a 'The British Museum: Floor over the King's Library', RCHME Report, NBR Index No. 93464

Guillery, P 1995b 'North Quay Transit Sheds, Royal Albert Dock, Newham, London', RCHME Survey Report NBR Index No. 33146

Guillery P 1999 'Warehouses and sheds: buildings and goods handling in London's nineteenth-century docks', *in* Jarvis A and Smith K (eds)

Albert Dock: Trade and Technology. Liverpool: National Museums & Galleries on Merseyside, 77–88

Guillery, P 2000 'Greenwich generating station'. *London's Industrial Archaeology* **7**, 3–12

Guillery P 2004 *The Small House in Eighteenth-Century London: A Social and Architectural History*. New Haven, CT: Yale University Press

Guillou, M Le 1972 'The South Staffordshire iron and steel industry and the growth of foreign competition (1850–1914)'. Part 1, West Midlands Studies **5**, 16–23

Gwilt, J 1826 *Rudiments of Architecture, Practical and Theoretical*. London: Priestley and Weale

Gwilt, J 1842 *An Encyclopaedia of Architecture: Historical, Theoretical, and Practical*. London: Longman, Brown, Green, and Longmans

Gwilt, J 1867 *An Encyclopaedia of Architecture, Historical, Theoretical and Practical*. London: Longmans, Green, and Co.

Hamilton, S B 1940–1 'The use of cast iron in building'. *TNS* **21** (1940–1), 139–55

Hamilton, S B, *et al* 1964 *A Qualitative Study of Some Buildings in the London Area*. London: HMSO [National Building Studies. Special Report no. 33]

Hanson, B 2004 'Pite, Arthur Beresford (1861–1934)', *Oxford Dictionary of National Biography*. Oxford: Oxford University Press

Harper, R H 1985 *Victorian Building Regulations: Summary Tables of the Principal English Building Acts and Model By-Laws 1840–1914*. London: Mansell

Harris J 1961 'Cast iron columns 1706'. *AR* **130** (July), 60–1

Hart, F, *et al* 1985 *Multi-Storey Buildings in Steel*, 2 edn. London: Collins

Hasluck, P N ed 1906 *Iron, Steel, and Fireproof Construction*. London: Cassell and Co.

Hay, G D and Stell, G P 1986 *Monuments of Industry: An Illustrated Historical Record*. Edinburgh: RCAHMS

Heathcote, C 1903 'A comparison of English and American methods in the erection of buildings' [read to the Institute of Builders], *BJ&AR* **18** (9 December), 251–54

Herbert, G 1978 *Pioneers of Prefabrication: The British Contribution in the Nineteenth Century*. Baltimore, MD: Johns Hopkins University Press [Series: The Johns Hopkins studies in nineteenth-century architecture]

Herbert, G 1999 'Architect-engineer relationships: overlappings and interactions'. *Architectural Science Review* **42** (2 June), 107–10

Hewitt, A S 1868 *The Production of Iron and Steel in its Economic and Social Relations*. Washington: Government Printing Office [In United States of America. Commission to the Paris Universal Exposition, 1867. Reports of the United States Commissioners, etc vol 2. 1870]

Heywood, J 1894 *John Heywood's Illustrated Manchester Ship Canal Route Guide*. Manchester: J. Heywood

Hill, C E W 1970 'The Victorian Grand Hotels of London' [unpublished typescript, April 1970, copy in English Heritage library]

Hill, H A 1909 'The influence on architecture of modern methods of construction' [unpublished typescript, awarded RIBA Silver Medal, 1909; RIBA Archive; X(079)E 693:72.036]

Hiorns, A H 1995 *Iron and Steel Manufacture*, 2 edn. London: Macmillan and Co.

Hitchcock, H-R 1949 'Victorian monuments of commerce'. *AR* **105** (February), 61–74

Hitchcock, H-R 1951 'Early cast iron façades'. *AR* **109** (February), 113–16

Hitchcock, H-R 1953 'Sullivan and the skyscraper', *JRIBA* **60** (July), 353–61

Hitchcock, H-R 1954 *Early Victorian Architecture in Britain* Vol 1: Text. London: Architectural Press [Yale historical publications. 9. History of art]

Hitchcock, H-R 1987 *Architecture: Nineteenth and Twentieth Centuries*, 4 edn. bibliography rev. New Haven, CT: Yale University Press [Series: Pelican history of art]

Hobson, GA and Wragge E 1900–01 'The metropolitan terminus of the Great Central Railway'. *PICE* **143** (1), 84–113

Hodson, H 1992 'The iron bridge; its manufacture and construction'. *IAR* **XV**, 36–44

Hogg, OFG 1963 *The Royal Arsenal: its Background, Origin, and Subsequent History Vol 2.* Oxford: Oxford University Press

Holden, C H and Holford, W G 1951 *The City of London: A Record of Destruction and Survival.* London: Architectural Press

Horner J 1896–1898 'Wrought-iron and steel in constructional work', *BN* [an article running in 37 parts from 25 September 1896 to 15 April 1898]

Horner J 1897 'Wrought-iron and steel in constructional work–XIII'. *BN* **72** (26 March)

Horner J 1898 'Wrought-iron and steel in constructional work – XXXVI', *BN* **74** (8 April), 482–3

Horner J G 1889 'The age of steel'. *The Quarterly Review* **169** (July), 132–61

Horsburgh, V D 1907 'On the influence of the use of iron and steel on modern architectural design'. *JRIBA* **14,** 3 ser (19 October), 689–701

Horsfield-Nixon, J 1908 'The Franco-British exhibition of science, arts and industries, London 1908'. *JRIBA* **15,** 3 ser (25 July), 546–56

Howard, D 1970 *London Theatres and Music Halls, 1850–1950.* London: Library Association

Hubert, H 1916a 'The great industries in Belgium before and during the war: no. I: iron and steel manufactures in Belgium and associated industries'. *The Engineer* 14 January, 27–9

Hubert, H 1916b 'The great industries in Belgium before and during the war: no. II: iron and steel manufactures in Belgium and associated industries'. *The Engineer* 21 January, 55–7

Hughes, Q 1964 *Seaport: Architecture & Townscape in Liverpool.* London: Lund Humphries

Humber, W 1868 *A Handy-Book For the Calculation of Strains in Girders and Similar Structures, and Their Strength; Consisting of Formulae and Corresponding Diagrams, with Numerous Details for Practical Application*, etc. etc. London.

Hurst, B L 1990 'The age of fireproof flooring', *in* Thorne, R (ed) *The Iron Revolution: Architects, Engineers and Structural Innovation, 1780–1880.* London: RIBA, 35–9

Hurst, B L 1996 'Concrete and the structural use of cements in England before 1890'. *PICE: Structures and Buildings, Special issue: Historic Concrete* **116** (3/4), 283–94

Hurst, B L 1998 'Edwin O. Sachs – engineer and fireman' in Wilmore, R (ed) Edwin O. Sachs: Architect, Stagehand, Engineer and Fireman – His Life and His Satellites. Summerbridge: Theatresearch, 120–31

Hurst, B L 1999 'An iron lineage' (Sutherland History Lecture), TSE **77** No. 10 (18 May 1999), 17–25 [discussion 26–8]

Hurst, B L 2009, 'Spanning tile creasing'. *Construction History Society Newsletter* **86** (December), 1–2

Hussey, C 1958 *English Country Houses: Late Georgian 1800–1840.* London: Country Life Limited

I'Anson, E 1864–5 'Some notice of office buildings in the city of London'. *TRIBA* **15**, 25–36

Ionides, J 1998 *Thomas Farnolls Pritchard of Shrewsbury: Architect and 'Inventor of Cast Iron Bridges'.* Ludlow: Dog Rose

Jackson, A A 1997 'The development of the steel framed building between 1875 and 1905 with reference to structural design and conservation issues'. Unpublished MA dissertation, University of York

Jackson, A A 1998 'The development of steel framed buildings in Britain 1880–1905'. *Construction History* **14**, 21–40

Jackson, F N and Dicksee, B 1913 'Modern steel building construction'. *JRIBA* 12 April, 413–35

Jackson, S 1964 *The Savoy: The Romance of a Great Hotel.* London: F. Muller

Jackson, S 1970 *The Great Barnato.* London: Heinemann

James Stewart & Company 1909 *Some Stewart Structures.* New York: W M Probasco

James Stewart & Company 1944 *A Century in Construction: A Historical Account with a List of Structures Recording an Achievement in Engineering and Construction Throughout the World During the Past Hundred Years by James Stewart & Company, 1844–1944.* New York: Marchbanks Press

Jefferson Smith, P 1997 'Office Buildings in the City: Royal Exchange Buildings (1841–45)' *in* Saunders, A L (ed) *The Royal Exchange.* London: London Topographical Society, 366–85

Jenkinson, D (ed) 1993 *Bedside Backtrack: Aspects of Britain's Railway History: An Anthology.* Penryn: Transport Atlantic Publishers

Jenkyns, P M 1984 *The Story of Sir Henry Bessemer: The Victorian Inventor and Engineer Who Lived on Denmark Hill 1863–1898.* London: Herne Hill Society

Jeremy, D J (ed) 1984 *Dictionary of Business Biography: A Biographical Dictionary of Business Leaders Active in Britain in the Period 1860–1980, Vol 1, A–C.* London: Butterworths

Jewett, R A 1967 'Structural antecedents of the I-beam, 1800–1850'. *Technology and Culture* **8** (1 January), 346–62

Jones, E 1985 *Industrial Architecture in Britain, 1750–1939.* London: Batsford

Joseph Westwood & Co., Limited, 1910 'Joseph Westwood & Co., Limited, Structural Engineers & Bridge Builders, Napier Yard & Millwall Yard, Millwall, London'.

Katona, L 1900 'Suggestions for the improvement of rolling-mills'. *JISI* 259–71

Kelly 1870 *Kelly's Directory of the Building Trades.* 1 edn. London

Kelly 1878 *Kelly's Directory of the Building Trades.* 4 edn. London

Kelly 1882 *Kelly's Directory of the Building Trades.* 5 edn. London

Kelly 1898 *Kelly's Directory of the Building Trades.* 7 edn. London

Kelsall, F 2001 'A report on the architectural history of nos 7–15 Fleet Street, City of London'. Unpublished report by The Architectural History Practice

Kennedy, A B W 1880a 'Mild steel and its application to building purposes'. *The Builder* 1 May 1880, 552–3, 8 May 1880, 582–3

Kennedy, A B W 1880b '"Mild steel" and its applications to building purposes'. *TRIBA* **30** (1879–80), 162–72 [discussion, 173–92]

Kent, L E and Kirkland, G W 1958 'Construction of steel-framed buildings'. *The Structural Engineer* Jubilee Issue, 102–10

Kerr, R M 1865 *On Ancient Lights.* London: J. Murray

Knight, S 1877 'The influence of business requirements upon street architecture', *TRIBA* **27** (1876/77), 15-32

Knowles, C C and Pitt, P H 1972 *The History of Building Regulation in London, 1189–1972 : With An Account of the District Surveyors' Association.* London: Architectural Press

Kohlmaier, G and Barna von S 1990 *Houses of Glass: A Nineteenth-Century Building Type.* Cambridge, MA: MIT [trans by John C Harvey]

Landau, S B and Condit, C W 1996 *Rise of the New York Skyscraper, 1865–1913.* New Haven, CT: Yale University Press

Landes D S 1969 *The Unbound Prometheus: Technological Change and Industrial Development in Western Europe from 1750 to the Present*. Cambridge: Cambridge University Press

Langham Hotel 1884 *The Langham Hotel Company's Guide to London*, 5 edn. London: Langham Hotel Company [revised and edited by W C Gordon, manager]

Langham Hotel 1904 *The Langham Hotel Guide to London*, 18 edn. London: Langham Hotel Company

Langton Cole, R 1899 'Notes, queries and replies: skeleton buildings'. *JRIBA* **6**, 3rd ser, 239

Larson, G R 1987 'The iron skeleton frame: interactions between Europe and the United States', *in* Zukowsky, J (ed) *Chicago Architecture 1872–1922: Birth of a Metropolis*. Munich: The Art Institute of Chicago and Prestel Verlag, 40–55

Lawford, G M (1889) 'Fireproof floors'. *Trans SE*, 43–70

Lawrence, D 1994 *Underground Architecture*. Harrow: Capital Transport

Lawrence, J C 1990 'Steel frame architecture versus the London building regulations: Selfridges, the Ritz, and American technology'. *Construction History* **6**, 23–46

Laxton, H 1857 *Examples of Building Construction, intended as an aide-memoire for the professional man and the operative*. London: Henry Laxton

Leacroft, R 1988 *The Development of the English Playhouse: An Illustrated Survey of Theatre Building in England from Medieval to Modern Times* [rev edn]. London: Methuen

Leboff, D 2002 *The Underground Stations of Leslie Green*. Harrow Weald, Middlesex: Capital Transport Publishing

Leonard, J 1997 *London's Parish Churches*. Derby: Breedon Books

Lewis, T H 1865 'Fire-proof materials and construction'. *TRIBA* **15**, (ser 1) 1864–5, 109–126

Ley, A J 2000 *A History of Building Control in England & Wales 1840–1999*. Coventry: RICS Books

Leyland, J 1901a 'Victualling the Navy: the Royal Victoria Yard, Deptford', *The Navy and Army Illustrated* 7 September, 614–16

Leyland, J 1901b 'Victualling the Navy: the Royal Victoria Yard, Deptford', *The Navy and Army Illustrated* 14 September, 638–40

Lingard, J and Lingard T 2007 *Bradshaw Gass & Hope: The Story of an Architectural Practice – The First One Hundred Years 1862–1962*. London: Gallery Lingard

Locker, F M 1980 'Full framed steel construction in Britain: the early years of confrontation with established attitudes and beliefs'. *Edinburgh Architecture Research* **7**, 29–63

Locker, F M 1984 'The evolution of Victorian and early twentieth century office buildings in Britain'. Unpublished PhD thesis, University of Edinburgh

London Trades Council 1867 *Report of the Various Proceedings Taken by the London Trades' Council and the Conference of Amalgamated Trades, in Reference to the Royal Commission on Trades' Unions, and Other Subjects in Connection Therewith*. London: J Kenny. [No7 in a volume of 13 pamphlets]

Loobey, P (ed) 2002 *Battersea Past*. London: Historical Publications

Mainstone, R 2001 *Developments in Structural Form*, 2 edn. Oxford: Architectural Press

Malcolm, I C 1990 'Dorman Long and Company, 1889–1967. Notes towards a history'. *The Cleveland Industrial Archaeologist* **20**, 31–57

Mangeot, S E 1939 'Queen Anne's mansions: the story of 'Hankey's Folly''. *A&BN* **13** (January), 77–9

Mansford, F H 1907 'Prudential assurance company's building, London'. *Architectural Record* 21 February, 135–47

Maskell, H P 1905 *Hints on Building a Church*. London: Church Bells Office

Matheson, E 1882 'Steel for structures'. *PICE* **69**(3) (1881–2), 1–79 [inc discussion]

Matheson, E and Grant, R C 1893 *Matheson and Grant's Handbook for Engineers*, 2 edn. London: E&F Spon

McWilliam, R C 2001 *BSI: The First Hundred Years 1901–2001*. London: Institution of Civil Engineers

Mende, M 2006 'The crucial impact of improvements in both steelmaking and rolling on 19th- and early 20th-century building construction', *in* Dunkeld, M *et al* (eds) *Proceedings of the Second International Congress on Construction History*, 3 vols. Ascot: Construction History Society, 2159–70

Metcalf, P 1972 *Victorian London*. London: Cassell

Middleton, G A T 1898 'Fire-resisting construction. – the regulations in force in London'. *The Engineering Magazine* **15** (April–September), 780–5

Middleton, G A T 1905–7 *Modern Buildings: Their Planning, Construction and Equipment* (Vol IV, 1906 Public Buildings, Steel Construction, Fire-Resisting Construction). London: Caxton Publishing

Middleton, G A T 1906 *Modern Buildings: Their Planning, Construction and Equipment* (Vol 4, Public Buildings, Steel Construction, Fire-Resisting Construction). London: Caxton Publishing

Milhous, J, *et al* 2001 *Italian Opera in Late Eighteenth-Century London. Vol 2 The Pantheon Opera and its Aftermath, 1789–1795*. Oxford: Clarendon Press

Milward, A S and Saul, S B 1977 *The Development of the Economies of Continental Europe, 1850–1914*. London: George Allen & Unwin

Misa, T J 1995 *A Nation of Steel: The Making of Modern America 1865–1925*. Baltimore, MD: Johns Hopkins University Press

Montgomery-Massingberd, H and Watkin, D 1980 *The London Ritz: A Social and Architectural*. London: Aurum

Montgomery-Massingberd, H and Watkin, D 1989 *The London Ritz: A Social and Architectural*, new edn. London: Aurum

Moreland, R 1884 'Constructional ironwork for buildings'. *Minutes of PICE* **77** (3 January), 281–96

Moreland, R 1893 'The use of steel for constructive purposes'. *BN* **17** (February), 227–8; **24** (February), 261–2; **3** (March), 320

Moreland, R & Son Ltd *c* 1895 *Steel Construction* [trade catalogue]

Moreland, R & Son Ltd *c* 1905 *Solid Steel Columns and Steel Construction* [trade catalogue]

Mornement, A and Holloway, S 2007 *Corrugated Iron: Building a New Frontier*. London: Frances Lincoln.

Morrison, K A 2003 *English Shops and Shopping: An Architectural History*. New Haven, CT: Yale University Press

Morrison, K A 2006 'Bazaars and bazaar buildings in regency and Victorian London'. *Georgian Group Jnl* **15**, 281–308

Mouchel, L G and Partners, Ltd 1920 *List of Works Executed in the United Kingdom 1897–1919*. London: L G Mouchel and Partners, Ltd

Munro, J 1891 *Heroes of the Telegraph*. London: The Religious Tract Society

Muthesius, S 1970 'The "iron problem" in the 1850s'. *Architectural History* **13**, 58–63

Muthesius, S 1972 *The High Victorian Movement in Architecture, 1850–1870*. London: Routledge & Keegan Paul

Nevill, R H 1911 *London Clubs: Their History and Treasures*. London: Chatto & Windus [new edn, revised, with additions by W Papworth. London: Longmans, Green]

Newby, F (ed) 2001 *Early Reinforced Concrete*. Aldershot: Ashgate [Studies in the history of civil engineering vol 11]

Newton, R H 1941 'New evidence on the evolution of the skyscraper'. *Art Quarterly* **IV**(Winter), 56–69

Norris, J V 1971 'Phoenix column made its mark in history'. *Daily Republican* 22 April

O'Dwyer, F 1997 *The Architecture of Deane and Woodward*. Cork: Cork University Press

Pacey, A J 1968 'Technical innovation in some late nineteenth-century railway warehouses'. *Industrial Archaeology* **5**(3), 364–72

Parissien, S 2008 *The Georgian House in America and Britain*, 2 edn. New York: Rizzoli

Parkinson-Bailey, J J 2000 *Manchester: An Architectural History*. Manchester: Manchester University Press

Payne, P L 1979 *Colvilles and the Scottish Steel Industry*. Oxford: Clarendon Press

Peach A 2004 'Alexander, Daniel Asher (1768–1846), architect and engineer' in Matthew H C G and Harrison B (eds), *Oxford Dictionary of National Biography*. Oxford: Oxford University Press

Peach, C S 1904 'Notes on the design and construction of buildings connected with the generation and supply of electricity know as central stations'. *JRIBA* **11**, 290

Pendred, H W 1883 'Designs, specifications, and inspection of ironwork'. *Trans SE* 1 October 1883, 87–112

Penfold, A (ed) 1980 *Thomas Telford Engineer*. London: Thomas Telford

'Pentagon' 1900 'Bovril and its new home in London'. *Grocery* **2** (2 February), 115–28

Perks, S 1905 *Residential Flats of All Classes, Including Artisans' Dwellings*. London: Batsford

Peters, T F 1996 *Building the Nineteenth Century*. Cambridge, MA: MIT Press

Pevsner, N 1957 *The Buildings of England: London 1: The Cities of London and Westminster*. Harmondsworth: Penguin

Pevsner, N 1976 *A History of Building Types*. London: Thames and Hudson

Phillips, W and Phillips T 1870 *Architectural Iron Construction*. London [trade catalogue]

Picton, J A 1880a 'The use of steel in architecture'. *The Builder* **24** (April), 496–8

Picton J A 1880b 'Iron as a material for architectural construction'. *TRIBA* **30**, 149–61 [discussion, 173–92]

Pike, W T (ed) 1908 *British Engineers and Allied Professions in the Twentieth Century*. Brighton: W.T. Pike [Pike's new century series; 24]

Pite, A B 1902 'Street architecture'. *BJ&AR* **16** (April), 129–33

Port M H 1995 *Imperial London. Civil Government Building in London 1850–1915*. New Haven, CT: Yale University Press

Port, M H 2009 'Founders of the Royal Institute of British Architects (act. 1834–1835)', *Oxford Dictionary of National Biography*, online edn, Oxford University Press, May 2009 [http://www.oxforddnb.com/view/theme/97265, accessed 26 October 2010]

Powers, A (ed) 1987 *End of the Line? The Future of London Underground's Past*. [Unpublished report produced by the Victorian Society and the Thirties Society]

Purdy, C T 1909 'The "New York Times" building'. *PICE* **178**, 185–205 [correspondence 215–19]

Rabun, J S 2000 *Structural Analysis of Historic Buildings: Restoration, Preservation, and Adaptive Reuse Applications for Architects and Engineers*. New York: Wiley

Reade, F T 1889 'The application of iron and steel to building purposes'. *The Architect* 22 November, 299–301

Rennison, R W (ed) 1996 *Civil Engineering Heritage: Northern England*, 2 edn. London: Thomas Telford

Richardson, R 1991 'Brompton Hospital (Hospital for Consumption and Diseases of the Chest)', RCHME Hospitals Survey Report, NBR Index No. 101058

Richardson, S 1999 'A Historic Building Record of Albion Works, Kings Cross', AOC Archaeology

Rinke, M and Schwartz, J (eds) 2010 *Before Steel: The Introduction of Structural Iron and its Consequences*. Zürich: Verlag Niggli AG

Rivington's Notes on Building Construction 1912 *Notes on Building Construction, Arranged to Meet the Requirements of the Syllabus of the Board of Education. Pt 4, Calculations for Building Structures*, 8 edn. London: Longmans, Green

Roscoe, ESA 1904 *Digest of the Law Relating to the Easement of Light*, 4 edn. London: Stevens and Sons

Rosenthal, E 2001 *Girders on the Veld: Structural Steel and its Story in South Africa*. Johannesburg: South African Institute of Steel Construction [first published 1981]

Rosoman, T 1980 'S. S. Teulon and his re-casting of two west London churches'. *Transactions of the London and Middlesex Archaeological Society* **31**

Royal Commission on Depression of Trade and Industry 1886 *Second Report of the Royal Commission on Depression of Trade and Industry*. London: HMSO

Royal Insurance Company 1912 *Old Lombard Street: Some Notes Prepared by the Royal Insurance Company Limited on the Occasion of the Opening of Their New Building in Lombard St, May 1912*. London: Causton & Sons

Ruddock, T 1979 *Arch Bridges and Their Builders, 1735–1835*. Cambridge; New York: Cambridge University Press

Ruskin, J 1849 *The Seven Lamps of Architecture*. London: Smith, Elder, and Co.

Ryland 1884 *Ryland's Iron, Steel and Allied Trades' Directory*. 2 edn. Birmingham

Ryland 1886 *Ryland's List of Merchant Exporters of Iron, Steel, Tin-Plates, Metals, Hardware, and Machinery*. 1 edn. Birmingham

Ryland 1887 *Ryland's Iron, Steel, Tin-Plate, Coal, and Allied Trades' Directory*. 3 edn. Birmingham

Ryland 1890 *Ryland's Iron, Steel, Tin-Plate, Coal, and Allied Trades' Directory*. 4 edn. Birmingham

Ryland 1895 *Ryland's List of Merchant Exporters of Iron, Steel, Tin-Plates, Metals, Hardware, and Machinery*. 3 edn. Birmingham

Ryland 1896 *Ryland's Iron, Steel, Tin-Plate, Coal, and Allied Trades' Directory*. 6 edn. Birmingham

S'international Architects 1989 'The Windsor Hotel Vol. 1: History of the Hotel'. Unpublished report, researched and written for S'international Architects by G D Marsh, June 1989

Sachs, E O and Woodrow E A E 1896–8 *Modern Opera Houses and Theatres*. London: B. T. Batsford, 3 vols. 1896–98

Sack, H 1889 'On universal rolling-mills for the rolling of girders and cruciform sections'. *JISI* 132–45

Saint, A 1976 *Richard Norman Shaw*. New Haven, CT: Yale University Press

Saint, A 2007 *Architect and Engineer: A Study in Sibling Rivalry*. New Haven, CT: Yale University Press

Saint, A *et al* (eds) 1982 *A History of the Royal Opera House, Covent Garden 1732–1982*. London: Royal Opera House

Salomons, E 1857 'Manchester art treasures exhibition building'. *MM* **66** (20 June 1857), 578–81

Saunders, A L (ed) 1997 *The Royal Exchange*. London: London Topographical Society [Series: London Topographical Society publication no. 152]

Schmiechen, J and Carls, K 1999 *The British Market Hall: A Social and Architectural History*. New Haven, CT: Yale University Press

Scott, W B 1925 'The revised British standard steel sections'. *TSE* **3**, new ser, 290–7

Scott, W B 1928 'Steelwork. A short history: ii'. *Architects Journal* 11 July, 55–7

Scott, W B 1929 'Some historical notes on the application of iron and steel to building construction'. *TSE* **7**(1), 4–12; **7**(3), 99–102

Scott, W B 1930 'Iron and steel' *in* Gilbert, W R (ed) *Modern Steelwork: A Review of Current Practice in the Employment of Structural Steelwork in Buildings and Bridges*. London: The British Steelwork Association

Scott, W B 1944 'First steel frame building'. *TSE* **22**(2), 98–9

Select Committee on Scientific Instruction 1868 *Report from the Select Committee on Scientific Instruction: Together With the Proceedings of the Committee, Minutes of Evidence, and Appendix*. London: HMSO [Series: British Parliamentary papers]

Service, A 1977 *Edwardian Architecture: A Handbook to Building Design in Britain, 1890–1914*. London: Thames and Hudson [Series: World of art library]

Service, A (ed) 1975 *Edwardian Architecture and its Origins*. London: Architectural Press

Shankland, E C 1897 'Steel skeleton construction in Chicago'. *Mins PICE* **128** (January), 1–27 [discussion 28–43; correspondence 43–57]

Shankland, E C 1898 'American and English steel construction'. *BN* **27** (May), 735–6

Sharman, J R 1906 'Recent examples of steel and concrete buildings in Scotland'. *C&CE* **1**, 199–205

Sharp, D 1993 'The 19th century', *in* Blanc, A, McEvoy, M and Plank, R (eds) *Architecture and Construction in Steel*. London: E&FN Spon, 14–31

Shaw, E 1960 *Peter Cooper & the Wrought Iron Beam*. New York: Cooper Union Art School [Series: CUAS 7]

Shipway, J S 1990 'The Forth railway bridge centenary, 1890–1990: some notes on its design'. *PICE* **88**(6), 1079–1107

Simmons, J 1978 *The Railway in England and Wales 1830–1914, I: The System and its Working*. Leicester: Leicester University Press

Skelton, H J 1891 *Economics of Iron & Steel*. London: Biggs & Co

Skelton, H J & Co 1908 *Structural Steel Handbook No. 13*, June

Skempton, A W 1959 'The evolution of the steel frame building'. *The Guild Engineer* **10**, 37–51

Skempton, A W 1961 'The boat store, Sheerness (1858–60) and its place in structural history'. *TNS* **32**, 57–78

Skempton, A W 1971 'Samuel Wyatt and the Albion mill'. *Architectural History* **14**, 53–73

Skempton, A W (ed) 2002 *A Biographical Dictionary of Civil Engineers in Great Britain and Ireland, Vol 1: 1500–1830*. London: Thomas Telford

Skempton, A W and Johnson, H R 1962 'The first iron frames'. *AR* (March), 175–86

Slater J 1886 'Some American methods of construction', *The Builder* **50** (20 March 1886), 434–5

Slater J 1887a 'New materials and inventions', *The Builder* **52** (7 May 1887), 703–06 and (14 May 1887), 741–42

Slater J 1887b 'New materials and inventions', *BN* **52** (6 May 1887), 672–73

Smith, D (ed) 2001 *Civil Engineering Heritage: London and the Thames Valley*. London: Thomas Telford

Smith, J 2000 'No. 7 Caledonian Road, King's Cross, London Borough of Islington'. English Heritage Architectural Investigation Report NBR BI No. 106340, September

Smith, S 1888 'Progress in construction'. *The Architect* **40** (23 November 1888), 293-4 (presidential address before the Society of Architects)

Smith, S 1994 'Craftsman and contractor: a re-appraisal of their role in the design and fabrication of mid-nineteenth century iron structures in Great Britain'. *Journal of Architectural and Planning Research* **11**(4), 311–38

Smith, T 1980 'Commercial Road goods depot'. *London's Industrial Archaeology* **2**, 9

Spain, G and Dromgoole, N 1970 'Theatre architects in the British Isles'. *Architectural History* **13**, 77–89

Sparrow, W S (ed) 1907 *Flats, Urban Houses and Cottage Homes: A Companion Volume to "The British Home of To-day"*. London, Hodder and Stoughton

Stamp, G 1979 *Temples of Power*. Burford: Cygnet Press

Stamper, J W 1989 'The Galerie des Machines of the 1889 Paris world's fair'. *Technology and Culture* **30** (April 1989), 332–53

Stannus, H 1882 'The artistic treatment of constructional iron-work'. *BN* **42** (27 January 1882), 100–1

Stansbie, J H 1907 *Iron and Steel*. London: Archibald Constable & Co. Ltd

Statham, H H 1898 'The esthetic treatment of engineering work'. *The Engineering Magazine* **14**, 103–9

Steiner, F H 1984 *French Iron Architecture*. Ann Arbor, MI: UMI Research [Studies in the fine arts. no. 3. Architecture]

Stratton, M 1993 *The Terracotta Revival: Building Innovation and the Image of the Industrial City in Britain and North America*. London: Victor Gollancz in association with Peter Crawley

Stratton, M 1999 'New materials for a new age: steel and concrete construction in the North of England, 1860–1939'. *IAR* **21**(1), 5–24

Summerson, J 1935 *John Nash: Architect to King George IV*. London: George Allen & Unwin Ltd

Summerson, J 1993 *Architecture in Britain, 1530–1830*. New Haven, CT: Yale University Press

Survey of London 1926 **10** *The Parish of St Margaret's, Westminster,* Part I. London: Batsford

Survey of London 1955 **25** *St George's Fields The Parishes of St George The Martyr, Southwark and St Mary, Newington*. London: The Athlone Press

Survey of London 1960a **29** *The Parish of St James Westminster,* Part I. London: The Athlone Press

Survey of London 1960b **30** *The Parish of St James Westminster,* Part I. London: The Athlone Press

Survey of London 1966a **33** *The Parish of St Anne Soho,* [Pt 1]. London: The Athlone Press

Survey of London 1966b **34** *The Parish of St Anne Soho,* [Pt 2]. London: The Athlone Press

Survey of London 1970 **35** *The Theatre Royal Drury Lane and The Royal Opera House, Covent Garden*. London: The Athlone Press

Survey of London 1975 **38** *The Museums Area of South Kensington and Westminster*. London: The Athlone Press

Survey of London 1977 **39** *The Grosvenor Estate in Mayfair,* Part I. London: The Athlone Press

Survey of London 1980 **40** *The Grosvenor Estate in Mayfair,* Part II. London: The Athlone Press

Survey of London 1983 **41** *Southern Kensington: Brompton*. London: The Athlone Press

Survey of London 1986 **42** *South Kensington: Kensington Square to Earl's Court*. London: The Athlone Press

Survey of London 1994a **43** *Poplar, Blackwall and the Isle of Dogs: The Parish of All Saints*. London: The Athlone Press

Survey of London 1994b **44** *Poplar, Blackwall and the Isle of Dogs: The Parish of All Saints*. London: The Athlone Press

Survey of London 2000 **45** *Knightsbridge*. London: The Athlone Press

Survey of London 2008a **46** *South and East Clerkenwell*. New Haven and London: Yale University Press

Survey of London 2008b **47** *Northern Clerkenwell and Pentonville*. New Haven and London: Yale University Press

Survey of London 2012 **48** *Woolwich*. New Haven and London: Yale University Press

Survey of London 2013 **49** Battersea Part 1: Public, Commercial and Cultural. New Haven and London: Yale University Press

Sutherland, J 2000 '19th-century iron and glass domes' *in* SAHBG 2000 Domes: papers read at the Annual Symposium of the Society of Architectural Historians of Great Britain, 111–29

Sutherland, R J M 1964 'The introduction of structural wrought iron'. *TNS* **36**, 67–84

Sutherland, R J M 1976 'Pioneer British contributions to structural iron and concrete: 1770–1855', *in* Ackerman, J S *et al* (eds) *The Garland Library of the History of Art: Vol 11, Nineteenth and Twentieth Century Architecture*. New York: Garland

Sutherland, R J M 1984 'The birth of stress: a historical review', *in* The Art and Practice of Structural Design: 75th anniversary international conference, 11–13 July 1984, Imperial College, London. London: Institution of Structural Engineers, , 5–15

Sutherland, R J M 1989 'The introduction of iron into traditional building', *in* Hobhouse, H and Saunders, A (eds) *Good and Proper Materials: The Fabric of London Since the Great Fire* [papers given at a conference organised by the Survey of London at the Society of Antiquaries on 21 October 1988, London Topographical Society 140, London: RCHME in association with the London Topographical Society, 48–58]

Sutherland, R J M 1989b 'Shipbuilding and the long-span roof'. *TNS* **60**, 107–26

Sutherland, R J M 1990 'The age of cast iron, 1780–1850: who sized the beams?', *in* Thorne, R (ed) 1990 *The Iron Revolution: Architects, Engineers and Structural Innovation, 1780–1880*. [Essays to accompany an exhibition at the RIBA Heinz Gallery, June–July 1990] London: RIBA, 24–33

Tanner, A W 1883 'The construction of theatres'. *The Builder* **44** (10 February 1883), 190–3

Tappin, S 2002 'The early use of reinforced concrete in India'. *Construction History* **18**, 93

Tarn, J N 1968 'The improved industrial dwellings company'. *Transactions of the London and Middlesex Archaeological Society* **22**(1), 43–59

Taylor, N 1965 'Ceramic extravagance'. *AR* **138** (November), 339–41

Thole, J 1992 *The Oxford and Cambridge Clubs in London*. London: The United Oxford and Cambridge University Club in association with Alfred Waller

Thorne, R 1987 'Paxton and prefabrication'. *Architectural Design* **57**(11/12), 22–8

Thorne, R (ed) 1990 *The Iron Revolution: Architects, Engineers and Structural Innovation, 1780–1880*. [Essays to accompany an exhibition at the RIBA Heinz Gallery, June–July 1990] London: RIBA

Thorne, R 1991 'Architects and engineers' *in* Worsley, G (ed) *Georgian Architectural Practice: Papers Given at the Georgian Group Symposium*

Thorne, R 2000 'Building the memorial: the engineering and construction history', *in* Brooks C (ed) *The Albert Memorial: the Prince Consort National Memorial, its Contexts and its Conservation*. New Haven, CT: Yale University Press, 134–59

Thorne, R 2001 'Inventing a new design technology: building and engineering' *in* MacKenzie, J M (ed) *The Victorian Vision: Inventing New Britain*. London: V&A Publications, 173–85

Thrupp, E C 1907 'Legal hindrances to modern methods of building construction'. *A&CR* **77** (19 April 1907), 257–60

Timbs, J 1866 *Club Life of London,* 2 vol. London: R. Bentley

Tredgold, T 1822 *A Practical Essay on the Strength of Cast Iron, etc.* London: J. Taylor

Trinder, B (ed) 1992 *The Blackwell Encyclopedia of Industrial Archaeology*. Oxford: Blackwell [Series: Blackwell reference]

Tucker, M T 1981 'London Docks: The "Skin Floor", An Appraisal'. Unpublished typescript, 1981, updated 1983 and 1986

Tucker, M T 2000 'London Gasholders Survey: The Development of the Gasholder in London in the Later Nineteenth Century Part A: General'. A report for Listing Branch and the Historical & Research Team, English Heritage

Turner, A W 1996 'Redpath Brown: Two Hundred Constructive Years'. Unpublished typescript

Turvey, R 1993–4 'London lifts and hydraulic power'. *TNS* **65**, 147–64

Tweedale, G 1984 'Brown, Sir John (1816–1896) iron and steel manufacturer', *in* Jeremy, D J (ed) *Dictionary of Business Biography: A Biographical Dictionary of Business Leaders Active in Britain in the Period 1860–1980*. London: Butterworths, 475–8

Tweedale, G 2004 'Brown, Sir John (1816–1896)', *in* Oxford Dictionary of National Biography. Oxford: Oxford University Press

Twelvetrees, W N 1900 *Structural Iron and Steel: A Text-Book for Architects, Engineering, Builders, and Science Students*. London: Whittaker & Co.

Twelvetrees, W N 1906 'Steel skeleton construction and the London building act'. *C&CE* **1** (March), 17–25

Twelvetrees, W N 1907 *Concrete-Steel Buildings: Being a Companion Volume to the Treatise on "Concrete-Steel"*. London: Whittaker and Co

Vaizey, J 1974 *The History of British Steel*. London: Weidenfeld and Nicolson

VCH Essex 1973 *The Victoria History of the County of Essex, **6***. Oxford: Oxford University Press

VCH Middlesex 1976 *The Victoria History of the County of Middlesex, **5***. Oxford: Oxford University Press

VCH Middlesex 1982 *The Victoria History of the County of Middlesex, **7***. Oxford: Oxford University Press

VCH Middlesex 1995 *The Victoria History of the County of Middlesex, **10***. Oxford: Oxford University Press

Verplanck, C 2001 'Louis Christian Mullgardt: an architect with a capital "A"'. *San Francisco Heritage News* **29**(5), 5–7

Walford, E 1892 *Old and New London: A Narrative of its History, its People, and its Places. Vol 5 The Western and Northern Suburbs*. London: Cassell

Walker, B M (ed) 1980 *Frank Matcham: Theatre Architect*. Belfast: Blackstaff Press

Walmisley, A T 1888 *Iron Roofs*, 2 edn. London: E&FN Spon

Walmisley, A T 1900 'The use of rolled joists in construction'. *The Builder* 1 December, 483–9

Walshaw, G R and Behrendt, C A J 1950 *The History of Appleby-Frodingham*. London: Appleby-Frodingham Steel Co.

Walter Jones 1893 'Illustrated Priced Catalogue Iron & Steel Roof Principles, Doors and Structural Work Generally' [trade catalogue in British Library Trade Literature Collection, MF 75]

Watkin, D, *et al* 1984 *Grand Hotel: The Golden Age of Palace Hotels: An Architectural and Social History*. London: Dent

Watson, G 1989 *The Smeatonians: The Society of Civil Engineers*. London: Thomas Telford

Watson, W 1892, *Paris Universal Exposition, 1889, Civil Engineering, Public Works, and Architecture*. Washington, DC: Government Printing Office

Weiler, J 1990 'The making of collaborative genius: royal engineers and structural iron 1820–1970', *in* Thorne, R (ed) *The Iron Revolution: Architects, Engineers and Structural Innovation, 1780–1880*. [Essays to accompany an exhibition at the RIBA Heinz Gallery, June-July 1990] London: RIBA, 41–7

Weinreb, B and Hibbert, C eds 1993 *The London Encyclopaedia*. 2 rev. edn. London: Papermac

Weinreb, B, *et al* (eds) 2008 *The London Encyclopaedia*. 3 rev. edn. London: MacMillan

Wermiel, S 1993 'The development of fireproof construction in Great Britain and the United States in the nineteenth century'. *Construction History* **9**, 3–26

Wermiel, S 2009 'Introduction of steel in US buildings, 1862–1920'. *Engineering History and Heritage* **162** (February), 19–27

Wermiel, S E 2000 *The Fireproof Building: Technology and Public Safety in the Nineteenth-Century American City*. Baltimore, MD: Johns Hopkins University Press [Series: Studies in industry and society]

Whitaker 1906 *Whitaker's Red Book of Commerce or Who's Who in Business*, 1 edn. London: J Whitaker & Sons.

Whitehead, J 2000 *The growth of Camden Town AD 1800–2000*, 2 edn. London: J. Whitehead

Whyte, I B 1999 'The neo-classical revival in turn of the century Britain'. *Edinburgh Architecture Research* **26** (September), 7–27

Wilkinson, R 1825 *Londina Illustrata. Vol 2, Including Theatrum Illustrata Graphic and Historic Memorials of Ancient Playhouses, Modern Theatres and Other Places of Public Amusement in the Cities and Suburbs of London & Westminster*. London: R. Wilkinson

Williams, M with Farnie, D A 1992 *Cotton Mills in Greater Manchester* [GMAU with RCHME]. Preston: Carnegie

Wilmore, D (ed) 1998 *Edwin O Sachs: Architect, Stagehand, Engineer & Fireman* [his life and his satellites: in celebration of the centenary of 'Modern opera houses and theatres'] Summerbridge: Theatresearch

Wilson, B 1999 '53 Northfield Road, Ealing, London W13'. Unpublished building survey report, April

Wolters J 1886–7 'On the manufacture of rolled joists in Belgium'. *PICE* **87**(1), 403–15 [a translation of a paper in the *Annuaire de l'Association des Ingénieurs sortis de l'École de Liége* **5** (1886)]

Wood, E 1898a 'The use of steel in buildings'. *Transactions Manchester Association of Engineers*, 165–78; discussion 178–87

Wood, E 1898b 'The use of steel in buildings'. *The Mechanical World* 2 December, 254–7

Woodbridge, G 1978 *The Reform Club, 1836–1978: A History from the Club's Records*. London: Reform Club

Woodley, R 1999 'Professionals: early episodes among architects and engineers'. *Construction History* **15**, 15–22

Woodrow, E A E 1892 'Some recent developments in theatre planning'. *BN* **62** (25 March 1892), 427–30 [a paper read before the Architectural Association on 18 March 1892]

Woodrow, E A E 1892–94 'Theatres' ... 'Theatres XLVII' *BN* **63** (15 July 1892), 63–5 (29 July 1892), 138–40 (5 August 1892), 168–70 (12 August 1892), 208–9 (19 August 1892), 237–9 (2 September 1892), 308–10 (16 September 1892), 382–4 (23 September 1892), 418–19 (30 September 1892), 449–50 (21 October 1892), 555–8 (28 October 1892), 590–2 (18 November 1892), 695–7 (2 December 1892), 765–6 (9 December 1892), 802–3 (30 December 1892), 932–4; **64** (27 January 1893), 122–4 (10 February 1893), 188–9 (10 March 1893), 330–1 (24 March 1893), 398–9 (14 April 1893), 500–1 (26 May 1893), 695–8 (23 June 1893), 830–1; **65** (14 July 1893), 37–9 (4 August 1893), 131–4 (18 August 1893), 198–200 (8 September 1893), 323–5 (22 September 1893), 365–8 (29 September 1893), 426–7 (10 November 1893), 608–9 (1 December 1893), 713–15 (29 December 1893), 851–2; **66** (19 January 1894), 76–8 (26 January 1894), 107–10 (23 February 1894), 247–9 (23 March 1894), 391–4 (20 April 1894), 526–9 (18 May 1894), 667–70; **67** (13 July 1894), 36–7 (20 July 1894), 64–5 (27 July 1894), 104–5 (24 August 1894), 243–6 (28 September 1894), 422–3 (26 October 1894), 566–9 (9 November 1894), 638–40 (23 November 1894), 708–9 (7 December 1894), 780–1 (28 December 1894), 912–13

Woodward C D 1972 *BSI: The Story of Standards*. London: British Standards Institution

Wordingham, C H 1901 *Central Electrical Stations: Their Design, Organisation, and Management*. London: Charles Griffin and Company

Yeomans, D T 1992 *The Trussed Roof: Its History and Development*. Aldershot: Scolar Press

Yeomans, D T 1997 *Construction Since 1900: Materials*. London: Batsford

Statutory Acts, Select committees etc

Statutes at Large 1774, 14 George III, c. 78; Fires Prevention (Metropolis) Act 1774; An Act for the further and better Regulation of Buildings and Party Walls, and for the more effectually preventing Mischiefs by Fire, within the Cities of London and Westminster, and the Liberties thereof, and other the Parishes, Precincts, and Places, within the Weekly Bills of Mortality, the Parishes of Saint Mary-la-Bonne, Paddington, Saint Pancras, and Saint Luke at Chelsea, in the County of Middlesex and for indemnifying, under certain Conditions, Builders and other Persons against the Penalties to which they are or may be liable for erecting Buildings within the Limits aforesaid contrary to Law

Public Statutes General 1844, 7 and 8 Victoria, c. 84; The Metropolitan Building Act 1844; An Act for regulating the Construction and the Use of Buildings in the Metropolis and its Neighbourhood

Public Statutes General 1855, 18 and 19 Victoria, c. 122; The Metropolitan Building Act 1855; An Act to Amend the Laws relating to the Construction of Buildings in the Metropolis and its neighbourhood

Public Statutes General 1860, 23 & 24 Victoria, c. 52; Metropolitan Building Act (Amendment) 1860; An Act to Amend The Metropolitan Building Act (1855)

Public Statutes General 1878, 41 and 42 Victoria, c. 32; The Metropolis Management and Building Acts Amendment Act 1878; An Act to amend the Metropolis Management Act, 1855, the Metropolitan Building Act, 1855, and the Acts amending the same respectively

Public Statutes General 1890, 53 & 54 Victoria, c. 218, The London County Council (General) Powers Act, 1890

Public Statutes General 1894, 57 & 58 Victoria, c. ccxiii., London Building Act, 1894; An Act to consolidate and amend the Enactments relating to Streets and Buildings in London

Public Statutes General 1905, 5 Edward 7, c. 209, London Building Acts (Amendment) Act, 1905; An Act to amend the Acts relating to buildings in London to confer various powers on the London County Council and for other purposes

The London County Council (General) Powers Act, 1908. Part III. Amendment of London Building Act 1894

The London County Council (General) Powers Act, 1909. Part IV. Amendment of London Building Acts

The Manchester Municipal Code: Being a Digest of the Local Acts of Parliament, Charters, Commissions, Orders, Bye-Laws, Regulations, and Public Instructions and Forms within the City of Manchester Vol. II (Manchester, 1895)

House of Commons 1831 *Second report from the Select Committee on Windsor Castle and Buckingham Palace; with the minutes of evidence and appendix: Buckingham Palace.* London: House of Commons paper No. 329 (14 October 1831)

House of Commons Select Committee on the Metropolitan Buildings and Management Bill 1874 Special report from the Select Committee on the Metropolitan Buildings and Management Bill: together with the proceedings of the Committee, minutes of evidence and plans. HMSO

INDEX

S